Pigmented Cereals and Millets
Bioactive Profile and Food Applications

Food Chemistry, Function and Analysis

Titles in the series:

1: Food Biosensors
2: Sensing Techniques for Food Safety and Quality Control
3: Edible Oil Structuring: Concepts, Methods and Applications
4: Food Irradiation Technologies: Concepts, Applications and Outcomes
5: Non-extractable Polyphenols and Carotenoids: Importance in Human Nutrition and Health
6: Cereal Grain-based Functional Foods: Carbohydrate and Phytochemical Components
7: Steviol Glycosides: Cultivation, Processing, Analysis and Applications in Food
8: Legumes: Nutritional Quality, Processing and Potential Health Benefits
9: Tomato Chemistry, Industrial Processing and Product Development
10: Food Contact Materials Analysis: Mass Spectrometry Techniques
11: Vitamin E: Chemistry and Nutritional Benefits
12: Anthocyanins from Natural Sources: Exploiting Targeted Delivery for Improved Health
13: Carotenoid Esters in Foods: Physical, Chemical and Biological Properties
14: Eggs as Functional Foods and Nutraceuticals for Human Health
15: Rapid Antibody-based Technologies in Food Analysis
16: DNA Techniques to Verify Food Authenticity: Applications in Food Fraud
17: Advanced Gas Chromatography in Food Analysis
18: Handbook of Food Structure Development
19: Mitigating Contamination from Food Processing
20: Biogenic Amines in Food: Analysis, Occurrence and Toxicity
21: Nutrition and Cancer Prevention: From Molecular Mechanisms to Dietary Recommendations
22: Health Claims and Food Labelling
23: Nutraceuticals and Human Health: The Food-to-supplement Paradigm
24: Nutritional Signalling Pathway Activities in Obesity and Diabetes
25: The Chemistry and Bioactive Components of Turmeric
26: Foodomics
27: Food Proteins and Peptides: Emerging Biofunctions, Food and Biomaterial Applications
28: Handbook of Antioxidant Methodology: Approaches to Activity Determination

29: Fats and Associated Compounds: Consumption and Human Health
30: Oral Processing and Consumer Perception: Biophysics, Food Microstructures and Health
31: Development of Trans-free Lipid Systems and their Use in Food Products
32: Advanced Spectroscopic Techniques for Food Quality
33: Berries and Berry Bioactive Compounds in Promoting Health
34: Metabolism of Nutrients by Gut Microbiota
35: Bio- and Nano-sensing Technologies for Food Processing and Packaging
36: Edible Fungi: Chemical Composition, Nutrition and Health Effects
37: Chemistry and Nutritional Effects of Capsicum
38: Pigmented Cereals and Millets: Bioactive Profile and Food Applications

Pigmented Cereals and Millets

Bioactive Profile and Food Applications

Edited by

Sneh Punia Bangar
Clemson University, USA
Email: snehpunia69@gmail.com

Sajid Maqsood
United Arab Emirates University, United Arab Emirates
Email: Sajid.m@uaeu.ac.ae

and

Anil Kumar Siroha
Chaudhary Devi Lal University, India
Email: anilsiroha@cdlu.ac.in

ROYAL SOCIETY
OF **CHEMISTRY**

Food Chemistry, Function and Analysis No. 38

Print ISBN: 978-1-83916-709-6
PDF ISBN: 978-1-83767-029-1
EPUB ISBN: 978-1-83767-030-7
Print ISSN: 2398-0656
Electronic ISSN: 2398-0664

A catalogue record for this book is available from the British Library

The Royal Society of Chemistry is a charity, registered in England and Wales, Number 207890, and a company incorporated in England by Royal Charter (Registered No. RC000524), registered office: Burlington House, Piccadilly, London W1J 0BA, UK, Telephone: +44 (0) 20 7437 8656.

Visit our website at www.rsc.org/books

Printed in the United Kingdom by CPI Group (UK) Ltd, Croydon, CR0 4YY, UK

Preface

The demand for functional foods increases daily due to the possible health benefits of consuming a healthy and balanced diet. Currently, food products prepared from pigmented crops such as wheat, rice, corn, and millets constitute a significant source of interest for health-conscious people. Pigmented grains demonstrate numerous health benefits, including anti-cancer, anti-diabetic, anti-inflammatory, and anti-heart disease activity. This book discusses pigmented grain's basic composition and functional components that are responsible for numerous health benefits. This book consists of 15 chapters, which are contributed by experts in their area of research.

The first chapter, contributed by Dangi, Chaudhary, Paul, Sharma, Dutta, and Razdan, focuses on pigmented wheat's nutritional and functional properties. This chapter elaborates on breeding methods, nutritional composition, health benefits, and product formulations from pigmented wheat. Chapter 2 deals with the composition and health benefits of pigmented rice. This chapter highlights the micro- and macronutrient content, various functional properties of flour and starch of pigmented rice, and their applications in various food products. Chapter 3 highlights the various nutritional properties and bioactive profile of pigmented maize, and was contributed by Dudeja, Gupta, Mankoo, and Singh. Many secondary metabolites, including phenolic acids, carotenoids, and flavonoids, contribute to maize pigmentation, with numerous health advantages also discussed in this chapter. Yadav, Guleria, and Kumar authored chapter 4, 'Pigmented Barley: Photochemical Composition, β-Glucan Content, and Applications.' This chapter highlights the phytochemical composition, health benefits, and industrial applications of pigmented barley. Various properties and food applications of β-glucan are also discussed in this chapter. Chapter 5 highlights the functional properties and bioactive diversity of the pigmented sorghum. Mehany and his

Food Chemistry, Function and Analysis No. 38
Pigmented Cereals and Millets: Bioactive Profile and Food Applications
Edited by Sneh Punia Bangar, Sajid Maqsood and Anil Kumar Siroha

Published by the Royal Society of Chemistry, www.rsc.org

co-workers contributed Chapter 6, which deals with chemistry, functionality, and technological aspects in food systems of pigmented pseudocereals. The nutritional benefits of the main pigmented pseudocereals, such as amaranth, buckwheat, and quinoa, are discussed in this chapter, focusing on their bioactive pigments. This chapter also covers the uses of colored pseudocereals in food applications. Millets exist in various colors, and the color of the kernel is an essential factor in determining market value and nutritional quality. The nutritional profile, carotenoids, anthocyanin content, and biological activity of colored millet are all discussed in Chapter 7. Sobti and her co-workers contributed Chapter 8, and this chapter highlights the comparison of qualities characteristics of pigmented and non-pigmented grains. Pigmented cereals have gained more attention due to their high nutritional value and bioactive compounds, which may provide health benefits.

The different processes for extracting pigments from various grains are addressed in Chapter 9. This chapter explored that cereal pigments can be extracted using traditional solvent-based or modern extraction methods, including ultrasonication, microwave-assisted extraction, and pulsed electric field extraction. Spectroscopic and chromatographic techniques can be used to quantify cereal colors. Chapter 10, 'Pigmented Cereals: Key Influencing Factors', is authored by Swain and her co-workers. Jasmeet and co-worker contributed Chapter 11. This chapter highlights the measurement methods and composition of anthocyanin in pigmented cereals. This chapter discusses the quantification and composition of anthocyanin molecules and which have been evaluated using high-performance liquid chromatography and ultra-performance liquid chromatography technologies. Chapter 12 highlights the chemical structures and biological properties of carotenoids. Various aspects of carotenoids, *i.e.*, metabolic pathways, human health, and carotenoids in various cereals grains, are explored in this chapter. The trends for 'functional foods' and a healthy diet are growing nowadays, owing to evidence of potential health advantages linked with frequent consumption of these items as part of a balanced and diverse diet. The dietary and health benefits of pigmented cereals are discussed in Chapter 13. Chapter 14 addressed the role of biotechnology in the production of pigmented cereals, and this chapter is contributed by Ishaq, Batool, Chen, Cai, and Asfandyar. Pigmented millet and cereals are considered a unique form of grains due to the relatively highest level of phenols and exhibit more remarkable antioxidant properties. Interest in the product formulations from these pigmented grains is increasing day by day. Chapter 15 will explore applications of pigmented cereals in formulations of various bakery products.

Sneh Punia Bangar
Sajid Maqsood
Anil Kumar Siroha

About the Editors

Sneh Punia Bangar, PhD, is a researcher at Clemson University, United States. Previously, she worked as Assistant Professor (C) in the Department of Food Science and Technology, Chaudhary Devi Lal University, Sirsa. Her research interests include extraction and functional characterization of bioactive compounds, starches, functional foods, and nanocomposites films and coatings. She has presented her research at various national and international conferences and has published more than 100 research papers/book chapters in national and international journals/books. To date, she has authored or co-authored more than 80 (published/accepted), one book series, 2 authored books (CRC, Taylor, and Francis), 5 edited books (CRC Taylor and Francis, Springer, Elsevier), one reference book (CRC, Taylor, and Francis), 35 book chapters, 20 conference proceedings/seminars, and acted as Guest Editor of three journal special issues (*Food Research International*, *International Journal of Food Science and Technology*, *Journal of Food Processing and Preservation*, *Foods and Polymers*).

Food Chemistry, Function and Analysis No. 38
Pigmented Cereals and Millets: Bioactive Profile and Food Applications
Edited by Sneh Punia Bangar, Sajid Maqsood and Anil Kumar Siroha

Published by the Royal Society of Chemistry, www.rsc.org

Sajid Maqsood is a Professor in Department of Food Science at College of Agriculture and Veterinary Medicine, UAEU. He has started his academic career in May 2011 at UAEU as a post-doctorate Fellow and is currently serving as Professor since 2021.

His research is dedicated and aligned exclusively on food and nutritional needs of UAE and the region. He has established an innovative research program at UAEU which focus on bioactive molecules in camel milk and dates and food wastes and by-products, plant-based proteins, alternative proteins, and exploring sustainable means of processing and diversifying their utilization through novel food product development. He has recently received grant on plant-based proteins and meat analogue. He is leading an active research group in Food Science Dept with multidisciplinary collaboration within UAEU and abroad.

He has published around 100 publications and edited 3 book and several book chapters.

Dr Sajid has received his PhD from Prince of Songkla University (Thailand) in the year 2010. He has received several awards for his research at UAEU and was listed in Top 2% scientist list released by Stanford University (2020–2021).

Anil Kumar Siroha, PhD, is presently working as Assistant Professor (C) in the Food Science and Technology Department at Chaudhary Devi Lal University, Sirsa. He received his doctorate in Food Science and Technology at Chaudhary Devi Lal University, Sirsa, India. His areas of interest include starch, starch modification, and the development of new products. He has published more than 25 research papers, 20 book chapters, two edited books, and one authored book. He also serves as a reviewer for various national and international journals. Dr Siroha is an active member of the Association of Food Scientists and Technologists (AFSTI) in Mysore, India. Dr Siroha also co-supervised more than 20 research projects of M.Sc. students in the Department of Food Science and Technology at Chaudhary Devi Lal University, Sirsa, Haryana.

Contents

Food Chemistry, Function and Analysis No. 38
Pigmented Cereals and Millets: Bioactive Profile and Food Applications
Edited by Sneh Punia Bangar, Sajid Maqsood and Anil Kumar Siroha

Published by the Royal Society of Chemistry, www.rsc.org

CHAPTER 1

Pigmented Wheat: Nutrition Scenario and Health Benefits

P. DANGI[a], N. CHAUDHARY*[b], A. PAUL[a], A. SHARMA[a], I. DUTTA[a] AND R. RAZDAN[a]

[a]Department of Food and Nutrition and Food Technology, Institute of Home Economics, University of Delhi, New Delhi, India; [b]Department of Food Science and Technology, College of Agriculture, Nagaur, Agriculture University, Jodhpur, India
*E-mail: chaudharynisha3@gmail.com

1.1 Introduction

Wheat (*Triticum aestivum*) is a major staple cereal crop that belongs to the grass family *Poaceae (Gramineae)*, and has an important place among crop species, playing a crucial role in the human diet. It is one of the primary sources of nutrition for about 2.5 billion people worldwide, and is used for making various wheat-based food products (pasta, crackers, noodles, pastries, *etc.*). Wheat grain is considered a major source of starch, protein, minerals, B-group vitamins, and dietary fibres.[1] It is believed that wheat was one of the first food crops ever grown. Wheat was first cultivated about 10 000 years ago, as a part of the agricultural revolution (*i.e.*, the shift to agriculture from hunting and gathering). For around 8000 years it has been the prime food source for some of the major civilizations of the world. Currently, wheat is the most commonly grown crop worldwide, with nearly 772 million metric tons being produced in more than 120 countries. Most of the world's wheat

Food Chemistry, Function and Analysis No. 38
Pigmented Cereals and Millets: Bioactive Profile and Food Applications
Edited by Sneh Punia Bangar, Sajid Maqsood and Anil Kumar Siroha

Published by the Royal Society of Chemistry, www.rsc.org

is produced by two countries: India and China. In India, which ranks second after China, production is estimated between 98 million to 105 million metric tons. It is also the second-most produced cereal behind maize and, in terms of food security, it is one of the dominant crops in India after rice.[2] Generally, amber-colored wheat varieties are consumed on a daily basis around the world. However, attention is increasingly being given to colored wheat varieties that are rich in anthocyanins, carotenoids, phenolic compounds, and other phytochemicals, owing to their numerous health benefits. Pigmented wheat came into existence in the 19th century by crossbreeding interspecies crosses of *Triticum aestivum* with distinct wild wheat species: *Triticum monococcum L. spp. aegilopoides, Thinopyrum ponticum*, and *Th. bessarabicum*.[3] Color variation in the wheat grain is dependent on genetic factors, growing conditions and technology used.

Wheat bran, covering the endosperm, comprises three tissue layers (the pericarp, testa, and aleurone layer), where pigments are localized and provide different colors.[4] Carotenoids (tetraterpenoids) are yellow, orange, and red colored pigments that are responsible for the color of the wheat grain located in the endosperm. Wheat grain contains a wide range of carotenoids, including lutein, β-carotene, β-apocarotenal, antheraxanthin, β-cryptoxanthin, zeaxanthin, triticoxanthin, taraxanthin (lutein 5, 6-epoxide), and flavoxanthin, whereas phlobaphenes (deoxyflavanoids) are responsible for the reddish brown color found in the outer layer of the kernel.[5] Anthocyanins are water-soluble flavonoids that give rise to the blue, black, purple, and red colors in the outer layer of kernel. However, genotype and environmental conditions are also responsible for the pigmentation in the wheat.[6] The purple color is due to anthocyanin accumulated in the pericarp, while the blue color is caused by anthocyanin in the aleurone layer.[7] In black wheat, anthocyanin is localized in both the pericarp and aleurone layers. A category of black wheat called Chinese black wheat contains high levels of polysaccharide and protein. It also contains more dietary fibre and less carbohydrate than conventional wheat. It has also been reported that black wheat (Qinhei No.1) contains more iron, zinc, manganese, copper, selenium, magnesium, potassium, and phosphorus, which are beneficial to health.[6] In purple wheat varieties, some of the anthocyanin present in high amounts include:

- cyanidin-3-glucoside (4 mg g^{-1}),
- peonidin-3-glucoside (2 mg g^{-1}),
- cyanidin-malonyl-glucose (1 mg g^{-1}),
- cyanidin-succinyl-glucose (1 mg g^{-1}),
- peonidin-malonyl-glucose,
- peonidin-succinyl-glucose,
- peonidin-malonyl-succinyl-glucose.

Blue wheat contains five to eight important anthocyanins, with delphinidin 3-rutinoside, delphinidin-3-glucoside and malvidin-3-glucoside the most prevalent pigments.[4] Purple wheat contains a lower level of anthocyanins than black and blue wheat varieties. There is a great variation in anthocyanin levels in pigmented wheat varieties, as is evident from studies, which found that: the total anthocyanin content of purple wheat varies from 10 to 305 mg kg^{-1} dry kernel weight; for blue wheat anthocyanin content ranges between 17 and 211 mg kg^{-1} dry kernel weight; and in black wheat it varies between 56 and 198 mg kg^{-1} dry wheat kernel.[7]

The protein content of pigmented wheat is 11.4–18.17%, which is higher than common wheat.[8] The percentage of total amino acids and essential amino acids present in pigmented wheat is approximately 8.88–18.91% and 7.31–18.13%, respectively.[9] Pigmented wheat varieties contain almost the same amount of fat and dietary fibres as conventional wheat.[3] Moreover, pigmented wheats are abundantly rich in vitamin B, including B1 (thiamine), B2 (riboflavin), B3 (niacin), B6 (pyridoxine), B9 (folate). It also contains (in smaller amounts) provitamin A (β-carotene), vitamin D (calciferol), and vitamin K (phylloquinone). Pigmented wheat also contains a large amount of zinc (108.54–142.68%), magnesium (5.31–40.63%) and iron (8.57–42.86%).[1] It has been found that flour yield is lower in colored wheat because of its lower volumetric weight than common wheat. As levels of anthocyanin is high in bran, bran-rich flour obtained from purple and blue wheat can be used to make food products such as biscuits, bread, *etc.* and in the extraction of natural colorants (dyes and fragrances).[4] The presence of carotenoids, total phenolic compounds (TPCs), total anthocyanin compounds (TACs), total dietary fibres (TDF) in the colored wheat exhibit antioxidant properties that basically helps in extending the shelf life of products and protects against degenerative diseases.[5] Pigmented wheat promotes a healthy lifestyle and helps in the prevention of several long-term diseases such as cardiovascular diseases (CVD), inflammation, cancer, obesity, gallstones, and metabolic syndrome.[10]

1.2 Transformation in the Consumption Patterns of Wheat

Wheat and wheat-based products are a good source of carbohydrates, proteins and other minerals and vitamins that constitute a healthy consumption pattern.[11] However, this crop still needs to be improved for its micronutrients and other functional components. Colored wheat makes up for the lack of certain micro-nutrients and functional compounds of the traditional wheat varieties. By virtue of being rich in bioactive substances such as phenolic acids, flavonoids and anthocyanins, it proves to be a good dietary supplement in the management of diseases such as diabetes, cardiovascular diseases, *etc.*[12] Pigmented wheat therefore quickly found a new place in the wheat world, and began to firmly establish itself as integral to human life. There is always a desire for a truly affordable and proportional diet in a single item in this busy civilization, which is unquestionably significant. While wheat is already considered one of the most cost-effective and essential food sources, and thus the introduction of pigmented wheat essentially began to meet this demand.[1] When compared to conventional wheat types, pigmented wheat is significantly rich in anthocyanins, flavonoids, carotenoids, phenolic acids, other bioactive compounds, and a variety of other necessary nutritive riches, as well as nutrients found in regular wheat, which is present in the bran and also in the flour, can lead to a bad effect on the consumers.[13] As a result, colorful ones enhance both diet and health, and they are being encouraged to be embraced by our society, which is highly crucial.

Over time, it has also been discovered that the natural colorants in colored wheat could take over and replace synthetic colors. These natural additives can be used in a wide range of products, including cosmetics, medicines, treatments, textiles, and several other things. They essentially give a much better texture, color, and other qualities than synthetic colorants, providing improved and appealing versions of foods and other products. Thus, pigmented wheat not only suits daily dietary needs but also facilitates management of health conditions and provision of alternative additives in the food industry. The multifaceted advantage of pigmented wheat would be key in encouraging its uptake among consumers. As a result, these characteristics make pigmented wheat an intriguing topic for food scientists and technologists to investigate, demonstrate, and utilize for the benefit of the world.[1]

Many clinical research and epidemiological studies suggest that regular consumption of pigmented wheat can reduce the risk of chronic diseases comparted with non-pigmented ones.[8] The pigments have antioxidant, anti-inflammatory, anti-microbial, anti-cancer, and anti-aging properties, provide protection to the nervous system and immune system, and provide the body with a wide range of functional assistance.[15] It even has various effects on platelets and blood vessels that reduce the risk of coronary heart diseases,[16] including preventing diseases like type 2 diabetes and obesity.[1] According to Sharma,[12] fruits like blueberries and blackberries are inherently high in anthocyanins, while mangoes are rich in carotenoids and contain high amounts of sugar. Therefore, high doses can pose a fundamentally high risk of high blood sugar. In pigmented wheat, the availability of pigments also provides a variety of essential nutrients, so the adoption of colored ones encourages switching to a low-sugar diet. It also contains some fairly unique properties that promote brain activity and even pigmented wheat like black wheat has a higher capacity than conventional wheat to scavenge free radicals in the colored seeds.[17]

Since colored wheat attracts consumers, it has begun to dominate the food industry. Due to various properties, pigmented wheat has attracted a lot of researchers, breeders, and industrialists to utilize it as commercial, functional, and nutraceutical food. Various innovative and colored products have been developed, such as noodles, bread, tortillas, muffins, and many other innovative foods, with some still under development.[12] Several agro-food companies have started their business, while many farmers have started new lives by filling their land with wheat of a specific color, gaining worldwide popularity.[18] For future and present generations, consumption patterns are shifting to now fixate on a diet that is preventive and nutritive in nature. This naturally has an impact on global consumption and production of pigmented wheat varieties over traditional ones due to their functional characteristics.

1.3 Types and Breeding of Pigmented Wheat

Although purple, blue, and black wheat are produced in small quantities, they show potential to be utilized in food, ornamentations, colorants, and other application. The welcoming of pigmented wheat on a large scale for their production and utilization is necessary because of the antioxidant properties, health benefits, and other features from the pigments providing an extra advantage over conventional wheat. However, these varieties of wheat did not originate from the local or wild crops, rather, they have been developed after much research on breeding and genetics over a long period.[19] The range of pigments generally agrees with the 'The Law of Homologous Series in Hereditary Variability,' which was drawn by N. I. Vavilov in 1920.[5] Although the pigments in wheat grains are quite low, they do influence the qualities of wheat products like bread, cookies, biscuits, pasta, and noodles.[19] The color gene expression of the wheat grain differs with the origin of the seed tissue. Moreover, in this grain, there are four main types of tissue *i.e.*, pericarp, seed coat, embryo, and endosperm.[5] In pigmented wheat, the purple color of wheat is due to the anthocyanins in the diploid pericarp layer while the blue color of the wheat is due to the anthocyanins in the aleurone layer, and black wheat develops with the genetic combination of the purple and the blue wheat.[17] Sometimes, a very dark grain can also emerge due to the genetic collaboration of purple pericarp, blue aleurone, yellow endosperm, and red grain.[5] In common black, blue and purple wheat, only forms of glucoside (mono-glycosylated) and rectinoside (di-glycosylated) were identified in every anthocyanin.[15] Further discussion on pigmented wheat (black, blue, and purple) follows.

1.3.1 Purple Wheat

The emergence of purple wheat has started to attract the scientific world. Purple wheat has been seen in variable shades, as growing and environmental conditions, light intensity and temperature hugely affect the amount of anthocyanins developed in the pericarp of the grain.[16] Ludwig Wittmack first reported purple-colored wheat in 1879. In around 1872/1873, Johann Maria, a famous botanist, collected some grains during a trip to East Asia and afterwards he handed these grains to the Museum of Agriculture, in Berlin. It was introgressed from Ethiopian sources to hexaploid bread wheat.[20] So, the purple wheat was first found in the tetraploid durum (*Triticum dicoccum*), in the regions of East Africa like Ethiopia and was genetically transferred to wheat.[19] Stable purple wheat was originated from the crosses of *Triticum dicoccum* var. *Arraseita percival* × *Triticum aestivum* L., and the color was much deeper. This variety inherited its color from the tetraploid wheat, *Triticum aestivum* var. *Arraseita percival* from Abyssinia containing a single dominant gene and hexaploid wheat, *Triticum aestivum* L. contains two incompletely dominant genes, and this led to the formation of purple-colored wheat.[15,19] The purple pigments are located in the diploid pericarp layers and since this pigment is present in the pericarp, its color does not segregate among the wheat kernels inside the spike.[17,21] In addition, due the localization of the purple color, the environmental effect on it is much stronger and the anthocyanins present in it are more prone to these environmental effects. Cyanidin 3-glucoside and peonidin 3-glucoside are the two major components seen in purple wheat, in which cyanidin 3-glucoside is the predominant one.[21] About 22 different types of anthocyanins could be found from purple wheat lines and even show higher peak intensities for the acylated anthocyanins.[15] This purple wheat provides many benefits, *e.g.*, anthocyanins from this wheat can also be utilized and crushed into larger pieces and spread over the surface of multigrain bread.[1]

1.3.2 Blue Wheat

The color of the blue wheat is due to the presence of anthocyanins in the aleurone layer.[16] Blue wheat and its varieties were developed using distant hybridization, and alien chromosomes into the wheat genome.[22] The blue color gene is transferred from three wild wheat species known as *Triticum monococcum* L. *spp. aegilopoids*, *Thinopyrum pouticum* (*Agropyron elongatum*), and *Th. bessarabicum*. However, these are also required for further study to determine the gene control and roles of different genes on the development of blue color in wheat.[17] Blue wheat was formed in the first half of the 20th century by crossing wheat with wheat's relatives. Originally, these crosses were carried out to make the crop disease resistant and for other factors like winter hardness, perennial habit, forage traits, and yield components, but this led to the formation of blue wheat. There is evidence that blue aleurone originates from einkorn wheat, including *Triticum boeoticum* boiss or *Triticum monococcum* L., and other types resulting from the interspecific crosses of *Agropyron* (syn. *thinopyrum*) and other wild wheat varieties.[20] Blue coloration is gene dependent and thus contributes to the different blue shades of wheat grain color.[17] Also, there are all over 22 different anthocyanins identified from blue wheat. The blue color is present in the aleurone layer around the seed, except the embryo. It was also noted that there are either two complementary genes or a single dominant gene to control the blue color of the wheat. The shade and stability of blue wheat is also caused by the different localization of anthocyanins in the wheat grain.[15] Unlike purple wheat, cyanidin 3-glucoside is the second most common in the blue wheat. However, anthocyanin present in maximum concentration still remains undefined for blue wheat.[21] Pelargonidin-3-glucoside was also not seen in the blue wheat.[15] The concentration of the color of blue wheat is also determined by light, temperature, growing, and environmental conditions, as with purple wheat.[21] If there is a lower yield, it is due to the negative impact of the genes linked to the blue aleurone chromosomes from the wheat species.[15] Blue wheat differs from other wheat varieties not in the composition of anthocyanins, but its presence of different layers that can be clearly seen during the cross section of the wheat kernel. Blue wheat contains mainly cyanidin, delphinidin, pelargonidin, malvidin, and petunidin glycosides.[5] This blue wheat also holds promises to develop wheat-based functional foods or natural colorants, based on the concentration of anthocyanins.[21] Since the blue color is present in the outer layers of the pigmented wheat, so wholemeal/bran-enriched flour can be utilized in many colored foods, like bakery products. However, blue wheat is more prone to thermal degradation, thus extra care has to be taken and also further studies are required to overcome this challenge.[5]

1.3.3 Black Wheat

Black wheat is a form of pigmented wheat that is developed by crossing purple and blue varieties of wheat and thus the black color is due to the anthocyanins present in both pericarp and aleurone layer. Black wheat was first originated in the laboratory of the Institute of Crop Genetics, Shanxi Academy of Agricultural Science, after 20 years, beginning in 1970.[17] Blue-grained hexaploid wheat was formed by crossing *Triticum aestivum* with *A. glaucum*, and the 'purple 12-1' was formed by crossing *Triticum aestivum* with *Elymus dasystachys*. Then the 'blue-purple 114' was developed by crossing blue-grained hexaploid wheat with 'purple 12-1'.[19] Then the first black wheat that was developed was named 'Black 76' and, gradually, this variety was formed by crossing 'purple 12-1' as the male parent (associated with the mutation in the pericarp) and 'blue–purple 114' as the female parent (as a result of wild introgression).[17,18] Just like blue wheat, black wheat can also be observed without any genetic disorder or affecting any chromosomes to affect its related yield.[15] Black wheat contains the highest parts per million (ppm) *i.e.*, 140 ppm (or 100–200 ppm), while regular wheat contains only 5 ppm.[23] It has the highest amount of anthocyanins among all the other colored varieties It contains 60% more iron and is thus more nutritious than any common wheat. There are twenty-six different anthocyanins that could be found in black wheat itself. Delphinidin-3-(6″-malonylglucoside) is the only compound that is seen in black wheat lines only. Due to the localization of the pigments in the outer pericarp layer, as in purple wheat, this black wheat also appear darker in color.[15] Among all the varieties of pigmented wheat, black wheat draws the most of the attention due to its high nutritional profile, good sensory attributes, and health-promoting activities. It is also used to produce various functional foods and colorants, and can produce products much better than the ones developed from the common wheat. No research has given a complete overview regarding the breeding line, composition, and functional properties of black wheat.[17] However, studies and experiments are still ongoing with the given information for the improved results, better economic value, nutritional and functional properties, and to make the best utilization of it.[15]

1.4 Nutritional Composition of Pigmented Wheat

Conventional wheat varieties have a protein concentration ranging from 10% to 18% of total dry matter, mainly affected by weather and nitrogen fertilizers.[24] Starch constitutes 60–75% of total dry weight in wheat grains.[25] However, it has been observed that pigmented wheat has 11.74–18.17% higher protein content, 8.88–18.91% higher total amino acids, and 7.31–18.13% higher essential amino acids. Even mineral content such as the amount of zinc, iron, and magnesium is raised by the respective percentages 108.54–142.68%, 8.57–42.86%, and 5.31–40.63% when compared with common wheat varieties.[9] The nutritional composition of pigmented wheat varieties is discussed in detail below (see Figure 1.1).

1.4.1 Protein

Black wheat is an abundant source of total essential amino acids, comprising 96 amino acids and their derivatives. The total amino acid content is 15.74%, while essential amino acid content is around 4.45%. The chief storage proteins of black wheat are constituted by alcohol soluble prolamins, glutenin and gliadin. The gluten index used to determine the end-quality of black wheat flour is 69.74, which is lower than that of conventional wheats with a gluten index of 98–99. However, it still falls in the optimal range

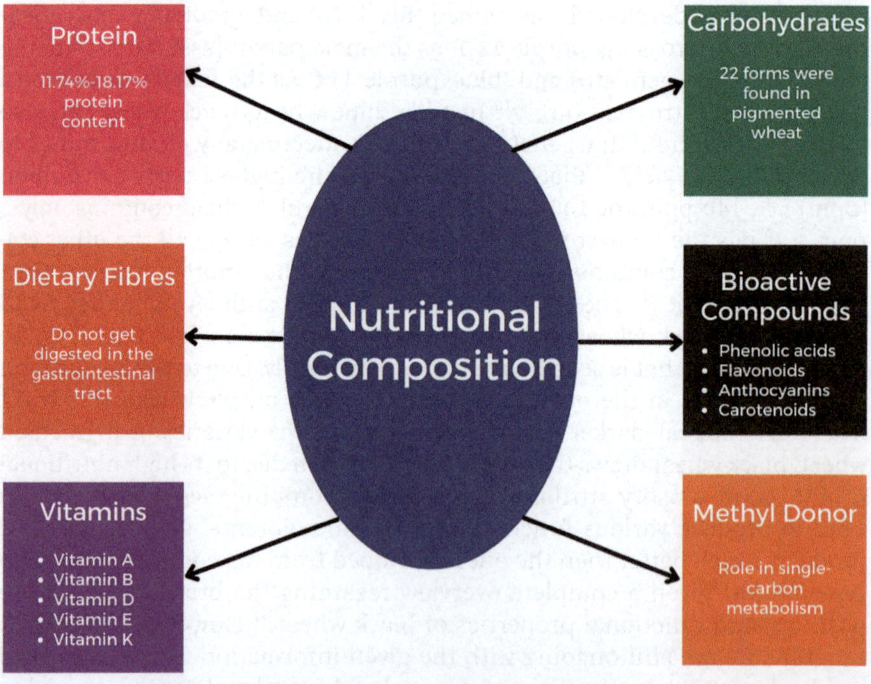

Figure 1.1 Nutritional composition of pigmented wheat.

for application in bread making and can be used to manufacture bread successfully. This paves the way for processing of black wheat to produce low-gluten food products that serve as an alternative for celiac disease patients. It has been suggested that it is critical to investigate wheat accessions in terms of total gluten content. As a result, considerable research, particularly for validation of black wheat's gluten nature, is required.[17] Purple wheat has a protein level that ranges from 29% to 55.37% more than the standard values, but its fat content also ranges from 7.69% to 15.38% higher than the standard values.[26]

1.4.2 Carbohydrates

In wheat, the primary carbohydrate is starch, which is a vital energy-providing component. It is present in the endosperm or flour section of the kernel. Free sugars, such as glucose, fructose, sucrose, raffinose and glucofructans and hemicelluloses are present in a small percentage of 1–2% by weight.[27] In the endosperm of a wheat grain containing a high amount of starch, crystalline and amorphous regions coexist, and there is a formation of semi-crystallized starch due to hydrogen bonding between radially oriented linear and branched macromolecules.[17] Starch is made up of two high-molecular weight polymers – amylopectin and amylose. Linear amylose makes up 20–30% of normal wheat starch, while branched amylopectin makes up 70–80%.[25] The distinctive physiochemical qualities of starch have allowed the production of various food items because of its properties. For the optimal production of bread, the usual amylose:amylopectin ratio present in wheat starch granules is vital.[27]

Twenty-two forms of carbohydrates were found in pigmented wheat varieties of red, blue, purple, and black.[28] More polysaccharides have been found in black wheat than in conventional wheat varieties. Around 75% of carbohydrates in black wheat are concentrated in its endosperm.[17] In the nutritional profiling of white, black, and blue wheat lines and advanced lines made by crossing the white and pigmented wheat varieties (purple, black, and blue), the carbohydrates and sugars are found to be lower in blue and white wheat lines, and higher in black donor and advanced lines.[29]

1.4.3 Dietary Fibres

Specific polysaccharides, such as cellulose, hemicellulose, pectin, *etc.*, are not digested in the gastrointestinal tract. These form the dietary fibre that boosts the volume of the food and thereby decreases the accessibility of starch granules by enzymes present in the digestive tract. In black wheat, total dietary fibres are around 1.15% of the total weight, and insoluble fibre contributes to a more significant fraction. This insoluble fibre offers volume to the body and the faeces, which helps prevent constipation and stimulate the intestinal transit.[17] However, the proportion of dietary fibre has been reported in purple wheat across multiple varieties (ZY4, ZY5, ZY11, ZY17-1, ZY17-2, ZY21 and ZY22) to be 11.30%.[26]

1.4.4 Bioactive Compounds

Bioactive compounds are phytochemicals extracted from natural food sources or their by-products, that can regulate metabolic functions and positively influence the overall health of an individual.[30] Pigmented wheat is abundant in anthocyanin and numerous other phytochemicals such as phenolic acids, flavonoids, carotenoids, *etc.*, thereby increasing their value as an advantageous food for the health-conscious crowd. Pigmented wheat kernels are a rich source of by-products like carotenoid and anthocyanin-rich pericarp (in purple wheat), aleurone layer (in blue wheat), and endosperm (in yellow wheat). Red wheat is a result of its outer layer containing phlobaphene. Black wheat that has been synthesized by crossing purple and blue wheat varieties has increased free radical scavenging quality and phenolic acid content.[5] Bioactive substances such as anthocyanins, phenols, flavonoids, *etc.*, of great value are found in black wheat.[17] Pigmented wheat varieties are abundant sources of vital bioactive compounds such as tannins, polyphenols, carotenoids, and anthocyanins. Thus, the different varieties have proven to fight against chronic diseases such as cardiovascular diseases, diabetes, inflammation, obesity, *etc.* Food manufacturers and researchers are in the process of tapping into this potential of pigmented wheat.[1]

In the red, purple, blue, and black pigmented wheat varieties, metabolite profiling has shown the presence of 174 flavonoids: 118 flavones and derivatives, 25 flavonols, 11 flavanones, 14 anthocyanins, 5 isoflavones, and 1 proanthocyanidin. In particular, tricin, chrysoeriol, apigenin, luteolin, quercetin, and their glycosides are present in greater concentrations.[28] Black, purple, and blue pigmented wheat lines and their antioxidant properties are linked with anthocyanins and phenolic compounds.[31]

1.4.5 Phenolic Acids

Phenolic acids are one of the most extensively found bioactive compounds, protecting plants from biotic (infections, wounds) and abiotic (nutrient deficits or excess, cold, and visible light) challenges. Furthermore, phenolics contribute to various food characteristics like bitterness, astringency, color, flavour, odour, and lipid oxidation resistance. Phenolic acid has anti-inflammatory, anti-bacterial, anti-proliferative, anti-carcinogenic, and antioxidative properties.[32] In common wheat, phenolic acids are present in soluble conjugate and insoluble form. Phenolic acids are sub-divided into hydroxybenzoic acid derivatives and hydroxycinnamic acid derivatives, which possess antioxidant properties. In metabolite profiling of red, yellow, blue, purple, and black pigmented wheat varieties, 53 phenylpropanoids, with 17 distinct phenolic acids and 36 derivatives, were found.

Hydroxycinnamic acids were present in the greatest concentration. The concentration of bound phenolic components is greater than those found in free form. Gentisic acid and protocatechuic acid were the most abundant phenolic acids and were found in higher concentrations in purple and black wheats. 4-methoxycinnamic, two cinnamic acid derivatives and 3,4,5-trimethoxycinnamic acids, have been identified and found to be much higher in yellow and blue wheats.[28]

However, black wheat comprises various phenolic acids like gallic, caffeic, gentisic, syringic, vanillic, *p*-hydroxybenzoic, ferulic acids, *p*-coumaric, and *o*-coumaric acid. Higher phenolic acid concentration is present in the bran than the wholemeal. Ferulic acid is the major phenolic acid present in black wheat. Higher soluble phenolic content is reported in black wheat than in white wheat. On comparing black wheat's phenolic acid with that of purple and common yellow varieties, bound total phenolic acids (TPC) content was 1.6 times higher in black wheat and free TPC was found to be 1.6 and 6 times more in black wheat than purple and yellow varieties, respectively.[17]

Phenolic acids have been detected exclusively in the outer layers of the wheat grain. A strong correlation between the antioxidant properties of phenolic acids – both free form and bound form – and total phenolic content has been depicted. Upon biochemical evaluation of the different bran fractions – outer bran, coarse bran, and shorts – of blue, purple, and black varieties of pigmented wheat, most phenolic acids were in bound form. These bound phenolic acids can be expected to promote health more effectively, since they may circumvent the upper gastrointestinal digestive process and can be absorbed directly into the plasma during intestinal digestion. The bran fractions with a higher concentration of phenolic acids can be obtained in debranning and utilized in the production of functional foods. Unfortunately, it requires changes by adding more bran in flour during processing.

The ferulic acid concentration in blue-pigmented wheat variety decreased gradually from outer bran to coarse bran to shorts. Coarse bran of black pigmented wheat and purple-pigmented wheat contained more ferulic acid than the other fractions. The highest concentration of total phenolic acids was noted in the outer bran of blue wheat, which amounted to 3459 $\mu g\ g^{-1}$, while the shorts of purple wheat have the lowest value of total phenolic acids of 1731 $\mu g\ g^{-1}$.[33]

1.4.6 Flavonoids

Flavonoids are polyphenols with great antioxidant properties and various medical benefits, and are generally found in the bran layer and germ.[1] They are one of the plant secondary metabolites that exhibit antioxidant, anti-inflammatory, anti-mutagenic, and anti-carcinogenic properties, along with the capacity to modulate key cellular enzyme functions. The general structure of flavonoids comprises a 15-carbon skeleton, made up of two phenyl rings A and B and a heterocyclic ring C. Six significant subtypes of flavonoids can be differentiated, although they have an overall structure of C6–C3–C6 carbon skeleton.[34] Depending on the structural distinctions, classification is made into different sub-groups such as flavones, isoflavones, anthocyanidins, chalcones, *etc.* Some of the flavonoids observed in wheat are tricin, quercetin, vitexin, kaempferol, *etc.*[17] Black wheat has been reported to have 174 flavonoids,[28] with the highest total flavonoid content (TFC) of 876 μg g^{-1} dry weight, whereas other wheat's TFC ranges from 694–841.81 μg g^{-1} dry weight.[35] The highest mean flavonoid content of black wheat may be a consequence of various flavonoids such as luteolin and its derivatives, anthocyanins, tricin, apigenin. Tricin derivatives, such as *O*-glucuronic acid with antioxidant properties have been reported to be concentrated in the bran and hull of certain types of black wheat.[17]

1.4.7 Anthocyanins

Anthocyanins are a part of the flavanol subclass of flavonoids and are present in the outermost layers of the wheat kernel. They play a pivotal role in the grain's color development.[17] These pigments are localized in specific kernel sections, which can be extracted to form anthocyanin-rich portions that can serve as either colorants or ingredients in functional foods[31].[28] After quantification of different types of wheat for 14 anthocyanins, black wheat was observed to contain 13 out of the 14 types of the anthocyanins in higher concentration when compared to other wheat varieties. In purple wheat, cyanidin 3-*O*-glucoside and peonidin 3-*O*-glucoside form the anthocyanin fraction and delphinidin 3-*O*-rutinoside and delphinidin 3-*O*-glucoside have been found in blue wheat.[36] The anthocyanin concentration has been found to be highest in black wheat, followed by purple, blue, red, and conventional yellow wheat.[28]

Generally, wheat with pigmented grain contains six anthocyanidins: cyanidin, delphinidin, malvidin, pelargonidin, peonidin, and petunidin. Cyanidin has been noticed to be effective against cardiovascular and liver diseases. It has improved metabolic systems and spatial memory and protected DNA cleavage, inhibition of invasion and mortality of tumor cells. Delphinidin is anti-inflammatory with bone protective and neuroprotective effects. It has an inhibitory effect on platelet activation, thrombosis and human glyoxalase. Malvidin is known to have anti-cancerous and anti-inflammatory activities in endothelial cells. It is known to be cytotoxic to human leukaemia cells and acts as an antioxidant. Pelargonidin is efficient in treating cancer and vascular diseases and has proven to prevent neuro-degenerative disorders. Peonidin has shown to prevent vascular and cardiac disorders and inhibit tumor cell growth with apoptotic effects on human breast cancer cells. Petunidin protects against liver cancer and reduces bone loss.[1]

1.4.8 Carotenoids

Plant pigments that give most fruits and vegetables their color are called fat-soluble carotenoids. Structurally, these compounds are 40-carbon isoprenoids having long conjugated polyene chains. Carotenoids are present in plant photosynthetic tissues, chromoplasts and chloroplasts. The biological activity of carotenoids is said to be affected by temperature and light. In cereals, the germ, bran, and endosperm contain carotenoids. The genotype of the cereal decides whether the carotenoid is present in an esterified or free form. The concentration of carotenoid is a determining factor for the color of wheat products and is used for analysing the quality of durum wheat. Some of the carotenoids in wheat such as alfa-carotene, beta-carotene, gamma carotene, and beta-cryptoxanthin, are converted to retinol, while lutein, lycopene, and zeaxanthin are functionally antioxidants. Other types of carotenoids that might be present in wheat are β-apocarotenal, triticoxanthin, lutein-5, 6-epoxide (taraxanthin), flavoxanthin, and antheraxanthin. Lutein is the most predominant carotenoid, followed by zeaxanthin, antheraxanthin, α-carotene, and β-carotene, with β-cryptoxanthin being a minor carotenoid, which is at the level of being non-detectable. In comparison, wheat flour during storage undergoes oxidation from either enzymatic or non-enzymatic mechanisms and causes carotenoid content to diminish.[17] Black wheat has been reported to show higher average total carotenoid content than red and white wheat varieties.[37] The carotenoid content of black wheat is at levels equivalent to those seen in ordinary wheat and other pigmented wheat such as blue, red, and purple varieties of wheat.[17]

1.4.9 Tocols

Tocols, commonly known as vitamin E (tocopherols and tocotrienols) are monophenols derived from 6-hydroxy-2-methyl-2-phytylchroman, which are used as food additives in the food and pharmaceutical sectors.[38] Tocopherol and tocotrienols are composed of a chromanol ring connected with a phytyl side chain. An unsaturated phytyl side chain distinguishes tocotrienols from tocopherols, which contains three double bonds. Based on the quantity and locations of methyl groups on the chromanol ring, tocopherols and tocotrienol has four isomers. Out of the isomers, α-tocopherol has long been known to be the most biologically active form of vitamin E when it comes to prevention of cell membrane oxidation.[25] Blue and purple-grained spring wheat varieties have an average tocol content of 32.68 mg kg^{-1} dry weight and 30.33 mg kg^{-1} dry weight, respectively. This was equivalent to the tocol levels of white-grained Novosibirskaya 67 spring wheat variety, *i.e.*, 32.00 mg kg^{-1} dry weight and red-grained Bohemia winter wheat variety, *i.e.*, 31.01 mg kg^{-1} dry weight.[5]

1.4.10 Vitamins

In comparison with standard wheat, most pigmented wheat varieties have been shown to have more levels of vitamin B (thiamine, riboflavin, niacin, pyridoxine, folate) and vitamin E (tocopherol/tocotrienol). Apart from this, minor amounts of β-carotene (precursor of vitamin A), vitamin D and vitamin K are also present.[39] Black wheat is an especially rich source of vitamin K. Black wheat has been reported to contain 19 vitamins; significantly, it contains 11.47 mg kg^{-1} of vitamin K in BGW 76 variety, which was 1.6 times greater than the vitamin K content of conventional wheat (Jinchun 9).[31] In black wheat, significant amounts of B3, B5, and E have been found.[9] All of these vitamins have a considerable impact on various bodily functions, which often overlap with each other. Vitamin B3 is concerned with the maintenance of healthy skin, DNA synthesis and energy metabolism. Vitamin E most significantly is concerned with prevention of haemolysis. The development of blood cells, hormone release, gland function, and food conversion to energy are linked to vitamin B5.[17] In terms of vitamin content, purple wheat varieties such as ZY17-1 and ZY22 outperform other wheats. More amounts of α-vitamin E, β + γ -vitamin E and total vitamin E compared with standard values were found in purple wheat.[26]

1.4.11 Methyl Donor

Principal methyl donors are betaine and choline, with wheat being the most abundant source of betaine.[40] Folate and B12 are members of the methyl-donor vitamin family, including choline and betaine. Methyl donors get their name from their role in single-carbon metabolism, which involves transferring methyl groups from amino acids like serine, glycine, and choline to a range of other substances like proteins, RNA, DNA, and intermediate metabolites.[41] Compounds like betaine, choline, dimethylglycine, and trigonelline have the ability to donate a methyl group to another molecule. Whole-wheat products contain increased concentrations of choline and betaine. Choline is required as a precursor of membrane phospholipids and acetylcholine, a neurotransmitter and through betaine. It also involved in donating methyl group to *S*-adenosylmethionine. Methyl donors, which are present as an interchangeable compound in wheat kernels, are present in much higher concentration in whole-wheat flour than refined flour.[42] In an animal model study, during the periods of gestation and early postnatal development, high choline intake showed improved cognitive function in adulthood and prevention of age-linked decline of memory and protection from neuropathological changes connected with Alzheimer's disease, *etc.*[43]

1.4.12 Anti-nutritional Factors

A major aim in nutrition is to diminish anti-nutritional factors, as they have been found to contribute towards micronutrient and mineral inadequacies, which negatively influences the nutritional properties that many cereals and legumes may otherwise offer. Anti-nutritional factors have been found to unfavourably affect the bioavailability of many nutrients.[44] Anti-nutrient factors found in wheat are tannins 1.43–1.84 mg g^{-1} dry basis (db), and phytic acid 7.95–8.00 mg g^{-1} db.[45] Phytic acid chelates with metal ions and decreases the availability of minerals, such as zinc.[44] Mineral bioavailability may also be affected by the elevated amounts of polyphenols.[46] In wheat that is a monocotyledon crop, the phytates present in the bran or aleurone layer can be removed through milling. Soaking the grains has also been an effective way to diminish anti-nutrient phytochemicals, such as phytate and tannins. Consumption of wheat after a period of soaking is recommended to ensure less impact of anti-nutrients.[44] Removal of phytate and other anti-nutritional factors in cereals is done during germination. Wheat flour that has been germinated and fermented shows decreased amounts of anti-nutrients.[46] A vital aim of many wheat-breeding programs to improve grain quality is the reduction of such anti-nutritional compounds. The current outlook on a balanced diet that promotes health, growth, and development comprises a diet that will supply sufficient calories, micronutrients, proteins, and low anti-nutritional constituents.[47]

1.5 Health Benefits of Pigmented Wheat

Pigmented wheat contains numerous phytochemicals, including various medical and health benefits like obesity, cardiovascular diseases, type 2 diabetes, cancer, *etc.* (see Figure 1.2).

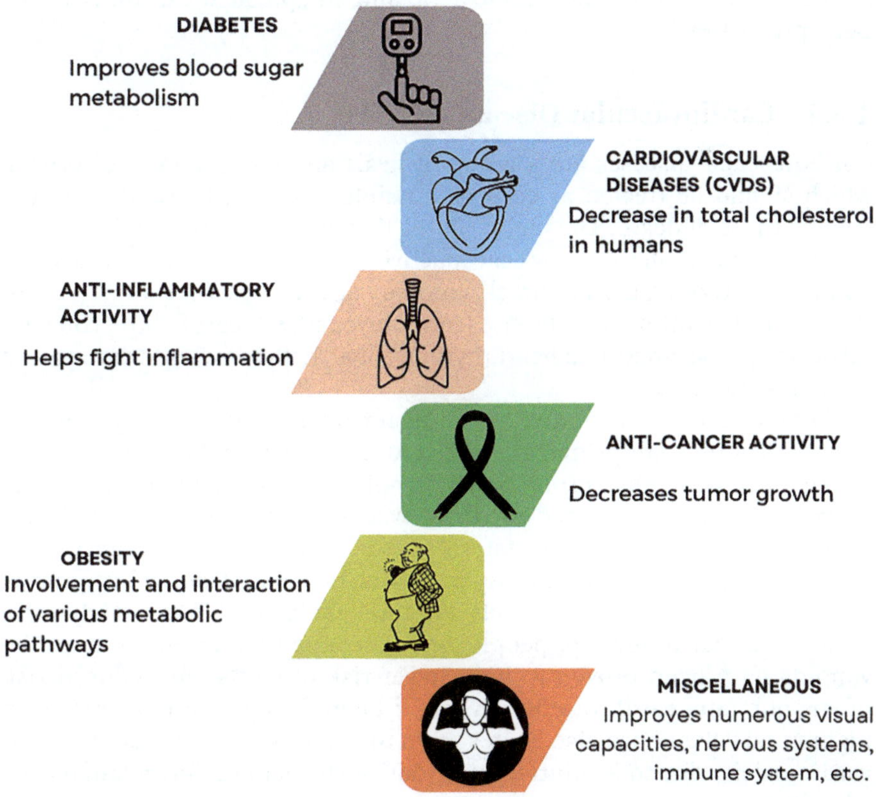

Figure 1.2 Health benefits of pigmented wheat.

1.5.1 Diabetes

Diabetes is a chronic disease that occurs when the pancreas does not produce enough insulin or when the body cannot effectively use the insulin it produces. Hyperglycemia, or raised blood sugar, is a common form of diabetes and may seriously damage the body's various systems, especially the nerves and blood vessels. In studies, the extended intake of whole-wheat was associated with a decreased risk and mortality from type 2 diabetes.[1] According to findings from Tyl and Bunzel,[48] blue wheat has been found to possess properties that decrease amylase-catalyzed starch absorption related to postprandial blood glucose levels, and is effective in curbing diabetes. The presence of high anthocyanin content in black wheat improves cholesterol levels and blood sugar metabolism in diabetic patients.[23] Compounds such as phenolic acids, tocopherol, and carotenoids are present in significant quantities across pigmented wheat varieties, which offer anti-diabetic effects. α-Tocopherol and β-carotene rich meals are also linked with decreased risk of type 2 diabetes and are consumed on a large scale by health-conscious people. Moreover, studies on rodents have shown that the presence of phenolic acids like gallic acid, syringic acid, ferulic acid, caffeic acid, and sinapic acid exhibit anti-diabetic properties.[1]

1.5.2 Cardiovascular Diseases (CVDs)

Cardiovascular diseases are a group of heart and blood vessels disorders, which should be treated as early as possible. Globally, these also are the leading cause of death. The major causes that lead to CVDs are tobacco use, unhealthy diets, obesity, physical inactivity, and excessive use of alcohol. Some of the CVDs are heart attack, coronary artery disease (*i.e.* narrowing of the arteries), heart muscle disease (cardiomyopathy), heart failure, congenital heart disease, abnormal heart rhythms, or arrhythmias, aorta disease and manfan syndrome.

Pigmented wheat contains a significant quantity of anthocyanin and carotenoids that are principally situated in the external aleurone layers. There is considerable proof that if wholegrain-based breakfast in the form of cereal is consumed for three weeks, then there is a significant decrease in total serum cholesterol up to 21.4%.[49] Whole bran of pigmented wheat contains arabinoxylan, which can be utilized to bring down the plasma, LDL-C fixations levels and total cholesterol.[50] In a study, after a mean 12-year follow up, people who consumed larger amounts of flavonoids have been proven to have lower risk of CVDs and reduced risk of any non-fatal cardiovascular events.[51] Phenolic compounds present in pigmented wheat have also been shown to suppress doxorubicin-induced cardiotoxicity due to inhibition of oxidative stress in the myocardial cells of rodent.[17]

1.5.3 Anti-inflammatory Activity

Anti-inflammation action is much displayed by flavonoids as their significant properties. Proanthocyanidins, anthocyanin, γ-oryzanol, and other bioactive mixtures have similarly been connected to anti-inflammatory activities. It has been reported that pigmented concentrates of wheat are quite convincing in reducing irritation, oxidative tension, and atherosclerotic wounds or injuries. In addition, the advancement of weight-related insulin barriers is mostly caused by inflammatory aggravations. Utilizing whole-wheat-based bread and fibre could chop down fundamental exacerbation among diabetic women as per clinical experts. Entire wheat grain usage can increase liver and inflammatory resilience in overweight and obese subjects with raised plasma cholesterol. The significant content of anthocyanin in black wheat varieties functionally acts as a safe-guard from lifestyle diseases.[1] Total anthocyanin content of wheat flour ranges from 6.61–95.04 mg kg^{-1} and follows in the order black > blue > purple > white. All colored wheat varieties had significantly higher TAC content, which helps to fight with inflammatory diseases.[52] Some bioactive compounds like proanthocyanins, anthocyanin, gamma-oryzanol *etc.*, are also beneficial for anti-inflammatory activities. Pigmented portions of wheat that offer antioxidant properties indicate a decrease in inflammation and oxidative pressure.[53] Wholewheat grain utilization can also improve liver and inflammatory resilience in overweight and fat subjects with raised plasma cholesterol.[54]

1.5.4 Anti-cancer Activity

Cancer is a disease in which some of the body's cells grow uncontrollably and spread to other parts of the body. Cancer can start almost anywhere in the human body, which is made up of trillions of cells. Usually, human cells grow and multiply (through a process called cell division) to form new cells as the body needs them. When cells grow old or become damaged, they die and new cells take their place. Sometimes this orderly process breaks down, and abnormal or damaged cells grow and multiply when they should not. These cells may form tumors, which are lumps of tissue. Moreover, cancer is one of the most common causes of death all around the globe.

The presence of naturally dynamic phytochemicals substantiate the role of pigmented wheat grain in probable prevention of breast cancer, colon cancer, liver cancer, ovarian cancer, and prostate cancer as upheld by epidemiological investigation of wholegrains.[1] Black wheat has also made an appearance and acts as a better option in the form of food supplements for all those suffering from this disease.[23]

1.5.5 Obesity

A disorder involving excessive body fat increases the risk of health problems. One of the primary casual factors of obesity is chronic intake of energy-dense or high-calorie foods. Obesity may cause insulin resistance, dyslipidemia, and hyperglycemia.[1] Recent studies have indicated that commonly consumed polyphenols and anthocyanin-rich diets suppressed co-morbidities. Whole grains are calorie deficit and give a feeling of satiety, accounting their role in body weight regulation. Many observational studies have shown an inverse linkage between whole grain intake and the risk of obesity.[12] The presence of naturally occurring antioxidants that develop in the grain help in filling the gap and therefore enriches pigmented wheat, especially black wheat with high nutritional value, highlighting its significant health benefits.[23]

1.5.6 Miscellaneous

According to the analysis of Tsuda,[55] an anthocyanin-rich eating regimen helps improve numerous visual capacities, for instance, decreasing eye exhaustion, improving nearsightedness, and upgrading the retinal bloodstream to lessen glaucoma. Black wheat is the solution to all the problems related to eyes among adults. Further, it was stated that it helps in the normal function of the cerebrums by decreasing age-related neurodegeneration and psychological decay. Black wheat contains anthocyanins and antioxidants than regular wheat, which controls antibodies and free radicals in our bodies that also boosts our immunity. Black wheat can also help in diminishing malnutrition.[23] Purple wheat containing anthocyanins have also been investigated for their antioxidant and anti-aging potential on *Caenorhabditis elegans* nematode experimental model.[56] Extracted anthocyanins from purple wheat were able to extend the life span of wild and oxidative stress mutants because of the ability to reduce oxidative stress.[12] Black wheat also significantly reduces fat pad and restricts body weight gain.[23] Pigmented wheat and its varieties contain diverse combinations of phytochemicals and are the main components known to impart protective and beneficiary effects to human body.[1]

1.6 Consumer Perceptions of Pigmented Wheat

Conventional wheat provides a sufficient amount of nutritional content to the consumer but nowadays, consumers are more concerned about their health and thus are looking for more nutrient-rich sources rather than only energy-giving foods. Pigmented wheat is getting more and more popular due to its amazing health benefits. Although pigmented wheat is produced in lesser yields than conventional wheat, it under higher demands for various purposes in the food industry.[4] There are many health claims for foods having sufficient dietary fibres in developed countries, and actually, it benefits humans for various health reasons. Pigmented wheat rich in dietary fibres help in lowering serum cholesterol level, insulin response and postprandial blood glucose.[17] Purple wheat noodles provide three-times more antioxidants than red grapes, and helps in lowering blood pressure and cholesterol, improving eyesight and prevent liver disease.[57] In bread wheat, commonly, white and red wheat are used, but purple and blue colored wheat is rare.[31] In terms of consumption-related traits, pigmented wheat does not contain any bitter compounds and enhances grain palatability for humans. Through the means of human sensory panels or highly reproducible, robust, and sensitive model systems or other "consumption" trials the consumption preference between or acceptance of pigmented wheat and conventional wheat varieties can be determined.[17] Currently, blue and purple corn are used for production of naturally colored blue tortillas.[58] In research studies, biscuit-making trials were performed from wholemeal flour and purple wheat flour; analysis was conducted on the basis of quality of characteristics purple flour consist of higher gluten index than that conventional flour. Higher TAC levels in purple wheat were detected, whereas in conventional wheat, no TAC level was observed, but somehow anthocyanin reduction was also observed. TPC level were slightly higher in purple than conventional wheat but on the other hand TYP level were less rich in purple than conventional wholemeal flour.[36] Some of the drawbacks were that TPC (Total phenolic compounds) present in flour gets degraded during dough mixing; this shows that flour contains higher antioxidant compounds compared with noodles and bread.[59] Blue wheat released in Australia has 25% less yield than the check cultivators.[15] Regular wheat has a 5 ppm concentration of anthocyanin, whereas black wheat grain is observed to have approximately 100–200 ppm of anthocyanin.[17] This attracts the consumer towards it usage, as it is a wiser and a healthier option. Black wheat is also more digestible than conventional wheat and poses many other health benefits; knowing about all these factors nowadays is attracting consumers.

1.7 Product Development From Pigmented Wheat

Pigmented wheat grains have varied processing qualities depending on their color, whether they are red, purple, or black, and the pigments (carotene, xanthophylls, anthocyanin, and phenolic compounds) contained in their seed coat.[17] Different structural compositions of pigmented wheat are given in Figure 1.1. Pigmented wheat has increased nutritional benefits, higher antioxidant activity and similar processing characteristics to regular wheat; it may be marketed to manufacture a variety of functional meals. Flours, breads, noodles, and pasta have been successfully formulated using pigmented wheat varieties. However, to overcome any shortcomings and to retain its original palatability, it is necessary to adopt effective processing technology for biologically active wheat cultivars.[16] The process parameters and circumstances must be optimized before they can be used in a commercial application.[3] The highest crude protein content has been found in flour produced from black grained wheat compared with flours produced from bread wheats. The protein and dough quality of purple and blue wheat has to be assessed prior to their utilization in bread making. Investigating the chemical changes occurring in bioactive compounds such as phenolic acids and anthocyanins during baking is vital.[31] Compared with known varieties used in bread-making such as Glenlea, black wheat has a higher crude protein content but similar HMW-glutenin subunits to those of Glenlea, black wheat can potentially be utilized in making bread. The antioxidant properties of purple wheat has been found to be the greatest, followed by the antioxidant property of whole grain bread and white bread.[31] However, before the production of bread, it is vital to determine the quality of wheat flour, as bread is the most common staple food, consumed globally.[16] Biscuits are consumed daily and are popular snack choices across countries. As one of the bakery items with a long shelf life, biscuits can be transformed by the addition of functional ingredients that are beneficial to health. Wholemeal flour made from purple wheat has been used to create biscuits, in which, although the anthocyanin profile decreased post-processing, it was similar to the values found in the assessment of wholemeal flour made from purple wheat (which is higher than conventional wheat). Structural differences due to higher gluten strength do act as a deterrent towards its popularity, but the creation of healthy products offering antioxidant properties is possible through purple wheat.[36] Indian unleavened flatbread has also been prepared using blue, purple and black wheat in which blue wheat had low dough extensibility, though the high nutritional content is present in the aleurone (outer pericarp) layer. Purple and black wheat were found to have sensory scores similar to conventional variety and are hard grained hence will form acceptable end products.[60] According to reports, black wheat possesses comparatively good properties and sensory characteristics, with substantial cooking losses and thus, black wheat flour can be used to make noodles. Noodles made from pigmented wheat flour have lower gumminess, hardness and chewiness, irrespective of milling method.[17] Chinese black grained wheat is suitable for noodle production after its combination with strong flours, as the gluten content present is heat-stable. The gluten content can be improved by breeding.[61] To achieve better health results, whole-wheat flour and partially debranned grain flour of black wheat can also be used for human consumption. There is a need for research to test the acceptability of noodles from pigmented wheat. Other products can also be developed, such as vinegar, breakfast cereals, muffin (including bran), antho-beer, soy sauce, and food colors using pigmented wheat. The use of bran fractions of pigmented wheat flour along with other cereal flours can improve the nutritional content with the formulation of various functional foods, but further studies are still required to understand consumer acceptance regarding palatability and the acceptability.[16]

1.7.1 Roasting of Pigmented Wheat Flour

When it comes to product development, we see continuous attempts to increase the efficacy of pigmented wheat processing lines. In order to do this, it becomes vital to study the properties and changes occurring in them by different processing methods. In a study conducted to examine the impact of roasting of pigmented wheat by three methods – sand roasting, pan roasting, and microwave roasting – significant variations in the functional properties, color and pasting properties were observed in the case of microwave roasting. Roasting affects the antioxidant content significantly and hence unroasted grains would be preferred over roasted grains when it comes to this aspect. To summarize a few of the findings, sand roasting has exhibited greatest reduction in total phenol content for black and yellow (conventional) wheat and greatest reduction in total phenol content for purple wheat was exhibited by microwave roasting. The lowest impact on total phenol content for yellow and purple wheat was seen in the case of pan roasting and microwave roasting for black wheat. Post-roasting, emulsifying capacity was also altered, and a reduction in the emulsifying capacity of black wheat (2–5.9%), purple wheat (4.52–6.86%), and conventional yellow wheat was observed (3.2–7.38%). Microwave roasting showed the most significant reduction, and sand roasting method the lowest reduction in the emulsifying capacity. The density of all flours was reduced considerably – lowest in the case of sand roasting and highest in case of microwave roasting – by all methods of roasting, due to moisture removal and void space creation in the cellular matrix or loss of integrity of the matrix. Percentage reduction in density for yellow wheat was 17.19–21.66%, purple wheat was 2.9–3.4%, and black wheat was 1.8–2.9%. Amylose content was boosted and amylopectin content saw a decline post roasting. The outlook for the future of pigmented wheat processing is the optimization of roasting such that we can retain the nutritive value of the pigmented wheat flour.[17]

1.8 Future Prospects

It has been proved that pigmented wheat is more nutritious than conventional wheat. The quantity and quality of proteins, minerals, dietary fibres, phytochemicals, and other bioactive compounds in pigmented wheats is far better than in conventional wheat varieties.[1] Although pigments have many beneficial features, their availability to most people is limited due to the low yield and high cost. Both farmers and researchers are trying to discover innovations as their consumption is not widely promoted.[17] Colored wheat is one of the underutilized crop due to its content, composition, dynamics, genotypic variation, and the products developed from it.[62] It is known that pigments can be more easily absorbed from cereals than any other food group.[21] In addition, colored grains can be easily stored and have a comparatively longer shelf life due to the stability in processed food products. So, it is necessary to consume in moderation and thus, the transfer of rich color pigmentation in wheat germplasm is done using breeding techniques in the hope of secured color, good yield, and some horticulture characters. Further studies are still required to analyze various activities related to the compounds from pigmented wheat.

As pigmented wheat contains higher concentrations of pigments and other nutrients, so the nutritional and health benefits can be easily acquired if added into different products.[22] However, it is necessary to find new processing technology that can keep the original attributes of pigments. This is because the conventional method damages them. Thus, the pearling process is encouraged but scientists still looking for more yield and better performance expecting to have in further generations, as it is known that different pigments have different processing technologies because of the different genetic backgrounds, and thus the results clash with each other.[8]

Further, significant progress has been noticed regarding identifying the gene responsible for the colored varieties. Most likely, RNA sequencing and SNP mapping will help identify and map the colored traits in the wheat kernel quickly.[22] It has been suggested that combining pigmented wheat with biofortification breeding could also play an important role in supplying additional nutrients left by it to form a complete diet in one food.[26] Even though the functional properties are limited, with the help of nutrient-controlling genes, modern breeding technologies, and genetic engineering, studies are continuously conducted to complete the demand for profit-oriented food products, providing promising tests for analysis.[17]

Farmers have an inadequate number of seeds, limited knowledge, and are unaware of the right processing methods, and are thus unable to facilitate the supply and demand of these pigmented varieties. Hence, their production cost is high, while the yield is low and that's the reason the general public cannot meet the expenses despite numerous advantages when compared with the regular ones.[18] At the same time, many people are unaware of how much nutrition these varieties contain. It is a great supplement to malnutrition as it is abundant in zinc, iron, manganese, copper, selenium, magnesium, potassium, phosphorous, and many amino acids (specifically black wheat)[26] but, ironically, poor people are unable to afford them as it costs twice, thrice, or more than the common varieties depending on the country. Thus, it is necessary to popularize these products with better research work into achieving a higher yield and the formation of better-performing products. There is also an expectation that the government will try to include pigmented wheat as a supplement in the mid-day meals by procuring it at a higher minimum support price (MSP).[18]

Colored wheat has attracted the attention of many, but the low yield is becoming the dragging point; hence, it is essential to provide a satisfactory yield along with the regional adaptation to the world.[29] Investigations are required to check the insects, pests, and disease resistance and the flexibility with the environment and the consumers. This is because the pigments degrade with heat (thermal processing techniques affect the anthocyanins content and thus the thermal stability of anthocyanins in foods is limited), change in pH, and several other environmental and chemical factors, which is a major challenge of applications of pigmented wheat in the food industry.[22] The research regarding these provides an expectation for the food industry to use colored wheat grains to develop better wheat and wheat-based products, enhancing their nutritive values.[1] Studies are still done as they can formulate into highly nutritive foods. These crops have huge sell prospects because of the potential for product development aside from the beneficial properties.[63]

1.9 Conclusions

Pigmented wheat is usually a rich source of pigments and bioactive compounds with better nutritional and functional properties. The health benefits of colored wheats are known because of their richness in dietary fibres, carbohydrates, proteins, phytochemicals, and other micronutrients present in the outer layer, compared with white ones. However, these characteristics mainly depend on the genotypes, crop processing method, storage conditions, and environment. It acts as a food and medicine because it reduces the risk of chronic diseases. It shows that colored wheat, in all senses and purposes, is a much better food source than the normal ones. Colored wheat is much healthier, as it increases the nutritional content, such as the developed products' anthocyanin content, than white one. A careful selection of appropriate fractionation processes is required, as most of the potent nutrients are present in the bran and germ layer. Thus, it depends on the processing method for the quality of the flour and the products developed from it. Although the pearling process is preferred over that of the roller-milling process for retaining the anthocyanins in the refined flour, further investigation is still needed. Pigmented wheat contains several nutrients and micronutrients and thus has the potential to tackle the issue of malnutrition persisting in the world. It can also benefit farmers by providing higher prices for higher yields, rather than giving higher production costs. So, there is a need to properly develop and utilize pigmented wheat and its products. Pigmented wheat has huge market potential and provides many additional features needed for commercial product development, laying the foundation for industrial use. However, research would be necessary to build out the supply chains, small- and large-scale tests could be modeled to evaluate the potential applications of pigmented wheat and whether it would compete with the normal varieties in terms of yields, function, and quality performance.

References

1. R. Gupta, M. Meghwal and P. K. Prabhakar, *Trends Food Sci. Technol.*, 2021, **110**, 240.
2. P. Giraldo, E. Benavente, F. Manzano-Agugliaro and E. Gimenez, *Agronomy*, 2019, **9**, 352.
3. P. Saini, N. Kumar, S. Kumar, P. W. Mwaurah, A. Panghal, A. K. Attkan, V. K. Singh, M. K. Garg and V. Singh, *Crit. Rev. Food Sci. Nutr.*, 2021, **61**, 3197.
4. F. E. Bokore, *Hulless Barley (Hordeum Vulgare L.) and Pigmented Wheat (Triticum L.): Genetic Diversity for Healthy Food*, Lap Lambert Academic Publishing GmbH KG, 2011.

5. J. Lachman, P. Martinek, Z. Kotíková, M. Orsák and M. Šulc, *J. Cereal Sci.*, 2017, **74**, 145.
6. S. Sharma, P. Kapoor, S. Kaur, A. Kumari, N. Sharma, A. Kumar, V. Chunduri and M. Garg, in *Physiological, Molecular, and Genetic Perspectives of Wheat Improvement*, ed. S. H. Wani, A. Mohan and G. P. Singh, Springer International Publishing, Cham, 2021, pp. 71–88.
7. A. Francavilla and I. J. Joye, *Nutrients*, 2020, **12**, 2922.
8. D. Giordano, M. Locatelli, F. Travaglia, M. Bordiga, A. Reyneri, J. D. Coïsson and M. Blandino, *Food Chem.*, 2017, **233**, 483.
9. S.-Q. Tian, Z.-C. Chen and Y.-C. Wei, *J. Cereal Sci.*, 2018, **83**, 63.
10. L. Dykes and L. W. Rooney, *Journal*, 2007, 105–111, https://nulifemarket.com/wpcontent/uploads/2016/02/CFWPhenolicCompoundsCerealGrainsTheirHealthBenefits.pdf.
11. S. Žilić, A. Serpen, G. Akıllıoğlu, V. Gökmen and J. Vančetović, *J. Agric. Food Chem.*, 2012, **60**, 1224.
12. S. Sharma, PhD thesis, Panjab University, 2020.
13. Y. Hemery, X. Rouau, V. Lullien-Pellerin, C. Barron and J. Abecassis, *J. Cereal Sci.*, 2007, **46**, 327.
14. L. Paznocht, Z. Kotíková, M. Šulc, J. Lachman, M. Orsák, M. Eliášová and P. Martinek, *Food Chem.*, 2018, **240**, 670.
15. M. Garg, M. Chawla, V. Chunduri, R. Kumar, S. Sharma and N. K. Sharma, *et al.*, *J. Cereal Sci.*, 2016, **71**, 138.
16. A. C. Zeven, *Euphytica*, 1991, **56**, 243.
17. S. Dhua, K. Kumar, Y. Kumar, L. Singh and V. S. Sharanagat, *Trends Food Sci. Technol.*, 2021, **112**, 780.
18. S. Kumari and L. Tzudir, *Biot. Res. Today*, 2021, **3**, 240–242.
19. T. Beta, W. Li and F. B. Apea-Bah, in *Flour and Breads and Their Fortification in Health and Disease Prevention*, ed. V. R. Preedy and R. R. Watson, Academic Press, 2 edn, 2019, p. 75.
20. S. N. Syed Jaafar, J. Baron, S. Siebenhandl-Ehn, T. Rosenau, S. Böhmdorfer and H. Grausgruber, *Plant Breed.*, 2013, **132**, 546.
21. E.-S. M. Abdel-Aal, J. C. Young and I. Rabalski, *J. Agric. Food Chem.*, 2006, **54**, 4696.
22. J. Diddugoda, D. Jeewani and M. D. Nishantha, *J. Agric. Vet. Sci.*, 2018, 230–235.
23. K. Kishor, *Just Agriculture*, March 2021, pp. 66–69.
24. I. V. Vrček, D. V. Čepo, D. Rašić, M. Peraica, I. Žuntar, M. Bojić, G. Mendaš and M. Medić-Šarić, *Food Chem.*, 2014, **143**, 522.
25. Z. Sramkova, E. Gregová and E. Šturdík, *Acta Chim. Slov.*, 2009, **2**, 115.
26. Z. F. Guo, Z. B. Zhang, P. Xu and Y. N. Guo, *Cereal Res. Commun.*, 2013, **41**, 293.
27. B. L. D'Appolonia and P. Rayas-Duarte, in *Wheat: Production, Properties and Quality*, ed. W. Bushuk and V. F. Rasper, Springer US, Boston, MA, 1994, p. 107, DOI: 10.1007/978-1-4615-2672-8_8.
28. X. Wang, X. Zhang, H. Hou, X. Ma, S. Sun, H. Wang and L. Kong, *Food Res. Int.*, 2020, **138**, 109711.

29. S. Sharma, V. Chunduri, A. Kumar, R. Kumar, P. Khare, K. K. Kondepudi, M. Bishnoi and M. Garg, *PLoS One*, 2018, **13**, e0194367.
30. C. M. Galanakis, in *Nutraceutical and Functional Food Components*, ed. C. M. Galanakis, Academic Press, 2017, pp. 1–14.
31. W. Li and T. Beta, in *Flour and Breads and Their Fortification in Health and Disease Prevention*, ed. V. R. Preedy, R. R. Watson and V. B. Patel, Academic Press, San Diego, 2011, pp. 59–67, DOI: 10.1016/B978-0-12-380886-8.10006-6.
32. A. Chandrasekara, in *Reference Module in Food Science*, 2018, DOI: 10.1016/B978-0-08-100596-5.22395-0.
33. J. Zhang, Y. Ding, H. Dong, H. Hou and X. Zhang, *J. Chem.*, 2018, 6459243.
34. A. N. Panche, A. D. Diwan and S. R. Chandra, *J. Nutr. Sci.*, 2016, **5**, e47.
35. P. B. Mellen, T. F. Walsh and D. M. Herrington, *Nutr. Metab. Cardiovasc. Dis.*, 2008, **18**, 283.
36. A. Pasqualone, A. M. Bianco, V. M. Paradiso, C. Summo, G. Gambacorta, F. Caponio and A. Blanco, *Food Chem.*, 2015, **180**, 64.
37. D.-Y. Ma, D.-X. Sun, Y. Zuo, C.-Y. Wang, Y.-J. Zhu and T.-C. Guo, *J. Integr. Agric.*, 2014, **13**, 1258.
38. D. Lu, Y. Yang, Y. Li and C. Sun, *Curr. Pharm. Anal.*, 2015, **11**, 66.
39. H. S. Balyan, P. K. Gupta, S. Kumar, R. Dhariwal, V. Jaiswal, S. Tyagi, P. Agarwal, V. Gahlaut and S. Kumari, *Plant Breed.*, 2013, **132**, 446.
40. P. R. Shewry and J. H. Sandra, *Food Energy Secur.*, 2015, **4**, 178.
41. J. A. Vaccaro and S. A. Naser, *Healthcare*, 2022, **10**, 61.
42. Y. Dror, E. Rimon and R. Vaida, in *Whole-wheat Bread for Human Health*, Springer International Publishing, Cham, 2020, pp. 1–20, DOI: 10.1007/978-3-030-39823-1_1.
43. J. K. Blusztajn, B. E. Slack and T. J. Mellott, *Nutrients*, 2017, **9**, 815.
44. M. Samtiya, R. E. Aluko and T. Dhewa, *Crit. Rev. Food Sci. Nutr.*, 2020, **2**, 6.
45. S. Singh, A. K. Gupta and N. Kaur, *Sci. World J.*, 2012, 485751.
46. M. K. Nadeem, F. Anjum, R. A. I. Amir, M. R. Khan, S. Hussain and M. S. Javed, *Pak. J. Food Sci.*, 2010, **20**, 54–61.
47. S. A. Mallick, K. Azaz, M. Gupta, V. Sharma and B. K. Sinha, *Indian J. Plant Physiol.*, 2013, **18**, 183.
48. C. E. Tyl and M. Bunzel, *Cereal Chem.*, 2014, **91**, 152.
49. A. Costabile, A. Klinder, F. Fava, A. Napolitano, V. Fogliano, C. Leonard, G. R. Gibson and K. M. Tuohy, *Br. J. Nutr.*, 2008, **99**, 110.
50. L. T. Tong, K. Zhong, L. Liu, J. Qiu, L. Guo, X. Zhou, L. Cao and S. Zhou, *Carbohydr. Polym.*, 2014, **112**, 1.
51. V. Ponzo, I. Goitre, M. Fadda, R. Gambino, A. De Francesco, L. Soldati, L. Gentile, P. Magistroni, M. Cassader and S. Bo, *J. Transl. Med.*, 2015, **13**, 218.
52. N. Sharma, V. Tiwari, S. Vats, A. Kumari, V. Chunduri, S. Kaur, P. Kapoor and M. Garg, *Molecules*, 2020, **25**, 5785.
53. Y.-P. Huang and H.-M. Lai, *J. Food Drug Anal.*, 2016, **24**, 564.
54. F. P. M. Hoevenaars, D. Esser, S. Schutte, M. G. Priebe, R. J. Vonk, W. J. van den Brink, J.-W. van der Kamp, J. H. M. Stroeve, L. A. Afman and S. Wopereis, *J. Nutr.*, 2019, **149**, 2133.

55. T. Tsuda, *Mol. Nutr. Food Res.*, 2012, **56**, 159.
56. X. Song, W. Zhu, Y. Pei, Z. Ai and J. Chen, *J. Cereal Sci.*, 2013, **58**, 400.
57. Koka, *Purple Power- the Powerful Health Properties of Purple Food*, accessed 9th February 2022.
58. M. Zanoletti, P. Abbasi Parizad, V. Lavelli, C. Cecchini, P. Menesatti, A. Marti and M. A. Pagani, *LWT--Food Sci. Technol.*, 2017, **75**, 663.
59. Y. Li, D. Ma, D. Sun, C. Wang, J. Zhang, Y. Xie and T. Guo, *Crop J.*, 2015, **3**, 328.
60. A. Kumari, S. Sharma, N. Sharma, V. Chunduri, P. Kapoor, S. Kaur, A. Goyal and M. Garg, *Molecules*, 2020, **25**, 5071.
61. W. Li, T. Beta, S. Sun and H. Corke, *Food Chem.*, 2006, **98**, 463.
62. A. Morgounov, Y. Karaduman, B. Akin, S. Aydogan, P. S. Baenziger and M. Bhatta, *et al.*, *Agronomy*, 2020, **10**, 86.
63. T. H. Gamel, A. J. Wright, M. Pickard and E.-S. M. Abdel-Aal, *Cereal Chem.*, 2020, **97**, 34.

CHAPTER 2

Pigmented Rice: Composition and Health Effects

ANIL KUMAR SIROHA[a], SNEH PUNIA BANGAR*[b] AND SHWETA SURI[c]

[a]Department of Food Science and Technology, Chaudhary Devi Lal University, Sirsa, Haryana, India; [b]Department of Food, Nutrition and Packaging Sciences, Clemson University, Clemson, SC, 29634, USA; [c]Department of Food Engineering, National Institute of Food Technology Entrepreneurship and Management, Sonipat, Haryana, 131028, India
*E-mail: snehpunia69@gmail.com

2.1 Introduction

Rice (*Oryza sativa* L.) is the world's most significant staple meal, and there are two major types: white (non-pigmented) and colored (pigmented) rice. Although the most commonly consumed type is white rice, some unique rice cultivars such as black rice, purple rice, red rice, and brown rice are also consumed. The color of these rices is attributed to the type of pigments found in the pericarp or bran section of the rice kernel. Black glutinous rice, the most prevalent variety of brown rice, is also known as 'black pearl' due to its high levels of protein, fat, carbohydrates, phenols, flavonoids, and anthocyanins.[1] The pigments are made up of anthocyanins, and their colors range from red to purple to black.[2] Rice is an essential grain because of its nutritional importance. Rice is also classified on the basis of amylose content as glutinous

Food Chemistry, Function and Analysis No. 38
Pigmented Cereals and Millets: Bioactive Profile and Food Applications
Edited by Sneh Punia Bangar, Sajid Maqsood and Anil Kumar Siroha

Published by the Royal Society of Chemistry, www.rsc.org

or non-glutinous, with subcategories of glutinous (0–5% amylose), low (6–18%) amylose content (AC), intermediate (19–23%) AC, and high (>23%) AC.[3] The viscosity of their kernels and method of edibility varies between glutinous and non-glutinous types.[4] According to the Food and Agricultural Organization of the United Nations,[5] global rice production was about 756 × 10^6 tonnes, with China being the world's top producer (213 × 10^6 tonnes), followed by India (178 × 10^6 tonnes). Rice grain is an important source of carbohydrates, protein, and other essential elements for billions of people worldwide, particularly in underdeveloped nations.[6,7] It is primarily a source of carbohydrates and minerals such as calcium, magnesium, phosphorus, and contains trace concentrations of copper, iron, manganese, and zinc.[8] Rice processing includes cleaning, hulling, and post-hulling processing such as whitening, polishing, and grading. Rice processing generates a variety of products such as husk, milled rice, and bran.[9] Paddy milling produces 70% rice (endosperm) as the main product and husk (20%), bran (8%), and germ (2%) as by-products.[10]

Antioxidants first gained attention as a 'miracle substance' in the 1980s and today research on antioxidant compounds is a major priority. The use of antioxidant supplements is increasing every day due to their positive effects.[11] White pericarp has lower antioxidant activity than red and black (purple) pericarp. A distinct subgroup of polyphenols can be found in the colored pericarp. Anthocyanins are found in black bran, while proanthocyanidins are found in red bran.[12] Brown rice had a total phenol level of 27.84 mg per 100 g, mainly ferulic acid and *p*-coumaric acid.[13,14] Anthocyanins (ACNs) are the pigment of black rice and proanthocyanidins (PA)/condensed tannins are the pigment of red rice.[15–17] According to Gao *et al.*,[18] red rice's primary bound phenolic acids are ferulic, *p*-coumaric acids and 2,5-dihydroxybenzoic acids, while black rice comprises protocatechuic and vanillic acids. The stomach and small intestine quickly absorbs the soluble phenolic acids and distributes them throughout the body, providing additional health advantages, such as suppressing the oxidation of low-density lipoprotein cholesterol and liposomes.[12,19] Phenolic acids are found primarily in rice bran, with trace levels in the endosperm of certain rice genotypes.[20]

Starch is the most prevalent biomolecule in rice grain, accounting for 72–82% (dry weight) of brown rice grain and nearly 90% of milled rice grain.[21] Starch is typically a mixture of amylose and amylopectin, amylose being a linear carbohydrate polymer of glucose residues linked *via* α-D-(1,4)-glucosidic linkages, and amylopectin a mixture of α-D-(1,4)-linked glucose residues having highly branched glucose units linked by α-D-(1,6)-glucosidic bonds. Rice starch granules are the smallest size fraction in cereal grains, with sizes ranging from 2 to 7 m.[22]

The physicochemical properties of starch determine the extent to which starch contributes to food quality, particularly its water absorption ability, thermal properties, and pasting behaviors.[23] The pasting and gelatinization characteristics of starch play an essential role in determining the eating and cooking quality of rice.[24] Because starches are unstable during heating, shearing, acidity, and subsequent storage, starch is often changed to increase its application in the food and non-food industries.[25] Starch is often modified through physical, chemical, and enzymatic processes to improve these qualities, and further investigated for specific purposes in food.[26] Physical treatments of starch modifications are preferred over chemical and enzymatic treatments, because people like to consume more natural foods, and chemicals are not used in the physical treatment. This chapter summarizes the nutritional makeup, functional properties, phytochemical constituents, and starch characteristics of pigmented rice.

2.2 Nutritional Composition

Pigmented rice contains good amounts of vitamins, minerals, fiber, and different phytochemicals, which have beneficial health effects.[27] The nutritional content of pigmented rice varies according to genetics and ecological conditions. Usually, colored rice is either semi-polished or unpolished; retaining most of the bran layer. Brown rice holds the bran layer (including micronutrients-vitamins, minerals, and fiber) because it is not polished or refined to make white rice. Different pigmented rice varieties are well known for their nutritional quotient. Brown rice is highly nutritious, containing fewer calories and more fiber and micronutrients *viz.* selenium, magnesium, manganese phosphorus, thiamin, niacin, and pyridoxine. The high nutrient content of brown rice may be due to the micronutrients attached to the rice's bran and outer husk. Rice bran and rice husks contain higher amounts of calcium, iron, and zinc. Red rice is a good source of minerals such as iron and zinc, whereas black rice is particularly rich in protein, fiber, and fat. Pigmented rice varieties like black and red rice scavenge free radicals and acquire most of their color through anthocyanin pigments, which are recognized for their antioxidant properties and therapeutic potential.[28]

2.2.1 Macronutrients

Pigmented rice flour contains high levels of protein (8.52–9.68%), ash (1.34–1.68%), fat (3.61–4.35%), and total dietary fiber (3.68–4.76%).[29] It was found that the amount of total dietary fiber ranged from 3.0 to 4.5% in red and black rice varieties grown in Thailand.[27] For macronutrient parameters, studies have found that red rice has more crude protein (10.49%) and crude fiber (2.71%) than white rice.[30] A comparative analysis of the proximate composition of Thai colored rice flour was conducted by Kraithong *et al.*[58] It was observed that ash, protein, and crude fat content varies from 1.14–1.44%, 7.27–7.54%, 2.05–3.60%, respectively, for rice berry flour and red jasmine rice flour. The moisture content of rice varies depending on the differences in genetic composition and the climatic situations in which the rice varieties are cultivated. The moisture level of pigmented rice varieties fluctuates from 7.1 to 13.1% (Table 2.1). It is the second biggest constituent of rice after

Table 2.1 Proximate composition of pigmented rice.

Rice variety	Moisture (%)	Protein (%)	Crude fiber (%)	Crude fat (%)	Crude ash (%)	Carbohydrate (%)	Ref.
Pigmented rice variety	9.3–13.1	7.16–10.85	0.28–0.61	1.15–3.19	0.82–1.5	70.75–81.29	27
Red (China)	11.90	9.72	2.52	2.35	1.37	74.66	
Red (Sri Lanka)	12.94	9.63	2.82	1.15	0.82	75.45	
Brown	7.1–11.6	5.9–11.0	0.22–0.95	1.6–2.8	0.43–2.34	71.31–84.75	128
Red	12.70	10.49	2.71	1.81	1.53	70.19	30
White	12.75	7.60	0.23	0.62	0.46	78.34	
Brown	11.40	8.75		3.33	0.83	74.67	63
Purple	12.05	7.77		3.73	1.79	74.38	
Black	11.12	5.57		3.05	1.38	78.24	
Black	11.8–13.5	7.4–11.3		3.0–3.7	1.63–1.88		132
Red	12.9–13.3	8.2–10.2		2.8–3.3	1.55–1.86		

starch. It has a big impact on the taste, as well as the nutritional value of the rice. Rice protein is superior to other grains because it has a better-balanced amino acid profile than wheat, corn, millet, and sorghum, due to lysine.[28] Red rice contains more protein than white rice.[30] Rice fat is an excellent source of essential polyunsaturated fatty acids such as linoleic acid, oleic acid, *etc.*; in fact, rice does not have any cholesterol content. Generally, the fat content is present in the bran portion (20% dry weight). Fat content in rice varies from 1.15 to 3.19% (Table 2.1). The presence of fiber in the rice provides bulk to the diet and acts as a laxative for the intestines. Arabinoxylan and β-glucan are the main constituents of water-soluble dietary fiber in rice. Furthermore, carbohydrates like rhamnose, xylose, glucose, galactose, and mannose are also contained in the water-soluble fiber. Insoluble fiber comprises cellulose, hemicellulose, insoluble β-glucan, and arabinoxylan. However, the amount of non-starch polysaccharides relies on the rice variety, milling quality, and water solubility.[31] Moreover, the ash content of rice varies between varieties, because of genetic factors and the mineral profile of the soil. The ash content signifies the mineral content of rice.[32] Minerals like calcium, iron, zinc, and magnesium are found in rice.

Processing, such as dry milling and wet processes can also influence the proximate content of rice flour. Wet milling tends to create rice flour with a higher carbohydrate content (and a lower content of other proximate parameters). This could be because of processes like soaking and grinding, where most of the sugars, soluble proteins, and non-starch components are combined with lipid to remove them.[33]

2.2.2 Micronutrients: Minerals & Vitamins

Pigmented rice varieties are considered a storehouse of essential vitamins and minerals such as iron, zinc, phosphorus, vitamin B-complex, *etc.*; however, it greatly depends on the rice variety and genotype. Red rice contains high concentrations of iron and zinc.[28] Several comparative studies of different colored rice varieties have shown that red rice genotypes have high iron levels in the range of 1.45 to 1.72 mg per 100 g,[34] then transgenic brown rice variety IR68144 has an iron content of 21 μg g.[35]

Other minerals found in red rice include calcium (5.74–9.55 mg per 100 g), sodium (13.72–19.54 mg per 100 g), potassium (19.48–29.00 mg per 100 g), magnesium (45.91–54.25 mg per 100 g), copper (0.27–0.36 mg per 100 g), zinc (0.46–0.65 mg per 100 g), and iron (0.25–1.72 mg per 100 g).[34] Besides, black rice varieties having the black color of bran, which covers the endosperm showed high micronutrient content *viz.* calcium (21.38 mg per 100 g), potassium (186.54 mg per 100 g), magnesium (107.21 mg per 100 g), sodium (10.19 mg per 100 g), copper (0.81 mg per 100 g), zinc (0.29 mg per 100 g) and iron (0.58 mg per 100 g).[36] However, the Korean variety of black rice showed higher levels of calcium (3215.50–4034.00 ppm), potassium (90.10–191.30 ppm) and magnesium (1202.00–1847.50 ppm).[37] The consumption of brown rice is observed to deliver up to 75% of the recommended daily intake of minerals such as zinc, copper, and iron, while white rice provides only 37%.[38]

Studies on the amount and concentration of vitamins in pigmented rice is rather scarce, however, research has revealed that black, as well as red rice, are good sources of carotenoids (vitamin A) and tocols (vitamin E). Pigmented rice grains are an excellent source of vitamin E, including tocopherols and tocotrienols. Among tocotrienols, β and γ-tocotrienols are abundant in rice.[39] Black rice contains total vitamin E content ranging from 28.54 to 51.33 mg kg^{-1}, while the total vitamin E content in red rice was 19.36–37.00 mg kg^{-1}.[40] However, rice processing, including dehulling and milling, significantly lowers the tocopherol content in the grain.[41]

Among carotenoids, lutein and β-carotene are most commonly found in pigmented rice varieties, while, zeaxanthin is observed in lower levels.[42] The total carotenoid content in black rice was found to be 2.2 mg per 100 g, while golden rice was observed to have a total carotenoid content of 3 to 16 mg per 100 g. The concentration of carotenoids also varies based on the genotype of pigmented rice. For instance, the black rice procured from different Asian countries showed high concentrations of all-*trans*-β-carotene,[43] while on the contrary, the Japanese black was found to have significant concentrations (around 94% of total carotenoids) of lutein and zeaxanthin, while β-carotene and lycopene were observed in minor quantities.[44]

2.3 Functional Properties

Functional properties are the basic physicochemical properties of grain that mimic the complex interactions between the composition, structure, and molecular conformation of the constituents and the environment with which they are associated.[45] These functional attributes of grains are influenced by either the genotype of the grain or other non-genetic factors. Conversely, it is vital to study the functional properties of the grain in order to optimize the design variables of the agricultural machinery used to produce, handle and store the grain. Therefore, it is important to define the functional properties of these grains.

2.3.1 Swelling Capacity

Swelling capacity is the ability of starch to bind to water *via* hydrogen bonds under certain circumstances, such as temperature and availability of water. The ratio of amylose to amylopectin mainly affects the functional properties of starch in cereals. The swelling property of starch in cereals is mainly due to its amylopectin content.[29] Swelling is recognized as a characteristic of amylopectin, while amylose acts as an inhibitor of swelling.[46] In addition, the swelling capacity of rice flour also depends on the protein and lipids, or channels in the rice flour granules. The packaging of starch granules with protein tends to form a rigid matrix, promote strong hydrogen bonds from the charged side chains, and slow water access to the starch granules. In addition, lipids form a complex with amylose, reducing charged molecules and obstructing swelling.[47–49]

2.3.2 Pasting Properties

The pasting properties of rice are important in the context of the sensory/organoleptic behavior of rice, and can therefore affect what recipes rice flour can be used in the future.[50] The pasting properties of flours and starches determine their cooking behaviour. The pasting behavior of starches/flours is influenced by the structure of starch, water content, shear rate, and temperature. Various researchers studied the pasting temperature of colored and uncolored rice varieties. Starches with low amylose content show greater paste temperatures, as higher thermal energy is needed to break polymer–polymer hydrogen bonds and form polymer–water–hydrogen bonds. On the contrary, starches with high amylopectin content have a greater ability to form hydrogen bonds and water molecules, leading to high viscosity.[51]

Of the three types of rice, the Umling ame red rice variety (90.5 °C) exhibited a substantial greater pasting temperature than the Lingkang taker ame red rice variety (83.7 °C) and white rice (77.7 °C), which indicates the minimum required temperature to cook the rice flour.[34] Usually, the rice starch is reported to have a pasting temperature ranging from 79.1 to 79.5 °C.[52]

The peak viscosity is another property of rice starch; that is, is the maximum viscosity that the gelatinized grain mix reaches all through heating in water, which indicates the water retention capacity of the mixture. White rice varieties have higher peak viscosity (2767.6 cP), followed by the two red rice varieties. This may be due to the damage caused to the starch throughout the dry milling procedure. Another pasting parameter of rice, *i.e.* final viscosity, varies depending on the amylose molecules and their amount.[53] The final viscosity of white rice is greater than that of colored rice.

Furthermore, breakdown value is another viscography parameter of rice starch. This indicates the thermal stability of starch at 95 °C.[54] For red rice varieties, breakdown viscosity was 17 cP, which was lower than their white counterparts. Moreover, the setback viscosity led to the reorganization of amylose molecules released from the starch granules that swelled throughout the cooling process. This is a criterion of starch gelling/retrogradation ability.[55]

Parboiling rice reduces its paste profile compared with raw rice. This is due to the increase in damaged starch, which absorbs moisture, and the decrease in peak viscosity caused by the resistance of starch granules to swell due to gelatinization during parboiling.[56]

2.3.3 Water Absorption/Solubility Properties

'Water absorption properties' means the ability of the grain to recombine with water in water-restricted settings. Previous studies on rice's functional attributes show that parboiling increases rice's water absorption index (WAI) and water solubility index (WSI).[56] The water absorption properties of pigmented rice are affected by the processing conditions. Thermal treatment for 8 to 12 hours led to an increase in the water absorption ability of red rice flour.[57] The moisture content of rice starch plays the leading role in influencing water absorption, as the starch molecular chains are weakened with the passage of water. The water absorption index of pigmented rice from Thailand varies from 5.44 to 7.14 g.[58] The higher water absorption index is provided by the numerous hydrophilic groups of starch molecules that give foods softness and viscosity. In addition, the negative charge of the phosphate group in amylopectin produces an increased ability to bind water. On the contrary, rice flour's high lipid content prevents the hydrophobic part from hydrating the starch granules, resulting in a low water absorption index.[47,59]

2.3.4 Hydration Properties

The rate of water uptake by the grains is typically dependent on temperature and soaking/immersion time. The hydration capacity of grains is a vital property that has an impact on the cooking quality and, consequently, the sensory attributes of the product.[30] Grains are moderately hydrated and their hydration behavior depends on their natural, as well as processed, conditions. The hydration properties of cereals have been extensively studied in the past. Several studies have described the importance of absorption kinetics. The rate of hydration is influenced by soaking time, the temperature of soaking, and solute concentration.[60]

The higher the parboiling temperature of rice, the higher the water absorption rate, with an analogous hydration curve pattern. It was observed that soaking rice at a temperature of 60 °C led to higher water absorption initially, which slowed down at a later stage. However, soaking rice grains at higher temperatures (70 °C and 80 °C) led to a sharp rise in hydration rate. This upsurge may occur due to the irreversible variations in starch granules caused by the gelatinization process.[56]

2.4 Starch and Starch Modification

Starch is the main carbohydrate constituent in rice, and is made up of amylose and amylopectin. Amylose is usually a linear molecule made up of glucose units joined by an α-1,4-glycosidic bond, while amylopectin is highly branched, with glucose monomers linked with an α-1,6-glycosidic bond.[29] The high-quality starch contained in rice is suitable for food manufacturing and processing. Examining the properties of rice starch is crucial to meeting the varying needs of consumers and processors for specific food processing.[37]

A study by da Silva *et al.*[61] characterized the pigmented rice starch (white, red, black) and observed a yield of starch of 35.74–47.05%, respectively. It was observed that starches are polygonal and angular in shape. Phenolic content in the rice starch was observed as 288.95 to 661.60 mg GAE per 100g, white starch showed the lesser value, whereas black rice starch had the highest value. Black rice starch also showed slightly higher crystallinity and thermal stability compared with other starches. Bhat and Riar[62] reported that starch granules showed irregular polyhedral morphology, with spherical granules that have a relatively smooth surface. Rice starch showed an A-type crystalline pattern, which is mainly found in cereal starches. The diameter of the analyzed starch particles varied from 5.139 µm to 8.453 µm. Peak viscosity (PV), breakdown viscosity (BV), final viscosity (FV), and pasting temperature (PT) of starches ranged from 3977–5359 cP, 1552–3205 cP, 3906–5464 cP and 75.25–87.20 °C, respectively. Tangsrianugul *et al.*[29] analyzed pigmented rice starch and flour for physicochemical and rheological properties. It was observed that the amylose content of pigmented rice starch varied from 2.77 to 12.09%. PT, setback, and FV increased, while breakdown and swelling power reduced with increased amylose content for flour and starch samples. The rheological behavior of starches was also observed and G' (storage modulus) of starches was observed to be higher than G'' (loss modulus), showing that starch pastes are more elastic than viscous. The $\tan\delta$ values of starches were observed to be less than 1, showing shear thinning behavior of starch paste. In pigmented rice, the starch granules differ in size and shape, leading to the formation of a cluster in the cell wall matrix, with a dispersion of small spherical aleurone particles. After the starch extraction process, starch granules come out from the cell wall matrix and the covered proteins plus lipids, resulting in a morphological change in the granules – from spherical to polygonal – however, some starch granules adhere to each other.[63] Usually, the size of starch granules in the rice varies with the rice cultivar, the cultivation conditions, and the physiology of the plant.[29] Native starches are less stable at high temperatures, pH, shear rate, and cold storage. Starches are generally modified using different physical, chemical, and enzymatic treatments to improve these properties. Dhull *et al.*[64] modified black rice starch using chemical and physical treatments. Starch was modified using octenyl succinic anhydride (OSA), heat moisture (HMT), ultrasonication, and gamma irradiation treatment. It was observed that swelling power, amylose content and solubility power were highest for OSA starch, ultrasonicated and gamma irradiated starch, respectively. After the modification process, significant effects were observed on the surface of starch granules. HMT-modified starch showed aggregation of rice starch granules, and grooves and cracks were observed for ultrasonication-modified starch, while rough granules, and fusion of granules were detected for OSA starch.

2.5 Phytochemical Diversity of Pigmented Rice and Other Bioactive Molecules

In recent years, colored or pigmented rice, *viz.* red rice, black rice, brown rice, purple rice, and wild rice varieties have attracted consumers who are seeking healthier foods. The therapeutic nature of colored rice is due to polyphenols having antioxidant, anti-inflammatory, anti-carcinogenic, anti-allergic, anti-atherogenic, and sugar-lowering effects.[65] The phytochemicals in colored rice can be categorized into different subclasses: phenols, carotenoids, nitrogen, alkaloids, and organosulfur compounds. Phenols are classified into phenolic acids, flavonoids, coumarins, and tannins. Anthocyanin, the main pigment compound in red and black rice, is a type of flavonoid.[28]

Pigmented rice is abundant in phenols, flavonoids, and antioxidant compounds. In comparison with white rice varieties, colored rice contains higher levels of polyphenols.[30] In general, the distribution of phenolic compounds in various rice fractions increases in the order of endosperm, whole grains, embryo, and bran. The rice bran holds the aleurone, an important tissue that hoards several phenolic compounds. Bran uses the polyphenolic components to further crosslink non-starch polysaccharides with other cell wall constituents, such as lignin, to enhance the physical integrity of the cell wall. In addition, phenols found in pigmented rice bran protect from invasive pathogenic microorganisms. The phenolic composition of different rice varieties differs because of genetic variation in the thickness of bran.[66]

Cinnamic acid is the primary precursor for synthesizing different phenolic acids in rice varieties, such as ferulic acid, *p*-coumaric acid, isoferulic acid, sinapic acid, and 2,5-dihydroxybenzoic acid.[67,68] Table 2.2 shows the composition of phenols and flavonoids in various pigmented rice types. The most common free phenols abundantly found in pigmented rice are

Table 2.2 Polyphenol composition of pigmented rice varieties.

Rice variety	Total phenol content (mg GAE per 100 g)	Total anthocyanin content (mg CYGE per 100 g)	Total flavonoid content (mg RE, QE, CE per 100 g)	Ref.
Sung Yod (red rice)	169.11 mg per 100 g	21.96 mg per 100 g	—	129
Hom Mali (white rice)	112.59 mg per 100 g	—	—	
Hom Nil (purple rice)	588.28 mg per 100 g	65.85 mg per 100 g	—	
Mali Dang (red rice)	943.53 mg per 100 g	18.56 mg per 100 g	—	
Mun Poo (red rice)	452.4 5 mg per 100 g	16.47 mg per 100 g	—	
Red rice	294.71 mg per 100 g	6.52 mg per 100 g		28
Red rice	143.38 mg per 100 g	—	120.0 mg per 100 g	30
White rice	24.26 mg per 100 g	—	166.23 mg per 100 g	
Red Thai rice	364.8 mg per 100 g	0.3–1.4 mg per 100 g	—	27
Black Thai rice	492.8 mg per 100 g	109.5–256.6 mg per 100 g	—	
Red rice	470.1 mg per 100 g	—		1
Black rice	1055.7 mg per 100 g	—		
Black rice	579.00 mg per 100 g	1.81 mg per 100 g		130

protocatechuic acid (0.17–143.23 mg per 100 g), ferulic acid (0.15–153.84 mg per 100 g), and vanillic acid (0.303–49.04 mg per 100 g). However, some of the phenols are found in lower quantities, *i.e.* chlorogenic acid, cinnamic acid, and ellagic acid.[67] Caffeic acid is found in smaller amounts in red rice but is not detected in brown rice varieties.[39] Black rice contains good amounts of phenolic acids *viz.* chlorogenic acid (11.26 mg per 100 g), ferulic acid (8.37 mg per 100 g), protocatechuic acid (143.23 mg per 100 g), *p*-hydroxybenzoic acid (15.38 mg per 100 g), sinapic acid (27.22 mg/100 g), vanillic acid (43.80 mg per 100 g).[69] A comparative study of the phenolic content of red and black rice varieties showed that red and black rice contain a free total phenol content of 0.173 mg GAE per g and 0.536 mg GAE per g, respectively.[70]

Several bioactive flavonoids are also found in pigmented rice, compared with uncolored/plain rice when analyzed by a metabolomics approach, based on ultra-high performance liquid chromatography triple with quadrupole mass spectrometry. The presence of higher transcript levels of flavonoid biosynthetic genes is the main reason for the higher flavonoid content of pigmented rice compared with their uncolored counterparts. Furthermore, the darker the color of the grain, the higher will be its flavonoid content.[71] The main flavonoids of colored rice grains are anthocyanins and proanthocyanidins. Other flavonoid-like compounds commonly seen were quercetin-3-*O*-glucoside, isorhamnetin-3-*O*-glucoside, quercetin-3-*O*-rutinoside, and methoxy-flavanol-3-*O*-glucoside.[39]

In black rice, the quantity of flavonoids in the free fraction is greater than that of flavonoids contained in the bound fractions.[72] Black rice contains good amounts of individual flavonoids, *viz.* epicatechin (6.35 mg per 100 g), kaempferol (1.23 mg per 100 g), quercetin (1.31 mg per 100 g), rutin (4.32 mg per 100 g), and cyanidin-3-*O*-glucoside (156.76 mg per 100 g).[69] Black rice contains higher levels of apigenin (0.20–1.49 mg per 100 g), catechin (0.241–17.00 mg per 100 g), kaempferol (0.375–94.62 mg per 100 g), and quercetin (0.05–17 mg per 100 g).[66] On the other hand, luteolin and quercetin were present in red rice below the detection limit. Of the seven flavonoids found in rice bran, tricin is the most important, accounting for about 77% of total flavonoid concentration.[73] Processing also affects the composition of bioactive compounds in the rice. Germination/sprouting leads to the enhancement of most nutrients, whereas the soaking of grains permits the water-soluble phenolic compounds to escape freely. However, anthocyanins are very sensitive to changes in temperature and also tend to oxidise easily.[59] Germinated rice has many proven therapeutic advantages in averting the risk of non-communicable diseases; for example, diabetes mellitus, cancer, cardiovascular disease, and Alzheimer's disease.[66,74,75] Twenty-one conjugated flavones have been identified in germinated brown rice, with flavones such as apigenin-6-*C*-glucoside-8-arbinoside and its two isomers being found in higher levels in the rice pericarp or testa region.[76] A drying temperature of 60 °C to 100 °C reduces the yield of head rice because of the development of cracks in the endosperm of black rice, which reduces the anthocyanin content in black rice. In contrast, a high drying temperature enhances the intestinal bioavailability of ferulic acid, but this has no influence on the anti-inflammatory properties of black rice.[77,78] Among polyphenolic acids found in rice, caffeic acid, ferulic acid, *p*-coumaric acid, and gallic acids are the most heat-labile polyphenols at drying temperatures.[66]

2.6 Heath Benefits of Pigmented Rice

Many health benefits of pigmented rice have been observed (Table 2.3), which are due to the higher polyphenolic content of pigmented rice.

2.6.1 Gluten-related Disorders

Rice is a gluten-free grain that contains the amino acid lysine as a limiting amino acid. As a result, it is often researched as a raw material for formulating foods directed toward celiac patients or those suffering from wheat-related illnesses.[79,80] Celiac disease (CD) is an immunological response to gluten consumption that causes damage to the upper small intestinal lining, resulting in malabsorption and medical complications.[81,82] Diarrhea is the most common symptom of CD, with malnutrition, growth retardation, anemia, and fatigue among some of the other symptoms of CD.[83] Currently, the most common way to manage CD is to adhere to a lifelong, rigorous

Table 2.3 Highlights of the health benefits of pigmented rice.

Source	Bioactivities	Key findings	Ref.
Brown rice	Anti-cancer activity	Rice bran may be one candidate of the active fraction suppressing DMH-induced early colon carcinogenesis in rats and may have the potential to be developed as a novel dietary supplement for chemoprevention of colon cancer	86
Black rice	Anti-cancer activity	Peonidin 3-glucoside and cyanidin 3-glucoside also exerted an inhibitory effect on the DNA binding activity and the nuclear translocation of AP-1. Furthermore, these compounds also exerted an inhibitory effect of cell invasion on various cancer cells (SCC-4, Huh-7, and HeLa).	90
Red rice	Prevention of celiac diseases	Gluten-free cake was prepared from extruded red rice flour. This gluten-free product prevents celiac diseases. The sensory quality of cake was acceptable.	131
Black, red rice	Anti-inflammatory activity	Extracts obtained from integral, black, red rice and bran showed that the expression of inducible Nitric Oxide Synthase (iNOS) was specifically reduced at both transcript and protein level by extracts, indicating the capacity of rice extracts to modulate the inflammatory response	94
Purple, red, white rice	Anti-inflammatory activity	Natural rice bran oil (NRBOs) inhibited NO and iNOS production by LPS/IFN-γ-stimulated RAW264.7 cells. Additionally, NRBO prepared from KGLP (rice cultivar) exhibited the highest inhibitory activity on NO and iNOS production	95
Brown, red, and purple rice bran	Anti-diabetic potential	Purple and red bran extracts inhibited α-glucosidase activity, however, only the red rice bran extracts inhibited α-amylase activity. These reduce starch digestion	104
Red rice	Anti-diabetic potential	These results indicate that taxifolin derivatives that inhibit the activity of carbohydrate hydrolyzing enzymes, resulting in reduced dietary carbohydrate absorption, which can potentially be used as a strategy to manage diabetes	106
Black rice	Anti-obesity	A high-fat diet supplemented with anthocyanin from black rice suppressed weight gain and ameliorated hypertriglyceridemia in rats without affecting their food consumption	133

gluten-free diet. This underlines the need to provide gluten-free foods that are safe, nutritionally balanced, digestible, and delicious.[84] A gluten-free diet helps lower the morbidity and mortality related to untreated CD.

2.6.2 Anti-cancer Activity

Pigmented rice varieties are high in bioactive chemicals, which provide several health advantages. Cancer is the biggest cause of malignancy-related deaths worldwide. Clinical trials have revealed that many anti-tumor medications have potential therapeutic benefits, but are very toxic to normal cells, thus impacting human health.[85]

People undergoing chemotherapy are urged to consume healthier foods that can diminish tumor cells to prevent the adverse effects of anti-tumor medications. Brown rice and rice bran dramatically decreased cyclooxygenase-2 expression in the middle colon of rats compared to white rice. Thus, rice bran was suggested as an active rice component for preventing colon 1,2-dimethylhydrazine early carcinogenesis.[86] Min *et al.*[12] examined the anthocyanin and oryzanol content of various colored rice bran and discovered that purple rice bran had the highest value when compared with other samples. According to Mazza,[87] anthocyanins have a high antioxidant capacity and act as an anti-cancer agent. γ-Oryzanol can inhibit tumor promotion[88] and tumor growth in tumor-bearing mice *via* induction of natural killer (NK) activity, macrophage inactivation, and angiogenesis inhibition.[89]

Chen *et al.*[90] discovered that anthocyanins from black rice had anti-cancer action. The molecular evidence related to the anti-metastatic properties of peonidin 3-glucoside and cyanidin 3-glucoside, important anthocyanins derived from black rice, demonstrated significant suppression of SKHep-1 cell invasion and motility. Finally, anthocyanins derived from black rice were shown to suppress the development of SKHep-1 cells *in vivo*.

2.6.3 Anti-inflammatory Activity

The build-up of oxidized lipid components (cholesterol and fatty acids) in the tissue cell membrane and biological fluids can induce inflammation.[91,92] Failure to promote resolution results in the creation of a persistent state of inflammation, which is a risk factor for a variety of illnesses, including cancer, diabetes, cardiovascular disease, and autoimmune disease.[93] Petroni *et al.*[94] discovered that colored rice cultivars exhibited increased antioxidant capabilities, which aid in preventing inflammatory disorders. The anti-inflammatory properties of lipopolysaccharide (LPS)-induced mouse RAW 264.7 macrophages pre-treated with extracts of integral, black, red rice, and bran revealed that extracts reduced the inducible Nitric Oxide Synthase (iNOS) expression at both the transcript and protein levels, indicating the ability of rice extracts to modulate the inflammatory response. Natural rice bran oil (NRBOs) suppressed NO and iNOS generation by LPS/IFN-stimulated RAW264.7 cells, according to Pattananandecha *et al.*[95] Furthermore, NRBO derived from KGLP (rice cultivar) had the most vital inhibitory efficacy against NO and iNOS generation. Pigmented NRBOs, particularly those grown in mountainous locations, and those with high levels of tocotrienols, tocopherols, and γ-oryzanol, may have a potential utility as a natural anti-inflammatory active ingredient in nutraceutical and cosmeceutical products.

2.6.4 Antioxidant Activities

Free radicals such as hydroxyl radical ($OH^{\bullet}$), superoxide anion ($O_2^{\bullet-}$), and hydrogen peroxide (H_2O_2) have been implicated in mediating a variety of clinical states and illnesses. Free radicals cause oxidative stress, which is linked to various illnesses, including diabetes, cardiovascular disease, respiratory disease, cancer, neurodegeneration, *etc.*[96] Bagchi *et al.*[97] compared the phenol content in products prepared from black rice, with those from white rice and observed 14 phenolic compounds in most of the products, including 4 hydroxybenzoic acid, 5 hydroxycinnamic acid, and 5 flavonoids, with higher levels in products prepared from black rice than those from white rice. Red rice, for example, is distinguished by the presence of pro-anthocyanidins, which commonly have catechin and epicatechin block units.[15] Furthermore, anthocyanins such as cyanidin-3-*O*-glucoside (Cy-3-glc) and peonidine-3-*O*-glucoside (Pe-3-glc) are found in black rice.[98]

2.6.5 Heart Diseases

Cardiovascular disease is a chronic condition caused by a variety of hereditary and environmental factors, such as smoking, saturated fat-based diets, and not enough physical activity. In developed countries, coronary heart disease and stroke are the two leading causes of death.[99] Oxidative stress, caused by free radicals or active oxygen, causes oxidized cholesterol and/or fatty acids. 7-ketocholerserol, the most common cholesterol oxidation product, and other products, such as aldehydes and short chain fatty acid peroxides, can change tissue cell membranes, causing the cells to have an inflammatory response.[100] Cardiovascular illness, for example, is significantly linked to increased levels of low-density lipoprotein (LDL) cholesterol and oxidized cholesterol.[101] These findings suggest that black rice pigment fraction (BRF) may have cardioprotective effects in patients with coronary heart disease (CHD) by improving plasma antioxidant status and inhibiting inflammatory factors.[102] According to Kushwaha,[103] the Fe in black rice is necessary for the production of healthy red blood cell production and digestion.

2.6.6 Hypertension

The use of flavonoid groups (anthocyanins, proanthocyanidins, flavonoids, flavones, flavonones, flavan-3-ols) regulate anthocyanins, which are most notably associated with hypertension prevention. Zhang *et al.*[100] stated that the anthocyanin extract from black rice bran has a significant effect on preserving cholesterol, but a smaller role in blocking fatty acid oxidation.

2.6.7 Anti-diabetic Potential

Diabetes mellitus is characterized by high blood glucose levels. Anthocyanins and proanthocyanidins are abundant in the bran layer of pigmented rice. The amount of phenolic, flavonoid, anthocyanin, and proanthocyanidin in five diverse rice bran extracts (1 brown, 2 red, and 2 purple) was determined.[104] Bran extracts were tested for their ability to inhibit activity of two major glucosidases (α-amylase and α-glucosidase), as they are necessary for starch digestion in humans. Red rice bran extracts inhibited only α-amylase activity, while all purple and red bran extracts inhibited α-glucosidase activity. Thermally treated grains exhibited significant activity against α-amylase, α-glucosidase, and glycation, whereas cooked grains lowered the estimated glycemic index (eGI) and improved resistant starch (RS). These data show that black rice is beneficial for the dietary control of metabolic illnesses such as diabetes and hyperlipidemia.[69] Although rice is a primary source of glycemic load in rice-consuming societies, black rice grain has been suggested for diabetes treatment in traditional Chinese medicine.[105] Yoon *et al.*[106] revealed that pigmented rice compounds such as taxifolin and its derivatives might inhibit sucrase, a carbohydrate hydrolyzing enzyme. This inhibitory impact affects dietary carbohydrate absorption and, as a result, decreases the volume of systemic blood glucose.

2.7 Recent Updates in the Utility of Pigmented Rice

The consumption of bakery items in India is rising daily. Consumers prefer bakery products because they are more readily available, more convenient, have a unique flavor, a long shelf life and are more cost effective in rural regions. As an evening snack, biscuits, buns, bread, cookies, and cakes are readily accessible baked and ready-to-eat goods.[107] Aside from the nutritional value of pigmented rice, its small starch granules, hypoallergenicity, and hydrophilicity are other important characteristics required to create a wide range of functional food and non-food uses. Rice is the most commonly utilized good in the gluten-free business since it can be used to make ready-to-eat product.[108] Consumers' interest in the nutritional features of food products has resulted in a need for better and healthier snacks than those already on the market.[109] Klunklin and Savage[110] made biscuits with purple rice instead of wheat flour. In terms of the nutritional content, rice flour changed the features of the biscuits. They were high in fibre, antioxidants, and antioxidant activities, but had slightly modified physical qualities such as hardness and spread ratio. Overall, the incorporation of purple rice in biscuits boosted crude protein content, protein digestibility, and bioactive components, while providing consumers with acceptable sensory ratings up to 50% substitution of purple rice. Papillo *et al.*[111] used polyphenol extract of Italian colored rice cultivars to make biscuits. When Artemide rice extract is microencapsulated with maltodextrins (MD) and gum arabic (GA) (50:50, w:w) as a coating agent, the polyphenols are preserved for at least 30 days at −20 °C. Model biscuits had a better polyphenolic profile with powdered rice extracts despite a partial deterioration during baking. Kim *et al.*[112] prepared the cookies by varying the proportion of black rice flour (0–20%), baking time (10–14 minutes), and sugar (sucrose, aspartame, and oligosaccharide). The spread ratio of the cookies was enhanced with black rice and sucrose. The sensory flavor got richer when the baking time and black rice flour were increased. Susmitha *et al.*[113] utilized the pigmented rice in the making of idli. In terms of protein (12.74–15.53%), amino acids (2.78–4.26%), total phenolic acids (0.25–0.49 mg g^{-1}), total flavonoids (0.36–0.71 mg g^{-1}), and antioxidant activity (0.16–0.46 mg phenolic acid per g), pigmented flour-based idli exhibited better than traditional white rice idli (WRI). Kong *et al.*[114] investigated the qualities of noodles made with different quantities (2%, 5%, 10%, and 15%) of black rice bran (BRB). The addition of BRB (5%, 10%, and 15%) considerably reduced the cohesiveness of the noodles' texture. Further, noodles with varying amounts of BRB were functionally examined, and the concentration of polyphenolics, flavonoids, and anthocyanins was higher than the control. The most promising plant sources of dark purple anthocyanins is black waxy rice. Nontasan *et al.*[115] prepared yoghurt using black rice bran. The colorant powder was dark purple and included 18.14 mg g^{-1} of γ-oryzanol, 37.10 mg g^{-1} of total phenolic, and 9.48 mg g^{-1} of anthocyanins. Sui *et al.*[116] used black rice extract powder (ABREP) to fortify bread. The 2% ABREP did not significantly ($p > 0.05$) improve bread quality; however, a 4% ABREP level resulted in reduced flexibility and increased bread density. Bread digestion rates with ABREP were shown to be lowered by 12.8%, 14.1%, and 20.5% with 1%, 2%, and 4% of ABREP, respectively. Mau *et al.*[117] used varying concentrations (10–100%) of black rice (BR) powders to make chiffon cake and observed improvement in the hardness, cohesiveness, adhesiveness, gumminess, and chewiness of the baked cakes. The total phenols, anthocyanins, and scavenging ability was enhanced with the increase in black rice powder levels. Lee *et al.*[118] prepared a cake utilizing a combination of ground black rice (GBR) and medium-grain brown rice (GMBR) and observed an increase in tempering moisture, heating temperature, and heating time, with an increase in the specific volume of black rice cakes. The elevated value of tempering moisture and heating duration decreased the hardness of puffed black rice cake, moreover, GMBR content had a greater impact than heating temperature. Blandino *et al.*[109] assessed the nutritional value of an extruded product made from brown, red, and black rice. The pigmented and brown rice exhibited no difference in terms of soluble and cell wall bound phenolic acid levels; however, red and black rice both had a greater antioxidant capacity after the extrusion. Black rice had the greatest anthocyanin and antioxidant capacity levels even after extrusion. However, red rice produced the best results when both biometric and textural features were considered. Black glutinous rice wine (BGRW) is a traditional rice wine produced in Guizhou, China.[119]

2.8 The Role of Biotechnology in Pigmented Rice

Biotechnology is used in various crops, including rice, to improve its qualitative and quantitative characteristics by transferring desired characteristics from genus/species barrier into the rice gene pool, modifying the target trait without affecting the non-target regions of rice genome.[120] Genetic changeability is the major part of any targeted breeding program. As a result, plant breeders are particularly interested in genetic variation among germplasm. The immature endosperm can produce geranylgeranyl diphosphate, which transforms into carotene phytoene *via* the expression of the enzyme phytoene synthase. Further, the complementation with Daffodil (*Narcissus pseudonarcissus*) phytoene synthase (PSY gene) and bacterial carotene desaturase (CrtI) (from *Erwinia uredovora*) increased β-carotene (vitamin A source for humans) by 4-fold in golden rice.[121,122] Raw data of nucleotides in various pigmented rice was stored in the European Nucleotide Archive (ENA) database and available *via* accession number PRJEB34340. By using this database, all expressed genes with specific characteristics (*i.e.* bioactive properties, folate, amylose content, aroma) of pigmented and non-pigmented rice cultivars may be identified. The data of potential gene markers and functional SNPs associated with functional traits may be utilized in the rice breeding program in the future.[123] Anthocyanins are responsible for the different colors of pigmented rice. MYB transcription factors, bHLH transcription factors, and a functional enzyme DFR regulate anthocyanin accumulation in rice tissues. OsC1 is the sole functionally described MYB TF in anthocyanin biosynthesis, and it is responsible for leaf, anther, and hull color. OsB2, Rc and Rb are a bHLH TF, which are essential for purple-hull and black rice development, red-pericarp creation and anthocyanin accumulation in rice leaves, respectively. Kala3 has been proposed as a MYB TF responsible for the black pericarp.[124]

The variance in pigmentation intensity has been interpreted to imply that the characteristic is polygenic, involving undiscovered genes.[125] While studying flour pigmentation, Tan *et al.*[126] discovered nine quantitative trait loci (QTL) in a recombinant inbred line (RIL) population. Three QTL were found to represent variation in the CIE 1976 color parameter L^* (lightness), two in a^* (red-green), and four in b^* (yellow-blue). 'Kasalath' (red pericarp) and 'Koshihikari', two rice cultivars, were crossed to produce a backcross RIL population. Sarif *et al.*[127] used agro-morphological features and simple sequence repeat (SSR) markers to investigate the genetic variability and diversity of 32 pigmented rice accessions. Quantitative traits (morphological, grain quality, and bioactive properties) and 34 SSR molecular markers were employed to examine cultivar identities and genetic diversity. Major quantitative features differed significantly among all rice accessions.

2.9 Conclusion

Pigmented rice is a rich source of antioxidants and phenolic compounds, compared with conventional rice cultivars (white rice). The health benefits of pigmented rice are mainly due bioactive compounds, especially anthocyanin content. Pigmented rice reduces the risk of various types of cancer, hypertension, anti-inflammatory activity, and has many more health benefits. Biotechnology plays an important role in producing pigmented rice of different colors. The vast collection of genetic material for pigmented rice is collected in gene banks, which help to further improve the nutritional value of pigmented rice. To improve the bioactive profile of food products, pigmented rice is preferred, and many researchers also prepared many extruded and bakery products from the pigmented rice. Starch properties of different pigmented cultivars are analysed by many researchers. Pigmented rice starches are good sources of polyphenolic compounds. To enhance the applications of starches, starch is modified using different treatments. But limited research has been conducted on modification of pigmented rice starches, so pigmented starch modification should be improved. More research should be conducted on the utilization of pigmented rice extract so that a more pure form of bioactive compound may be collected. Limited research has been conducted on pigmented rice starch and modification of pigmented rice starch. More research should be conducted on pigmented rice starch and product preparation from pigmented rice.

References

1. Y. Shen, L. Jin, P. Xiao, Y. Lu and J. Bao, *J. Cereal Sci.*, 2009, **49**(1), 106.
2. R. Yawadio, S. Tanimori and N. Morita, *Food Chem.*, 2007, **101**(4), 1616.
3. C. J. Bergman, K. R. Bhattacharya, and K. Ohtsubo, *Rice Chemistry and Technology*, in ed. E. T. Champagne, AACC, MN, USA, 2004.
4. Y. Gao, X. Guo, Y. Liu, M. Zhang, R. Zhang, A. M. Abbasi and R. H. Liu, *LWT–Food Sci. Technol.*, 2018, **96**, 19.
5. FAO (Food and Agricultural Organization of the United Nations), The statistical division, https://www.fao.org/faostat/en/#data/QCL, 2020, accessed 23 March 2022.
6. Y. Huang, C. Tong, F. Xu, Y. Chen, C. Zhang and J. Bao, *Food Chem.*, 2016, **192**, 873.
7. M. Parengam, K. Judprasong, S. Srianujata, S. Jittinandana, S. Laoharojanaphand and A. Busamongko, *J. Food Compos. Anal.*, 2010, **23**, 340.
8. D. K. Verma and P. P. Srivastav, *Trends Food Sci. Technol.*, 2020, **97**, 355.
9. K. Gul, B. Yousuf, A. K. Singh, P. Singh and A. A. Wani, *Bioact. Carbohydr. Diet. Fibre*, 2015, **6**(1), 24.
10. V. Van Hoed, G. Depaemelaere, J. V. Ayala, P. Santiwattana, R. Verhé and W. De Greyt, *J. Am. Oil Chem. Soc.*, 2006, **83**(4), 315.
11. S. Sen, R. Chakraborty and P. Kalita, *Trends Food Sci. Technol.*, 2020, **97**, 265.

12. B. Min, A. M. Mcclung and M. H. Chen, *J. Food Sci.*, 2011, **76**(1), C117.
13. I. L. Massaretto, M. F. Madureira Alves, N. V. Mussi de Mira, A. K. Carmona and U. M. Lanfer Marquez, *J. Cereal Sci.*, 2011, **54**(2), 236.
14. X. Ni, K. Zhong, L. Tong, L. Liu, X. Zhou and S. Zhou, *Food Sci.*, 2018, **39**(16), 105.
15. A. Gunaratne, K. Wu, D. Li, A. Bentota, H. Corke and Y. Z. Cai, *Food Chem.*, 2013, **138**, 1153.
16. Z. Hou, P. Qin, Y. Zhang, S. Cui and G. Ren, *Food Res. Int.*, 2013, **50**, 691.
17. P. Y. Lin and H. M. Lai, *Food Chem.*, 2011, **127**, 86.
18. Y. Gao, X. Guo, Y. Liu, M. Zhang, R. Zhang, A. M. Abbasi and R. H. Liu, *LWT–Food Sci. Technol.*, 2018, **96**, 19.
19. Y. F. Shao and J. S. Bao, *Food Chem.*, 2015, **180**, 86.
20. Y. Shao, F. Xu, X. Sun, J. Bao and T. Beta, *J. Cereal Sci.*, 2014, **59**(2), 211.
21. J. Bao, Rice starch, in *Rice Chemistry and Technology*, ed. J. Bao, Elsevier Inc. in cooperation with AACC International, 4th edn, 2019, p. 55.
22. G. E. Vandeputte and J. A. Delcour, From sucrose to starch granule to physical behaviour: a focus on rice starch, *Carbohydr. Polym.*, 2004, **58**, 245.
23. K. Xu, Z. She, H. Wang, X. Liu, Y. Zhang, C. Chi and H. Zhang, *Int. J. Food Sci. Technol.*, 2021, **56**(8), 4081.
24. M. Obadi and B. Xu, *Food Hydrocolloids*, 2021, **112**, 106286.
25. C. L. Lin, J. H. Lin, J. J. Lin and Y. H. Chang, *Int. J. Biol. Macromol.*, 2019, **135**, 472.
26. S. I. Rafiq, S. Singh and D. C. Saxena, *Food Hydrocolloids*, 2016, **57**, 103.
27. R. Sompong, S. Siebenhandl-Ehn, G. Linsberger-Martin and E. Berghofer, *Food Chem.*, 2011, **124**(1), 132.
28. T. S. Rathna Priya, A. R. L. Eliazer Nelson, K. Ravichandran and U. Antony, *J Ethn. Foods*, 2019, **6**, 1–11.
29. N. Tangsrianugul, R. Wongsagonsup and M. Suphantharika, *Int. J. Biol. Macromol.*, 2019, **137**, 666.
30. R. Raghuvanshi, A. Dutta, G. Tewari and S. Suri, *J. Rice Res.*, 2017, **10**, 49–53.
31. V. M. Lai, S. Lu, W. H. He and H. H. Chen, *Food Chem.*, 2007, **101**(3), 1205.
32. A. O. Oko, B. E. Ubi, A. A. Efisue and N. Dambaba, *Int. J. Agric. For.*, 2012, **2**, 16–23.
33. P. Y. Chiang and A. I. Yeh, *J. Cereal Sci.*, 2002, **35**(1), 85.
34. D. Samyor, S. C. Deka and A. B. Das, *J. Food Process. Preserv.*, 2016, **40**(2), 174.
35. S. Majumder, K. Datta and S. K. Datta, *Agronomy*, 2019, **9**(12), 803.
36. R. Thomas, R. Bhat and Y. T. Kuang, *Int. Food Res. J.*, 2015, **22**(3), 1148.
37. M. Y. Kang, J. H. Kim, C. W. Rico and S. H. Nam, *Int. J. Food Prop.*, 2011, **14**(6), 1241.
38. M. I. Hashmi and J. S. Tianlin, *Int. J. Agric., For. Plant.*, 2016, **2**, 31.
39. E. G. N. Mbanjo, T. Kretzschmar, H. Jones, N. Ereful, C. Blanchard, L. A. Boyd and N. Sreenivasulu, *Front. Genet.*, 2020, **11**, 1–18.

40. B. Shammugasamy, Y. Ramakrishnan, H. M. Ghazali and K. Muhammad, *J. Sci. Food Agric.*, 2015, **95**(4), 672.
41. M. Zubair, F. Anwar, M. Ashraf and M. Uddin, *Int. J. Mol.*, 2012, **13**(4), 4608.
42. S. Lakshmi, G. Goudar, M. Singh, H. S. Dhaliwal, P. Sharma and T. Longvah, *J. Food Meas. Charact.*, 2021, **15**(3), 2774.
43. M. Frei and K. Becker, *J. Sci. Food Agric.*, 2005, **85**(14), 2380.
44. G. Pereira-Caro, S. Watanabe, A. Crozier, T. Fujimura, T. Yokota and H. Ashihara, *Food Chem.*, 2013, **141**(3), 2821.
45. M. Siddiq, R. Ravi, J. B. Harte and K. D. Dolan, *LWT–Food Sci. Technol.*, 2010, **43**(2), 232.
46. R. F. Tester and W. R. Morrison, *Cereal Chem.*, 1990, **67**(6), 551.
47. A. Aprianita, T. Vasiljevic, A. Bannikova and S. Kasapis, *J. Food Sci. Technol.*, 2014, **51**(12), 3669.
48. K. O. Falade and A. S. Christopher, *Food Hydrocolloids*, 2015, **44**, 478.
49. S. Yu, Y. Ma, L. Menager and D. W. Sun, *Food Bioprocess Technol.*, 2012, **5**(2), 626.
50. S. Graham-Acquaah, J. T. Manful, S. A. Ndindeng and D. A. Tchatcha, *J. Food Process. Preserv.*, 2015, **39**(6), 2286.
51. S. L. Rivero Meza, I. Louro Massaretto, P. Sinnecker, M. Schmiele, Y. K. Chang, J. A. Noldin and U. M. Lanfer Marquez, *Int. J. Food Sci. Technol.*, 2021, **56**(7), 3218.
52. K. Huaisan, J. Uriyapongson, P. Rayas-Duarte, I. Alli and V. Srijesdaruk, *Int. J. Food Prop.*, 2009, **12**(1), 145.
53. W. Ding, Y. Wang, W. Zhang, Y. Shi and D. Wang, *Int. J. Food Sci. Technol.*, 2014, **50**(3), 744.
54. I. Lee, G. J. We, D. E. Kim, Y. S. Cho, M. R. Yoon, M. Shin and S. Ko, *LWT–Food Sci. Technol.*, 2012, **48**(2), 164.
55. A. A. Karim, L. C. Toon, V. P. L. Lee, W. Y. Ong, A. Fazilah and T. Noda, *J. Food Sci.*, 2007, **72**(2), C132.
56. S. A. Mir and S. J. D. Bosco, *Food Nutr. Sci.*, 2013, **4**(3), 282–288.
57. F. Pratama, *Agric. Eng. Int.: CIGR J.*, 2018, **20**(3), 227.
58. S. Kraithong, S. Lee and S. Rawdkuen, *J. Cereal Sci.*, 2018, **79**, 259.
59. L. Wang, J. Guo, R. Wang, C. Shen, Y. Li, X. Luo, Y. Li and Z. Chen, *Cereal Chem.*, 2016, **93**(6), 593.
60. V. Singh, K. H. Vishwanathan, K. N. Aswathanarayana and Y. M. Indhudhara Swamy, *J. Food Sci. Technol.*, 2010, **47**(1), 34.
61. L. R. da Silva, C. W. P. de Carvalho, J. I. Velasco and F. M. Fakhouri, *Int. J. Biol. Macromol.*, 2020, **156**, 485.
62. F. M. Bhat and C. S. Riar, *Food Chem.*, 2019, **297**, 124984.
63. C. K. Reddy, L. Kimi, S. Haripriya and N. Kang, *Rice Sci.*, 2017, **24**(5), 241.
64. S. B. Dhull, S. Punia, M. Kumar, S. Singh and P. Singh, *Starch/Staerke*, 2021, **73**(1–2), 2000098.
65. F. F. Paiva, N. L. Vanier, J. D. J. Berrios, J. Pan, F. de Almeida Villanova, G. Takeoka and M. C. Elias, *J. Food Compos. Anal.*, 2014, **35**(1), 10.

66. R. J. N. Tiozon, K. J. D. Sartagoda, A. R. Fernie and N. Sreenivasulu, *Crit. Rev. Food Sci. Nutr.*, 2021, DOI: 10.1080/10408398.2021.1995697.
67. Y. Shao, Z. Hu, Y. Yu, R. Mou, Z. Zhu and T. Beta, *Food Chem.*, 2018, **239**, 733.
68. H. Zhang, Y. Shao, J. Bao and T. Beta, *Food Chem.*, 2015, **172**, 630.
69. H. Aalim, D. Wang and Z. Luo, *Food Res. Int.*, 2021, **141**, 109898.
70. Y. Pang, S. Ahmed, Y. Xu, T. Beta, Z. Zhu, Y. Shao and J. Bao, *Food Chem.*, 2018, **240**, 212.
71. X. Yu, T. Yang, Q. Qi, Y. Du, J. Shi, X. Liu, H. Zhang, Z. Zhang and N. Yan, *Food Chem.*, 2021, **344**.
72. Y. P. Huang and H. M. Lai, *J. Food Drug Anal.*, 2016, **24**(3), 564.
73. P. Goufo and H. Trindade, *Food Sci. Nutr.*, 2014, **2**(2), 75.
74. S. B. Patil and M. K. Khan, *J. Food Sci. Technol.*, 2011, **48**(6), 661.
75. P. Pramai, N. A. A. Hamid, A. Mediani, M. Maulidiani, F. Abas and S. Jiamyangyuen, *J. Food Drug Anal.*, 2018, **26**(1), 47.
76. M. Galland, S. Boutet-Mercey, I. Lounifi, B. Godin, S. Balzergue, O. Grandjean, H. Morin, F. Perreau, I. Debeaujon and L. Rajjou, *Plant Cell Physiol.*, 2014, **55**(9), 1646.
77. G. H. Lang, I. Lindemann, S. da, J. T. Goebel, C. D. Ferreira, T. Acunha, S. dos and M. de Oliveira, *J. Food Sci.*, 2020, **85**(6), 1717.
78. T. Qiu, Y. Sun, X. Wang, L. Zheng, H. Zhang, L. Jiang, X. Zhu and H. Xiong, *Food Biosci.*, 2021, **41**, 100841.
79. I. F. Bolarinwa, P. T. Lim and K. Muhammad, *Food Res.*, 2018, **3**(3), 199.
80. P. Nitisuk, J. Uriyapongson and I. Alli, *Food Appl. Biosci. J.*, 2019, **7**(1), 51.
81. L. Shan, O. Molberg, I. Parrot, F. Hausch, F. Filiz and G. Gray, *et al.*, *Science*, 2002, **297**, 2275.
82. H. R. P. Green and C. Cellier, *N. Engl. J. Med.*, 2007, **357**, 1731.
83. H. R. P. Green and B. Jabri, *Lancet*, 2003, **362**, 383.
84. C. M. Rosell, F. Barro, C. Sousa and M. C. Mena, *J. Cereal Sci.*, 2014, **59**(3), 354.
85. S. P. Bangar, K. Dunno, M. Kumar, H. Mostafa and S. Maqsood, *J. Funct. Foods*, 2022, **89**, 104937.
86. S. C. Li, T. C. Chou and C. K. Shih, *Food Res. Int.*, 2011, **44**, 209.
87. G. J. Mazza, *Ann. Ist. Super. Sanita*, 2007, **43**, 369.
88. K. Yasukawa, T. Akihisa, Y. Kimura, T. Tamura and M. Takido, *Biol. Pharm. Bull.*, 1998, **21**, 1072.
89. S. P. Kim, M. Y. Kang, S. H. Nam and M. Friedman, *Mol. Nutr. Food Res.*, 2012, **56**, 935.
90. P. N. Chen, W. H. Kuo, C. L. Chiang, H. L. Chiou, Y. S. Hsieh and S. C. Chu, *Chem.-Biol. Interact.*, 2006, **163**(3), 218.
91. H. Bays, L. Mandarino and R. A. DeFronzo, *J. Clin. Endocrinol. Metab.*, 2004, **89**, 463.
92. G. Paolisso, P. Gualdiero, D. Manzella, M. R. Rizzo, M. R. Tagliamonte and A. Gambardella, *et al.*, *Am. J. Cardiol.*, 1997, **80**, 932.
93. R. Medzhitov, *Nature*, 2008, **454**, 428.

94. K. Petroni, M. Landoni, F. Tomay, V. Calvenzani, C. Simonelli and M. Cormegna, *Univers. J. Agric. Res.*, 2017, **5**, 312.
95. T. Pattananandecha, J. Sirithunyalug, B. Sirithunyalug, K. Thiankhanithikun, C. Khanongnuch and C. Saenjum, *J. Pharm. Nutr. Sci.*, 2019, **9**, 205.
96. A. Phaniendra, D. B. Jestadi and L. Periyasamy, *Indian J. Clin. Biochem.*, 2015, **30**, 11.
97. T. B. Bagchi, K. Chattopadhyay, M. Sivashankari, S. Roy, A. Kumar, T. Biswas and S. Pal, *J. Cereal Sci.*, 2021, **100**, 103263.
98. M. Hiemori, E. Koh and A. E. Mitchell, *J. Agric. Food Chem.*, 2009, **57**, 1908.
99. P. Wongsa, *Recent Adv. Rice Res.*, 2020, **4**, 19.
100. X. Zhang, Y. Shen, W. Prinyawiwatkul, J. M. King and Z. Xu, *Food Chem.*, 2013, **141**, 111.
101. Y. Cui, D. J. Watson, C. J. Girman, D. R. Shapiro, A. M. Gotto, P. Hiserote and M. B. Clearfield, *Am. J. Cardiol.*, 2009, **104**(6), 829.
102. Q. Wang, P. Han, M. Zhang, M. Xia, H. Zhu, J. Ma, M. Hou, Z. Tang and W. Ling, *Asia Pac. J. Clin. Nutr.*, 2007, **16**, 295.
103. U. K. S. Kushwaha, *Black Rice: Research, History and Development*, Springer, 2016.
104. S. M. Boue, K. W. Daigle, M. H. Chen, H. Cao and M. L. Heiman, *J. Agric. Food Chem.*, 2016, **64**, 5345.
105. G. F. Deng, X. R. Xu, Y. Zhang, D. Li, R. Y. Gan and H. B. Li, *Crit. Rev. Food Sci. Nutr.*, 2013, **53**, 296.
106. K. D. Yoon, J. Y. Lee, T. Y. Kim, H. Kang, K. S. Ha, T. H. Ham, S. N. Ryu, M. Y. Kang, Y. H. Kim and Y. I. Kwon, *J. Agric. Food Chem.*, 2019, **68**(3), 742.
107. H. C. R. Sangma and S. Parameshwari, *Mater. Today: Proc.*, 2021, DOI: 10.1016/j.matpr.2021.07.257.
108. C. G. Dalbhagat, D. K. Mahato and H. N. Mishra, *Trends Food Sci. Technol.*, 2019, **85**, 226.
109. M. Blandino, A. Bresciani, M. Loscalzo, F. Vanara and A. Marti, *J. Cereal Sci.*, 2022, **103**, 103347.
110. W. Klunklin and G. Savage, *J. Food Qual.*, 2018, 1–8.
111. V. A. Papillo, M. Locatelli, F. Travaglia, M. Bordiga, C. Garino, M. Arlorio and J. D. Coïsson, *Food Chem.*, 2018, **269**, 603.
112. Y. S. Kim, G. H. Kim and J. H. Lee, *J. Korean Soc. Food Sci. Nutr.*, 2006, **35**, 499.
113. T. Susmitha, T. B. Bagchi, B. S. Deb, T. Biswas, T. Adak, H. Banerjee and S. Pal, *Food Chem. Adv.*, 2022, **1**, 100021.
114. S. Kong, D. J. Kim, S. K. Oh, I. S. Choi, H. S. Jeong and J. Lee, *J. Food Sci.*, 2012, 77(3), C303.
115. S. Nontasan, A. Moongngarm and S. Deeseenthum, *APCBEE Proc.*, 2012, **2**, 62.
116. X. Sui, Y. Zhang and W. Zhou, *Food Chem.*, 2016, **196**, 910.
117. J. L. Mau, C. C. Lee, Y. P. Chen and S. D. Lin, *LWT–Food Sci. Technol.*, 2017, **75**, 434.

118. J. C. Lee, J. D. Kim, F. H. Hsieh and J. B. Eun, *Int. J. Food Sci. Technol.*, 2008, **43**, 1078.
119. C. Zhao, W. Su, Y. Mu, L. Jiang and Y. Mu, *Food Res. Int.*, 2020, **138**, 109800.
120. M. Lema, *Mod. Concept Dev. Agron.*, 2018, **2**(1), 120.
121. P. K. Burkhardt, P. Beyer, J. Wünn, A. Klöti, G. A. Armstrong, M. Schledz, J. Lintig and I. Potrykus, *Plant J.*, 1997, **11**, 1071.
122. X. Ye, S. Al-Babili, A. Kloti, J. Zhang, P. Lucca, P. Beyer and I. Potrykus, *Science*, 2000, **287**, 303.
123. R. A. Zainal-Abidin, Z. Zainal, Z. A. Mohamed-Hussein, N. Abu-Bakar, M. S. F. Ab Razak, S. Simoh and Y. S. Sew, *Data Brief*, 2020, **30**, 105432.
124. E. Grotewold, P. Athma and T. Peterson, *Proc. Natl. Acad. Sci. U. S. A.*, 1991, **88**, 4587.
125. T. H. Ham, S. W. Kwon, S. N. Ryu and H. J. Koh, *Plant Breed. Biotechnol.*, 2015, **3**, 160.
126. Y. F. Tan, M. Sun, Y. Z. Xing, J. P. Hua, X. L. Sun and Q. F. Zhang, *et al.*, *Theor. Appl. Genet.*, 2001, **103**, 1037.
127. H. M. Sarif, M. Y. Rafii, A. Ramli, Y. Oladosu, H. M. Musa, H. A. Rahim, Z. M. Zuki and S. C. Chukwu, *Biotechnol. Biotechnol. Equip.*, 2020, **34**, 747.
128. G. N. Devi, G. Padmavathi, V. R. Babu and K. Waghray, *J. Rice Res.*, 2015, **8**, 23.
129. J. Ratseewo, F. J. Warren and S. Siriamornpun, *Food Chem.*, 2019, **298**, 124949.
130. S. Saikia, H. Dutta, D. Saikia and C. L. Mahanta, *Food Res. Int.*, 2012, **46**, 334.
131. A. B. Das and S. Bhattacharya, *LWT–Food Sci. Technol.*, 2019, **102**, 197.
132. V. Melini, G. Panfili, A. Fratianni and R. Acquistucci, *Food Chem.*, 2019, **277**, 119.
133. Y. Yang, M. C. Andrews, Y. Hu, D. Wang, Y. Qin, Y. Zhu, H. Ni and W. Ling, *J. Agric. Food Chem.*, 2011, **59**, 6759.

CHAPTER 3

Pigmented Maize: Nutritional Properties and Bioactive Profile

ISHA DUDEJA[a], MUSKAN GUPTA[b], RAMANDEEP KAUR MANKOO[a] AND ARASHDEEP SINGH*[b]

[a]Department of Chemistry, Punjab Agricultural University, Ludhiana, Punjab, India; [b]Department of Food Science and Technology, Punjab Agricultural University, Ludhiana, Punjab, India
*E-mail: arash.pau@gmail.com

3.1 Introduction

Maize is widely distributed and consumed worldwide, especially in developing countries. Because of this, some researchers have focused their efforts on the generation of new varieties of pigmented maize through cross-hybridization, to obtain new varieties of grains with a greater antioxidant capacity.[1] Such improved maize varieties could function as bioactive agents and potential candidates to be included in the prevention and integral management of several chronic degenerative diseases, reducing their incidence.

Maize is pigmented mainly due to a huge number of secondary metabolites, such as phenolic acids, carotenoids, and flavonoids. The different expression of these pigments imparts to maize tissues different pigments, from yellow-orange to dark purple-blue, ivory, and cream. Pigments are mostly concentrated in the thick pericarp or aleurone layers of kernels and

Food Chemistry, Function and Analysis No. 38
Pigmented Cereals and Millets: Bioactive Profile and Food Applications
Edited by Sneh Punia Bangar, Sajid Maqsood and Anil Kumar Siroha

Published by the Royal Society of Chemistry, www.rsc.org

in maize cobs. The most prevalent colors are red and blue/purple and at least 59 species have been identified, many of which correlate to pigmented grain variants.[2] Anthocyanins are mainly responsible for imparting different pigments to maize grains. The amount of anthocyanins in different pigmented maize cultivars has been found to range from 2.50 to 696.07 mg CGE per kg d.m.[3] Anthocyanins can be accumulated in seeds, maize stalks, cobs, and leaves, and their different chemical structures and contents lead to various pigmented shades, from pink to dark blue. Anthocyanins in pigmented grains can be beneficial dietary elements due to their antioxidative properties, anti-inflammatory potential, phase II enzyme catalysts, anti-cell proliferating, and hypoglycemic effects.[4,5] In general, the aleurone layer and, to a lesser extent, the starchy endosperm of the seeds contain a significant amount of pigment.[6]

Carotenoids are lipid-soluble isoprenoid pigments that play vital roles in photo protection, provide substrates for the synthesis of abscisic acid, and are essential nutritional components of the human diet, as chemopreventive agents and sources of vitamin A.[7] As lipid-soluble pigments, carotenoid extracts are primarily used in lipid-based products, but some water-soluble formulations are available. Major carotenoids present in various maize varieties are β-carotene and lutein (see Table 3.1). Phlobaphenes are water-insoluble 3-deoxyflavanoid pigments found in the maize cob and pericarp. They are derived from the polymerization of the flavan-4-ols apiforol and luteoforol, ratios of which are controlled by the purple aleurone 1 (*Pr1*) gene. During pericarp and cob maturation, flavan-4-ol content decreases with the development of pigment, resulting in orange to brick-red pigments. Functional *Pr1* plants produce mainly luteoforol, resulting in brick-red phlobaphenesin pericarp and cob. In contrast, plants homozygous for recessive *Pr1* are incapable of producing luteoforol and therefore accumulate apiforol instead, leading to orange phlobaphene pigments.

Little is known about the enzymes or mechanisms responsible for the polymerization reaction, but several of the biosynthetic genes early in the pathway are shared in the formation of flavan-4-ols and anthocyanins. Starch, proteins, lipids, carbohydrates, anthocyanins, salicylic acid, resins, saponins, potassium, sodium salts, sulphur, phosphorus, and other phenolic chemicals are all abundant in pigmented maize. The outer layer of the grains containing a large amount of anthocyanins can be removed to obtain natural dyes, and the remaining elements can be used to make flour, ethanol, feed, food additives, and many other substances. Maize grains are a raw material rich in carbohydrates, mainly starch, the content of which ranges from 58.0% to 71.5%. The protein content in the blue varieties is higher compared with those of the yellow and white maize varieties, from 9.1% to 13.1%. The lipid content of the blue varieties ranges from 4.5% to 6%.[13] The main fatty acids found in maize grains are unsaturated, including linoleic and oleic acids (about 12% each). Saturated fatty acids, such as stearic and palmitic acids are also present in smaller amounts. The lipid content ranges from 4.3% to 6.9%.

Table 3.1 Amount of anthocyanins and carotenoids reported in differently pigmented maize varieties.

Class	Compounds	Units	Amount	Maize variety	Reference
Anthocyanins					
	Cyanidin-3-*O*-glucoside	mg per 100 g	17.44	Yellow maize	8
	Cyanidin-3-*O*-glucoside	mg per 100 g	21.25	Red maize	
	Cyanidin-3-*O*-glucoside	mg per 100 g	96.3	Blue maize	
	Cyanidin-3-(6″-malonylglucoside)		—	Blue maize extract	9
	Cyanidin 3-glucoside	mg per 100 g	2.0 ± 0.5	Blue maize flour	10
	Cyanidin-3-glucoside	mg per 100 g dry weight, DW	310.04	Peruvian purple maize (*Zea mays* L.)	11
	Cyanidin-3-glucoside (M)		1×10^{-6}	Pigmented Mexican maize	12
	Pelargonidin-3-glucoside (M)		1×10^{-7}		
Carotenoids					
	β-carotene	μg g^{-1} extract	0.39	White maize	8
			0.49	Yellow maize	
			1.01	Red maize	
			0.18	Blue maize	
	β-carotene	mg kg^{-1} d.m.	0.21	White maize	3
	Lutein	mg kg^{-1} d.m	0.08	Lemon yellow maize	
	β-carotene	mg kg^{-1} d.m.	0.70	Yellow maize	
	Lutein	mg kg^{-1} d.m	5.91		
	β-carotene	mg kg^{-1} d.m.	2.42	Orange maize	
	Lutein	mg kg^{-1} d.m	11.14		
	β-carotene	mg kg^{-1} d.m.	1.15	Red-yellow maize	
	Lutein	mg kg^{-1} d.m	9.59		
	β-carotene	mg kg^{-1} d.m.	0.60	Red maize	
	Lutein	mg kg^{-1} d.m	13.89		
	Lutein	mg kg^{-1} d.m	0.01	Light blue maize	
	β-carotene	mg kg^{-1} d.m.	1.68	Dark blue maize	
	Lutein	mg kg^{-1} d.m	5.34		
	β-carotene	mg kg^{-1} d.m.	1.40	Multi-pigmented maize	
	Lutein	mg kg^{-1} d.m	4.63		

A variety of phenolics other than anthocyanins have been detected in pigmented maize extracts. These compounds have been widely reported to contribute potential health-promoting properties to animal and human cells. Phenolic acids from purple maize were recovered using alcohol and acetone[14] and identified by high performance liquid chromatography-mass spectrometery (HPLC-MS).[3] The *p*-coumaric acid and ferulic acid contents for dark-hued Bolivian purple maize were reported to be 607.5 and 154.2 mg per 100 g DW, respectively; much higher than yellow cultivars, which are reported to contain 366.7 and 132.9 mg per 100 g DW, respectively.[15] A more recent study showed ferulic acid to be the most abundant phenolic acid (39.9 mg kg^{-1} DW for native and 33.3 mg kg^{-1} DW for hybrid maize), followed by diferulic and coumaric acids in Mexican purple maize.[16] Most of the phenolic acids are in conjugated or bound forms in purple maize; just a small portion is in free soluble form that can be easily extracted without hydrolysis treatment. More than 9 different phenolic acids have been found in purple maize, *i.e.* protocatechuic acid, vanillic acid, syringic acid, 2,4,6-trihydroxybenzoic acid, *p*-coumaric acid (also called *p*-hydroxycinnamic acid), caffeic acid, ferulic acid, chlorogenic acid, and *p*-hydroxyphenyl acetic acid, and also their derivatives.[3,15]

In addition to phenolic acids, other functional compounds found in purple maize include flavonoids. The total flavonoids content in purple maize ranged from 307.42 to 337.51 mg kg^{-1} DW, while the yellow-hued maize contained less at 248.64 to 281.20 mg kg^{-1} DW.[3] Similar to phenolic acids, flavonoids in purple maize are recovered by organic solvent mixture.[3,14] Rutin, hirsutrin, morin, kaempferol, quercetin, naringenin, hesperitin, and their derivatives are the most commonly reported flavonoids recovered from purple maize.[17,18]

Physical qualities such as grain size, density, hardness, chemical composition, and environmental factors such as climate, soil type, production practices, and lastly, individual genetic elements of each variety, all contribute to the variety of pigmented maize types. Each variety of pigmented maize has a specific observable property, such as color, and a distinct biological activity based on the quantity and profile of secondary metabolites, which increases its functional food potential. Pelargonidin, for example, is the most prevalent anthocyanidin in light red grains,[19] whereas cyanidin derivatives predominate in blue grain, which also includes grains with blue, purple, and black pigment.[17,20] Cyanidin derivatives predominate in magenta red kernels, but they also contain derivatives of pelargonidin and peonidin.[21] The primary anthocyanins identified in maize are cyanidin 3-*O*-glucoside and pelargonidin 3-*O*-glucoside, which have been described as malonylated in a reasonably high proportion (approximately 40%) as cyanidin 3-*O*-(6″-malonyl-glucoside) and cyanidin 3-*O*-(3″, 6″-dimalonyl-glucoside). Some landraces' cobs might have up to three different pigmented kernels.[22] The pigment is found in the aleurone layer in blue grain types, in the pericarp in light red grain varieties, and in the pericarp and aleurone layer in magenta/red grain varieties.[23] Pigmented maize grains include both acylated and non-acylated anthocyanins. In comparison with other cereals, pigmented maize has a high quantity of acylated anthocyanins, which boosts its chemical stability under extreme pH and temperature conditions.[21] Cyanidin-3-*O*-glucoside, pelargonidin-3-*O*-glucoside, and peonidin 3-*O*-glucoside were the acylated anthocyanins found in maize grain. One or more acyls may bond to the sugar moiety (aliphatic acids such as malic, malonic, or succinic acid). Cinnamic acids (p-coumaric, caffeic, ferulic, or synaptic) acting on glucose and rhamnose can also produce acyl radicals.[24]

The procedures involved in extraction involve the use of aqueous-organic solvent mixtures, generally acidified to improve the yields, at temperatures lower than 50–60 °C with different reaction times.[25] The extraction of anthocyanins from purple maize cob was carried out using a food-friendly 50% aqueous ethanol acidified with 0.01% 6 N HCl for 45 minutes, and double washing of the remaining pigmented insoluble powder, after extract filtration. This extract contained a high level of polyphenols and monomeric anthocyanins.[26] Recently, a new approach for water-based anthocyanin recovery from purple maize was investigated, which involved single-stage, two-stage, three-stage, and two-stage countercurrent water-based extractions, with recoveries of 48, 68, 77, and 66%, respectively. The extraction process can be improved by using ultrasound with 95% ethanol with 0.1 M citric acid, and was successfully scaled up for the extraction of C3G (45% total anthocyanins) and cyanidin-3-*O*-(6″-*O*-malonylglucoside) (40% of total anthocyanins) from purple bran. Lower percentages of pelargonidin-3-*O*-glucoside, pelargonidin-3-*O*-(6″-*O*-malonylglucoside), peonidin-3-*O*-glucoside, and cyanidin-3-*O*-(3″,6″-di-*O*-malonylglucoside) were also obtained. The use of 55% ethanol for extraction from Mexican red maize cob allowed the extraction of 5-CQA, *p*-coumaric acid and caffeic acid 4-*O*-hexoside, and three glycosylated apigenin (hexoside, pentosylhexoside, and 6-*C*-pentosyl-8-*C*-hexoside), luteolin-*O*-rutinoside, and scopoletin.[27] Ultrasound-assisted extraction was also the most efficient procedure to extract anthocyanins (1 : 10 solvent-to-solid ratio, ethanol/water/lactic acid mixture, 80 : 19 : 1, for 20 min, 100 W) from Cahuacintle maize and husk. Additionally, an enzymatic-assisted extraction using xylanases, celluclast, and depol, was used, since the residual material obtained after the extraction exhibited an intense pigmentation. The optimized pH and temperature conditions gave an extract richer in cyanidin-3-*O*-(6″-*O*-malonyl) glucoside than the maceration or microwave-assisted extraction methods.[28]

Another exciting approach, *i.e.* ohmic heating extraction, was applied to a new purple maize single cross-hybrid from Thailand, and the yield of total anthocyanins extraction was compared to that obtained by using other extraction methods based on the use of water as solvent and operating in optimized conditions (conventional extraction applied for 5 min at 70 °C, ultrasound-assisted extraction for 5 min at 80 °C, 37 KHz, and microwave-assisted extraction for 2 min, 2.45 GHz, 700 W). This new approach has two main advantages: it does not affect the nutritional properties differently from conventional extraction, and it leads to higher yield in less time. Microwave-assisted extraction was the best method, and it was scaled up, giving a total anthocyanin yield of 0.25%, and producing a dark purple powder, which created a deep red pigmented solution after dissolution in water.[29]

By optimizing reaction conditions using response surface methodology (RSM),[30] experiments were performed to set up the best conditions (at 40 °C for 8 h, with 1 : 15 in solvent volume, and a 33% solid : liquid ratio) for extracting 3-*O*-glycosidic derivatives of cyanidin, pelargonidin, peonidin, and the malonylglycosidic derivatives of cyanidin and peonidin in the highest amounts from a purple maize kernel. In these conditions, the best yield extraction was obtained for other polyphenols, such as hyrsutrin and its methoxylated derivative, hydroxy benzoic (2,4,6-trihydroxy benzoic, protocatechuic, and vanillic acids) and hydroxycinnamic (ferulic and *p*-hydroxycinnamic acids) acid derivatives. By using this approach, the same anthocyanins in addition to pelargonidin-3-*O*-(6″-*O*-malonyl)-glucoside, malvidin-3-*O*-(6″-*O*-malonyl-phydroxybenzoyl)-glucoside, and malvidin-3-*O*-(6″-*O*-dimalonyl)-glucoside were extracted from a purple maize produced in China, without any degradation. RSM was also adopted to optimize the ultrasound-assisted extraction of anthocyanins and zein, a prolamin protein representing about 40% of total protein content, which is widely used in different industrial fields for its peculiar characteristics. The highest yield of anthocyanins (450 mg kg^{-1} maize) was obtained using a 1 : 26 solid (g) : liquid (mL) ratio, 74% ethanol (v/v) at 70 °C for 90 min, with an ultrasound power of 105 W, after different purification steps involving precipitation with ammonium sulphate, followed by the separation from zein using a cross-flow ultrafiltration, and final purification by AB-8 macroporous resins.[31]

Purple maize cob could be a good and economical source of anthocyanins, with its general content higher than kernel. Thus, it could represent a relatively cheap material for extracting natural pigment.

In the last decade, the demand for natural pigments has increased, replacing synthetic substances, which are considered less safe for food purposes; therefore, studies to check the stability of pigments have become mandatory. Anthocyanins are one of the most studied classes of natural pigments, together with anthraquinones, betalains, and carotenoids. One of the main issues regarding the use of pigmented maize extract is its stability. Anthocyanins are red-purple pigmented in acidic conditions, and following the deprotonation and quinoidal form generation, due to pH increase, they generate a purple-blue pigment; moreover, metal chelation and co-pigmentation can generate a blue pigmented, even in acidic conditions.

3.2 Pigment Accumulation and Their Biosynthesis in Pigmented Maize

Pigment development is a crucial stage in the growth of maize kernels. Maize kernels lose their pigments as they adapt to new growth environments, primarily due to human selection and, as a result, the majority of farmed maize is yellow.[32] Demand for dark-pigmented maize kernels has recently increased due to growing insight into the nutritional relevance of pigments in the prevention of chronic diseases. As a result, research into the formation of pigment in a variety of crops, including maize kernels, has focused on historic flavonoid-rich landraces.[33] Flavonoids, specifically anthocyanins and phlobaphenes, are responsible for the pigments found in maize kernels. Anthocyanin compounds cause blue, purple, red, and black pigments, whereas phlobaphenes cause red and purple hues.[3,34]

Pigment formation in maize is linked to several genes. There are two gene-controlled pathways – anthocyanin and carotenoid – involved in pigment formation. Each pigment's synthesis necessitates the expression of specific genes, including structural (enzyme-coding) and regulatory genes (transcription activators and coding for proteins that operate the transcription of structural genes). Commonly, such genes are referred to as pigmented genes. As cited by Ford,[35] *Pr1*, C1, R1 and Y1 are the four important genes that account for pigment synthesis in maize. Red aleurone 1 (*Pr1*) characterizes the recessive phenotype's kernel pigment, and identifies a unique structure of the seed engaged in its expression, the aleurone. This structural gene encodes an enzyme flavonoid 3′-hydroxylase that produces the purple pigment in the anthocyanin pathway. The next structural gene, White1 or (Y1), has a name that also explains its recessive phenotype. It encodes phytoene synthetase, an enzyme involved in the carotenoid pathway of synthesis. The other two genes, pigmented aleurone 1 (C1) and pigmented 1 (R1) encode transcription activators that interact to regulate numerous structural genes in the anthocyanin pathway. Several genes influence the phenotype, and the sum of their effects is referred to as genetic background. Other than the *Pr1*, C1, and R1 genes, several other genes are required for the expression of anthocyanins, as this is a multistep metabolic pathway. No anthocyanin production is observed when either recessive alleles – r1 or c1 – are present in homozygous conditions. Hence, the kernel pigment is either yellow or white, depending on the kind of carotenoid pathway genes that are present. Thus, the genetic background includes genes from both the anthocyanin and carotenoid pathways.

In a different study,[32,36] it was found that the r1 (red1)/b1 (booster1) transcription factor family, which belongs to the bHLH (basic helix-loop-helix) transcription factor class; the c1 (pigmentless 1)/pl1 (purple plant1)/p1 (pericarp pigmented1) transcription factor family, which belongs to the MYB-like transcription factors; and the WD40 factor PAC1 all play roles in pigment forming pathways (pale aleurone pigment 1). Each member of this family manifests in a tissue-specific or plant-wide manner. The allelic combination of the MYB and bHLH loci greatly influences maize pigmentation patterns.[37] The allelic combination of the regulatory pl1/B1 genes, for example, results in the dark blue hue seen in some tropical maize grains.[38] This allelic combination can stimulate the production of cyanidin-3-*O*-glucosides, primarily in the maize pericarp and specific tissues. The r1/c1 combination is required to form pelargonidin-3-*O*-glucoside in the aleurone layer.[36,39]

A ternary complex of transcription factors, comprising the MYB factor C1 (pigmentless 1) or PL1 (purple leaf 1), the bHLH factor R1 (red pigment 1) or B1 (booster 1), and the WD40 factor PAC1, regulates the anthocyanin pathway in maize (pale aleurone pigment 1). Although each transcription factor appears to be controlled individually, anthocyanin pigmentation requires the expression of all three. In the absence of R1 or B1, however, the MYB member P1 (pericarp pigment 1) controls the production of phlobaphene. These transcriptional factors create precursors of different anthocyanins and phlobaphenes by regulating three common flavonoid biosynthetic structural genes, chalcone synthase (CHS), chalcone isomerase (CHI), and dihydroflavonol 4-reductase (DFR). Furthermore, epigenetics play an essential role in regulating the complex spatial and temporal pigmentation in maize due to the notably uncompact structure of the promoters of the R1, B1, and P1 genes.[32,33] Maize gets its blue hue from anthocyanins in the aleurone layer, which is controlled by anthocyaninless 1 (a1), anthocyaninless 2 (a2), bronze 1 (bz1), bronze 2 (bz2), pigmentless 1 (c1), pigmentedless 2 (c2), defective kernel 1 (dek1), red aleurone (pr), pigmentless (r), and viviparous 1 (v1) (vp1). Every factor requires the presence of a dominant allele for deep purple pigment. The blue pigment can be diminished by the presence of dilution elements, such as C2-Idf. The R locus alleles (R and r) have dosage-dependent expression. R/R/r genotypes are purple, but R/r/r genotypes are mottled purple.[6]

3.3 Bioactive Compounds in Pigmented Maize

Pigmented maize is a rich source of carbohydrates, minerals, and bioactive compounds.[17] The bioactive compounds found in pigmented (blue, red, and purple) maize are anthocyanins, *p*-hydroxycinnamic acids, flavonoids, and, to a lesser extent, carotenoids, phytoesterols, vitamin E, lignans, policosanols, and xylans. The presence of the maximum number of anthocyanins and phenolic compounds in the aleurone and pericarp layer of grains is primarily responsible for giving maize varieties their distinct pigments (black, blue, red and purple). When comparing pigmented maize to pigmented rice, barley, and wheat grains, the former has the highest anthocyanins. Maize has become more diverse, as a result of changes in geographical regions and environmental conditions. The differences in pigment could be due to various anthocyanin derivatives in varying concentrations. It was observed that the pigment characteristics in 15 Mexican variants *i.e.*, L*,

C* and h ranged from 25.13 to 63.64, 2.41 to 33.58 and 14.36 to 359.19. Also, the percentage of carbohydrates, protein and moisture ranged from 71.30–74.88%, 9.72–12.57% and 7.98–9.67%, respectively. The presence of phosphorous, potassium, magnesium, zinc, and iron was also reported.[40]

Anthocyanins are a family of polyphenolic pigments found throughout nature. Generally, they are present in the kernel, cob, husk, silk, leaves, and stem of maize plants. Anthocyanins are found in most abundance in the kernel, but they can also be found in the pericarp and aleurone layer. The pericarp can be transparent, orange, red, or brown, whereas the aleurone component can be clear, red, or purple.[41] Of the various anthocyanins discovered to date, the three most common anthocyanins present in pigmented maize are: cyanidin 3-glucoside (most abundant), pelargonidin, and peonidin 3-glucoside, as revealed by HPLC and liquid chromatography-mass spectrometry (LC-MS). The concentration of anthocyanins in different pigmented maizes varies according to geographical and environmental factors. In one study,[42] extraction of anthocyanin from purple maize cob powder was done using deionized water, acidified water, acidified ethanol, and 70% acetone. The extraction was carried out for 1 hour at various temperatures, *i.e.* at room temperature (RT), and at 50 °C, 75 °C, and 100 °C. It was observed that the maximum anthocyanin content was present in 70% acetone and acidified boiling water extract. The anthocyanins reported were cyanidin-3-glucoside, pelargonidin-3-glucoside, peonidin-3-glucoside, cyanidin-3-maloylglucoside, cyanidin-3-maloylglucoside, pelargonidin-3-maloylglucoside, peonidin-3-malonylglucoside and cyanidin-3-dimaloylglucoside.

Cyanidin 3-glucoside was the most abundant anthocyanin in black and red rice, and blues, purplish, and reddish maizes. The pelargonidin 3-glucoside and delphinidin 3-glucoside were reported as predominant anthocyanins in pinkish maize and blue pigmented wheat, respectively.[19] The amount of anthocyanin in several varieties of blue pigmented maize ranges from 2.50 to 696.07 mg CGE per kg,[3] while in a few varieties, it can reach up to 1052 mg Cy3glu per kg cyanidin-3-*O*-glucoside per kg of sample.[43] In 57 native maize accessions and three commercial varieties grown in diverse cities across Mexico, Central and South America, cropping site, accessions, and location-accession all had a substantial impact on the total flavonoid and phenolic content of maize. The content of anthocyanin in blue maize was higher than in red maize. Also, significant differences in total phenolic, total flavonoids, and antioxidant activity were detected across the different grain pigments (*i.e.* blue, red, yellow), and significant interactions were observed between the localities and grain pigments[44] (see Table 3.2). In pigmented maize, cyanidin and pelargonidin have been identified as significant anthocyanins. Cyanidin 3-*O*-(6″-malonyl-glucoside) and cyanidin 3-*O*-(3,6-dimalonyl-glucoside) are the most common glycosylated forms. Some landraces' cobs may feature three-pigmented kernels.[22] The pigment could be found in the aleurone layer, pericarp, or both. Pelargonidin and cyanidin derivatives were the most abundant anthocyaninins in light and blue maize, respectively.[19,20] The majority of the analogs identified in magenta red kernels were

Table 3.2 Phenolic acid content in different pigmented maize.

Color	Source	Solvent	Time	Content	Reference
Red and purple	Kernel flour	MeOH:CHCl3 (1:1)	3 h	69.4 to 212.8 mg GAE per 100 g DW	58
Purple	Kernel flour	Methanol	20–72 h	1903 mg chlorogenic acid per 100 g DW	59
Yellow	Kernel flour	Methanol	2 h	1170–1640 μg GAE per g	60
White	Kernel flour	Methanol	2 h	903–1332 μg GAE per g	
Purple	Kernel flour	Methanol	2 h	1223–1843 μg GAE per g	
Blue	Kernel flour	Methanol	30 min	307.57 GAE per 100 g DW	61

cyanidin analogs; however, pelargonidin and peonidin analogs were also found in a few cases.[21]

Pigmented maize contains both acylated and non-acylated forms of anthocyanins. In Mexican pigmented maize, a significant proportion of acylated anthocyanins was reported. Cyanidin-3-*O*-glucoside, pelargonidin-3-*O*-glucoside, and peonidin 3-*O*-glucoside were among the acylated anthocyanins found in maize grain.[21,45] In nine Bolivian purple maize (*Z. mays* L.) varieties, the soluble and insoluble-bound fractions of phenolic compounds were investigated using HPLC-DAD and HPLC-ESI-MSn. Anthocyanin was found in differing amounts across all of the varieties. Ferulic acid and p-coumaric acid (non-anthocyanin phenolics) concentrations in dry weight (DW) ranged from 132.9 to 298.4 mg per 100 g and 607.5 mg per 100 g, respectively. The principal anthocyanins found were cyanidin-3-glucoside and its malonate derivative. The total amount of phenolic compounds varied from 311.0 to 817.6 mg gallic acid equivalents (GAE) per 100 g DW, while the total monomeric anthocyanin content varied from 1.9 to 71.7 mg cyanidin-3-glucoside equivalents per 100 g DW. As minor constituents, several dimalonylated monoglucosides of cyanidin, peonidin, and pelargonidin were present.[15] The total amount of phenolic compounds in 100 g DW ranged from 311.0 to 817.6 mg GAE, while the total monomeric anthocyanin content ranged from 1.9 to 71.7 mg cyanidin-3-glucoside equivalents per 100 g DW. As minor constituents, several dimalonylated monoglucosides of cyanidin, peonidin, and pelargonidin were confirmed.[15] The chemical profile of whole kernels of yellow, red, blue, and purple maize varieties revealed clear discrimination in their chemical profile. The concentration of acylated anthocyanins was higher than that of glycosilated anthocyanins. The purple kernel maize variety was enriched in total phenolics, flavonoids, and proanthocyanidin, whereas yellow and red maize has more carotenoid and total tocol content.[46] Phytochemical screening of blue maize (BM) (*Z. mays* L.) flour revealed that the total phenolic content (164 ± 14 mg gallic acid per g dry matter) was higher than that of white maize (127.7 mg gallic acid per g dry matter). Camelo-Méndez *et al.*[47] determined the anthocyanin content to be 2.0 ± 0.5 mg cyanidin 3-glucoside per 100 g. Flour prepared from four different types of Italian pigmented maize (orange to red and dark red) revealed that insoluble-bound phenols and flavonoids account for roughly 70–80% of the total.

The amount of anthocyanins was less, whereas the amount of red-brick phlobaphenes was significantly greater. Local varieties had higher levels of zeaxanthin and -carotene, while commercial varieties had lower levels of lutein. The highest amount of zeaxanthin was found in all local varieties. The nano variety contained the most zeaxanthin, -carotene, and -cryptoxanthin of any local variety.[48]

At the milk and mature stages, HPLC-DAD-ESI/MS analysis of kernels from 12 waxy maize varieties revealed that acylated derivatives contributed 67.1–88.2% and 46.2–83.6% of the total anthocyanin contents, respectively. Cyanidin-3-glucoside and its derivatives were identified as the most prominent. During the maturation of all maize varieties, the amount of monomeric anthocyanin increased. The highest proportion of anthocyanins was detected in the kernels of a purplish black waxy maize variety (KKU-WX111031).[49] The determination of anthocyanin composition during grain filling in purple maize kernels revealed that anthocyanins increased as the grain pigment darkened. Quantitatively, cyanidin-3-β-Oglucoside was the major compound, with concentrations ranging from 57.0–409.1 mg kg^{-1} fresh weight in raw kernels and 1027.6 mg kg^{-1} in dry seeds, followed by pelargonidin-3-β-*O*-glucoside and malvidin-3-β-*O*-glucoside. Maize's anthocyanin content was determined by its physicochemical properties and amino acid content. Rather than increasing in a linear fashion, the amount of anthocyanin increased in a stepwise manner.[50]

The total anthocyanin content in the white grain of the inbred line of sweet maize CE401 and the purple grain of the population F4 (CE401 Chornosteblova) was 1174.5 and 2951.4 mg kg^{-1}, respectively, according to differential spectrophotometry. The concentration of anthocyanins in grains increased in a linear relationship with the intensity of pigment. In both white and purple grains, the amount of glucosidic anthocyanins was twice as high as non-glycosidic anthocyanins. In both the glucosidic and non-glucosidic forms, peanidin-3-glucoside and pelargonidin were dominant.[51] The acylated anthocyanins cyanidin-3-(6″-succinylglucoside) (Cy-Suc-Glu) and cyanidin-3-(6″-disuccinylglucoside) (Cy-diSuc-Glu) were found to be the most abundant in 15 Mexican blue pigmented maize, accounting for 52.1% and 15.6%, respectively, of total anthocyanin content. Cyanidin-3-glucoside, pelargonidin-3-glucoside, pelargonidin-3-(6″-malonylglucoside), and cyanidin-3-(6″-malonyglucoside) are also important anthocyanins. It was also[52] discovered that raw blue maize has a similar anthocyanin profile. The concentration of anthocyanins, a type of flavonoid, in purple maize ranged between 1.5% and 6.0%. The percentages of carbohydrates, proteins, minerals, and vitamins (Band C) in purple maize were 80%, 11%, 2%, and 2%, respectively.[53] Eight anthocyanin compounds were isolated from blue maize extracts (BCE) and its yellow variant. The principal anthocyanins are pelargonidin, cyanidin-3-glucoside, and various glycoside derivatives. Compared with saturated fatty acids, the proportion of unsaturated fatty acids (linoleic acid and oleic acid) (stearic and palmitic acid) was lower. Among sterols, β-sitosterol and campesterol were present in larger amounts than stigmastanol and stigmasterol.[54]

A significant difference between phytic acid and total phenolic compounds was observed for genetic variation in 192 Turkish maize landraces. The phytic acid concentration ranged from 0.82 to 4.87 mg g^{-1}, and the total phenolic content ranged from 0.03% to 1.99%.[55] Phytochemical analysis of the pericarp, aleurone layer, kernel, and maize cob of 52 Mexican purple maize populations with varying degrees of pigmentation found that the total anthocyanin (TA) content of purple maize was higher than that of blue and red maize. TA per 100 g aleurone layer, pericarp, and kernel biomass ranged from 0.0044 to 0.0523 g, from 0.2529 g to 2.6452 g, and from 0.2529 to 2.6452 g, respectively.[56] Research concluded that cooking at a lower temperature of standard sugary sweet maize (*Z. mays* saccharata Sturt.) resulted in no change in anthocyanin content, whereas cooking at a higher temperature resulted in a change in anthocyanin content without structural modifications in molecules.[57]

Seven anthocyanins were found in blue maize extracts during chromatographic studies. Cyanidin-3-(600-malonylglucoside) was reported as the principal anthocyanin.[9] In another study, cyanidin peonidin and pelargonidin anthocyanin glycosides were isolated from rose maize. HPLC profiling revealed cyanidin-3-(6-malonyl-glucoside), a malonic derivative of cyanidin, as the most abundant anthocyanin.[62] HPLC profiling of hydrolyzed and non-hydrolyzed extracts of pigmented Mexican maize showed only minor differences in composition. Major anthocyanins detected by MALDI-ToF MS were cyanidin-3-glucoside (1×10^{-6}) and pelargonidin-3-glucoside (1×10^{-7} M) in the hydrolyzed and non-hydrolyzed extracts, respectively.[12] The nixtamalization process reduced the anthocyanin content of creole maize races before and after tortilla formation. Prior to processing, the number of anthocyanins in blue and red varieties was higher than in white and yellow varieties. When blue maize was processed into tortillas and masa, a loss of 64% and 83% in anthocyanin content was observed,[8] respectively, whereas another study reported the detection of more anthocyanins in blue maize (cyanidin 3-*O*-glucoside, equivalent to 620.9 mg kg^{-1}) than in other pigmented maize *via* an extrusion process. It also resulted in the formation of anthocyanin derivatives, which changed the anthocyanin content.[63,64] Another study discovered that uncooked maize had a higher proportion of glycosylated anthocyanins and a lower proportion of acylated anthocyanins.[52] The variation in the anthocyanins and the total phenolic content was observed during high-energy milling (HEM), as compared to traditional nixtamalization (TN) in four creole maize flours obtained from maize varieties grown under rainy temporal conditions in the semi-desert climate of Mexico. Total anthocyanins and phenolics were higher in flour obtained by HEM than by traditional nixtamalization.[65]

Four native maize varieties (Arrocillo, Cónico, Peruano, and Purepecha) analysed using HPLC revealed that the anthocyanin profile was similar in all the samples and total anthocyanin content varied from 54 to 115 mg per 100 g of sample. Cyanidin-3-glucoside, pelargonidin-3-glucoside, peonidin-3-glucoside, cyanidin-3-(6″ malonylglucoside), and cyanidin-3-(3″,6″-dimalonylglucoside) were the main anthocyanins identified.[66] HPLC analysis confirmed the presence of ferulic acid, quercetin, sinapic acid, gallic acid and protocatechuic acid in white, yellow and purple pigmented maize. Total phenolic content (TPC) varied from 903 to 1843 mg GAE per g. Purple accessions had the highest concentrations of phenolics. Ferulic acid was the most prevalent phenolic acid in all accessions.[60] In fifteen pigmented maize varieties of maize, total phenolic, total flavonoid, total anthocynanin and condensed tannins content varied from 349.39 to 485.71 mg GAE per 100 g, 24.00 to 105.75 mg CE per 100 g, 1.38 to 74.52 mg C3GE per 100 g and 33.70 to 158.55 mg CE per 100 g, respectively. The percentage composition of bound fractions was higher in all cases. Ferulic acid and coumaric acid were the main compounds identified.[40] Total phenolic content (TPC) present in five blue hybrid maize genotypes (503 × 27, 503 × 67, 613 × 27, 611 × 8, and 631 × 27) was estimated using the Folin–Ciocalteu method. The extracts were prepared by macerating maize kernels with organic solvent. It was observed that the TPC ranged from 10 390 to 13 313 mg GAE per kg of maize.[43] In another study, TPC and TAC of blue maize and tortillas were compared. It was determined that the blue maize contained 287.3 mg GAE per 100 g phenolic compound while the tortilla contained 70.3 mg GAE per 100 g, which is comparatively less than the content in former. The TAC present in blue maize and tortilla were 70.50 and 27.8 mg C3G per 100 g, respectively. It was inferred that the nixtamalization process, which uses high temperatures and alkaline conditions, may cause a decrease in the concentration of polyphenols and anthocyanins in tortilla. 4 red and 7 purple genotypes were collected and analysed for TPC, TAC, and TFC. Observations revealed total phenolic content and ferulic acid in the maize samples ranging from 69.4 to 212.8 mg GAE per 100 g DW and 7.4–42.6 mg FA per 100 g DW, respectively. TFC and TAC values ranged from 0.07 to 12.19 mg CE per 100 g DW and 3.89 to 34.17 mg CGE per 100 g DW, respectively.[58]

3.4 Pigment Extraction From Pigmented Maize

Carotenoids, anthocyanins, flavonoids, and chlorophylls are some of the pigments present in plants. Anthocyanins are the main pigments in pigmented maize, along with trace amounts of flavonoids and carotenoids. To extract these pigments from pigmented maize varieties, several extraction methods such as solvent extraction (stirred vessel, soxhlet, and ultrasonication), pressurized liquid extraction (PLE), supercritical fluid extraction, and subcritical water extraction, among others, have been developed in recent years. The total content of each pigment (anthocyanin, phenols, flavonoids, *etc.*) is usually determined using a spectroscopic method, and the total content is calculated by taking some standard compound and reporting it as an equivalent of that standard compound (see Table 3.3). However, HPLC combined with mass spectrometry can be used to identify and confirm these individual pigments (MS). In brief, reverse phase chromatography is performed with a non-polar C-18 column as the stationary phase and polar solvents such as acetonitrile and dilute trifluroacetic acid (0.1%) as the mobile phase. UV/DAD detectors are commonly used in the detection process. Individual component identification can also be accomplished with the help of Fourier transform ion cyclotron resonance (FT-ICR) and LC-MS. Using these methods, a previously unknown anthocyanin was discovered (cyanidin 3-disuccinylglucoside) in blue maize.[67]

The majority of these pigments, particularly anthocyanins, are water soluble, but a mixture of water and the organic solvent is recommended to increase extraction efficiency, because most of these molecules are bound to carbohydrates due to ionic interactions. The conventional extraction technique is solvent extraction, which involves extracting pigments with a mixture of water and polar organic solvents, such as methanol, acetone, and ethanol, in specific proportions (40–100% solvent). To increase the ionization of the extracting solution and thus break ionic interactions between the anthocyanins and carbohydrates, the solution is acidified with HCl (0.01 to 1%) or citric acid (1%).[20] Since these pigments are extremely sensitive to high temperatures, pH levels, and light,[68] these parameters are considered consciously during the extraction procedure. The extraction of these pigments can be enhanced by increasing the extraction time and temperature, but because anthocyanins are unstable, they are more prone to degradation under these conditions. In general, a temperature of 50–60 °C is maintained during solvent extraction, with varying time intervals for different matrices. Following extraction, the extracts are concentrated under vacuum at a low temperature (45 °C or less) and stored in the dark in a freezer (−4 °C to −20 °C) to await further analysis. In a previous study, anthocyanins were extracted using a conventional method from both seed and cob,[69] and the anthocyanin content was found to be higher for cob than seeds, indicating their use as an antioxidant and a natural food pigment in the food industry.

Scientists[70] reported the extraction of carotenoids and anthocyanins, as well as their monomeric content, in 49 different genotypes of waxy maize kernels. Extraction was achieved by immersing 2.0 g of maize samples in 25 mL of an acetone : water (70 : 30) mixture that had been acidified with HCl (0.01%). At room temperature, the flasks were shaken for about 2 hours. After shaking, the samples were filtered, evaporated, diluted with 70% acetone again, and extracted with chloroform (15 mL). The acetone layer was recovered and kept for subsequent pH differential analysis of monomeric anthocyanin content.[42]

Tristimuluspigmentedimetry was used[71] to determine the anthocyanin content of purple maize. The highest yield of anthocyanins was obtained at 70 °C with a solid : liquid ratio of 1 : 25 and an extraction time of 73 minutes. The presence and confirmation of three anthocyanins (pelargonidin-3-glucoside, cyanidin-3-glucoside, and peonidin-3-glucoside) were determined

Table 3.3 Effect of different extraction conditions on anthocyanin content.

Color of corn	Source	Time	Temperature	Solvent	Total anthocyanin content/yield	Type of anthocyanin detected	Reference
Purple	Kernel	73 minutes	70 °C	1 *n*HCl–95% ethanol (15 : 85, v/v)	6.02 mg g^{-1}, 55.8 ± 1.5 mg per 100 g	Cyanidin-3-glucoside(Cy3Glu), pelargonidin-3-glucoside(Pr3G) and peonidin-3-glucoside(P3G)	71
		24 hours	4 °C	Methanol	55.8 ± 1.5 mg per 100 g	Cyanidin-3-glucoside, pelargonidin-3-glucoside, peonidin-3-glucoside	69
	Cob powder	1 hour	Room temperature, 50 °C, 75 °C and 100 °C	1. Deionized water 2. 0.01%-HCl-acidified water 3. 0.01%-HCl-acidified ethanol 4. 70% aqueous acetone.	0.68, 0.94, 0.83 and 0.52 g per 100 g dry weight (DW) at RT, 50 °C, 75 °C and 100 °C respectively 0.60, 0.84, 0.77 and 0.63 g per 100 g DW at RT, 50 °C, 75 °C and 100 °C respectively 0.14 g per 100 g DW at RT 0.98 g per 100 gm DW at RT	Anthocyanins in 70% acetone and acidified boiling water: Cyanidin-3-glucoside, pelargonidin-3-glucoside, peonidin-3-glucoside, cyanidin-3-maloylglucoside, cyanidin-3-maloylglucoside, pelargonidin-3-maloylglucoside, peonidin-3-malonylglucoside and cyanidin-3-dimaloylglucoside	42
		3 hour	60 °C	60% aq. Ethanol containing 1% citric acid	34 g per 100 g DW	Cyanidin-3-glucoside, pelargonidin-3-glucoside and peonidin-3-glucoside	45

	Pericarp (genotypes)	—	—	Water	12.8 to 93.5 mg Cy3Glu equivalents per g	Three glycosides (Cy3Glu, Pr3G, and P3G), three acylated glycosides (Cy3Glu-Mal, Pr3G-Mal, and P3G-Mal), and a flavanol-anthocyanin dimer (catechin-cyanidin-3,5-diglucoside).	77
Blue	Kernel Flour (hybrid and native genotypes)	30 minutes	50 °C	96% ethanol and 1 N HCl 85 : 15 v/v	646 to 1052 mgCy3Glu per kg	—	16
	Kernel (various accessions and locations)	60 minutes	4 °C	Acidified methanol	Average 0.43 g kg^{-1} DW	Cyanidin 3-glucoside (most abundant),pelargonidin and peonidin 3-glucoside, Succinyl and disuccinyl glycosidic forms of cyaniding, Cyanidin 3-disuccinylglucoside	67
Rose	Kernel flour	20 minutes	RT	Methanol : water : HCl	—	Cyanidin 3-glucosides, Peonidin 3-glucosides Pelargonidin 3-glucosides, Cyanidin-3-(6-malonyl-glucoside) (major)	62
Purplish-red	Kernel flour (different varieties)	24 hours	RT	Methanol–acetic acid–water (10 : 1 : 9, v/v/v)	54–115 mg per 100 g DW	Cyanidin-3-glucoside, pelargonidin-3-glucoside, and Peonidin 3-glucosides in varying proportions among different varieties	66

using HPLC-MS. One study[72] also reported the effect of nixtamalization on anthocyanin, total phenolic, and carotenoid content in five different varieties of maize (white, yellow, red, blue and high carotenoid). Blue maize has the highest anthocyanin content when compared with red maize. Overall, it was discovered that lime-cooking significantly reduced pigment concentrations (phenols, anthocyanin and carotenoids). Extraction of carotenoids, anthocyanins, and phenolics from three types of waxy maize (black, white, and yellow) was done.[73] The extraction was carried out using the solvent extraction method with acetone at a concentration of 80%. To achieve the highest yield, the extraction procedure was repeated three times. Soluble free residues were flushed with nitrogen and treated with 4 N NaOH for one hour with continuous shaking for the extraction of bound pigments. The mixture was shaken before being acidified with hydrochloric acid to a pH of 2 and extracted with ethyl acetate. The supernatant was used for carotenoid, phenolic, and carotenoid content analysis. The extract was further acidified and heated to break the anthocyanin-glycoside bond for anthocyanin determination. The highest carotenoid content was found in yellow maize (22.5–39.61 μg of dry weight) whereas anthocyanins were major constituents (0.09–276.11 mg of cyanidin 3-glucoside equivalents per 100 g of dry weight) in black maize, while both the pigments were present in the least quantities in white maize. In addition, it was also observed that the carotenoid and anthocyanin content of yellow and white maize decreases with maturity, whereas it increases in black maize. According to recent reports, individual anthocyanins in a mixture can now be identified using HPLC coupled with MS. The anthocyanin content of waxy maize pigmented varieties at maturity was investigated.[49] The identification and quantification of individual anthocyanins was accomplished using HPLC-DAD-ESI/MS analysis. The highest quantity of anthocyanins detected were cyanidin-3-glucoside and its derivatives. Furthermore, acylated anthocyanins in milk and mature stages accounted for 67.1–88.2% and 46.2–83.6% of the total contents, respectively. Anthocyanin content and antioxidant activity were also found to have a positive correlation.

As sprouts have health benefits, their consumption is encouraged. This is attributed to their increased nutritional value, primarily due to the increased phytochemical content of the seeds during the sprouting or seedling stage. A study compared the phytochemical content of yellow, white, and violet waxy and field maize varieties. The extracts of maize seeds, sprouts, and seedlings were prepared using a solvent extraction method with methanol as the solvent. The extracts of maize seeds, sprouts, and seedlings were prepared using a solvent extraction method, with methanol as the solvent. The extracts were acidified further with HCl (1%) and shaken for about two hours. The filtered extracts were concentrated and used to determine pigment concentrations (phenols, anthocyanins, carotenoids, *etc.*). It was discovered that seedlings had the highest concentration of these pigments when compared with seeds and sprouts.[74] Waste utilization is always encouraged, and in the maize crop cob, silk, and husk are waste materials that can be used to increase the crop's production efficiency. In one such study,[75] the anthocyanin content of three varieties of purple waxy maize was determined in the cob, kernel, silk, and husk. The sample (0.5 g) was extracted with methanol containing 1% citric acid. All of the contents were thoroughly mixed and stored at 4 °C in the dark for 24 hours. The mixture was filtered and centrifuged after extraction. The obtained supernatant was stored at −20 °C for further analysis.

The pH differential method was used to determine the total anthocyanin content. It was discovered that there was more anthocyanin in the maize in the seed stage than in the edible stage. Although seeds possessed the highest anthocyanin content, maize silk and maize cob contain significant amounts of anthocyanins, which can be used in the food industry. Previously reported methods for anthocyanin identification (HPLC-PDA-MS) and quantification (pH differential, total anthocyanin contents, and HPLC) were compared.[26] The experiment was conducted on 14 different varieties of purple maize cob, using both intact and hydrolyzed samples. The quantification of anthocyanins and the quantification of pigments was discovered to be highly dependent on the extraction method used (anthocyanins and other pigments).[76] Furthermore, three fractionation methods were used[78] to compare the anthocyanin distribution in blue and purple maize coproducts. Total monomeric anthocyanin contents were ten times higher in purple maize than in blue maize, with the highest concentrations in the pericarp of the purple variety and the aleurone layer of blue maize.

Organic solvents were used in conventional plant pigment extraction techniques. However, these are now being replaced with more environmentally friendly and cost-effective solvents. In one previous study,[79] new solvents such as supercritical carbon dioxide, ethanol, and water were used for the sequential extraction of plant pigments from purple maize (cob and pericarp). Higher anthocyanin yields were observed in ethanol, whereas water extraction yielded higher phenolic and flavonoid content. Furthermore, sequential extraction techniques proved to be superior in terms of increasing extraction yield and producing extracts with different chemical compositions.

Many studies have been conducted to compare pigmented cob's yield and extract composition using both conventional and more innovative extraction procedures. In one study, extracts of Peruvian purple maize pericarp (*Z. mays* L.) were prepared using four methods: two conventional low-pressure methods (stirred vessel and Soxhlet) and two new high-pressure methods (supercritical fluid extraction using CO_2 as a solvent and EtOH–H_2O (70 : 30, v/v) as a co-solvent and pressurized liquid extraction (PLE) with EtOH–H_2O (70 : 30, v/v) as a solvent). Extraction yields of phenolic compounds and anthocyanins were higher with high pressure methods at higher temperatures (60 °C) than with low pressure methods[80](see Table 3.3).

3.5 Pigmented Maize-based Food Products

Color is considered to be one of the most vital parameters for assessment of the sensory quality of foods from a market standpoint. Nowadays, plant-based natural colorants have gained substantial interest due to their perceived safety and potential health benefits compared with chemical alternatives. However, natural pigments have limited stability upon exposure to light, temperature, oxygen, and pH.[81] Secondary metabolites such as anthocyanins/flavonoids, particularly in tandem with carotenoids and betalains, are a principal family of photosynthetic pigments that have found wide application in improving the functional properties of food due to their antioxidant qualities and potential health benefits, such as preventing risk of developing cardiovascular diseases, cancer, and diabetes.[7] Apart from being safe and nutritionally beneficial, natural pigments such as anthocyanins are now considered a promising alternative to coloring additives used in food products.[82] Different pigmented maize variants are being investigated for their potential in manufacturing various products *viz.* popcorn, tortilla chips, extruded snacks, expanded products, pasta, breads, cakes, muffins, *etc.* to improve the color, nutritional, and medicinal characteristics, owing to their high anthocyanin and fibre content.

3.5.1 Tortillas and Tortilla Chips

Traditionally, tortillas are prepared using normal maize. A common food in many countries, it is considered an important source of calories in many rural and suburban areas, where it supplies about 50% of people's daily energy intake.[83] However, nowadays, pigmented maize is also of great interest in preparing maize-based foods, due to the anthocyanin content and its perceived health benefits, which act as natural antioxidants and prevent cell and molecular damage by scavenging free radicals.

Nixtamalization, or the alkaline cooking of maize grains is the initial step in the manufacturing of maize-based food products such as tortillas, maize chips, tacos *etc.* Although anthocyanins and carotenoid losses were significantly higher in blue than yellow maize grains upon processing them into masa and tortillas, pigmented creole maize varieties possess antioxidant and antimutagenic activities after nixtamalization.[8] Apart from nixtamalization, lime-cooking extrusion is a promising alternative to produce dough suited for enchiladas, with comparable properties to those created using standard methods.[84]

Tortillas, tortilla chips, and other Hispanic foods made with masa and nixtamalized flour have high protein, calcium, and dietary fibre. In comparison with yellow Mexican maize varieties, in red maize grains, nixtamalization and lime-cooking extrusion was found to have a negative impact on total carotenoid content and oxygen radical absorbance capacity. However, lutein concentration was affected to lesser extent in conventional and traditional tortillas.[85] In artisanal blue maize tortillas, protein, fat, ash content, calcium, insoluble dietary fibre, and soluble dietary fibre content increased significantly compared with commercially produced white maize tortillas. In addition to their enhanced nutritional potential, the bioactive profile of tortillas was also improved, as a significant rise in anthocyanins and total phenolic content was observed, with subsequent increased antioxidant potential in blue handmade tortillas.[86] Blue maize tortillas have been found to exhibit high fat and protein content than white tortillas, and they have a softer texture due to the reduced total starch content and diminished interaction among starch chains. They also have a lower glycemic index and moderate amount of resistant starch content, reducing the tendency to starch retrogradation.[87]

Owing to the potential health benefits of pigmented maize, producers can switch to these varieties to produce nixtamalized maize flours for manufacturing products such as tlacoyos, quesadillas, and fajitas, which can result in a greater visual appeal of the product, as well as increased nutraceutical properties.[88] Blue maize has been widely studied to prepare tortillas and tortilla chips.[8,88–90] To further improve maize-based products' functional characteristics, several modifications to the traditional nixatamalization process have been investigated. The fractionated method of nixtamalization was used,[88] in which endosperm was separated from the pericarp, tip and germ were separated and then nixtamalized to prepare tortillas. The study found that the functional characteristics of blue maize tortillas from nixtamalized flour prepared with calcium hydroxide for 45 min was on a par with handmade tortillas produced from blue maize. The alteration in the nixtamalization process also aided in reducing process time, water consumption, promoting the utilization of whole grain, and reducing the generation of polluting effluents. Another study[90] produced blue maize tortillas by nixtamalization with calcium salts (NCS) and reported higher indigestible starch content and low resistant starch content than its counterpart, which was prepared using the conventional nixtamalization method. Also, it showed very low glycemic load and relatively high antioxidant properties, indicating a potential health benefit.

Red maize has also been widely investigated for the preparation of tortillas, due to their higher anthocyanins and carotenoids content, compared with blue maize.[8,90,92] Red pigmented maize can be successfully used to prepare tortillas through both the conventional nixtamalization process and the lime-cooking extrusion process. Due to the good amount of bioactives such as anthocyanins, carotenoids, and other compounds, consumers have greater awareness of the synergistic effect of these foods on human health.[8] Cereal products made from pigmented grains are attractive to consumers. Black and purple maize have also been utilized to prepare tortillas, which retain 34.1% and 24.9% of their anthocyanin content, respectively. Tortilla chips prepared from black maize retain more anthocyanin content (28.6%) than those prepared from purple maize. Also, they have a smoother texture and a sweeter taste compared with those prepared from white or yellow maize.[87,91]

3.5.2 Bread, Muffins, and Cake

Bread is one of the most commonly consumed wheat-based food products around the globe. It is considered an ideal food for fortifying essential micro- and macronutrients and bioactive compounds, as it can act as a vehicle for their delivery.[92] Substituting anthocyanin-rich pigmented maize grain flour against white wheat flour can considerably improve the functional and nutritional characteristics of bread.[3] Bread prepared by using whole grain blue maize flour and dark-red popped maize flour with 30% of wheat flour has enhanced bioactive compounds *viz.* total phenolic content, anthocyanins, and phenolic acids.[93] However, anthocyanin-rich maize flour did not influence the enhancement of bread's antioxidant properties, which are attributed to the synergistic and antagonistic interactions that emerge from the coexistence of multiple antioxidant components in food, but Maillard's reaction was found to have a substantial impact in increasing antioxidative properties in wheat and mix-breads. Anthocyanins from maize flour may significantly contribute to the beneficial properties of bread, including its appealing color, and overcoming thermal deterioration during baking.

Muffins are popularly consumed at breakfast, and as an evening snack. Colored maize has great potential to be used in developing functional gluten-free muffins, as zein protein-based doughs (functionally modified) had similar rheological properties as wheat-based doughs at higher temperatures, which otherwise is not able to form viscoelastic networks upon hydration as wheat gluten.[94,95] Several research findings have been in favor of using maize doughs in gluten-free product lines due to its suitability and required rheological characteristics. Gluten-free eggless muffins without the addition of hydrocolloids were prepared successfully. They utilized purple corn flours, as well as white and yellow germplasm to produce these muffins exhibiting low specific volume.[60] As well as having the appropriate physical characteristics, muffins have improved functional characteristics, as they exhibit higher levels of phenolic compounds and antioxidant activities. Although purple maize muffins did not appear as appealing to the sensory panelists in contrast to muffins made from yellow maize accessions, the formulations look promising as they were eggless and can help to mitigate the rise in egg allergies.

A blend of purple maize and rice flours in a ratio of 71.1 : 28.9 was also used as an alternative to wheat flour for the production of cakes. Purple maize cakes with maize silk were found to have a high anthocyanin content and high antioxidant activity, as well as the best textural score. Furthermore, its acidification with 3% fumaric acid retained the color and prevented phenolic degradation, making this a highly nutritional food.[96] Hence, it suggests that pigmented maize flours appear to be a promising alternative to wheat flour for bakery products. Similarly, polvornes, a traditional Mexican bread treat that is both nutritious and energetic, could be produced using pigmented maize. In comparison with wheat flour, polvorone from blue maize were more acceptable in terms of overall sensory attributes, including flavor and color. Also, these have significantly high phenolic and anthocyanin content. The availability of anthocyanins emphasized healthy positive characteristics, since the only deterioration observed during uncooked flour processing is the de-acylation of certain acylated anthocyanins.[97]

3.5.3 Cookies and Crackers

Cookies could easily be fortified with functional compounds among bakery products to promote health and reduce disease risk. Refined wheat flour used in such products can be partially replaced by anthocyanin-rich ingredients.[98,99] Acylated anthocyanins can indeed be used as a replacement of synthetic colors. Dark red popping corn, blue popping corn and blue standard corn flour was used[100] to make functional cookies. It demonstrated that the addition of citric acid prevented the degradation of phenolic compounds and antioxidant activity during baking in colored maize cookies by stabilizing anthocyanins. It lowers the pH of total anthocyanins in maize cookie dough and in the final cookies, causing acylation of sugar residues or flavylium cation. The addition of citric acid caused an increase in total flavonoids and anthocyanins of 60% and 70%, respectively, compared to ones prepared without citric acid. Hence, citric acid is a useful organic acid ingredient of bakery food, as it stabilizes the anthocyanins in maize cookies. However, low pH also inhibits the Maillard reaction, exhibiting a loss in antioxidant activity in blue popping corn and blue standard corn cookies. Cookies were also prepared from blue, dark blue and dark red popcorn maize varieties[101] and the availability of multiple polyphenolic compounds in these colored maize cookies had a significant effect on dicarbonyl creation and eradication. This is because organic polyphenol compounds could reduce glyoxal, methylglyoxal, and diacetyl content through the thermal process of baking, particularly those containing ammonium bicarbonate, which creates an alkaline environment. Due to the ability of their phenolic components to trap C2, C3, or C4-dicarbonyl molecules, colored maize flour can be a source of natural dietary antiglycation agents. In order to combat food and nutrition shortages in developing countries such as the Philippines, there has been increased interest in adopting indigenous crops as alternative food sources and, in this manner indigenous pigmented maize (*Z. mays* L.), called *camotes*, can be successfully used for the development of crackers by replacing all-purpose flour. The study suggested that crackers made with 80:20 camotes:all-purpose flour were comparable to those prepared with 100% all-purpose flour in terms of sensory acceptability. Also, the incorporation of maize flour enhances the nutritional content, *i.e.* protein, dietary fibre, lysine, tryptophan, zinc, antioxidant properties, phenolic compounds, and flavones. Therefore, they can be used as a nutritionally dense alternative in developing countries, to combat food and nutrition scarcity.

3.5.4 Pasta

Pasta, a wheat-based foodstuff, is widely consumed, and can be purchased ready to eat or prepared quickly at home. Celiac patients are increasingly demanding gluten-free products that are comparable to gluten-containing products in terms of organoleptic and nutritional properites.[102] The creation of gluten-free pasta with bioactive components is an important strategic area for the food manufacturing industry, as there has been a considerable growth in the production of food aimed at celiac disease (CD) sufferers in recent years.[103] Clients with CD appreciate gluten-free pasta, although there are only a few gluten-free items containing bioactive constituents like phytonutrients and soluble dietary fibre (especially resistant starch) that significantly contribute to the possible health beneficial properties of gluten-free eating.[104]

Antioxidant capacity and anti-cancer characteristics caused by the presence of anthocyanins can aid in boosting the nutraceutical value of gluten-free pasta.[89] The addition of 50–75% whole blue maize flour to regular maize flour resulted in an increase in total anthocyanins, phenolic content, and antioxidant capacity, increased protein content and dietary fibre, and a subsequent increase in slowly digestible starch and resistant starch content, while starch hydrolysis decreased. The incorporation of blue maize flour also led to a decrease in glycemic index. Therefore, consumption of pigmented maize-based food can contribute to a reduction in diet-related metabolic disorders.[102,104] Blue maize flour can be considered functional ingredients that could be used to develop functional food products with essential antioxidative potential, substantial levels of ingestible carbohydrates, a reduced GI, and therapeutic benefits for the overall population, such as lowering the risk of nutrition metabolic disorders. Another study investigated the preparation of composite pasta from blue maize flour, chickpea, and plantain flour. Composite pasta was found to have reduced firmness, gumminess, and a dark color, but stronger adhesiveness over white maize-based pasta. It has an adequate gruel solid loss of 9–11%. Extrusion and cooking resulted in maximum overall phenolic content retention *i.e.*, 80% and 70%, respectively. Extrusion and subsequent boiling of blue maize-containing pasta resulted in a minor drop in phenolic content and antioxidant properties, revealing that the network formed during manufacturing shielded the phytochemical compounds from heat degradation. In this regard, gluten-free pasta made with blue maize flour may provide a nutraceutical benefit.[104]

3.5.5 Yoghurt and Milk

Anthocyanins from high-anthocyanin-content colorful maize grains (Arrocillo, Peruano, Purepecha, and Cónico) have much potential for use as natural colorants in yoghurt.[66] Anthocyanin methanol extract was taken from four different cultivars and infused in yoghurts. The Peruano and Arrocillo extracts exhibited a darker red tone in comparison with that of Cónico and Purepecha extracts. However, after 5–10 days, the color degraded to a slightly pale yellow under refrigerated conditions. Another study used purple corm pigment to prevent lipid oxidation in milk and preserve its quality. The inclusion in milk samples of 0.3% purple maize pigment comprising pelargonidin, peonidin, cyanidin, malvidin, petunidin, and delphinidin resulted in a 7-day delay in the oxidation of unsaturated free fatty acids. Pelargonidin and petunidin severely impacted C14: 1*n*-5, C17: 1*n*-7, C18: 2*n*-6, C20: 2*n*-6, C20:

3*n*-3, and C20: 4*n*-6 degradation, whereas cyanidin and total anthocyanins directly impacted C14: 1*n*-5, C16: 1*n*-7, C17: 1*n*-7, and C20: 4*n*-6 degradation. This shows that pigment from purple maize has the potential to prevent lipid oxidation and maintain the concentration of unsaturated fatty acids in milk during its storage.[105]

3.5.6 Beverages

Flores-Calderón *et al.*[106] proposed a modified 'Sendechó' beverage, which added hops and brewer's yeast to guajillo chili and blue corn malt (the conventional flavors) in an attempt to produce beer with higher antioxidant activity and phenolic content, resulting in a beer with Sendechó-like character traits, as well as corn beer-like character traits, due to the ale fermentation process. The effect of adding caramel malt, varying hops, and guajillo to prepare different blue corn malt beers was evaluated. Eight blue maize malt beers were analysed for titratable acidity, bitterness units, total reducing sugars, alcohol, *cis*- and *trans*-iso-acids concentration, total anthocyanins and phenolics, and radical scavenging activity at three separate stages of the process: boiling wort, fermentation (green beer), and maturation (mature beer). Except for pH, total acidity and alcohol level, other physiochemical parameters were particularly affected in mature beer. There was a significant drop in polyphenolic concentration and antioxidant activity. Low-alcohol beers were produced with pH, total acidity, bitterness units, iso-acids level, and total reducing sugars comparable to barley beers. The maximum anthocyanin concentration and antioxidant activities were found in beers made with 85% maize malt and 15% caramel malt.

Another beverage, chicha morada, has been prepared using the Peruvian Andean maize variety, which is typical of a limited Peruvian zone. The beverage comprises extract of purple maize pigments and fruits such as pineapple, quince, apple, and peach, which act as a starch source and add taste to the product. A dessert, Mazamorra Morada, is also made using a cooking technique and adding dried or fresh fruits with this variety of purple maize. This cultivar had significant nutritional and health benefits, and its flour can be utilized in a range of dishes (pasta, bread, and cakes) to provide color and taste to processed goods.[107]

3.5.7 Extruded and Expanded Snacks

Extrusion is an extensively used method for production of variety of maize based products, such as expanded snacks and other ready-to-eat foods. Conventional nixtamalization and lime-cooking extrusion are also very common methods of maize processing. Nixtamalization involves adding $Ca(OH)_2$ to maize and water at 100 °C to generate a solution boiled for 20–40 minutes then steeped for at least 16 hours as lime-cooking extrusion is a quicker and more accurate option. Extrusion is a method of making nixtamalized items, but is not as common as other methods. Maize is suitable for expanded snacks manufacture as it exhibits good expansion characteristics. Therefore, colored maize has also been studied as it may also possess good physical and technological characteristics for second- and third-generation extruded snack production when used with added calcium hydroxide.[108] Colored maize such as blue and purple were widely used to produce extruded third-generation snacks from whole blue maize and corn starch mixture in ratio of 65 : 35.[41] The extrusion process in pigmented maize results in a loss of anthocyanins of 50–60% as anthocyanins lose their stability at elevated temperature and pH levels. Changes in specific processing parameters can result in the creation of a bluish quinoidal base or a colorless carbinol pseudo base that also affects the color of goods manufactured with pigmented maize and changes their visual effect. Calcium hydroxide aids in maintaining the maximum possible total anthocyanin concentration, strong blue and purple coloring, and maximum expansion of the snacks during extrusion.[109] Nixtamalized blue maize can also be used to develop expanded extruded snacks at 16 of feed moisture, barrel temperature of 130 °C and particle size index of 83.97. Compared with raw maize, it showed a higher total anthocyanin (211.1 mg kg^{-1}) and Cyanidin 3-glucoside (11.3%). Researchers[109] also used the blue maize to develop second-generation snacks using lime-cooking extrusion instead of the traditional nixtamalization process. These products have been found to have enhanced functional benefits with sensory attributes on a par with those prepared with lime-cooking extrusion, due to the anthocyanic chemicals found within these grains.

Snacks are food products that may temporarily satisfy a person's appetite, is ready to eat, lightweight, tiny, and easily accessible.[110] They are consumed worldwide by people of all social classes, ages, and genders, and the snack industry is flourishing, offering a broad variety of snacks; primarily containing potato and maize derivatives.[111] Extrusion is commonly used to prepare third-generation (3G) snacks, commonly referred as intermediate snack foods or pellet. These are produced at low pressure to prevent expansion and then dried to make a vitreous pellet.[112] This is followed by necessitating yet another expansion step, which can be accomplished using cooking oil, hot air, or electromagnetic treatment, which are required to give the snack a porous structure by boiling the water in the pellet and producing vapors that lead to expansion.[113] Regular maize is generally used for the production of snack foods, due to its expansion and nutritional properties and, hence, colored maize has similar properties, with the added benefit that anthocyanin and polyphenol chemicals can be used to manufacture snack foods.[17]

The third-generation extruded snacks from blue-colored maize by single screw extrusion of blue maize[111] were prepared at a barrel temperature of 120 °C to 126 °C and a feed moisture percentage of 23.8–25.2%, which could be further enhanced by microwave processing to produce 3G snacks. The product formed exhibited good expansion, optimum hardness, and enhanced nutritional and functional properties. The total anthocyanin content was 71.09 mg kg^{-1} and improved dietary fibre content.

Roasting is another well-known dry thermal technique for crispy snack production. However, extreme heat can destroy chemical components like phenolics and anthocyanins, reducing their antioxidative action considerably. Researchers[114] have suggested roasting with the intensification of vaporization by decompression to the vacuum (IVDV). This is a new texturizing method that can be employed as a pre-treatment for roasting purple maize grains, which involves subjecting wet grains to a saturated steam pressure accompanied by decompression to the vacuum, preceded by roasting in relatively mild conditions. This helps to texturize purple maize grains, whilst also conserving as much of their own antioxidant content as possible.[114] The effect of the incorporation of colored maize on product characteristics is summarized below (see Table 3.4).

Table 3.4 Effect of incorporation of colored maize on product characteristics.

Food products	Pigmented maize used	Product characteristics	Reference
Tortilla	Red colored Mexican maize with conventional nixtamalization and lime-cooking extrusion technique	• Reduction in carotenoid content and radical scavenging activity • 72.7% and 60% lutein retained in traditional and extruded tortillas in comparison with raw grains	85
Homemade tortillas	Blue maize	• Higher antioxidant potential, fibre, calcium, and total polyphenol concentration than white tortilla • 4.5 times ferulic acid concentration than white tortilla	86
Tortilla	Nixtamalized blue maize with calcium salts	• High indigestible starch and less resistant starch content • Low glycemic index • High antioxidant potential	90
Tortilla	Blue colored maize	• Higher fat and protein content • Softer texture • Lower total starch content. • Lower glycemic index in comparison with white tortillas • Reduced starch retrogradation tendency	87
Tortillas	Black and purple maize	• Smooth texture • Sweet taste • 34.1% and 24.9% anthocyanin retention in black and purple maize tortilla, respectively • Higher anthocyanin retention in tortilla chips made from black maize (28.6%) than purple maize	87 and 91

(*continued*)

Table 3.4 (*continued*)

Food products	Pigmented maize used	Product characteristics	Reference
Bread	Whole grain blue and dark red popped maize flour with 30% wheat flour	• Enhanced total phenolic content, anthocyanin content, antioxidant activity • Reduced color degradation during baking	93
Gluten-free, eggless Muffins	Purple maize	• Improved phenolic content and antioxidant activities • Good height • Low specific volume • Lower sensory acceptability in comparison with yellow maize	60
Cake	Purple maize flour: Rice flour in ratio 71.1 : 28.9 upon acidification with 3% fumaric acid	• Maximum retention of total anthocyanin content, total phenolic content, ABTS-radical scavenging ability and ferrous chelating ability	6
Polvorones	Blue maize	• Higher overall sensory acceptability • Higher total phenolic (837.1 mg gallic acid per kg) and anthocyanin content (373.0 mg cyanidin-3-glucoside per kg)	97
Cookies	Dark red popping corn, blue popping corn and blue corn flour	• Increased phenolic content and improved stabilization of anthocyanins due to addition of citric acid; aiding acylation of flavylium cation	100
	Blue, dark blue and dark red popcorn maize	• Anthocyanin caused significant reduction in glyoxal, methylglyoxal, and diacetyl content	101
Crackers	*Camotes* (Indigenous pigmented maize in Philippines): All purpose flour in 80 : 20.	• Enhanced protein content, dietary fibre, lysine, zinc • Enhanced phenolic compounds, flavones and antioxidant properties	115
Gluten-free Pasta	Whole blue maize flour	• Increase in total phenolic content, anthocyanin, antioxidant activity • Low GI due to increase in slowly digestible starch and resistant starch • Reduced firmness, gumminess, but higher adhesiveness • Darker color	102 and 104
Yoghurt	Yoghurt milk infused with 1 mg of methanolic anthocyanin extract	• Delayed oxidation of unsaturated fatty acids by 7 days	105

Table 3.4 (*continued*)

Food products	Pigmented maize used	Product characteristics	Reference
Sendecho	Blue maize malt	• Significant reduction in polyphenolic content and antioxidant activity in mature beer	106
Expanded extruded snacks	Nixtamalized blue maize	• Higher total anthocyanin and Cyanidin 3-glucoside in comparison with raw maize	109
Third-generation expanded snacks	Whole blue colored maize	• High total anthocyanin content of 71.09 mg kg^{-1} • Good expansion • Optimum hardness	111
Roasted maize with IVDV pre-treatment	Purple maize grains	• Prevent destruction of antioxidant due to extreme heat	114

3.6 Conclusion

Pigmented maize is an abundant source of bioactive components such as anthocyanin, polyphenolics, carotenoids, and flavonoids, which are attributed to various health benefits offered by it. The different expression of these pigments imparts different pigments to maize tissues, from yellow-orange to dark purple-blue, as well as ivory and cream. Pigments are generally mostly concentrated in the thick pericarp or in the aleurone layers of kernels, and also in maize cobs. Owing to the potential health benefits of pigmented maize, production companies can switch to these varieties for production of maize flours for manufacturing various food products such as tortillas, cookies, cakes, *etc.*, which may result in greater visual appeal of the product, as well as increased nutraceutical properties.

References

1. A. Tiessen, E. A. Cubedo-Ruiz and R. Winkler, *Am. J. Plant Sci.*, 2017, **8**, 502.
2. J. M. Magaña-Cerino, H. A. Peniche-Pavía, A. Tiessen and C. M. Gurrola-Díaz, *Pol. J. Food Nutr. Sci.*, 2020, **70**, 85.
3. S. Zilic, A. Serpen, G. Akıllıoğ, V. Gökmen and J. Vančetović, *J. Agric. Food Chem.*, 2012, **60**, 1224.
4. T. Tsuda, F. Horio, K. Uchida, H. Aoki and T. Osawa, *J. Nutr.*, 2003, **133**, 2125.
5. M. Urias-Peraldi, J. A. Gutierrez-Uribe, R. E. Preciado-Ortiz, A. S. Cruz-Morales, S. O. Serna-Saldivar and S. Garcia-Lara, *Field Crops Res.*, 2013, **141**, 69.

6. F. J. Betrán, A. J. Bockholt, and L. W. Rooney, *Blue Corn*, ed. A. R. Hallauer, CRC Press, Boca Raton, Florida, USA, Specialty Corns, 2nd edn, 2000.
7. Y. Tanaka, N. Sasaki and A. Ohmiya, *Plant J.*, 2008, **54**, 733.
8. S. Mendoza-Díaz, M. C. Ortíz-Valerio, E. Castaño-Tostado, J. D. Figueroa-Cárdenas, R. Reynoso Camacho, M. Ramos-Gómez and G. Loarca-Piña, *Plant Foods Hum. Nutr.*, 2012, **67**, 442.
9. G. A. Camelo-Méndez, E. Agama-Acevedo, M. M. Sanchez-Rivera and L. A. Bello-Pérez, *Food Chem.*, 2016, **211**, 281.
10. G. A. Camelo-Méndez, E. Agama-Acevedo, J. Tovar and L. A. Bello-Pérez, *J. Cereal Sci.*, 2017, **76**, 179.
11. L. G. Ranilla, A. Christopher, D. Sarkar, K. Shetty, R. Chirinos and D. Campos, *J. Food Sci.*, 2017, DOI: 10.1111/1750-3841.13973.
12. A. Castañeda-Ovando, C. A. Galán-Vidal, L. Pacheco, J. A. Rodriguez and M. E. Paez-Hernandez, *Food Anal. Methods*, 2010, **3**, 12.
13. C. Mutlu, S. Arslan-Tontul, C. Candal, O. Kilic and M. Erbas, *J. Food Sci.*, 2018, **83**, 53.
14. J. Moore, Z. Hao, K. Zhou, M. Luther, J. Costa and L. Yu, *J. Agric. Food Chem.*, 2005, **53**, 6649.
15. E. C. Montilla, S. Hillebrand, A. Antezana and P. Winterhalter, *J. Agric. Food Chem.*, 2011, **59**, 7068.
16. D. A. Urias-Lugo, J. B. Heredia, M. D. Muy-Rangel, J. B. Valdez-Torres, S. O. Serna-Saldívar and J. A. Gutiérrez-Uribe, *Plant Foods Hum. Nutr.*, 2015, **70**, 193.
17. R. Pedreschi and L. Cisneros-Zevallos, *Food Chem.*, 2007, **100**, 956.
18. F. Ramos-Escudero, A. M. Muñoz, C. Alvarado-Ortíz, A. Alvarado and J. A. Yáñez, *J. Med. Food*, 2012, **15**, 206.
19. E. S. M. Abdel-Aal, J. C. Young and I. Rabalski, *J. Agric. Food Chem.*, 2006, **54**, 4696.
20. X. Zhao, C. Zhang, C. Guigas, Y. Ma, M. Corrales, B. Tauscher and X. Hu, *Eur. Food Res. Technol.*, 2009, **228**, 759.
21. Y. Salinas-Moreno, G. Salas-Sanchez, D. Rubio-Hernandez and N. Ramos-Lobato, *J. Chromatogr. Sci.*, 2005, **43**, 483.
22. L. Barrientos-Ramírez, H. E. Ramírez-Salcedo, M. F. Fernández-Aulis, M. A. Ruíz-López, A. Navarro-Ocaña and J. J. Vargas-Radillo, *Interciencia*, 2018, **43**, 188.
23. Y. Salinas-Moreno, F. Cruz-Chavez, S. Diaz-Ortiz and F. Castillo-Gonzalez, *Rev. Fitotec. Mex.*, 2012, **35**, 33.
24. H. Wang, G. H. Cao and R. L. Prior, *J. Agric. Food Chem.*, 1997, **45**, 304.
25. H. Hong, M. Netzel and T. O'Hare, *Food Chem.*, 2020, **319**, 126515.
26. F. Lao and M. M. Giusti, *Food Anal. Methods*, 2016, **9**, 1367.
27. M. Hernández, J. Ventura, C. Castro, V. Boone, R. Rojas, J. Ascacio-Valdés and G. Martínez-Ávila, *Molecules*, 2018, **23**, 1425.
28. F. Fernandez-Aulis, L. Hernandez-Vazquez, G. Aguilar-Osorio, D. Arrieta-Baez and A. Navarro-Ocana, *J. Food Sci.*, 2019, **84**, 954.

29. W. Piyapanrungrueang, W. Chantrapornchai, V. Haruthaithanasan, U. Sukatta and C. J. Aekatasanawan, *J. Food Process. Preserv.*, 2016, **40**, 1125.
30. B. Huang, Z. Wang, J. H. Park, O. H. Ryu, M. K. Choi, J. Y. Lee, Y. H. Kang and S. S. Lim, *Nutr. Res. Prac.*, 2015, **9**, 22.
31. T. Jiang, S. Zhan, S. Li, Z. Zhu, J. He, J. M. Lorenzo and F. J. Barba, *J. Sci. Food Agric.*, 2018, **98**, 4919.
32. C. Y. Hu, Q. L. Li, X. F. Shen, S. Quan, H. Lin, L. Duan, Y. F. Wang, Q. Luo, G. R. Qu, Q. Han, Y. Lu, D. B. Zhang, Z. Yuan and J. X. Shi, *Sci. Rep.*, 2016, **6**, 1.
33. K. Petroni, R. Pilu and C. Tonelli, *Planta*, 2014, **240**, 901.
34. K. Morohashi, M. I. Casas, M. L. Falcone Ferreyra, M. K. Mejía-Guerra, L. Pourcel and A. Yilmaz, *et al.*, *Plant Cell*, 2012, **24**, 2745.
35. R. H. Ford, *Am. Biol. Teach.*, 2000, 181–188.
36. M. Sharma, M. Cortes-Cruz, K. R. Ahern, M. McMullen, T. P. Brutnell and S. Chopra, *Genetics*, 2011, **188**, 69–79.
37. L. Y. Shen and J. F. Petolino, *Mol. Breed.*, 2006, **18**, 57.
38. C. Lago, E. Cassani, C. Zanzi, M. Landoni, R. Trovato and R. Pilu, *Plant Breed.*, 2014, **133**, 210.
39. Q. Li, P. Somavat, V. Singh, L. Chatham and E. G. de Mejia, *Food Chem.*, 2017, **231**, 332.
40. P. A. Rodriguez-Salinas, F. Zavala-Garcia, V. Urias-Orona, D. Muy-Rangel, J. B. Heredia and G. Nino-Medina, *Arabian J. Sci. Eng.*, 2020, **45**, 95.
41. R. O. Navarro-Cortez, E. A. Palazuelos, J. Z. Zazueta-Morales, J. C. Rosas, C. A. GÃmez-Aldapa and F. S. Aguirre-Tostado, *Int. J. Food Process. Technol.*, 2014, **1**, 10.
42. P. Jing, V. Noriega, S. J. Schwartz and M. M. Giusti, *J. Agric. Food Chem.*, 2007, **55**, 8625.
43. D. A. Urias-Lugo, J. B. Heredia, S. O. Serna-Saldivar, M. D. Muy-Rangel and J. B. Valdez-Torres, *CyTA–J. Food*, 2015, **13**, 336.
44. R. Martínez-Martínez, A. M. Vera-Guzmán, J. L. Chávez-Servia, E. N. A. Bolaños, J. C. Carrillo-Rodríguez and A. Pérez-Herrera, *Interciencia*, 2019, **44**, 549.
45. S. dePascual-Teresa, C. Santos-Buelga and J. Rivas-Gonzalo, *J. Sci. Food Agric.*, 2002, **82**, 1003.
46. S. Suriano, C. Balconi, P. Valoti and R. Redaelli, *LWT*, 2021, **144**, 111257.
47. G. A. Camelo-Méndez, E. Agama-Acevedo, J. Tovar and L. A. Bello-Pérez, *J. Cereal Sci.*, 2017, **76**, 179.
48. A. Capocchi, S. Bottega, C. Spano and D. Fontanini, *Int. J. Food Sci. Nutr.*, 2017, **68**, 515.
49. B. Harakotr, B. Suriharn, R. Tangwongchai, M. P. Scott and K. Lertrat, *Food Chem.*, 2014, **164**, 510.
50. J. T. Kim, G. Yi, I. M. Chung, B. Y. Son, H. H. Bae, Y. S. Go, J. Y. Ha, S. B. Baek and S. L. Kim, *ACS Omega*, 2020, **5**, 15702.
51. A. O. Psolova, K. V. Derkach and T. M. Satarova, *Agrology*, 2019, **2**, 128.

52. S. Mora-Rochín, N. Gaxiola-Cuevas, J. A. Gutiérrez-Uribe, J. Milán-Carrillo, E. M. Milán-Noris, C. Reyes-Moreno, S. O. Serna-Saldivar and E. O. Cuevas-Rodríguez, *Food Sci. Technol.*, 2016, 563–569.
53. J. Guillén-Sánchez, S. Mori-Arismendi and L. M. Paucar-Menacho, *Sci. Agropecu.*, 2014, **5**, 211.
54. A. J. Smorowska, A. K. ˙Zołnierczyk, A. Nawirska-Olszanska, J. Sowinski and A. Szumny, *J. Food Qual.*, 2021, 8813613.
55. F. Kahriman, F. Akta, G. Pinar, U. Songur and C. O. Egesel, *Acta Fytotech. Zootech.*, 2021, **24**, 16.
56. C. G. Mendoza-Mendoza, M. C. Mendoza-Castillo, A. Delgado-Alvarado, F. J. Sánchez-Ramírez and T. A. Kato-Yamakake, *Maydica*, 2020, **65**, M21.
57. C. Lago, M. Landoni, E. Cassani, S. Atanassiu, E. Cantaluppi and R. Pilu, *Maydica*, 2020, **59**, 2014.
58. G. Loarca-Piña, M. Neri, J. D. D. Figueroa, E. Castaño-Tostado, M. Ramos-Gómez, R. Reynoso and S. Mendoza, *J. Food Sci. Technol.*, 2019, **56**, 3177.
59. B. A. Cevallos-Casals and L. Cisneros-Zevallos, *J. Agric. Food Chem.*, 2003, **51**, 3313.
60. S. Trehan, N. Singh and A. Kaur, *J. Food Sci. Technol.*, 2018, **55**, 2334.
61. L. L. Méndez-Lagunas, M. Cruz-Gracida, L. G. Barriada-Bernal and L. I. Rodríguez-Méndez, *J. Food Sci. Technol.*, 2020, **57**, 4688.
62. L. B. Ramírez, H. E. R. Salcedo, M. F. Fernández Aulis, M. A. RuízLópez, A. N. Ocaña and J. J. Vargas Radillo, *Interciencia*, 2018, **43**, 188.
63. A. Escalante-Aburto, B. Ramírez-Wong, P. I. Torres-Chávez, J. López-Cervantes, J. D. D. Figueroa-Cárdenas, J. M. Barrón-Hoyos and R. Gutiérrez-Dorado, *Molecules*, 2014, **19**, 21066.
64. A. Escalante-Aburto, B. Ramirez-Wong, P. I. Torres-Chavez, J. M. Barron-Hoyos, J. D. D. Figueroa-Cardenas and J. Lopez-Cervantes, *Rev. Fitotec. Mex.*, 2013, **36**, 429.
65. K. Y. Amador-Rodríguez, F. Martínez-Bustos and H. Silos-Espino, *Plant Foods Hum. Nutr.*, 2019, **74**, 241.
66. Y. S. Moreno, G. S. Sánchez, D. R. Hernández and N. R. Lobato, *J. Chromatogr. Sci.*, 2005, **43**, 483.
67. A. N. Nankar, B. Dungan, N. Paz, N. Sudasinghe, T. Schaub, F. O. Holguin and R. C. Pratt, *J. Sci. Food Agric.*, 2016, **96**, 4542.
68. S. Haggard, D. Luna-Vital, L. West, J. A. Juvik, L. Chatham, M. Paulsmeyer and E. G. de Mejia, *Food Res. Int.*, 2018, **105**, 286.
69. Z. Yan and W. Zhai, *Innov. Food Sci. Emerg. Technol.*, 2010, **11**, 169.
70. B. Harakotr, B. Suriharn, M. P. Scott and K. Lertrat, *Euphytica*, 2015, **203**, 237.
71. Z. Yang, Z. Chen, S. Yuan, W. Zhai and X. Piao, *Int. J. Food Sci. Technol.*, 2009, **44**, 2485.
72. C. De La Parra, S. O. Serna Saldivar and R. H. Liu, *J. Agric. Food Chem.*, 2007, **55**, 4177.
73. Q. P. Hu and J. G. Xu, *J. Agric. Food Chem.*, 2011, **59**, 2026.

74. W. Chalorcharoenying, K. Lomthaisong, B. Suriharn and K. Lertrat, *Int. Food Res. J.*, 2017, **24**, 552.
75. S. Simla, S. Boontang and B. Harakotr, *Aust. J. Crop Sci.*, 2016, **10**, 675.
76. F. Lao and M. M. Giusti, *Food Anal. Methods*, 2016, **9**, 1367.
77. Q. Zhang, D. Luna-Vitala and E. G. de Mejia, *J. Funct. Foods*, 2019, **54**, 175.
78. T. C. Li, W. Zhang, H. Y. Yang, Q. Dong, J. Ren, H. H. Fan, X. Zhang and Y. B. Zhou, *Sci. Rep.*, 2019, **9**, 2485.
79. Y. M. Monroy, R. A. Rodrigues, A. Sartoratto and F. A. Cabral, *J. Supercrit. Fluids*, 2016, **107**, 250.
80. Y. M. Monroy, R. A. Rodrigues, A. Sartoratto and F. A. Cabral, *Braz. J. Chem. Eng.*, 2020, **37**, 237.
81. A. Bakowska-Barczak, *Polish J. Food Nutr. Sci.*, 2005, **14/55**, 107–116.
82. M. M. Giusti and R. E. Wrolstad, *Biochem. Eng. J.*, 2003, **14**, 217.
83. O. N. Campas-Baypoli, E. C. Rosas-Burgos, P. I. Torres-Chavez, B. Ramımez-Wong and S. O. Serna-Sald´ıvar, *Starch-Stärke*, 1999, **5**, 173.
84. J. Milan-Carrillo, R. Gutierrez-Dorado, E. O. Cuevas-Rodríguez, J. A. Garzon-Tiznado and C. Reyes-Moreno, *Plant Foods Hum. Nutr.*, 2004, **59**, 35.
85. A. B. Corrales-Bañuelos, E. O. Cuevas-Rodríguez, J. A. Gutiérrez-Uribe, J. E. M. Milán-Noris, C. Reyes-Moreno, J. Milán-Carrillo and S. Mora-Rochín, *J. Cereal Sci.*, 2016, **69**, 64.
86. C. Colín-Chávez, J. J. Virgen-Ortiz, L. E. Serrano-Rubio, M. A. Martínez-Téllez and M. Astier, *Curr. Res. Food Sci.*, 2020, **3**, 189.
87. J. P. Hernández-Uribe, E. Agama-Acevedo, J. J. Islas-Hernández, J. Tovar and L. A. Bello-Pérez, *J. Sci. Food Agric.*, 2007, **87**, 2482.
88. A. Cortes-Gomez, E. San Mart´ın-Mart´ınez, F. Mart´ıınez-Bustos and G. M. V´azquez-Carrillo, *J. Food Eng.*, 2005, **66**, 273.
89. J. Aguayo-Rojas, S. Mora-Rochın, E. O. Cuevas-Rodrıguez, S. O. Serna-Saldivar, J. A. Gutierrez-Uribe, C. Reyes-Moreno and J. Milan-Carrillo, *Plant Foods Hum. Nutr.*, 2012, **67**, 178.
90. L. A. Bello-Pérez, P. C. Flores-Silva, G. A. Camelo-Méndez, O. Paredes-López and J. D. D. Figueroa-Cárdenas, *Cereal Chem.*, 2015, **92**, 265.
91. L. X. Lopez-Martınez, K. L. Parkin and H. S. Garcia, *Int. J. Food Sci. Technol.*, 2012, **47**, 1321.
92. A. A. A. El-Megeid, I. J. A. AbdAllah, M. F. Elsadek and Y. F. A. El-Moneim, *World J. Dairy Food Sci.*, 2009, **4**, 107.
93. M. Simic, S. Zilic, O. Simuruna, B. Filipcev, D. Skrobot and J. Vancetovic, *Polish J. Food Nutr. Sci.*, 2018, **68**, 299–308.
94. B. A. Bugusu, O. Campanella and B. R. Hamaker, *Cereal Chem.*, 2001, **78**, 31.
95. D. P. Erickson, O. H. Campanella and B. R. Hamaker, *Trends Food Sci. Technol.*, 2012, **24**, 74.
96. H. Kokkaew and T. Pitirit, *Int. Food Res. J.*, 2016, **23**, 2390.
97. M. G. Vázquez-Carrillo, L. A. Aparicio-Eusebio, Y. Salinas-Moreno, M. O. Buendía-Gonzalezand and D. Santiago-Ramos, *Plant Foods Hum. Nutr.*, 2018, **73**, 321.

98. S. Mildner-Szkudlarz, R. Zawirska-Woitasiak, W. Obuchowski and M. Goslinski, *J. Food Sci.*, 2009, **74**, S362.
99. A. Pasqualone, A. M. Bianco, V. M. Paradiso, C. Summo, G. Gambacorta and F. Caponio, *Food Res. Int.*, 2014, **65**, 385.
100. S. Zilic, T. Kocadağlı, J. Vančetović and V. Gökmen, *LWT–Food Sci. Technol.*, 2016, **65**, 597.
101. T. Kocadağlı, S. Žilić, N. G. Taş, J. Vančetović, D. Dodig and V. Gökmen, *Eur. Food Res. Technol.*, 2016, **242**, 51.
102. G. A. Camelo-Méndez, J. Tovar and L. A. Bello-Pérez, *J. Food Sci. Technol.*, 2018, **55**, 2739.
103. T. Matthias, S. Neidhöfer, S. Pfeiffer, K. Prager, S. Reuter and M. E. Gershwin, *Cell. Mol. Immunol.*, 2011, **8**, 121.
104. G. A. Camelo-Méndez, M. G. Ferruzzi, G. A. González-Aguilar and L. A. Bello-Pérez, *Food Eng. Rev.*, 2016, **8**, 76.
105. X. Z. Tian, Q. Lu, P. Paengkoum and S. Paengkoum, *J. Dairy Sci.*, 2020, **103**, 7808.
106. A. M. D. Flores-Calderón, H. Luna, H. B. Escalona-Buendía and J. R. Verde-Calvo, *J. Inst. Brew.*, 2017, **123**, 506.
107. R. Salvador-Reyes, P. Silva and M. T. Clerici, *Food Res. Int.*, 2020, **130**, 108934.
108. J. Zazueta-Morales, F. Martínez-Bustos, N. Jacobo-Valenzuela, C. Ordorica-Falomir and O. Paredes-Lopez, *J. Food Sci.*, 2002, **67**, 3350.
109. A. Escalante-Aburto, B. Ramirez-Wong, P. I. Torres-Chavez, J. D. Figueroa-Cardenas, J. Lopez-Cervantes, J. M. Barron-Hoyos and I. Morales Rosas, *CyTA–J. Food*, 2013, **11**, 29.
110. M. Hurtado, B. Escobar and A. M. Estévez, *Arch. Latinoam. Nutr.*, 2001, **5**, 303.
111. I. L. Camacho-Hernández, J. J. Zazueta-Morales, J. A. Gallegos-Infante, E. Aguilar-Palazuelos, N. E. Rocha-Guzmán, R. O. Navarro-Cortez and C. A. Gómez-Aldapa, *CyTA–J. Food*, 2014, **12**, 320.
112. P. Hollingsworth, Third-generation snacks take aim at popcorn market, *Food Technol.*, 2001, **55**, 20.
113. C. Boischot, C. I. Moraru and J. L. Kokini, *Cereal Chem.*, 2003, **80**, 56.
114. R. Mrad, E. Debs, R. Saliba, R. G. Maroun and N. Louka, *J. Cereal Sci.*, 2014, **60**, 397.
115. Z. G. Sales, C. B. Juanico, E. I. Dizon and W. A. Hurtada, *Nutritional Status, Dietary Intake and Body Composition*, 2018, vol. 24, p. 617.

CHAPTER 4

Pigmented Barley: Phytochemical Composition, β-Glucan Content, and Applications

ROSHANLAL YADAV[a], PRIXIT GULERIA*[b] AND ASHWANI KUMAR[c]

[a]Department of Food Technology, Bhaskaracharya College of Applied Sciences, Dwarka, New Delhi, India; [b]Department of Food Technology, Maharshi Dayanand University, Rohtak, Haryana, India; [c]Department of Post Harvest Technology, College of Agriculture and Forestry, Rani Lakshmi Bai Central Agricultural University, JJhansi, Uttar Pradesh, India
*E-mail: prixit.guleria@gmail.com

4.1 Introduction

Recently, minor bioactive compounds and carotenoids, phenols, flavonoids, and anthocyanins have attracted massive interest from food manufacturers and researchers because of their disease-preventing and health-promoting properties. Various epidemiological studies confirmed that long-term consumption of plant-derived polyphenols could defend against different diseases such as brain and immune disorders, cardiovascular diseases, cancers, osteoporosis, diabetes, and numerous neurodegenerative diseases.[1,2] In this respect, pigmented cereals may be considered a natural source of bioactive

Food Chemistry, Function and Analysis No. 38
Pigmented Cereals and Millets: Bioactive Profile and Food Applications
Edited by Sneh Punia Bangar, Sajid Maqsood and Anil Kumar Siroha

Published by the Royal Society of Chemistry, www.rsc.org

compounds in the diet because they are staple foods and were utilized to develop new value-added food products.[2] Among all the cereals, barley is the most cultivated crop globally (12% of total cereal cultivated), ranking fourth among cereal grains after wheat, rice, and maize, and it is the most historical plant grown globally.[3] This cultivated barley crop has diverse uses; most of the portions, *i.e.* about 65% is generally used in animal feed, 33% is used for malting, while only 2% is directly used for the consumption of human beings.[4]

Regarding genetic variations, barley is one of the most genetically varied cereal grains, including spring or winter types, colored or uncolored, with shells or hull-less, malting or feed varieties, and two-row or six-row. With practical genetic engineering, barley cultivars can now grow whole pigmented kernels because of carotenoids, anthocyanins, phenolic acids, flavans, and flavonols, and other secondary phytochemical compounds that can be synthesized in plants through their metabolism. In barley, different grain color is mainly caused by diverse levels of polyphenolic compounds in the pericarp and aleurone layer, which may vary from light yellow to purple, blue, black, and violet.[5] The nutritional composition of barley grains is presented in Table 4.1.

4.2 Phytochemicals in Pigmented Barley

Phytochemicals are the non-nutritive chemical compounds produced by plants, and they have protective or disease-preventing properties. The chief phytochemicals present in pigmented or colored barleys are anthocyanins,

Table 4.1 Nutritional composition of barley.

Parameters	Quantity	Ref.
Starch	54–75%	150
Protein	10–20%	150
Lipid	2–3%	150
Fiber	Total: 11–34% Soluble: 3–20% β-Glucan: 2–10% Arabinoxylan: 3.8–6.05%	150
Vitamin B^1, mg	0.31	151
Vitamin B^2, mg	0.10	151
Vitamin B^3, mg	5.2	151
Vitamin B^6, mg	0.56	151
Vitamin B^9, µg	50	151
Iron, mg	6.0	151
Zinc, mg	3.3	151
Magnesium, mg	91	151
Tocols (µg g^{-1})	61.49[a]	152
Unsaturated fatty acids (mg per 100 g)	1505.32[a]	153
Sterols (mg per 100 g)	71.24[a]	154

[a]Mean value.

proanthocyanidins, and flavonols.[6] In addition to these phytochemicals, a small number of different phytochemicals, such as phenolic acids, folate, sterols, tocols, and alkylresorcinols are also present in pigmented barley.[4] These phytochemicals are vital components that could contribute to the beneficial properties of barley, and are randomly disbursed in barley, with higher concentrations in the outer layer than around the endosperm. Barley has a wide range of phenolic acids, especially derivatives of benzoic acid (C6–C1) and cinnamic acid (C6–C3) (Figure 4.1).[7] Moreover, flavonols and anthocyanins in barley generally exist in the form of glycoside derivatives, consisting of cyanidin-3-glucoside, penidin-3-glucoside, and delphinidin-3-glucoside.[6]

4.2.1 Flavonoids

Flavonoids are a cluster of naturally occurring plant substances with variable phenolic structures, and these belong to a large family of secondary metabolic compounds. To date, more than 10 000 different types of flavonoids have been reported.[8] In cereal crops, the most plentiful flavonoids composed of 15 carbons have a core C6–C3–C6 skeleton structure, which is composed of two aromatic rings connected by a 3-carbon bridge (Figure 4.2),[9] except for the aurones that have a C6–C2–C6 skeleton (Figure 4.3).[10] As per the chemical modifications (methylation, hydroxylation, and glycosylation) of the central C-ring, these flavonoids can be separated into idiosyncratic structural classes, including flavanols (leucocyanidin, leucopelargonidin, and leucodelphinidin), flavonols (kaempferol, quercetin, and myricetin), flavanones (flavanone, pentahydroxy naringenin, and eriodictyol), isoflavones

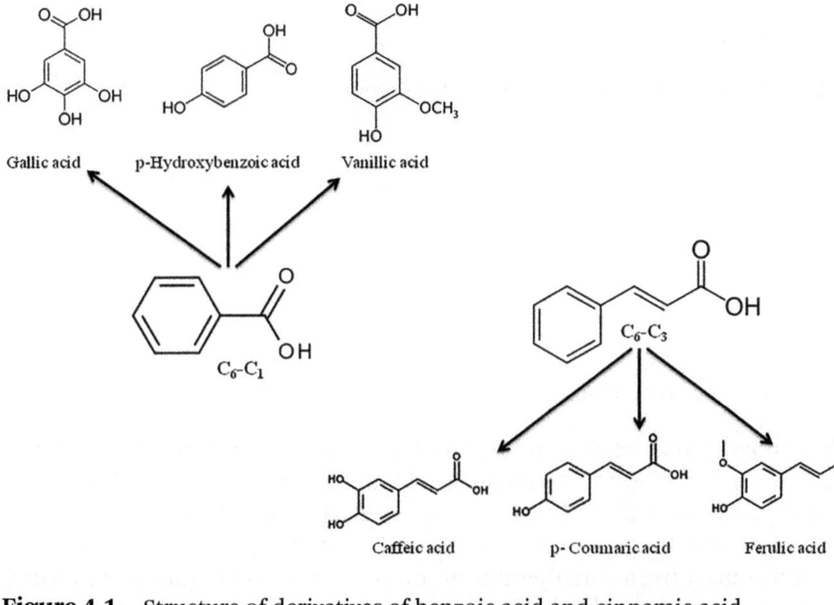

Figure 4.1 Structure of derivatives of benzoic acid and cinnamic acid.

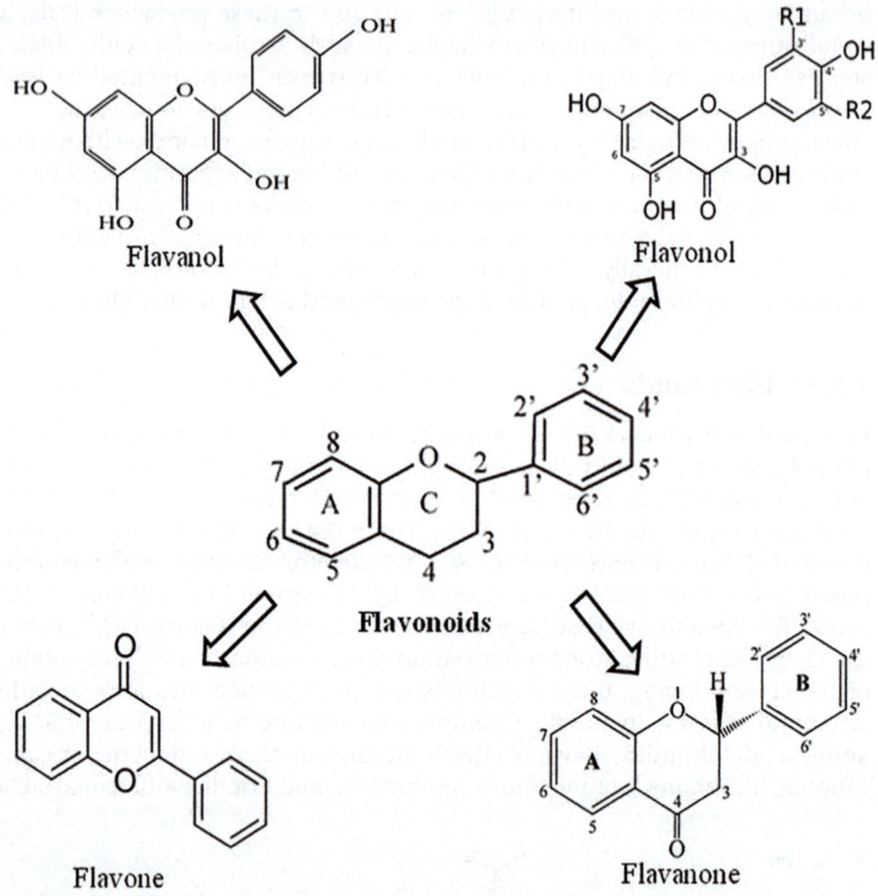

Figure 4.2 Generic structure of major flavonoids.

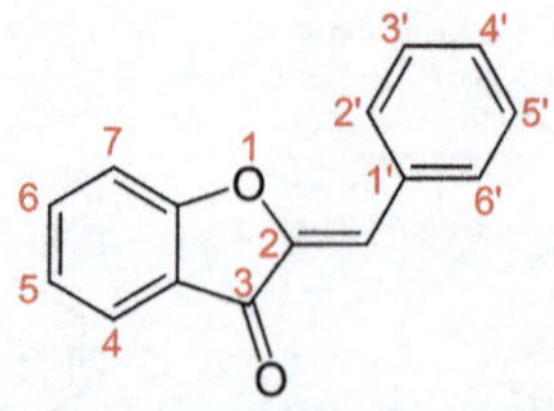

Figure 4.3 Structure of aurones.

and anthocyanins (delphinidin, cyanidin, and pelargonidin).[11] Among most cereal crops, barley grain is a staple food that can be used directly for human consumption and indirectly *via* livestock feed. Because of this reason, particular interest has been paid to growing and cultivating more varieties of barley grain containing a considerable amount of flavonoid composition, which may benefit human health.[12] The flavans (a type of flavonoid) constitute 58%

to 68% of the overall phenolic content of barley.[13] The intensity of the color of barley grain is directly proportional to the concentration of flavonoids in it, and this was proved in a study where purple and black kernels showed more flavonoids than pink and red-colored kernels. Irrespective of a specific type of flavonoids, the total flavonoid content in different varieties of colored barley ranged from 11.23 to 24.13 mg catechin equivalents per 100 g.[8] Moreover, in a comparative study, it was observed that the purple pigmented barley has a higher concentration of total flavonoids (57.3 mg per 100 g) than the yellow-colored barley (21.0 mg per 100 g).[14] This study supports the idea that pigmentation in grains is mainly due to the availability of different flavonoids and their concentration. Flavonols are a kind of flavonoid located in the outer pericarp, and are responsible for barley grain pigmentation. They commonly exist in cereal grains as glycoside derivatives including cyanidin-3-glucoside, delphinidin-3-glucoside, and penidin-3- glucoside.[12,15,16] On the other hand, the plentiful flavanols in cereals are formed by the proanthocyanidins in their monomeric, dimeric (procyanidin B3, prodelphinidin B3), and trimeric (procyanidin C2, prodelphinidin C2, other prodelphinidin isomers) forms.[8,17] These three forms of flavanols *i.e.* monomeric, dimeric, and trimeric contribute approximately 58–68% of the total phenolic composition in pigmented/colored barley, and among these forms, trimeric flavanols are the most abundant flavanols in barley.[12] Kim *et al.*[18] observed the flavonoid content between 62.0 and 300.8 mg g^{-1}.

Furthermore, a higher concentration of flavonoids was observed in the unhulled barley group (75.9 mg g^{-1}) compared with the hulled group (56.29 mg g^{-1}). As flavonoids are a broad class of phenolic compounds, and proanthocyanidin is one of them, its concentration also takes part in the occurrence of color. Proanthocyanidin (a type of flavonoid) concentration in pigmented/colored barley generally varied from 15.8 mg g^{-1} to 131.8 mg g^{-1}. Moreover, the purple and blue groups (83.0 mg g^{-1}) have considerably higher proanthocyanidin content than the black barley group (55.3 mg g^{-1}).[18] It can therefore be concluded that blue and purple barley grains hold the most flavonoid content among all the colored varieties of barley grain.[5]

4.2.2 Anthocyanins

Anthocyanins are water-soluble vacuolar pigments, and are the most studied polyphenols in pigmented barley. They are derivatives (polyhydroxylated or methoxylated) of glycosides of flavylium cation (Figure 4.4), which are mainly responsible for developing colors ranging from orange and red, to blue and purple.[19] As most of the color in barley is due to the presence of these compounds, these anthocyanins are abundantly present in purple, blue, and black colored genotypes of barley.[8,12,20,21] In colored barley grain, chiefly, these compounds are generally present either in the pericarp or in the aleurone layer, which causes the purple and blue color in the kernel.[20] In pigmented barley grain, the anthocyanin concentration generally varied

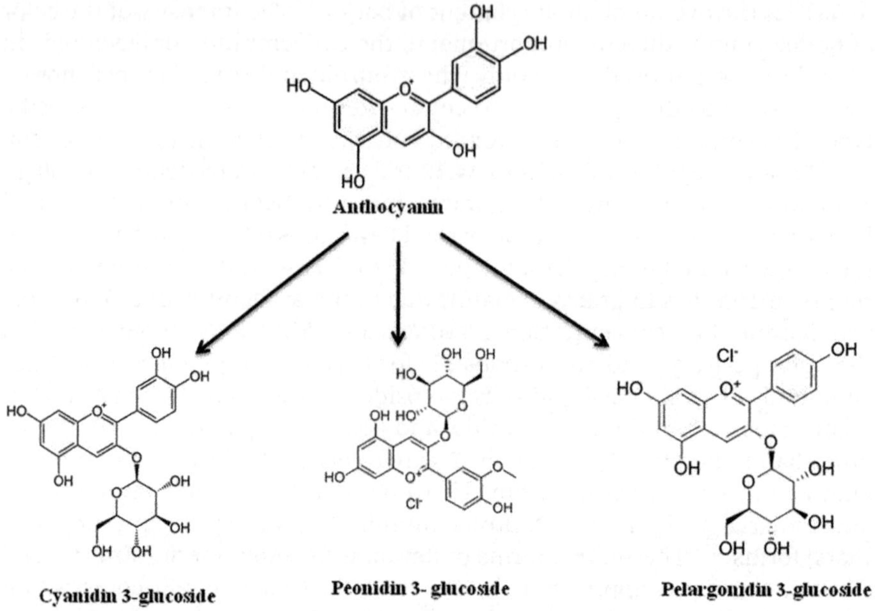

Figure 4.4 Structure of anthocyanin and its derivatives.

from 5.98 to 35.03 mg per 100 g.[8] In another study of two-rowed and six-rowed black barley, the total anthocyanin content was reported in 2.43 to 2.96 mg per 100 g.[22] Additionally, in blue pigmented barley, the total anthocyanin content was 3.46 mg per 100 g.[12] The aglycones of anthocyanins including delphinidin, pelargonidin, petunidin 3-glucoside, cyanidin, cyanidin 3-glucoside, and pelargonidin glycosides are mainly present in colored/pigmented barley grains.[12,23] The most widespread anthocyanins are cyanidin 3-glucoside (214.8 mg g^{-1}), followed by peonidin 3-glucoside and pelargonidin 3-glucoside in purple colored barley, and these three types of anthocyanins contribute to about 50–70% of their total anthocyanin content. In the same study, blue and black colored barley possessed a significant amount of delphinidin 3-glucoside reported (167.6 mg g^{-1} and 36.0 mg g^{-1}, respectively). Different studies on anthocyanin composition show that blue and purple pigmented barley (320.5 mg g^{-1}) possess a higher proportion of anthocyanins than black colored barley grain (49.0 mg g^{-1}).[20,24,25] This also indicates that the different types and concentration of anthocyanin compounds in barley grain is responsible for developing different colors. Pelargonidin is a type of anthocyanin that is abundantly available in purple barley, whereas cyanidin and delphinidin are other anthocyanin compounds are largely present in blue-black and yellow barley cultivars.[25] The black color in barley grain is possibly a result of co-pigmentation between anthocyanins and melanin-like pigments (brown/black polymers of high molecular weight), which often cover the purple pigmentation.[26–28] Moreover, the development of the black color in pericarp and the lemma of barley is may be due to this melanin-like pigment,[29] which has the tendency to overlap with a purple or blue shade that generally occurs due to the anthocyanin content.

4.2.3 Phenolic Acid Composition

The phenol-carboxylic acid or phenolic acids are types of aromatic acid compounds. In barley grain, these phenolic acids belong to one of two significant two groups, *i.e.* benzoic acids (protocatechuic acid, *p*-hydroxybenzoic acid, and vanillic acids) and cinnamic acid (caffeic acid, *p*-coumaric acid, and ferulic acids) and their derivatives, which are available in both bound and in free form.[30] Phenolic acid (free, conjugated, and bound) composition in barley is presented in Table 4.2. Because of their characteristics, these phenolic acids may promote human health,[31] and influence the color and flavor of grains.[32] Phenolic acids are the primary phytochemicals in pigmented/colored barley, and are chiefly concentrated in the outermost layers of the barley kernel, while lower concentrations can be found in the endosperm.[33,34] Since both insoluble and soluble phenolic acids are available in barley grain, the chief composition of phenolics in barley grain is the insoluble bound form, followed by the conjugated and free forms.[35] In the absence of free phenolic acids in barley grain, the bound phenolic acids are ester-linked to the cell walls of grains, and for their release from the cell matrix, it requires hydrolysis like acid, alkali, and enzymatic hydrolysis.[1] Free phenolic acids, including ferulic acid (4-hydroxy-3-methoxy cinnamic acid) and *p*-coumaric acid (4-hydroxycinnamic acid), are the primary low-molecular-weight phenolic acids present in the outer layers (husk, pericarp, testa, aleurone) and endosperm of pigmented barley grain and are usually represented as a small segment of the total phenolic acid.[34] On the other hand, the bound form of phenolic acids, including *p*-hydroxybenzoic, vanillic, and sinapinic acid are present in significant amounts in pigmented barley.[8,13,17] Because of the various forms of phenolic acids present in pigmented barley, numerous studies have been conducted to identify and determine the concentrations of

Table 4.2 Phenolic acid (free, conjugated and bound) composition in barley.

Phenolic acid	Free form ($\mu g\ g^{-1}$)	Conjugated form ($\mu g\ g^{-1}$)	Bound form ($\mu g\ g^{-1}$)	Ref.
p-Hydrobenzoic acid	ND	5.8–26.7	0.5–5.4	155
2,4-Dihydroxybenzoic acid	0.04–2.62	6.8–61.8	11.1–74.4	155
Vanillic acid	1.45–4.71	8.9–30.2	0.5–7.5	156 and 157
Syringic acid	0.45–3.74	2.2–10.0	0.0–3.0	158
Sinapic acid	ND	12.4–24.4	8.9–17.8	159
Ferulic acid	1.32–5.87	21.7–42.5	104.3–365.4	160
p-Coumaric acid	0.57–7.01	1.7–13.1	2.7–109.7	161
o-Coumaric acid	0.27–1.31	1.2–3.2	2.7–4.7	161

different phenolic acids. In pigmented or colored barley, a plentiful amount of phenolic acids (ferulic acid and *p*-coumaric acid) was reported in a study conducted by Siebenhandl *et al.*[20] on two different varieties of black colored barley, and *p*-coumaric acid was reported in the highest concentration, followed by ferulic acid. Similarly, in another study performed on 28 different varieties of colored (black, blue, and yellow) barley, the ferulic acid (149–413 μg g^{-1}) was measured as the prominent and most plentiful phenolic acid, which contributed approximately 68% of total phenolic acids in pigmented/colored barley; moreover *p*-coumaric acid (15–374 μg g^{-1}) was the second most plentiful phenolic acid in colored barley grain. Furthermore, the lower composition of some other phenolic acids like *p*-hydroxybenzoic acid, syringic acid, sinapic acid, vanillic acid, 2,4-dihydroxybenzoic acid, and *o*-coumaric acids were also measured in colored barley grains.[34,36]

Similarly, Abdel-Aal *et al.*[37] also reported ferulic acid as the predominant phenolic acid in free and bound extracts of pigmented barley grain, followed by *p*-coumaric acid. These previous studies indicate that many phenolic acids are available in pigmented barley in different concentrations. Pigmented barley has higher concentrations of phenolic acids, such as ferulic acid, when compared with non-pigmented barley, which may be due to the difference in climacteric conditions, environmental, and genetic behavior.[38]

4.2.4 Carotenoids

Carotenoids, often termed tetraterpenoids, are yellow, orange, and red organic pigments generally produced by plants and sometimes by other microorganisms. Depending on the genotype of cereals, carotenoids mainly occur naturally either in esterified forms (mostly with linoleic acid and palmitic) or in free form.[35,39] Some essential carotenoids, such as zeaxanthin and lutein, are abundantly available in colored barley (Figure 4.5). The most cultivated and consumed yellow barley grain contains a smaller (2.25 μg g^{-1}) amount of carotenoids than purple grain (4.54 μg g^{-1}), which is almost twofold higher than uncolored barley grain.[40] This indicates that the concentration of total carotenoids directly relates to the occurrence and strength of color. Recently, various studies have found that barley is a moderate source of carotenoids. A study on barley harvested in Czechia in 2014 and 2015 quantified five carotenoids, namely antheraxanthin, lutein, zeaxanthin, α- and β-carotene, in yellow barley, and among these, zeaxanthin was reported as being the most abundant carotenoid in barley.[41] As a diverse range of carotenoids is available in pigmented barley, it is essential to identify and quantify their profile. During the analysis of the carotenoid profile of pigmented/colored barley, the zeaxanthin was measured as the major carotenoid (49.7% of TCC; 1.26 μg per g DW), followed by lutein (41.5%; 0.94 μg per g DW) and antheraxanthin (7.6%; 0.18 μg per g DW).[41] In a study, Ndolo and Beta[38] also reported that zeaxanthin is the primary carotenoid, and its concentration in barley was almost twofold higher than that of lutein.

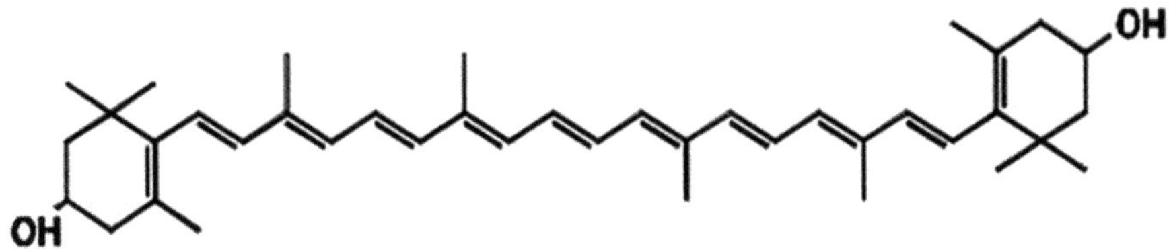

(a) Zeaxanthin

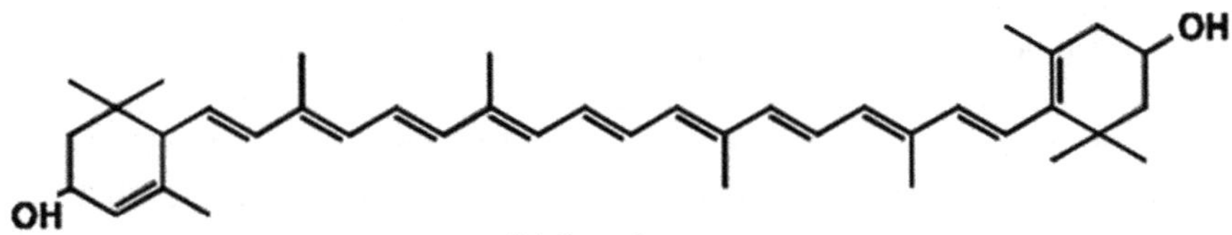

(b) Lutein

Figure 4.5 Structure of carotenoids (a), zeaxanthin, and (b) lutein.

4.3 Health Benefits and Application of Pigmented Barley

Pigmented barley grain is an excellent source of bioactive phytochemicals, such as polyphenolic compounds, including flavonoids, phenolic acids, tocols, anthocyanins, and proanthocyanidins, which all possess antioxidant activity and could be used as dietary supplements to improve human health.[13,42] Because of different bioactive compounds, pigmented/colored cereal grains and their products possess superior antioxidant activity than non-pigmented/uncolored cereal grains.[43–45] Therefore, pigmented barley grains can be used in functional food, as nutraceutical food ingredients, and natural food colorants.[171]

4.3.1 Flavonoids

Flavonoids are the most abundant compounds in pigmented barley, and they may promote human health by enhancing immunity and reducing the risk of various chronic diseases. Different epidemiological investigations have established a positive correlation between a high dietary intake of flavonoids and a reduction in degenerative diseases such as type 2 diabetes,[46] cardiovascular disease,[47] cancer,[48] and atherosclerosis. The flavonoids majorly contribute to the scavenging of free radicals; therefore, they can prevent the damage caused by free radicals. The uncontrolled generation of free radicals can cause atherosclerosis, neoplasia, and neurodegenerative disease.[49] In addition, some individual subgroups of flavonoids could reduce the risk of specific types of cancers. In a study, Chang *et al.*[50] reported that the moderate intake of flavonols and flavones through diet might reduce the risk of breast cancer, especially in post-menopausal women. Flavones and luteolin in pigmented barley have been shown to help prevent sunburn.[51]

4.3.2 Anthocyanins

Various pigmented barley colors (blue, purple, and black) contain a good quantity of anthocyanins, which contribute to their antioxidant activity.[23] The anthocyanins present in pigmented barley can scavenge free radicals, which may prevent oxidative stress in human cells.[52,53] The antioxidant activity of anthocyanins can have anti-obesity,[54] anti-inflammatory,[10,52] hypoglycemic,[55,56] and anti-cancer effects.[16] Anthocyanins therefore have considerable potential in the food industry as functional foods,[57] safe and effective food colorants,[58] and nutraceuticals.[59] In previous studies, a strong correlation between the concentration of anthocyanin and free radical scavenging activity of pigmented barley was reported. The purple barley grain possesses significantly higher total anthocyanin content and consequently exhibited much higher antioxidant activity, superoxide radical scavenging capacities and DPPH free radical scavenging capacity than blue, black, and yellow barley grain.[18] All these studies indicate that the concentration and types of anthocyanin are much more responsible for their antioxidant activities.

4.3.3 Phenolic Acids

The potential health benefits of phenolics – such as acting as antioxidants, free radical scavengers, and immune enhancers – have also been established. In previous studies, phenolic compounds were effective free radical scavengers, metal chelators, and active antioxidants.[60,61] Similarly to flavonoids and anthocyanin, the radical scavenging ability of phenolics is directly proportional to their concentration, and the effectiveness may vary concerning their types. The free radical scavenging capacity in different pigmented barley samples is broadly diverse. During investigation of different colored barley, Kim *et al.*[8] concluded that a significant amount of total phenol content is responsible for their antioxidant activity. Similarly, pigmented barley, mainly purple barley, has the greatest antioxidant activity due to its higher polyphenolic content.[18] Different phenolic acids, chiefly homogentisic acid, chlorogenic acid, 3,4-dimethoxybenzoic acid, rutin, and protocatechuic acid, are responsible for scavenging free radicals able to stop further chain reactions.[100] Cereal grains, including pigmented barley, possess outstanding resistance against various diseases and insect damage, because of the superior level of phenolic acids present.[62,63] Moreover, an augmentation in the actions of phenolic-related enzymes and the accumulation of phenolic compounds also showed significant resistance of cereals to biotic stresses.[64] In other words, it can be concluded that the metabolism of phenolic compounds[65,66] often synchronizes the plant's resistance to biotic and abiotic stresses.

4.3.4 Carotenoids

There is a lot of evidence to indicate that carotenoids may act as antioxidant compounds that defend cells against reactive oxygen species, with the effect that they possess potent health benefits. Carotenoids can prevent sunburn,[67] liver cancer,[68] and improve the human body's immune system. Some important types of carotenoids like zeaxanthin and lutein are abundantly available in colored barley, which is chiefly accountable for the coloration of the retina's macula lutea (yellow spot): the area of maximal visual acuity. Therefore, dietary zeaxanthin and lutein are supposed to protect against age-related macular degeneration and cataract[69]. Sometimes, lutein and zeaxanthin may act together with other carotenoids and phytochemicals and protect against various cardiovascular risks, cancer, and other health-related disorders[70]. Further, carotenoids have shown anti-inflammatory properties that take part in intracellular signaling cascades by influencing transcription factors.[71] Carotenoids with an unsubstituted β-ionone ring possess vitamin A activity.[72] However, humans cannot synthesize carotenoid compounds; therefore, these carotenoids need to be included in the diet, and barley may act as a good source of these carotenoids.[73]

4.4 Food Applications of Barley and its Industrial By-products

Barley is an important crop, which is used in various food applications. It is rich in nutrients such as protein, amino acids, minerals, vitamins, and dietary fiber.[74] Due to the superior quality and mouth-feel of food products prepared from other cereals like wheat and rice, the use of barley in foods decreased, especially in the 19th and 20th centuries.[75] When used in food, barley grains are firstly abraded to produce pot or pearled barley, and then further processed to produce grits, flakes, and flour. Pearled barley, grits, or flour are used for making various types of traditional dishes in many countries and regions such as Poland, Tibet, Japan, Russia, and India.[76] Barley seeds are the main raw material in beer-making in the beverage industry. Both raw barley and malted barley are used, often in combination with adjunct grains, such as rice and corn.[77] Highland barley flour is also produced by wheat flour milling equipment and mixed with wheat flour to make cakes,[78] biscuits,[79] cookies,[80] bread,[81] rusks,[82] and muffins.[83] It is also used for making yellow alkaline noodles,[84] and pasta products.[85] Dhingra and Jood,[86] made wheat bread with the addition of 15–20% barley flour and found that the bread produced the best results in terms of flavor, appearance, and texture, however, there was a decrease in loaf volume, dull brown color, and hard crumb texture.

On the other hand, Klamczynski and Czuchajowska,[87] reported comparable loaf volumes of chemically leavened bread baked from barley flour of regular starch composition to that of wheat flour bread. The hardness of cooked white salted wheat noodles was not affected with the addition of 15–50% barley flour with starch, but it was found to decrease with the addition of waxy barley flour.[88] Newman *et al.*[89] prepared barley flour muffins, which were slightly inferior in volume, density, and moisture compared with wheat flour muffins. Ames *et al.*[90] also reported favorable consumer acceptance of barley flour tortillas compared with wheat flour, particularly in textural properties. The use of protein isolated from brewers' spent grain has the potential for use in the making of bakery products to increase their nutritional and functional properties. Zong *et al.*[91] studied the effects of the addition of spent grain protein in cookies and found an acceptable score for sensory, flavor, textural characteristics, and increased nutritional values. To increase the nutritional value in terms of protein, dietary fiber, and minerals, Brewer's spent grains (BSG) is a beneficial raw material for food product formulation. Waters *et al.*[92] added 10% BSG and fermented BSG in bread and found that the dough displayed improved handling characteristics, and that softness, and staling were improved in both cases. While sweetness was decreased and acidulous flavor was increased. In another study, Ktenioudaki, *et al.*[93] added 10% BSG in a crispy snack product to increase its fiber content. They found that this formulation almost doubled the product's fiber content without changing the product's acceptability.

4.5 β-Glucan of Barley

Recently, barley has attracted much interest from companies producing healthy foods. This is due to its bioactive compounds, including tocols (vitamin E) and β-glucan (soluble fiber).[30] β-Glucan is a linear chain of the β-glucopyranosyl unit, in which about 70% is linked (1→4) and about 30% (1→3) linkages are present.[94,95] The β-glucan that is present in barley is a water-soluble polysaccharide, which is considered soluble dietary fiber. On a commercial level, β-glucan is mainly derived from barley, oats, mushrooms, and a few microorganisms. A linear chain of D-glucose linked in the β (1-3) position with various sized D-glucose branches linked to the main chain by β (1-6) linkages are found in mushrooms and microorganisms.[96] In cereals, β-glucan presents 1% in wheat grains, 3–7% in oats and 5–11% in barley.[97] The presence of β-glucan in barley usually ranges from 2% to 10%, but some varieties have been reported to contain up to 20% β-glucan. The high content of (1→3)(1→4)-β-D-glucan (β-glucan) barley makes it different from other cereals.[98] The major difference between soluble and insoluble β-glucan is its structural difference, which greatly influences its water solubility. The molecular weights of soluble β-glucans are greater than those of the insoluble glucans and the soluble β-glucans have a greater ratio of β-(1→4) linkages as well as of cellotriosyl units.[99] β-Glucan content of 3.7–7.9% water-soluble and 10.8–21.7% water-insoluble has been reported in different barley cultivars.[100] Similarly, Sharma *et al.*[101] reported 1.95 to 3.07 and 2.0 to 3.0% soluble and insoluble β-glucan content in eight different

Indian hulled barley cultivars. Soluble, insoluble and total β-glucan content of different cultivars of barley is reported in Table 4.3. The presence of β-glucan gives barley the status of an important functional grain, which provides various health benefits. It is also widely used in many food industries due to its capacity to form a gel and increase the viscosity of aqueous solutions. It is also used for enhancing the appearance and texture of salad dressings, various types of gravy and ice cream desserts. To develop calorie-reduced food products β-glucan is also used as a fat mimetic.[102] In some food industries, the flow behavior and gelling properties of β-glucan create some technological problems, such as slow filtration of solutions or slurries, low yield, retardation of malting, decrease in yield during mashing, increase in viscosity of wort and beer, and formation of haziness and gumminess precipitates in the stored beer. Therefore, the malting industry prefers only barley with lower levels of β-glucan.[103] The use of β-glucan in the cosmetics and pharmaceutical industries has been limited due to its high molecular weight and viscosity.[39,104] Many researchers have reported in their studies that the application of physical, chemical, and enzymatic methods can alter the molecular structure of β-glucan, which significantly impacts its solubility, viscosity, and other rheological properties.[105–107]

4.5.1 Extraction of β-Glucan

Various types of processing methods are used for barley β-glucan extraction. Different methods change the molecular structure of β-glucan content, which affects solubility and molecular weight distribution and due to these, the health benefits of β-glucan may also be affected.[69] Several difficulties occur during the extraction of β-glucans from cereals. Normally, four methods are commonly used to extract β-glucan from barley grains: hot water extraction, alkaline extraction, acid extraction, and enzyme extraction.[23.] Among these methods, the order of barley β-glucan extraction rate is hot water > enzyme > acid > alkaline.[108] The high purity of β-glucan can be extracted by only the hot water method. A simple extraction process involves the three stages: (a) inactivation of endogenous enzymes, (b) extraction of β-glucans with a suitable solvent, (c) a purification–isolation stage, because other chemicals present in cereals like proteins, saccharides and starch can be extracted with β-glucan.[109] Extraction of β-glucan from barley with different operating conditions is shown in Table 4.4. A simple process of dry milling can produce β-glucans obtained from barley fractions and sieving using either abrasion milling[110] or hammer milling,[111] but this process requires a long sifting time, and the yield of β-glucans-enriched product is also low (less than 5%). Different researchers use different procedures for β-glucan extraction, but all of them can be divided into two categories, as Vasanthan and Temelli report:[112] (a) procedures carried out in dry conditions and (b) performed in wet conditions. Dry processing procedures include dry milling and sieving techniques.[113] The final product contains 15.2% β-glucan.[114] Wet procedures are complex and have two or three stages. In this procedure, cereal bran or flour

Table 4.3 Soluble, insoluble and total β-glucan content from different barley cultivars.

β-Glucan (%)	Barley type	Cultivars	Growing country/region	Ref.
Total β-glucan				
3.37 to 6.23	Hull-less	Dolma, HL-335, Karan-16	Hisar, India	162
8.0	High amylose (hull-less)	92 550 648	Saskatoon, Canada	121
3.91 to 5.93	Normal	Persefoni, Kos, Thessaloniki, Athinaida, Dimitra, Triptolemos	Thessaloniki, Greece	163
3.49 to 4.98	Spring	Alexis, Triumph, Nevada, Cooper, Caminant, Texana, Trebon, Cork, Delibes, Polygena, Mentor, Mie, Reggae, Anni	Jutland, Denmark	164
2.64 to 8.05	Two-rowed (hulled)	Adagio, Alexis, Astoria, Barke, Bodega, Bombay, Cellar, Clarine, County, Hanka, Jersey, Labea, Ludine, Madou, Meteor, Otis, Orchidea, Romina, Riviera, Sabel, Scarlett, Silvana, Svenja, Tiffany, Vanessa, BA, BB, BC, BD	Acquapendente, VT, Italy	165
8.0 to 8.4	Six-rowed (low amylose starch – waxy	Saechal, Hichal, Jinjuchal	Korea (2008)	166
4.84	Waxy (hull-less)	Merlin	The USA	167
4.60 to 6.23	Hull-less	Dolma, DL-88	Hisar, and Palampur, India	168
3.2 to 4.6	NM	NM	Turkey	169
5.16	Commercial	NM	Spain	170
3.5 to 5.3	Two rowed	$n = 50$	Anttila (Southern Finland)	171
2.8 to 4.3	Six rowed	$n = 68$	Anttila (Southern Finland)	171
Soluble β-glucan				
2.06 to 5.39	Hull-less	Dolma, DL-88	Hisar, and Palampur, India	168
3.75 to 7.96	NM	Bear USA, CI 5 DEU, CI 15 DEN, Hyproly, KM 1910, KM 2037, Oriflame, Peregrine, Tibet White	Piešťany, Slovak Republic	100
2.0	High amylose (hull-less)	92 550 648	Saskatoon, Canada	121
1.16 to 1.96	Normal (hull-less)	Falcon, Condor, CDC Gainer, Bear, AC Hawkeye, AC Bacon, CDC Dawn, SB90354, SR9310	Manitoba & Saskatoon, Canada	172

1.52 to 2.02	High amylose (hull-less)	Genotypes: 92-55-06-54, 92-55-06-48, SB94893, SB94897	Saskatoon, Canada	172
2.22 to 3.55	Waxy (hull-less)	CDC Candle, SB94917, SB93965, SR93135, SH96054, SH96076, SH96090, SH96093, SB95134		
2.16 to 3.07	Zero amylose waxy (hull-less)	Genotypes: SB94785, SB94792, SB94794		
2.6 to 3.1	High amylose starch	Whole grain Karmose, shifted Karmose	Moss, Norway	173
Insoluble β-glucan				
10.89 to 21.70	NM	Bear USA, CI 15 DEN, Hyproly, KM 1057, KM 1910, KM 2037, Oriflame, Peregrine, Tibet White	Piestany, Slovak Republic	100
1.3 to 1.4	Spring	NM	Montana & Scandinavia	174
2.5 to 5.1	Two-rowed (waxy, normal, high amylose)	Cindy, Magdalena, Kinnan, Iver, Tyra, Karmose, SES 2/11	Norwegian University of Life Sciences, Norway. Temperature; 9 °C, 12 °C, 15 °C, 18 °C, 21 °C	175
0.1 to 0.9	Two-rowed (hulled normal)	Justina, Olve, CDC Dolly	Bjorke Research farm, Ilseng, Norway	176

is used as the initial raw material, treated with a solvent (usually water, aqueous solvent at basic pH, or semi-alcoholic solution), obtaining an aqueous extract. This aqueous extract also contains starch, proteins, and fats other than β-glucan, and due to this a purification step is required. In this step, β-glucans are precipitated with alcohol, or separated by other methods, for instance, freezing–thawing the extract.[115] The precipitate obtained by this procedure is dried, and this contains 20–70% of β-glucan.

Basic steps in barley β-glucan extraction:[109]

Bran + solvent (As per methods and extraction conditions)

↓

Centrifuge at 6000 × g for 15 min

↓

Combine supernatants, where necessary,

Adjust to pH 4.5 with 2M HC1

↓

Centrifuge at 6000 × g for 15 min, discard pellet

↓

Add Termamyl® (1 ml/ 100 ml of supernatant), incubate at 96 °C for 1 hr with shaking

↓

Centrifuge at 6000 × g for 15 min, discard pellet

↓

Add ethanol to 50% concentration, let stand overnight at 4 °C

↓

Centrifuge at 6000 × g for 15 min

↓

Re-suspend pellet in water, wash two times with 50% ethanol,

Centrifuge, homogenize pellet in water and then freeze-dry

Table 4.4 Extraction of β-glucan from barley with different operating conditions.

Solvent	Extraction yield (%)	Extraction time and temperature	Barley	Ref.
Acidic media	44.7	1 h at room temperature	HL-W-4.7%	177
Water, pH 10	61	0.5 h at 45 °C	HL-N-4.5%	46
Water	16.6	1.5 h at 40 °C	HL-W-4.7%	178
NaOH, 1 N	73–77	1 h at ambient temperature	HL-N-5.6% HL-W-11.3%	109
Water	85	0.5 h at 40–55 °C	HL-N-5.7%	179
Water	53.3	2 h at 38 °C	H–W-6.1% HL-W-5.8%	22
Water	52.7	0.5 h at 25 °C	H–W-6.9%	172

4.5.2 Food Applications of Barley β-Glucan and its Fractions

The polysaccharide nature of β-(1,3) glucans has many food applications, such as being used as edible films, as a feed for domestic animals, and making low-calorie food. Other foods that can be made using barley β-glucan include: cookies, pumpkin pudding, layer cake, dip for pot crust, taro custard, saute, and meat products.[116] The main aim of adding barley or barley constituents (β-glucan) to foods is to increase the total and soluble fibre content in foods, improving their physiological efficacy and providing numerous health benefits.[117] The addition of β-glucan also alters the processing and handling of foods and food products' texture, color, flavor, and taste.[94] The addition of β-glucan in foods also provides various characteristics like improved viscosity, water-holding capacity, oil-binding capacity, emulsion stabilization, and improvement in organoleptic characteristics.[118] The addition of barley ingredients with high fiber content results in a dramatic increase in the water absorption capacity of dough.[94,119] β-Glucan is a major and most important component of the soluble dietary fiber and it influences the gelation ability and rheological properties of food gels.[120] Izydorczyk *et al.*[121] reported that adding barley β-glucan into wheat flour significantly increased dough strength, as indicated by peak dough resistance during mixing, mixing stability, and work input during mixing and elastic modulus. Brennan and Cleary,[122] found that in their studies, the addition of soluble barley β-glucan to dough reduced the dough resistance to extension and the addition of water could not overcome this. They also observed that adding barley soluble β-glucan to wheat flour changes starch pasting properties, lowering gel viscosity and increasing gelatinization temperature. Glucans can act as capsules or 'tablet-like' ingredients in the pharmaceutical and nutraceutical industries, as they release their content even further after ingestion. Cereal β-glucan can be used as soluble dietary fiber and used as a stabilizer in making low-fat food products, such as low-fat ice creams,[123] soft brined cheeses,[102] and low-fat sausages.[124] Another food application of β-glucan includes its uses in the making of soups, sauces, and beverages.[125] Much research was done on the incorporation of β-glucan in cereal-, meat-, and dairy-based products, but further investigation is required to determine how to use this ingredient in other products.[120] White bread made from refined wheat flour will always lack dietary fiber. However, there is currently high demand for high-fiber white bread. Such bread is possible with the use of 'Glucagel', which is commercially available, and is extracted from barley. The addition of Glucagel increases starch availability for its digestion and helps reduce hyperglycemia and hyperinsulinemia, which is important in the control of diabetes.[122] The manufacturer used cereal β-glucans as a bioactive ingredient to provide health benefits, as well as acting as stabilizers and texturizers.[126] Prebiotic sausages are also made with a combination of glucan, starch, and resistant starch to provide various health benefits.[127] Brennan, *et al.*[128] observed that ready-to-eat snacks made with β-glucan fractions from barley and mushrooms also had an improved glycemic response after consuming them. A cake prepared with 1 g of β-glucan had the similar attributes to the control sample.[129] β-Glucan is a very important functional bioactive component in the production of many dairy products. Yogurt prepared with β-glucan results in faster proteolysis action, which lowers the release of large peptides and provides more free amino acids.[130] Kaur and Riar,[131] found that the addition of barley β-glucan (0.5, 1, 1.5, and 2%, w/v) full-fat yogurt during storage significantly increased the separation, viscosity, texture profile, and sensory characteristics. Nowadays, β-glucan is also added as a structure-making agent in gluten-free bakery products due to its health benefits. Karp *et al.*[132] added oat β-glucan (with an optimized percentage of 2.63%) to gluten-free yeast-leavened cake and found good texture, volume, and sensory characteristics. Similarly, Wang *et al.*[130] prepared biscuits with the addition of β-Glucan (5.2 g per 100 g) in barley flour and found a positive response from all customers in terms of sensory quality, compared with the control sample. β-Glucan is widely used in the making of bread and has also been widely tested in terms of its effect on bread firmness, bread loaf volume, rate of staling, and consumer acceptance. All these parameters depend on certain conditions, including the molecular weight of β-glucan, its concentration and source of β-glucan.[133] β-Glucan was also used as a fat substitute in various meat products, such as beef patties, sausages, and low-fat beef burgers, having 2.2% β-glucan, and exhibited improved textural, nutritional, and cooking properties.[134]

4.5.3 Health Benefits of Barley β-Glucan

The healthy properties of barley are mainly attributed to its high dietary fiber content, *i.e.* β-glucan, and it is almost 10-times higher in barley than in other cereals.[135] Researchers have documented several applications of β-glucan against various diseases and disorders, including: reducing the onset of colorectal cancer,[136] increasing stool bulk and preventing constipation,[137] decreasing the glycemic index,[138,139] decreasing postprandial blood glucose levels and raising insulin,[138] preventing insulin resistance,[126] decreasing serum cholesterol levels,[16,140] preventing coronary heart disease,[141] producing short-chain fatty acids,[142] and increasing the growth of beneficial gut microflora,[9,143] hypertension, stroke and type 2 diabetes.[144] Based on the source of β-glucan, barley β-glucan was more effective in regulating glucose and insulin responses than oat β-glucan.[139] Based on various studies, the Food and Drug Administration has also recommended a daily diet containing 3 g of β-glucan to achieve its health benefits.[2] Delaney *et al.*[145] reported an increase in fecal-neutral cholesterol in their studies and a decrease in aortic cholesterol levels in hamsters fed with 8 g per 100 g β-glucan from barley or oats. Jenkins *et al.*[138] reported that 4 units of the glycemic index were decreased by taking of 1 g of β-glucan per 50 g of carbohydrates. A diet containing barley can be used by diabetic patients to prevent insulin resistance.[144]. In treating burn injuries, β-glucan was also used as an active therapeutic agent. Delatte *et al.*[146] reported that the application of poly(lactic-*co*-glycolic acid) membranes containing β-glucan helps in wound healing and is also used as a skin substitute,[147] the flexible hydroxyapatite/glucan composite was used as a bone-substituting material,[148] and carboxymethylated β-glucan was used in arthritis treatment.[149]

4.6 Conclusion

Pigmented barley is one of the most ancient crops containing phytochemicals, such as β-glucan, polyphenols (phenolic acids, flavonoids, and anthocyanins), polysaccharide (arabinoxylan), phytosterols (β-sitosterol and campesterol), and tocols (β-tocotrienol, α-tocotrienol, β-tocopherol, and α-tocopherol). It is an adequate source of minerals (Cu, Zn, Fe, Se, and Mg), vitamins (E and B), and dietary fiber. Hence, cereal flour can be added, *i.e.* wheat, rice, and other flours, for well-balanced nutrition. As a result of the special chemical composition of barley, it has considerable health benefits. These beneficial effects involve anti-inflammatory, anticancer, antidiabetic, antibacterial, anti-obesity, antioxidant, antifatigue, antiaging, and cholesterol-lowering activities. Barley also contains β-glucan, an essential water soluble polysaccharide in barley, which ranges between 2% to 10%, but in some varieties has been reported up to 20% β-glucan. β-Glucan also has various food applications and is widely used as an emulsion stabilizer, thickening agent, fat substitute, and bioactive ingredient in the food industry due to its water solubility, viscosity, gelation property, and health benefits. β-Glucan is also beneficial against various diseases and disorders and reduces the onset of colorectal cancer, increases stool bulk, prevents constipation, decreases glycemic index, decreases postprandial blood glucose levels, raises insulin, and prevents coronary activity, heart disease, and type 2 diabetes. All these unique properties of pigmented barley make it an essential crop in the making of functional foods and, due to this, the future of barley in food production is very promising.

References

1. K. B. Pandey and S. I. Rizvi, *Oxid. Med. Cell. Longevity*, 2009, **2**, 270.
2. FDA, *Federal Regulations*, 1997, vol. 62, p. 3584.
3. D. Schulte, T. J. Close, A. Graner, P. Langridge, G. Muehlbauer, T. Matsumoto, G. Muehlbauer, K. Sato, A. H. Schulman, R. Waugh, R. P. Wise and N. Stein, *Plant Physiol.*, 2009, **149**, 142.
4. E. Zannini, A. Mauch, S. Galle, M. Ganzle, A. Coffey, E. K. Arendt, J. P. Taylor and D. M. Waters, *J. Appl. Microbiol.*, 2013, **115**, 1379.
5. Y. Liu, J. Liu, Z. Pu, J. Wang, Y. Zheng, Y. Li and Y. Wei, *Biotechnol. Lett.*, 2013, **35**, 1765.
6. L. Gong, in *Bioactive Factors and Processing Technology for Cereal Foods*, ed. J. Wang, B. Sun and R. Tsao, Springer, Singapore, 2019, p. 55.
7. F. Shahidi and M. Naczk, *Food Phenolics*, Technomic Publishing, Basel, Switzerland, 1995.
8. M. J. Kim, J. N. Hyun, J. A. Kim, J. C. Park, M. Y. Kim, J. G. Kim, S. J. Lee, S. C. Chun and I. M. Chung, *J. Agric. Food Chem.*, 2007, **55**, 4802.
9. V. Mussi de Mira, R. M. Cerdeira Barros, M. A. Schiocchet, J. A. Noldin and U. M. Lanfer-Marquez, *Cienc. Tecnol. Aliment.*, 2008, **28**, 994.
10. Y. Lin, C. Y. Li and I. F. Hwang, *Food Chem.*, 2008, **109**, 771.
11. L. Lepiniec, I. Debeaujon, J. M. Routaboul, A. Baudry, L. Pourcel, N. Nesi and M. Caboche, *Annu. Rev. Plant Biol.*, 2006, **57**, 405.
12. M. Zydorczyk, S. J. Symons and J. E. Dexter, *Whole Grain Foods in Health and Disease*, American Association of Cereal Chemists, AACC International, St. Paul, MN, 2002.
13. P. Goupy, M. Hugues, P. Boivin and M. J. Amiot, *J. Sci. Food Agric.*, 1999, **79**, 1625.
14. R. Andersson, K. Autio and P. Aman, *J. Cereal Sci.*, 1999, **30**, 183.
15. D. C. Knievel, E. M. Abdel-Aal, I. Rabalski, T. Nakamura and P. Hucl, *J. Cereal Sci.*, 2009, **50**, 113.
16. S. Y. Kang, N. P. Seeram, M. G. Nair and L. D. Bourquin, *Cancer Lett.*, 2003, **194**, 13.
17. F. Dvorakova, P. Guido, Z. Dostalek, M. Skulilova, M. Moreira and A. A. Barros, *J. Inst. Brew.*, 2008, **114**, 27.
18. C. Lee, D. Han, B. Kim, N. Baek and B. K. Baik, *Int. J. Food Sci. Technol.*, 2013, **48**, 984.
19. L. Jackman and J. L. Smith, *Natural Food Colorants*, Springer, M. A Boston, 1996, pp. 244–309.
20. S. Siebenhandl, H. Grausgruber, N. Pellegrini, D. Del Rio, V. Fogliano, R. Pernice and E. Berghofer, *J. Agric. Food Chem.*, 2007, **55**, 8541.
21. N. I. Yao, J. L. Jannink and P. J. White, *Cereal Chem.*, 2007, **84**, 471.
22. C. Collar and A. Angioloni, *Eur. Food Res. Technol.*, 2014, **3**, 2128.
23. G. Goudar, P. Sharma, S. Janghu and T. Longvaha, *Int. J. Biol. Macromol.*, 2020, **162**, 1204.
24. G. Bellido and T. Beta, *J. Agric. Food Chem.*, 2009, **57**, 1022.
25. B. Jende-Strid, *Hereditas*, 1993, **119**, 187.

26. A. Riley, *Int. J. Biochem. Cell Biol.*, 1997, **29**, 1235.
27. K. Pandey and M. R. Dhakal, *Curr. Sci.*, 2001, **80**, 933.
28. D. Pugh, P. Balachandran, H. Lata, F. E. Dayan, V. Joshi, E. Bedir, T. Makino, R. Moraes, I. Khan and D. S. Pasco, *Int. Immunopharmacol.*, 2005, **5**, 637.
29. U. Lundqvist, J. D. Franckowiak and T. Konishi, *Barley Genetic Newsletter*, 1996, vol. 26, p. 516.
30. M. Bonoli, V. Verardo, E. Marconi and M. F. Caboni, *J. Agric. Food Chem.*, 2004, **52**, 5195.
31. L. Dykes and L. W. Rooney, *Cereal Foods World*, 2007, **52**, 105.
32. J. Yu, T. Vasnathan and F. Temelli, *J. Agric. Food Chem.*, 2001, **49**, 4352.
33. A. Brandolini, L. Castoldi, L. Plizzari and A. Hidalgo, *J. Cereal Sci.*, 2013, **58**, 123.
34. E. Idehen, Y. Tang and S. Sang, *J. Food Drug Anal.*, 2017, **25**, 148.
35. J. U. Ziegler, S. Wahl, T. Würschum, C. F. H. Longin, R. Carle and R. M. Schweiggert, *J. Agric. Food Chem.*, 2015, **63**, 5061.
36. E. Sullivan, E. Arendt and E. Gallagher, *Trends Food Sci. Technol.*, 2013, **29**, 124.
37. S. M. Abdel-Aal, T. M. Choo, S. Dhillon and I. Rabalski, *Cereal Chem.*, 2012, **89**, 198.
38. V. U. Ndolo and T. Beta, *Food Chem.*, 2013, **139**, 663.
39. E. Mellado-Ortega and D. Hornero-Mendez, *J. Cereal Sci.*, 2015, **62**, 15.
40. L. Lachman, K. Hejtmánkova and Z. Kotíkova, *J. Cereal Sci.*, 2013, **57**, 207.
41. L. Paznocht, Z. Kotíkova, M. Sulc, J. Lachman, M. Orsak, M. Eliasova and P. Martinek, *Food Chem.*, 2018, **240**, 670.
42. P. Van Hung, *Crit. Rev. Food Sci. Nutr.*, 2016, **56**, 25.
43. S. M. Abdel-Aal, A. A. Abou-Arab, T. H. Gamel, P. Hucl, J. C. Young and I. Rabalski, *J. Agric. Food Chem.*, 2008, **56**, 1171.
44. R. Hirawan, W. Diehl-Jones and T. Beta, *J. Agric. Food Chem.*, 2011, **59**, 12330.
45. V. Mussi de Mira, R. M. Cerdeira Barros, M. A. Schiocchet, J. A. Noldin and U. M. Lanfer-Marquez, *Cienc. Tecnol. Aliment.*, 2008, **28**, 994.
46. R. S. Bhatty, *Cereal Chem.*, 1993, **70**, 73.
47. G. Li, Y. Zhu, Y. Zhang, J. Lang, Y. Chen and W. Ling, *J. Acad. Nutr. Diet.*, 2013, **113**, 786.
48. E. Angst, J. L. Park, A. Moro, Q. Y. Lu, X. Lu, G. Li, J. King, M. Chen, H. A. Reber, A. Howard, V. L. W. Go, E. Guido and O. J. L. Hines, *Pancreas*, 2013, **42**, 223.
49. K. E. Heim, A. R. Tagliaferro and D. J. Bobilya, *J. Nutr. Biochem.*, 2002, **13**, 572.
50. H. Chang, Q. Xie, X. Zhang, J. Peng, J. Zhu and M. Mi, *PLoS One*, 2013, **8**, 54318.
51. L. Verschooten, K. Smaers, S. V. Kelst, C. Proby, D. Maes, L. Declercq, P. Agostinis and M. Garmyn, *J. Invest. Dermatol.*, 2010, **130**, 2277.
52. T. Tsuda, M. Watanabe, K. Ohshima, S. Norinobu, S. Choi and S. Kawakishi, *J. Agric. Food Chem.*, 1994, **42**, 2407.

53. H. Nam, S. P. Choi, M. Y. Kang, H. J. Koh, N. Kozukue and M. Friedman, *Food Chem.*, 2006, **94**, 613.
54. R. L. Prior and X. Wu, *Free Radical Res.*, 2006, **40**, 1014.
55. J. Jurgonski and Z. Zdunczyk, *Plant Foods Hum. Nutr.*, 2008, **63**, 176.
56. M. H. Grace, D. M. Ribnicky and P. Kuhn, Vaccinium angustifolium Aiton, *Phytomedicine*, 2009, **16**, 406.
57. M. Karaaslan, M. Ozden, H. Vardin and H. Turkoglu, *LWT–Food Sci. Technol.*, 2011, **44**, 1065.
58. F. J. Francis and P. C. Markakis, *Crit. Rev. Food Sci. Nutr.*, 1989, **28**, 273.
59. E. C. Wildman, *Handbook of Nutraceuticals and Functional Foods*, CRC Press LLC, Florida USA, 1st edn, 2001.
60. K. Nascimento, J. A. Sattler, L. F. L. Macedo, C. Gonzalez, I. Melo, E. Araujo and L. B. A. Muradian, *LWT–Food Sci. Technol.*, 2018, **91**, 85.
61. M. Shen, K. Li, H. Jing and L. Zheng, *J. Food Sci. Technol.*, 2018, **83**, 221.
62. M. Cabrera, O. Munoz, G. E. Zuniga, L. J. Corcuera and V. H. Argandona, *Phytochemicals*, 1995, **39**, 1023.
63. L. Long, M. J. Gorecki, M. Renton, J. K. Scott, L. Colville and D. E. Goggin, *Biol. Rev. Biol. Proc. Cambridge Philos. Soc.*, 2014, **5**, 601.
64. M. Mohammadi and H. Kazemi, *Plant Sci.*, 2002, **162**, 491.
65. M. Tattini, C. Galardi, P. Pinelli, R. Massai, D. Remorini and G. Agati, *New Phytol.*, 2004, **163**, 547.
66. V. Lattazioa, A. Cardinali, A. Ardinali, C. Ruta, F. I. Morone, V.M.T Lattazioa, V. Linsalata and N. Ciccod, *Environ. Exp. Bot.*, 2009, **65**, 54.
67. W. Stahl and H. Sies, *Biochim. Biophys. Acta*, 2005, **1740**, 101.
68. H. Nishino, M. Murakoshi, H. Tokuda and Y. Satomi, *Arch. Biochem. Biophys.*, 2009, **483**, 165.
69. M. K. Beatty, D. P. Henson and M. Boulton, *Surv. Ophthalmol.*, 2000, **45**, 115.
70. M. M. Calvo, *Crit. Rev. Food Sci. Nutr.*, 2005, **45**, 671.
71. A. Kaulmann and T. Bohn, *Nutr. Res.*, 2014, **34**, 907.
72. E. Fernandez-Garcia, I. Carvajal-Lerida, M. Jaren-Galan, J. Garrido-Fernandez, A. Perez-Galvez and D. Hornero-Mendez, *Food Res. Int.*, 2012, **46**, 438.
73. R. Fernandez-Orozco, L. Gallardo-Guerrero and D. Hornero-Mendez, *Food Chem.*, 2013, **141**, 2864.
74. P. Zhang, Y. Yu, Y. Qin, Y. Zhou, R. Tang, Q. Wang and X. F. Huang, *J. Nutr. Biochem.*, 2019, **65**, 54.
75. C. W. Newman and R. K. Newman, *Cereal Foods World*, 2006, **51**, 4.
76. B. K. Baik and S. E. Ullrich, *J. Cereal Sci.*, 2008, **48**, 233.
77. S. I. Mussatto, *J. Sci. Food Agric.*, 2014, **94**, 1264.
78. Z. Ruan, C. Zhang, D. Sun-Waterhouse, B. S. Li and D. D. Li, *Food Bioprocess Technol.*, 2019, **12**, 1232.
79. M. Martinez-Subira, M. P. Romero, E. Puig, A. Macia, I. Romagosa and M. Moralejo, *LWT–Food Sci. Technol.*, 2020, 109582.
80. X. Q. Deng, Z. F. Pan, Q. Li, G. B. Deng, H. Long, N. Tashi and M. Q. Yu, *Cereal Chem.*, 2020, **97**, 39.

81. J. Liu, Q. Li, H. Zhai, Y. Zhang, X. Zeng, Y. Tang and Z. Pan, *J. Food Sci.*, 2020, **85**, 3141.
82. S. Punia, K. S. Sandhu and M. Kaur, *J. Food Sci. Technol.*, 2020, **57**, 3782.
83. M. S. Jacobs, *Cereal Chem.*, 2006, **80**, 645.
84. D. Hatcher, S. Lagasse, J. Dexter, B. Rossnagel and M. Izydorczyk, *Cereal Chem.*, 2005, **82**, 60.
85. G. A. Al Shehry, *Aust. J. Basic Appl. Sci.*, 2017, **11**, 98.
86. S. Dhingra and S. Jood, *Int. J. Food Sci. Technol.*, 2004, **39**, 213.
87. A. P. Klamczynski and Z. Czuchajowska, *Cereal Chem.*, 1999, **76**, 530.
88. B. K. Baik and Z. Czuchajowska, *Food Sci. Technol. Int.*, 1997, **3**, 423.
89. R. K. Newman, C. F. McGuire and C. W. Newman, *Cereal Foods World*, 1990, **35**, 563.
90. N. Ames, C. Rhymer, B. Rossnagel, M. Therrien, D. Ryland, S. Dua and K. Ross, *Cereal Foods World*, 2006, **51**, 23.
91. X. Y Zong, M. H. Bian, L. Li, L. Qiang and H. Luo, *J. Sichuan Univ., Eng.*, 2012, **4**, 14.
92. D. M. Waters, F. Jacob, J. Titze, E. K. Arendt and E. Zannini, *Eur. Food Res. Technol.*, 2012, **235**, 767.
93. A. Ktenioudaki, E. Crofton, A. G. M. Scannell, J. A. Hannon, K. N. Kilcawley and E. J. Gallagher, *J. Cereal Sci.*, 2013, **57**, 384.
94. M. S. Izydorczyk and J. E. Dexter, *Food Res. Int.*, 2008, **41**, 850.
95. P. Sharma and H. S. Gujral, *LWT–Food Sci. Technol.*, 2014, **55**, 301.
96. P. Guleria, S. Kumari and N. Dangi, *Int. J. Enhanced Res. Sci., Technol. Eng.*, 2015, **4**.
97. A. Skendi, C. G. Biliaderis, A. Lazaridou and M. S. J. Izydorczyk, *J. Cereal Sci.*, 2003, **38**, 15.
98. A. M. Andersson, C. Elfverson, R. Andersson, S. Regner and P. Aman, *J. Sci. Food Agric.*, 1999, **79**, 979.
99. L. Johansson, P. Tuomainen, M. Ylinen, P. Ekholm and l. Virkki, *Carbohydr. Polym.*, 2004, **58**, 267.
100. A. Gajdosova, Z. Petrulakova, M. Havrlentova, V. Cervena, B. Hozova, E. Sturdık and G. Kogan, *Carbohydr. Polym.*, 2007, **70**, 46.
101. P. Sharma, H. S. Gujral and C. M. Rosell, *J. Cereal Sci.*, 2011, **53**, 25.
102. P. Volikakis, C. G. Biliaderis, C. Vamvakas and G. K. Zerfiridis, *Food Res. Int.*, 2004, **37**, 83.
103. J. Yu-Lai, R. A. Speers, A. T. Paulson and J. S. Robert, *MBAA TQ*, 2004, **41**, 231.
104. K. Zhong, Q. Zhang, L. Tong, L. Liu, X. Zhou and S. Zhou, *Ultrason. Sonochem.*, 2015, **23**, 75.
105. A. Khan, A. Gani, F. A. Masoodi, F. Amin, I. A. Wani, F. A. Khanday and A. Gani, *Carbohydr. Polym.*, 2016, **140**, 442.
106. Y. J. Chang, S. Lee, M. A. Yoo and H. G. Lee, *Food Chem.*, 2006, **54**, 3815.
107. L. De Souza, J. Bartz, R. Z. Eda, P. D. De Oliveira, W. S. V. Da Silva, G. H. Alves and A. R. G. Dias, *Food Chem.*, 2015, **178**, 243.
108. G. Maheshwari, S. Sowrirajan and B. Joseph, *J. Food Sci.*, 2017, **82**, 1535.
109. R. S. Bhatty, *J. Cereal Sci.*, 1995, **22**, 163.

110. B. E. Knuckles and M. M. Chiu, *Cereal Chem.*, 1992, **69**, 198.
111. S. H. Yoon, P. T. Berglund and C. E. Fastnaught, *Cereal Chem.*, 1995, 72, 187.
112. T. Vasanthan and F. Temelli, *Food Res. Int.*, 2008, **41**, 876.
113. J. Kiryluk, A. Kawka, H. Gasiorowski, A. Chalcarz and J. Aniola, *Nahrung*, 2000, **44**, 238.
114. T. Vasanthan and R. Bhatty, *Cereal Chem.*, 1995, **72**, 379.
115. R. Morgan, *WO pat.*, Patent International Publication Number WO98/13056, 1998.
116. G. Inglett, C. Carriere, S. Maneepun and P. Tungtrakul, *Int. J. Food Sci. Technol.*, 2004, **39**, 1365.
117. S. Brennan and L. J. Cleary, *J. Cereal Sci.*, 2005, **42**, 1.
118. A. Andersson, A. M. Lampi, L. Nyström, V. Piironen, L. Li, J. L. Ward, K. Gebruers, C. M. Courtin, J. A. Delcour, D. Boros and A. Fras, *J. Agric. Food Chem.*, 2008, **56**, 9767.
119. M. S. Jacobs, M. S. Izydorczyk, K. R. Preston and J. E. Dexter, *J. Sci. Food Agric.*, 2008, **88**, 558.
120. A. Lazaridou and C. G. Biliaderis, *J. Cereal Sci.*, 2007, **46**, 101.
121. M. S. Izydorczyk, A. Hussain and A. W. MacGregar, *J. Cereal Sci.*, 2001, **34**, 251.
122. S. Brennan and L. J. Cleary, *Food Res. Int.*, 2007, **40**, 291.
123. G. El-Nagar, G. Glowers, C. M. Tudorica, V. Kuri and C. S. Brennan, *Int. J. Dairy Technol.*, 2002, **55**, 89.
124. A. Morin, F. Temelli and L. McMullen, *J. Food Sci.*, 2002, **67**, 2391.
125. M. Lyly, K. Roininen, K. Honkapaa, K. Poutanen and L. Lahteenmaki, *Food Quality Preference*, 2007, **18**, 242.
126. J. A. Nazare, S. Normand, A. O. Triantafyllou, A. B. De La Perriere, M. Desage and M. Laville, *Mol. Nutr. Food Res.*, 2009, **53**, 361.
127. S. R. Amini, H. Hosseini, D. Bondarianzadeh, F. J. Colmenero and R. Khaksar, *LWT–Food Sci. Technol.*, 2015, **62**, 704.
128. A. Brennan, E. Derbyshire, B. K. Tiwari and C. S. Brennan, *Plant Foods Hum. Nutr.*, 2013, **68**, 78.
129. M. J. Lee, N. Y. Lee, Y. K. Kim, J. C. Park, I. D. Choi, S. G. Cho, J. G. Kim, H. K. Park, K. H. Park, K. J. Kim and H. S. Kim, *Food Sci. Biotechnol.*, 2011, **20**, 55.
130. L. Wang, F. Ye, L. Feng, F. Wei and G. Zhao, *J. Texture Stud.*, 2017, **48**, 562.
131. R. Kaur and C. S. Riar, *J. Food Sci. Technol.*, 2020, **57**, 41.
132. S. Karp, J. Wyrwisz and M. A. Kurek, *J. Food Sci. Technol.*, 2020, **57**, 3628.
133. K. M. Andrzej, M. Małgorzata, K. Sabina, O. K. Horbanczuk and E. Rodak, *Food Sci. Technol. Int.*, 2020, **26**, 53.
134. H. Afshari, H. Hosseini, R. Khaksar, M. A. Mohammadifar, Z. Amiri, R. Komeili and A. M. Khaneghah, *Food Technol. Biotechnol.*, 2015, **53**, 436.
135. L. Geng, M. Li, S. Xie and D. Wu, *Mol. Plant Breed.*, 2021, **41**, 6.
136. G. Dongowski, M. Huth, E. Gebhardt and W. Flamme, *J. Nutr.*, 2002, **132**, 3704.
137. H. S. Odes, H. Lazovski, I. Stern and Z. Madar, *Nutr. Res.*, 1993, **13**, 979.

138. L. Jenkins, D. J. A. Jenkins, U. Zdravkovic, P. Wursch and V. Vuksan, *Eur. J. Clin. Nutr.*, 2002, **56**, 622.
139. L. N. Granfeldt and I. Bjoarck, *Eur. J. Clin. Nutr.*, 2008, **62**, 600.
140. K. N. Smith, K. Queenan, W. Thomas, G. Fulcher and J. Slavin, *FASEB J.*, 2004, **18**, 149.
141. W. Jinshui, M. R. Cristina and C. B. De-Barbera, *Food Chem.*, 2002, **79**, 221.
142. E. Wisker, M. Daniel, G. Rave and W. Feldeim, *Br. J. Nutr.*, 2000, **84**, 31.
143. S. Crittenden, S. Karppinen, M. Ojanen, R. Tenkanen, J. Fagerstrom, M. Matto, T. Saarela, S. Matiila and K. Poutanen, *J. Sci. Food Agric.*, 2002, **82**, 781.
144. J. Beck, S. M. Tosh, M. J. Batterham, L. C. Tapsell and X. F. Huang, *Mol. Nutr. Food Res.*, 2009, **53**, 1343.
145. R. J. Delaney, T. A. Nicolosi, T. Wilson, S. Carlson, G. H. Frazer, R. Zheng, R. Hess, K. Ostergren, J. Haworth and N. Knutson, *J. Nutr.*, 2003, **133**, 468.
146. S. J. Delatte, J. Evans, A. Hebra, W. Adamson, H. B. Othersen and E. P. E. Tagge, *J. Pediatr. Surg.*, 2001, **36**, 113.
147. H. L. Kim, J. H. Lee, M. H. Lee, B. J. Kwon and J. C. Park, *Tissue Eng., Part A*, 2012, **18**, 2315.
148. A. Belcarz, G. Ginalska, T. Pycka, A. Zima, A. Ślósarczyk, I. Polkowska, Z. Paszkiewicz and W. Piekarczyk, *Open Life Sci.*, 2013, **8**, 534.
149. A. Anusuya and M. Sathiyabama, *Int. J. Biol. Macromol.*, 2014, **70**, 440.
150. J. Wang, B. Sung and R. Tsao, *Bioactive Factors and Processing Technology for Cereal Foods*, Springer Singapore, 1st edn, 2019.
151. L. F. Calinoiu and D. C. Vodnar, *Nutrients*, 2018, **11**, 1615.
152. U. Tiwari and E. Cummins, *Trends Food Sci. Technol.*, 2009, **20**, 511.
153. L. Wang, R. K. Newman, C. W. Newman, L. L. Jackson and P. J. Hofer, *Plant Foods Hum. Nutr.*, 1993, **43**, 9.
154. O. J. R. Richard, S. B. Racette, O. Alfred and F. S. William, *Am. J. Clin. Nutr.*, 2002, **75**, 1000.
155. K. T. Chung, C. I. Wei and M. G. Johnson, *Trends Food Sci. Technol.*, 1998, **9**, 168.
156. J. M. Brimson, N. Onlamoon, T. Tencomnao, P. Thitilertdecha and P. S. M. Clerodendrum, *Biomed. Pharmacother.*, 2019, **118**, 109319.
157. J. M. Brimson, N. Onlamoon, T. Tencomnao, P. Thitilertdecha and P. S. M. Clerodendrum, *Biomed. Pharmacother.*, 2019, **118**, 109319.
158. C. Srinivasulu, M. Ramgopal, G. Ramanjaneyulu, C. M. Anuradha and K. Suresh, *Biomed. Pharmacother.*, 2018, **108**, 547.
159. C. Chen, *Oxid. Med. Cell. Longevity*, 2016, 3571614.
160. M. Srinivasan, A. R. Sudheer and V. P. Menon, *J. Clin. Biochem. Nutr.*, 2007, **40**, 92.
161. G. Garrait, J. F. Jarrige, S. Blanquet, E. Beyssac, J. M. Cardot and M. Alric, *J. Agric. Food Chem.*, 2006, **54**, 2944.
162. S. Jood and S. Kalra, *Nahrung*, 2001, **45**, 35.
163. A. Brennan, E. Derbyshire, B. K. Tiwari and C. S. Brennan, *Plant Foods Hum. Nutr.*, 2013, **68**, 78.

164. J. P. Nielsen and L. Munck, *J. Cereal Sci.*, 2003, **38**, 173.
165. G. Panfilli, A. Fratianni, T. Di-Criscio and E. Marconi, *Food Chem.*, 2008, **107**, 84.
166. M. J. Lee, N. Y. Lee, Y. K. Kim, J. C. Park, I. D. Choi, S. G. Cho, J. G. Kim, H. K. Park, K. H. Park, K. J. Kim and H. S. Kim, *Food Sci. Biotechnol.*, 2011, **20**, 55.
167. H. Koksel, G. H. Ryu, A. Basman, H. Demiralp and P. K. W. Ng, *Nahrung*, 2004, **48**, 19.
168. S. Kalra and S. Jood, *J. Cereal Sci.*, 2000, **31**, 141.
169. A. Demirbas, *Food Chem.*, 2005, **90**, 773.
170. C. Collar and A. Angioloni, *Eur. Food Res. Technol.*, 2014, **3**, 2128.
171. M. Lehtonan and R. Aikasalo, *Cereal Chem.*, 1987, **64**, 91.
172. M. S. Izydorczyk, A. Hussain and A. W. MacGregor, *J. Cereal Sci.*, 2000, **34**, 251.
173. A. K. Rieder, S. Holtekjolen, S. Sahlstrom and A. Moldestad, *J. Cereal Sci.*, 2012, **55**, 44.
174. P. Aman and H. Graham, *J. Agric. Food Chem.*, 1987, **35**, 704.
175. A. Nilssen, S. Sahlstrom, S. H. Knutsen, A. K. Holtekjolen and A. K. Uhlen, *J. Cereal Sci.*, 2008, **48**, 670.
176. K. Holtekjolen, H. H. R. Olsen, E. M. Færgestad, A. K. Uhlen and S. H. Knutsen, *LWT–Food Sci. Technol.*, 2008, **41**, 2085.
177. R. S. Bhatty, A. W. McGregor and B. G. Rossnagel, *Cereal Chem.*, 1991, **68**, 221.
178. L. Saulnier, S. Gevaudan and J. F. Thibault, *J. Cereal Sci.*, 1994, **19**, 171.
179. F. Temelli, *J. Food Sci.*, 1997, **62**, 1194.

CHAPTER 5

Pigmented Sorghum: Functional Properties and Bioactive Diversity

TAHA MEHANY(0000-0001-7960-7558)*[a,b], YOUNES M. RASHAD[c], BABATUNDE OLAWOYE[d], ILARIA CACCIOTTI[e], ESUA OKON JOHNSON[f,g], OYEKEMI OLABISI POPOOLA[d], ZHONG HAN[g] AND WALEED M. E. FEKRY[h]

[a]Food Technology Department, Arid Lands Cultivation Research Institute, City of Scientific Research and Technological Applications, 21934 New Borg El-Arab City, Alexandria, Egypt; [b]Department of Chemistry, University of La Rioja, C/Madre de Dios 51, 26006 Logroño, La Rioja, Spain; [c]Plant Protection and Biomolecular Diagnosis Department, Arid Lands Cultivation Research Institute, City of Scientific Research and Technological Applications, 21934 New Borg El-Arab City, Alexandria, Egypt; [d]Department of Food Science and Technology, Faculty of Engineering and Technology, First Technical University, Ibadan, Nigeria; [e]Department of Engineering, INSTM RU, University of Rome "Niccolò Cusano", Via Don Carlo Gnocchi 3, 00166, Roma, Italy; [f]Department of Agricultural and Food Engineering, University of Uyo, Uyo, 520101, Nigeria; [g]School of Food Science and Engineering, South China University of Technology, Guangzhou 510641, China; [h]Department of Plant Production, Arid Lands Cultivation Research Institute, City of Scientific Research and Technological Applications, 21934 New Borg El-Arab City, Alexandria, Egypt
*E-mail: tmehany@srtacity.sci.eg

Food Chemistry, Function and Analysis No. 38
Pigmented Cereals and Millets: Bioactive Profile and Food Applications
Edited by Sneh Punia Bangar, Sajid Maqsood and Anil Kumar Siroha

Published by the Royal Society of Chemistry, www.rsc.org

5.1 Introduction

Cereal grains are humans' main food source.[1] Sorghum, commonly called *Sorghum bicolor* L. Moench (Figure 5.1), is a monocotyledonous herbaceous plant that belongs to the same family *(Poaceae)* as maize (*Zea mays* L.) and barley (*Hordeum vulgare*), and is very similar to corn in physiology, genomic, shape, and food applications.[2] It is the fifth most vital grain crop worldwide after rice, wheat, corn, and barley, and it serves as the main food for approximately 750 million inhabitants of semi-arid areas of Asia, Australia, Africa, and America, owing to its characteristics, such as drought tolerance.[3,4] In 2018, worldwide production of sorghum grains reached about 58 million tons. Figure 5.2 illustrates the worldwide sorghum production share, and Figure 5.3 shows the top 10 sorghum-producing countries in tons, according to FAOSTAT.[5] In contrast to C3 plants like wheat and rice, sorghum is a C4 photosynthetic plant, and thus is highly efficient in assimilating carbon and

Figure 5.1 Plant of sorghum. Source: personal archive of Taha Mehany.

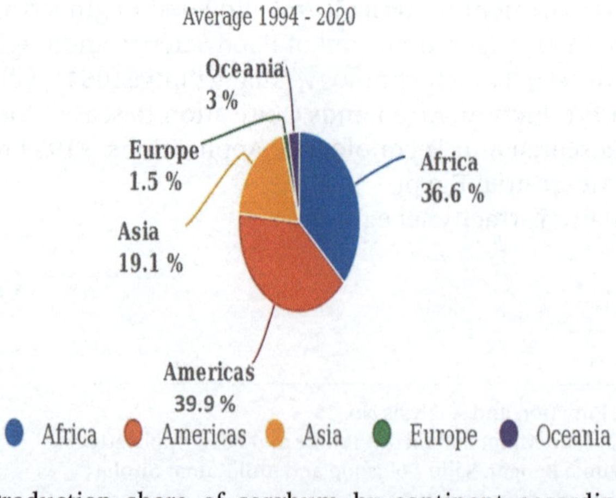

Figure 5.2 Production share of sorghum by continent according to FAOSTAT (2022).[5]

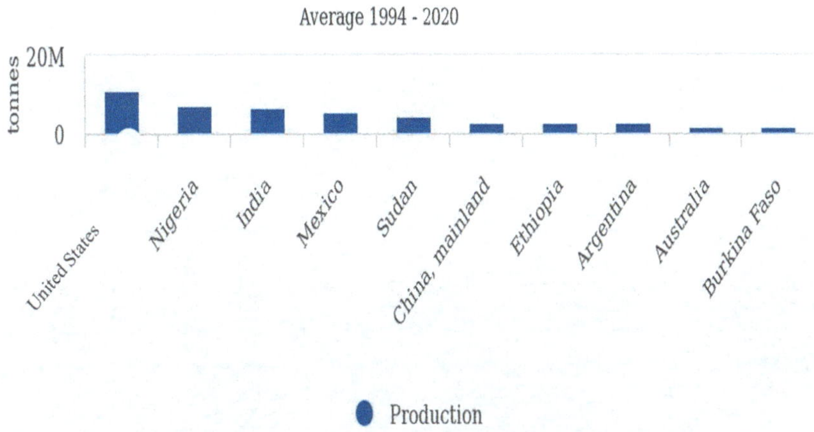

Figure 5.3 Top 10 sorghum-producing countries in tonnes according to FAOSTAT (2022).[5]

accumulating biomass at elevated temperatures.[6,7] To address the challenge of climate change, a shift toward more climate-resilient cereal crops such as sorghum is of great importance.[6] Among them, sorghum is well known as an adapted crop that can be produced even in a warming climate and with growing water shortages. In the semi-arid countries of Asia and Africa, sorghum plays an integral role as a feed and food source rich in proteins, dietary fiber, minerals, and vitamins.[6]

In addition to its agronomical features, sorghum possesses attractive attributes that are highly relevant to modern food habits, particularly considering the growing chronic diseases related to extra calorie intake and poor diet worldwide. Sorghum is grown not only as cereal, but also for sugar (sweet stalk), forage, and cellulosic biomass for biofuel manufacturing.[7] Moreover, there is demand for sorghum consumption as boiled grains, beverages, pasta, or semolina. However, sorghum impairment is often characterized by the presence of pigments in the testa and seed coat. For example, in some sorghum genotypes, the testa is occasionally partial, non-visible, or even lost, whereas it is highly colored in others. Sorghum varies in grain color, and can be naturally white, tan, red, pale orange, dark brown or brownish-red.[4] Nevertheless, the chief commercially available grains are white, black, and red (Figure 5.4). The color of the seed coat or pericarp (testa) is a genetic character that is controlled by the Y and R genes. The sorghum pericarp widely varies in color, depending principally on the composition of its polyphenol profile, increasing sorghum's diversity and the desirability of its products.[4]

Among key cereal grains, sorghum is exceptionally rich in bioactive diversity, owing to polyphenols and other bioactive compounds that are connected with a reduced risk of nutrition-linked chronic diseases such as type 2 diabetes, cardiovascular disease, and cancer.[8] Sorghum presents high levels of bioactive components that are not common to other grains,

Figure 5.4 Mature panicles of different types of sorghum in the field. Black (left), white (middle), and red (right) sorghum. Reproduced from ref. 4, with permission from Elsevier, Copyright 2017.

with proven potential health benefits to humans.[7] Various bioactive compounds, *e.g.* phenolic acids, anthocyanins, flavonoids, among others, were found in sorghum grains in considered amounts, while concentrated tannins are the most widely abundant plant secondary metabolites in the sorghum, playing a vital role in defense mechanisms against insects and birds.[3] Indeed, these bioactive ingredients, especially polyphenols, are widely studied and applied in both the food and pharmacological industries, due to their many potential health benefits, including antioxidant, anti-inflammation, immune-regulation and antimicrobial effects.[9] From a nutritional viewpoint, sorghum grains present better functional and nutritional properties compared with other common cereals. This has been ascribed to its higher quantity of slowly digestible and resistant starch that helps in reducing postprandial hyperglycemia in humans, and the calories it provides in a typical daily food intake. Sorghum grains have wide techno-functional benefits because of their aptitude to modify starch and protein functionality, which can be utilized to produce novel bioactive constituents or improve the quality of food products.[6–8,10,11] However, Gerard *et al.*[7] reviewed the key issues that limit extensive sorghum use in food as follows: (1) the challenges associated to the sorghum endosperm functionality; (2) the moderately limited clinical evidence to verify the unique health benefits of the sorghum; (3) the plentiful availability of alternate cereals with desirable organoleptic characteristics; (4) the lack of optimal processing standardization technologies; and (5) a fairly under-developed sorghum food value chain. Teferra and Awika[6] addressed the key challenges related to sorghum utility as a food ingredient due to its lower endosperm functionality and the relatively low digestibility of its protein. Progress in the use of genetics to alter the functionality and structure of kafirin protein in sorghum means that solving the challenges of using sorghum grains in food systems may be achievable. These improved sorghum lines have demonstrated enhanced food use quality and protein digestibility.[6] The aim of this chapter is to: (1) review the chemical composition, functional properties and food applications of PS; (2) provide an overview of innovative technologies for the development of functional food from PS; (3) highlight the uniqueness of PS colorants and polyphenols, with specific consideration for their bioactive flavonoids and anthocyanins.

5.2 Nutritional Properties of Pigmented Sorghum

Sorghum belongs to a group of cereal crops that consists of red, white, brown, black, or yellow endosperms. Despite the color differences, most of the varieties include starch, proteins, lipids, fibers and ash, with a value of approximately 75%, 12%, 4%, 3%, and 2%, respectively.[12] Other shared nutrient sources include various lipids, vitamins, minerals and amino acids.

5.2.1 Carbohydrates

Sorghum composition is similar to that of maize, except for its lower lipid content. The major component of sorghum is starch, and the dietary fiber in the grain cell wall makes up to 75% of this. The starch-free polysaccharide content in sorghum is responsible for its ability to lower cholesterol levels and improve bowel function in humans. The total soluble sugar of sorghum ranges from 0.7% to 4.2%, while reduction of sugars ranges from 0.05% to 0.53%.[2] Starch content ranges between 32.1 and 72.5 g 100 g in some sorghum varieties, and mainly comprises amylopectin (81.0–96.5%) and amylose (3.5–19.0%).[13] The rheological properties of starch in sorghum, such as gelatinization, gelling and retrogradation, as well as its digestibility, are usually affected by the amylose : amylopectin ratio.[14] Compared with other cereals, sorghum is the least digestible due to the strong interaction between the starch granules and the proteins, together with tannins. Generally, most of these starch granules are slowly digestible (30.0–66.2%), while the remainder are rapidly digestible starch or resistant starch with values from 15.3% to 26.6%, and from 16.7% to 43.2%, respectively.[15] Soluble fibers, with a value between 10% and 25%, and insoluble fibers, between 75% and 90% are the non-starch polysaccharides in sorghum (6.0–15.0 g per 100 g).[16]

5.2.2 Proteins

Sorghum proteins are mainly categorized as prolamins (79% of the entire proteins), while the other ones consist of globulins, albumins, and glutelin.[17] The majority of sorghum prolamins consist in kafirins that are present in three major categories, *i.e.* α (66–84%), β (8–13%) and γ (9–21%).[12]

The endoplasmic reticulum is responsible for sorghum kafirin storage in the spherical protein bodies: the peripheral proteins house the γ-kafirins, while the inner region the α-kafirins.[18] As a result of intracellular disulphide bond formation, sorghum kafirins offer opposition to peptidase, and thus, sorghum protein digestibility after cooking is lower compared with that of cereals such as maize and wheat. However, in tannin-rich varieties, protein digestibility can be reduced to about 50% by the binding of kafirins with tannins.[19] In addition, other endogenous factors contribute to low digestibility, as well as protein's organizational structure in grain and exogenous factors (interconnection between proteins and non-proteins constitutes [*e.g.* lipids, starch, phytic acid and non-starch polysaccharides]). When the sorghum is cooked in heat (wet), proteins' digestibility is reduced, while sorghum digestibility may be doubled when processing methods such as germination and fermentation are applied. There have been recent efforts to enhance the digestibility of sorghum's protein by reducing the amount of kafirins, especially the β and γ forms, as well as improving the albumin and glutelin.[17] Genetically modified varieties are characterized by *in vitro* digestibility values between 23% and 102%, which is higher than those of the control varieties. Sorghum proteins are excellent sources of nonpolar amino acids (alanine, leucine and proline) and glutamic acid, which is lysine, the major limiting amino acid.[20] In addition, sorghum protein may be inadequate as source of some amino acids such as cysteine, threonine, methionine, isoleucine and valine.[21] Varieties obtained through breeding usually have a lysine content that is 52–115% greater than conventional varieties. The digestion of α kafirins in the intestine usually takes place after that of other nutrients due to their overabundance, also their nutritive quality is usually reduced by their indigestibility.[17] β and γ kafirins are excellent sources of cysteine, forming diffusive bonds, which are presumed to block the α kafirins' accessibility to hydrolysis. Therefore, in modified sorghum varieties, higher digestibility can be linked to an increase in the α kafirin enzyme vulnerability, as a result of differences in the protein body morphology and a decline in disulfide bond formation between β and γ kafirins.[22] Interestingly, compared with the main prolamins of barley (hordein), rye (secalin) and wheat (gliadin), the kafirins usually do not activate autoimmune reactions or allergic responses in human beings. Thus, people with celiac disease can safely consume sorghum.

5.2.3 Lipids

The lipid levels in sorghum, are particularly high in comparison with most other cereals.[23] The sorghum lipid content is (1.2–3.07 g per 100 g) and is mainly made up of saturated fatty acids between 83% and 88%.[24] In the majority of sorghum varieties, monounsaturated fatty acids are usually lower than polyunsaturated ones. Linoleic, oleic, palmitic and linoleic acids are the main fatty acids present in sorghum, at levels of 4.5–5.1%, 32.2–42%, 12.4–16% and 1.4–2.8%, respectively.[12] However, these lipids are supplied in proportions different within the grain. Akin *et al.*[25] demonstrated that the highest lipid levels are contained in the germ (76.2%), whereas the endosperm and pericarp contain 13% and 10.6% of lipids, respectively. The lipids surrounding the sorghum kernels' surface have a distinctive compositional profile. Several studies have shown that the surface lipids comprise wax, as well as oil, at a ratio of 2:1 or 3:1. The wax and oil between them contain aldehydes, hydrocarbons, wax esters, alcohols, sterylesters, monoacylglycerols, diacylycerols, triacylylcerols, free sterols and free fatty acids, regardless of endosperm color.[26] The wax fraction comprises fatty aldehydes, policosanols, fatty acids, hydrocarbons, wax esters and triglycerols with values of 44–55%, 37–44%, 0.1–0.5%, 1.4% and 0.3–0.4%, respectively.[24] Triacyclglycerols represent the most rapidly supplied group for the oil fraction, while oleic, linoleic and palmitic acids are the most richly supplied fatty acids. In contrast to corn oil, sorghum oil possesses higher oleic and stearic acid levels and lower palmitic, myristic and linoleic acid levels, making it a less saturated oil. Much attention has been focused on the health benefits of sorghum's bioactive agents, especially the policosanols and phytosterols available in wax and oil fractions. There is a structural similarity between phytosterols and cholesterols, while policosanols form a combination of long-chained aliphatic alcohols. Ratnavathi and Komala[27] reported that the total levels of phytosterols present in sorghum have a range of value between 46 and 51 mg per 100 g. Also, the extractable phytosterol level is affected by sorghum's physical state (*i.e.* whole grain, ground kernel, dry distiller's sorghum), whereas the total composition stays the same. Mehmood *et al.*[28] showed that all the lipids acquired out of the above physical states comprised stigmasterol, campesterol and sitosterol, with dry distiller's sorghum possessing the highest levels, because the carbohydrate fraction is not present any longer.

5.2.4 Minerals and Vitamins

Several minerals, such as potassium, phosphorus and zinc are present in sorghum.[29] However, little is known about bioavailability in the sorghum. For example, the bioavailability of Zn in sorghum is between 9.7% and 17.1%, and Fe bioavailability is between 6.6% and 15%.[22] Similarly, there is a scarcity of information on sorghum's vitamin content. It is known that sorghum contains some Vitamin B-complexes (thiamine and riboflavin, in addition to pyridoxine) and the fat-soluble vitamins D, E and K.[29] An inconsistent level of vitamin E and its end products in sorghum has been detected.[30] Table 5.1 shows the nutritional composition of sorghum grains according to USDA[31] data regarding the composition of cereal grains and pasta category.

5.3 Phytochemical Diversity of Pigmented Sorghum and Other Bioactive Molecules

Among the major cereals, sorghum shows a distinct characteristic that is not shared with other cereals and is associated with the presence of various bioactive compounds that promote health, including tannins, polyphenols and

Table 5.1 The nutritional composition of sorghum grains according to USDA (2018).[31]

Nutritional items	Components	Amount (g per 100 g)
Macronutrients	Water (g)	12.4
	Energy (kcal)	329
	Protein (g)	10.6
	Total lipid (fat) (g)	3.46
	Ash (g)	1.43
	Carbohydrate (g)	72.09
	Fiber, total dietary (g)	6.7
	Sugars, total including NLEA (g)	2.53
	Starch (g)	60
Minerals	Calcium (mg)	13
	Iron (mg)	3.36
	Magnesium (mg)	165
	Phosphorus (mg)	289
	Potassium (mg)	363
	Sodium (mg)	2
	Zinc (mg)	1.67
	Copper (mg)	0.284
	Manganese (mg)	1.605
	Selenium (μg)	12.2
Vitamins	Total ascorbic acid (C) (mg)	0
	Thiamin (mg)	0.332
	Riboflavin (mg)	0.096
	Niacin (mg)	3.688
	Pantothenic acid (mg)	0.367
	Vitamin B-6 (mg)	0.443
	Folate, total (μg)	20
	α-tocopherol (E) (mg)	0.5
	Phylloquinone (K) (μg)	0
Amino acids	Tryptophan (g)	0.124
	Threonine (g)	0.346
	Isoleucine (g)	0.433
	Leucine (g)	1.491
	Lysine (g)	0.229
	Methionine (g)	0.169
	Cystine (g)	0.127
	Phenylalanine (g)	0.546
	Tyrosine (g)	0.321
	Valine (g)	0.561
	Arginine (g)	0.355
	Histidine (g)	0.246
	Alanine (g)	1.033
	Aspartic acid (g)	0.743
	Glutamic acid (g)	2.439
	Glycine (g)	0.346
	Proline (g)	0.852
	Serine (g)	0.462
Fatty acids	Total saturated (g)	0.61
	Total monounsaturated MUFA (g)	1.131
	Total polyunsaturated PUFA (g)	1.558
	Fatty acids, total trans (g)	0.004

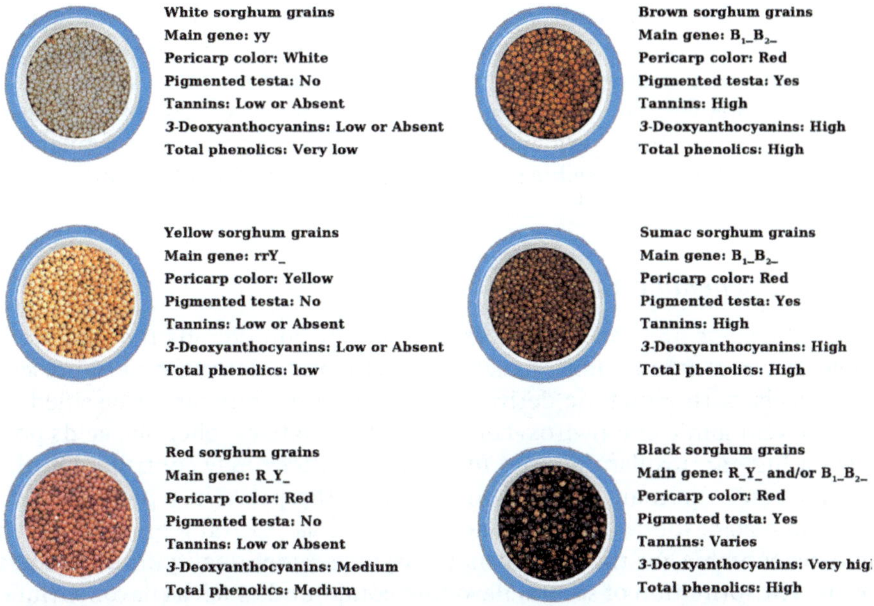

Figure 5.5 The genetics, classification, and physical characterization of sorghum grains. Reproduced from ref. 10 with permission from Elsevier, Copyright 2021.

flavonoids. A diverse, high level of bioactive lipids and polyphenols is present in sorghum in comparison with other types of cereals.

The health benefits of sorghum, such as fat accumulation reduction, lipid and glucose metabolism and insulin sensitivity, have been associated with its phytochemicals, especially polyphenols. In different sorghum varieties, bioactive compounds such as anthocyanin-derivatives; 3-deoxyanthocyanidins have been mainly identified in black pericarp.[32] Sorghum's bioactive components are located in the bran (testa and pericarp, as well as the aleurone tissues). These compounds co-exist with the polysaccharide cell wall, which is richly supplied in cereal bran, significantly contributing to health advantages characteristic of whole-grain intake.[33] However, the benefits obtained from cereals can be diminished by processing food that may remove the majority of the bran. Pigmented sorghum and its byproducts are essential sources of bioactive substances such as 3-deoxyanthocyanidins, phenolic acids, proanthocyanins and carotenoids, which provide the color of the pigmented sorghum. Figure 5.5 shows sorghum grains' genetics, classification and physical characterization.[10]

5.3.1 Polyphenols

Polyphenols' free radical scavenging ability and their modulation of various signaling pathways have been widely studied in terms of disease prevention. In sorghum, the dominant polyphenols consist of flavonoid derivatives or phenolic acids. Polyphenols perform an essential function in providing natural protection for the plant against pests and pathogens. For instance, in sorghum, the presence of condensed tannin increases bird predation tolerance. Moreover, the wide variety sorghum pericarp colors (from whitish cream to red, to lemon yellow, as well as black) is a function of polyphenol's diversity, making it a possible natural means for food diversification in terms of sensory visual appeal.[4]

5.3.1.1 Phenolic Acids

Phenolic acids are well-characterized phenolic compounds that are available in cereal grains. Both benzoic and cinnamic derivatives are present in cereals, with cinnamic derivatives being more abundant. Classified as hydroxycinnamic and hydroxybenzoic acid derivatives, phenolic acids possess a high antioxidant activity *in vitro*, hence they may contribute to the promotion of human health.[34] In sorghum, the prevailing phenolic acid is ferulic acid, followed by flavonoids, as well as natural pigments, which are accountable for the appropriate enzyme's functionality in the plant's cells. The formation of several flavonoid compounds such as flavones (luteolin and apigenin), flavonols (kaempferol and quercetin), and naringenin result from their three-ring structure. Just like in other types of cereal grains, phenolic acids (70–95%) in sorghum are covalently bonded to the polysaccharide cell wall. Phenolic acids that are not covalently bound to the polysaccharide cell wall are relatively small molecules of carbohydrates, especially glycerol.[35]

Phenolic acid's covalent connection to the polysaccharide cell wall greatly contributes to the effective biological action of components of whole grain in the human diet, and to absorption in the upper gastrointestinal tract by their forms, which are unbound. In some varieties of sorghum, the amount of the phenolic acid ranges between 135.5 and 479 μg g^{-1}.[36] Protocatechuic and ferulic acids constitute the major components, with values ranging from 15.3 to 178.2479 μg g^{-1} and from 120.5 to 173.5 μg g^{-1}, respectively, whereas *p*-hydroxybenzoic, cinnamic, caffeic, vanillic, syringic, *p*-coumaric acids are present in smaller amounts, *i.e.* 6.1–164 μg g^{-1}, 9.8–15.0479 μg g^{-1}, 13.6–20.8479 μg g^{-1}, 14.8–21.5479 μg g^{-1}, 15.7–17.5479 μg g^{-1} and 41.9–71.9479 μg g^{-1}, respectively.[37] Lignin or arabinoxylans chains usually bind to phenolic acids in sorghum, in addition to cereals.[38] The hydrolysis of human digestive enzymes, decreasing their bioavailability in humans by these phenolic acids is not possible, but fermentation occurs in the colon by microbial inhabitation.[39] Several methods for improving sorghum phenolic acids' bioavailability are emerging, and the processing and microbial action can help improve this bioavailability. The collected results suggest that the fermentation using specific strains of probiotics, as well as processes of cooking, can significantly increase the phenolic acid content, leading to their improved availability.[40] Figures 5.6 and 5.7 show the phenolic acids and flavoinds contents, respectively, in red sorghum compared with white sorghum (mg kg^{-1}), according to Przybylska-Balcerek *et al.*[41]

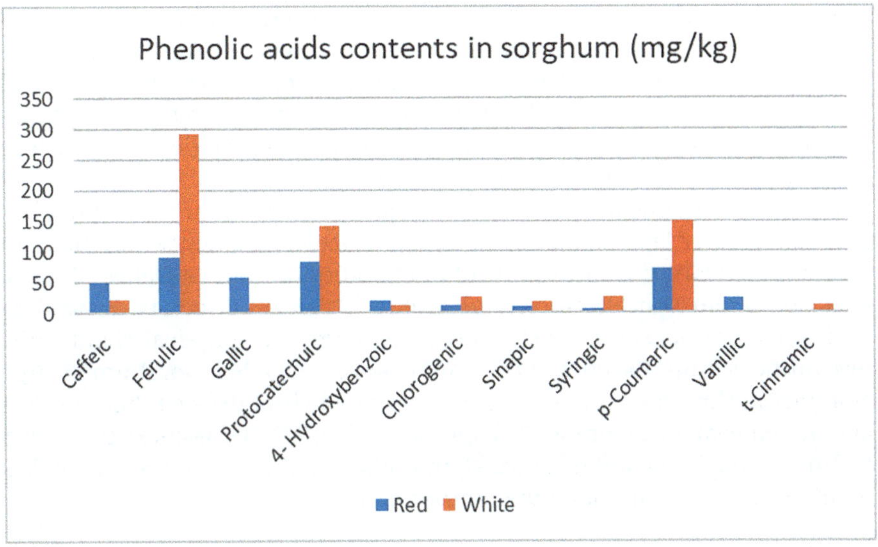

Figure 5.6 Phenolic acids content in red sorghum compared with white sorghum (mg kg^{-1}).[41]

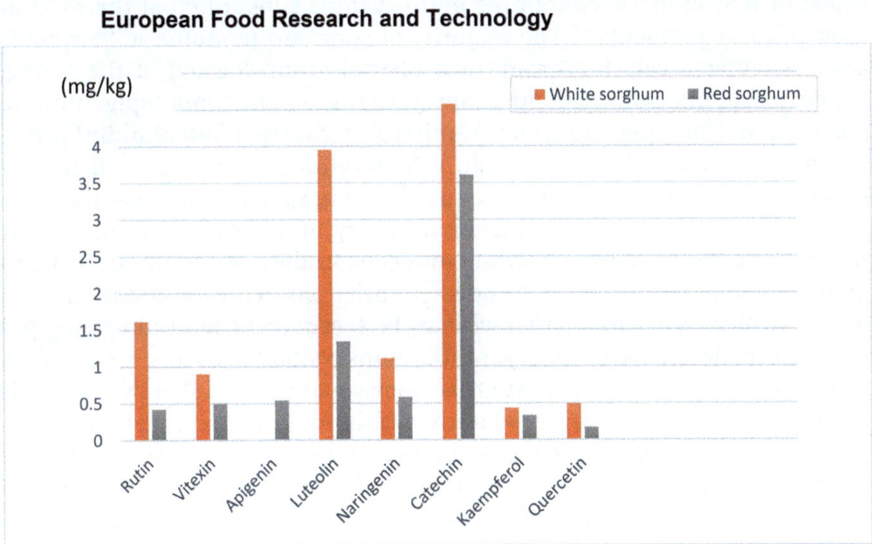

Figure 5.7 Flavonoid contents in red sorghum compared with white sorghum (mg kg^{-1}).[41]

5.3.1.2 Flavonoids

The importance of flavonoids cannot be overemphasized; they are crucial due to their distinct characteristics and, as a result of their significant influence on food's sensory attributes (color and flavor [bitterness], as well as astringency). The condensed tannin, a polymeric flavonoid, can also significantly impact the digestibility of macronutrients, or their bioavailability. Meaningful levels of monomeric, in addition to polymeric flavonoids are present in sorghum, as well as some less important millets such as pearl and teff millet.[42] Tannin levels as high as 33 mg g^{-1} are found in some sorghum, which may exist simultaneously with flavonoids (monomeric) levels such as 3-deoxyanthocyanins. Hence, in grains, flavonoids may play an important role in promoting health and other characteristics relevant to food quality. Sorghum's outer layer is inhabited by most flavonoids, hence variations in sorghum pericarp thickness and color, in addition to the presence of testa, affect flavonoid concentration and profile.[43,44] Genetic factors, as well as environmental ones, are a determinant of sorghum's physical characteristics. Anthocyanins, flavones and flavanones are available in sorghum in large amounts. Anthocyanins in sorghum are associated with 3-deoxyanthocyanidins and are equivalent to almost 79% of flavonoids.[45,46] The non-existence of a hydroxyl group at position carbon-3 differentiates their structure and makes them more consistent than other anthocyanins.[47]

5.3.1.2.1 Anthocyanins. Sorghum's stability is a function of its unsubstituted C-3. In addition, sorghum presents a distinct set of pigments called 3-deoxyanthocyanins, in contrast to other types of grains like red rice, purple wheat or purple maize containing anthocyanins substituted at their C-3 as their primary pigments.[47] The majority of sorghum monomeric flavonoids possess either a catechol group or a phenol group located at the B-ring; hence, in the phenylpropanoid pathway, they are naringenin (apigeninidin, apigenin and proapigenidin) or eriodictyol derivatives (luteolinidin, proluteolinidin, procyanidin and luteolin). In sorghum, there is less identification of some flavonoids (galloyl groups) found in specific plant commodities and cereals. Anthocyanin composition in sorghum kernels has continually revealed great variations, according to recent studies. For example, 381 sorghum genotypes were analyzed and large variations in the anthocyanins content were observed.[48] In another work by Petti *et al.*,[49] 13 sorghum genotypes were also analyzed and a 3-deoxyanthocyanins content of 68 μg g^{-1} was found. Aside from sorghum kernel, extensive analysis of the leaf sheath – which is colored and a potential source of anthocyanin for food and other uses – has been performed. Six (6) colored genotypes of the sorghum leaf sheath were analyzed for their anthocyanin contents, and values between 14 and 35 CGE (cyanidin-3-glucoside equivalents) mg g^{-1} (ds) were found.[50] Hence, materials from the sorghum leaf, which is usually handled as waste material, can be used for anthocyanin's extraction and further applications. Anthocyanins from the kernels and leaves of several sorghum types have been identified and quantified.[51] The deoxyanthocyanidins are the unique kinds of anthocyanins obtainable in sorghum, but not found in other types of cereals.[41] The pyrano-apigeninidin 4-vinylyphenol, having a symmetrical structure, was detected in sorghum leaf sheath.[52] Other types of anthocyanins, namely apigenin-7-0-methylflavene dimmer and apigenin-flavene dimmer arkek were found in the sorghum leaf sheath at levels of 2.8 mg g^{-1} and 7.7 mg g^{-1}, respectively. These showed great stability under several conditions, such as an extensive pH range.[51] The anthocyanins' composition in terms of genetic diversity has also been recorded: 13 genotypes of the sorghum kernels contained 5-methoxyluteolinidin (0–154 μg g^{-1}), luteolinidin (0–282 μg g^{-1}), 7-methoxyapigenidin (0–137 μg g^{-1}), together with apigeninidin (0–166 μg g^{-1}); 6 genotypes of the sorghum leaf sheath contained malvidin and luteolinidin, as well as apigeninidin, at levels of 0.6–1.0 mg g^{-1}, 0.4–2.4 mg g^{-1} and 17–46 mg g^{-1}, respectively.[50] The different kinds of anthocyanins present in sorghum are in agreement with results reported by Escribano-Bailón *et al.*[53]

5.3.1.2.2 Flavones and Flavanones. Flavones and flavanones are the major (monomeric) flavonoids present in the sorghum.[44,54] They present a similarity in the sharing of 4-quinone structures, whereas their structural difference is in their carbon-to-carbon bonding. Flavones possess a C2=C3 double bonding and, hence, are relatively planar and conjugated, while flavanones exhibit a chiral property at C as a result of the C2–C3 structure, which is saturated.[33] The structural difference is significant; experimental evidences suggest that a determining factor of the flavonoids power is the C2=C3 structure, which prevents various types of diseases.[55,56]

Flavones. Flavones possibly exist as a compound of biological importance among all phenolic compounds available in cereals. Zamora-Ros *et al.*[57] thought that this is because most of the commonly eaten plant foods, such as legumes, fruits, vegetables, as well as tubers, are deficient in flavones. The majority of cereal grains are composed of flavones (primarily as *C*-cylycosides of luteolin and apigenin) as the main flavonoid components.[33] A similar flavone level with wheat is exhibited by sorghum, which contains the highest flavone level. In comparison with 3-deoxyoanthocyanins, a large proportion of flavones in sorghum exist as glycosides. Glycosides dominate luteolin derivatives of sorghum, while glycosides dominate its apigenin derivatives.[56] There is a structural disparity of biological importance between sorghum flavones and that of other cereals. In cereals grain, flavones are mostly the *C*-glycosides, while sorghum flavones are primarily *O*-glycosides. In similar sorghum varieties, both apigenin and luteolin are usually available in meaningful quantities, while most other types of cereals mainly possess apigenin glycosides.[58]

Compared with flavones, flavanones are usually uncommon and present in meaningful quantities. Sorghum contains higher amounts of flavanones than other cereal grains: it presents approximately 1800 μg of flavanones in the case of some phenotypes possessing a lemon-yellow pericarp.[56] Only specific sorghum phenotypes possess flavanones, indicating an extreme genetic component in their accumulation.[33] Sorghum flavones mainly exist as *O*-glycosides, with their glycones having relative limitations.[56] The *O*-glycoside flavanones are responsive to low pH and readily hydrolyze in their aglycones, suggesting their easy accessibility in the upper gastrointestinal tract.[56]

5.3.1.3 Tannins (Proanthocyanidins)

Tannins are identified as secondary metabolites that are usually found in several species of plants. They are regarded as phenolic compounds that play roles as a defense mechanism against predators.[59] Tannins are not present in cereals like wheat, rice and maize, but are readily available in varieties of pigmented sorghum testa.[60] Sorghum tannins vary according to the type, content and individual oligomer and polymer distribution, and can be classified in Type I, Type II and Type III. Sorghum tannins are usually condensed and composed of catechins, oligomers or polymers.[42] Awika *et al.*[61] reported that sorghum tannins possess a high molecular weight, as well as a degree of substitution, whose value is greater than 10 (69–81%). The tannin content in sorghum is between 0.2 and 48 mg g^{-1}, while the sorghum supplied with black testa has the highest tannin content. Mkandawire *et al.*[62] also observed that the tannin content and bioactivity in the sorghum can be influenced by the cultivation season. In spite of their anti-nutritional outcome, tannins' radical-scavenging effectiveness is 15–30 times higher than that of simple phenolics,[63] hence an extensive study of tannins has been conducted to determine their health-promoting value.[64] The study ascribed sorghum's functional importance to oligomers. The anti-inflammatory, immunomodulatory, antioxidant, anticancer, anti-thrombotic, vasodilatory antiradical, anti-UV action and cardioprotective ability of tannins' oligomers have been reported to contribute almost 19% of tannin's antioxidant ability in food.[65] In sorghum, tannin's digestibility can be enhanced by processing. Depolymerization can be achieved by drying at 121 °C for 30 mins or 95 °C for 20 mins. Moreover, tannin's bioavailability can be increased by thermal treatments, which help to reduce the content of all these compounds, maintaining or even increasing the tannin-rich sorghum varieties' functional ability. Indeed, reducing the polymeric tannins in a grain may lead to an increase in its nutritional value, which is also associated with a higher protein and starch digestibility.

5.3.2 Peptides

When proteins are digested, they form peptides, which have specific biological activity.[66] Sorghum's use in food is limited because its proteins are not digestible. Therefore, studies involving the improvement of sorghum proteins' digestibility are emerging. Peptides with enhanced antioxidant activity are produced by enzymatic hydrolysis through *in vitro* studies. Proteins' hydrolysis is related to their antiviral and antihypertensive activities, and is linked to kafirin, the major storage protein present in sorghum. Thus, diets supplemented with sorghum karifin extracts exhibit an enhanced lipid metabolism and an increase in serum antioxidant potential. Through *in vivo* studies, it has been demonstrated that peptides produced from kafirin are associated with improved antioxidants and serum lipid activity. Karifin extracted from sorghum could be utilized in the prevention of atherosclerosis, as well as other generated diseases.[67]

5.3.3 Other Bioactive Ingredients

Carotenoids are another group of bioactive compounds in sorghum. They protect the plant's cells against damage from reactive oxygen species (ROS) by inhibiting free radical activity. Xanthophylls are a group of different carotenoids in sorghum, and include lutein and zeaxanthin, characterized by yellowish-orange color, being products of the β-carotene metabolic pathway. Stilbenes also belong to a family of phenolic compounds originating from the phenylpropanoid pathway, with several implications for plant resistance against diseases, as well as for human health.[68] A correlation exists between the grain's color and the stilbenes total content: traces of *trans*-piceid are present in white sorghum varieties without *trans*-resveratrol, while both classes are available in red sorghum. Other bioactive ingredients include polycosanols and phytosterols, and are related to the sorghum lipid fraction having physiological benefits.[69]

5.4 Functional and Health Properties of Pigmented Sorghum

5.4.1 Health Benefits of Pigmented Sorghum's Bioactive Compounds

Several studies have shown that sorghum and sorghum components have a distinctive health-protective ability when consumed, suggesting that sorghum possesses bioactive components that are beneficial to health.[4] Similar outcomes have also been detected after consuming plant foods such as fruits and vegetables, and this suggests sorghum possesses bioactive components that can be beneficial to health. Sorghum's ability to prevent chronic diseases can be largely ascribed to its polyphenol impact and endosperm's slow digestibility. Both *in vitro* and animal studies demonstrated that the fat-soluble compounds or phenolics isolated from sorghum benefit the modulation of its microbial community, and also have potential to protect against non-communicable diseases, like reduction of chronic inflammation, as well as oxidative stress, prevention of certain kinds of cancer, diabetes, hypertension and obesity, reduction in insulin resistance, improvement of glucose and lipid metabolism, suppression of fat accumulation, and improvement of colon health, as a result of gut microbiota modulation. Factors that may influence the desired results, such as food composition, degree of processing, dose and exposure time, are necessary for exploration, as these parameters have an impact on manufacturing, as well as on the health benefits.[70]

5.4.1.1 Anticancer Action

Sorghum's anticancer properties have been demonstrated against different types of cancer.[71] Phenolic compounds, mostly directly the estrogenic 3-deoxyanthocyanidins, act against cancer cells, promoting their apoptosis and growth inhibition, as well as against metastases of the skin melanoma, esophagus, colon, breast, bone marrow and liver.[72] In addition, sorghum tannins may possess anticancer activity, as several studies have revealed that tannins extracted from foods affect the regulatory enzymes that block the pathways of signal transduction and induce apoptosis. For this reason, tannins have attracted a lot of attention from the cancer treatment field. *In vivo* studies concerning pigmented sorghum's anticancer effects have also been performed.[73]

5.4.1.2 Reduction in Oxidative Stress

Chronic and extreme free radical production is important in the growth of non-contagious diseases.[74] There is a stronger relationship between sorghum consumption and decline in oxidative stress. *In vitro* studies have demonstrated the antioxidative stress activities of components isolated from sorghum. These benefits are characteristics of sorghum phenolic compounds, which are mostly observed in red or black sorghum.[75] Indeed, they ae able to monitor the expression of enzymes (phase II) involved in the defense system against oxidative stress, by the continuous action of converting highly reactive electrophilic species to harmless and excretable metabolites.[76] Black sorghum varieties, due to their high 3-deoxyanthocyanidins and their rich profile, may exert a vigorous action on NADH (1,4-dihydronicotinamide adenine dinucleotide): quinone oxidoreductase.[76] The influence of tannin-free sorghum on oxidative stress has also been demonstrated.[70]

5.4.1.3 Anti-obesity and Anti-inflammation Properties

Obesity diffusion has generated interest in identifying alternative healthy and dietary carbohydrates as functional ingredients in order to control blood glucose and insulin levels, among these alternatives, sorghum is a great option due to its slower digestion compared to other cereals. It has been found that tannin-rich sorghum reduces weight gain in some animals such as pigs, rats, poultry and rabbits.[77] This reduction in weight gain is objectionable in animals that are slaughtered, but it is highly beneficial against obesity in humans. Indeed, sorghum's polymeric tannins can naturally modify the starch and form resistant starch by strongly interacting with amylose.[78] Mkandawire *et al.*[62] further reported that sorghum tannins could result in starch digestion inhibition due to saccharase and amylase enzyme inhibition. Furthermore, there is a correlation between obesity and chronic inflammation,[79] since the adipocytes play a vital role as inflammation mediators.

5.4.1.4 Anti-diabetes

It has been reported that sorghum may contain resistant starch up to 65%, depending on the variety, with brown sorghum having the highest amount.[80] Major studies on sorghum-resistant starch have concentrated on its content and processing effect. Shen *et al.*[81] suggested that consuming isolated sorghum-resistant starch regulated the colon microbiota, increased *Bifidobacteria* and *Lactobacillus*, and decreased *Escherichia coli*. A reduction in blood triacylglycerol was observed after the intake of sorghum-resistant starch, suggesting its ability to prevent and treat diabetes. Pelpolage *et al.*[82] demonstrated that resistant starch derived from whole sorghum grain induces the production of short-chain fatty acids during colon fermentation, suggesting that it may play a role as a prebiotic.

5.4.1.5 Dyslipidemia

In vitro studies have demonstrated that sorghum phenolic fractions and lipids regulate the parameters connected to dyslipidemia and the adverse event of cardiovascular diseases. In particular, sorghum lipid fraction may result in the gut microbiota alteration affecting the cholesterol absorption, due to the presence of phytosterol as a significant bioactive component in sorghum lipid fraction, which can restrain cholesterol absorption. In addition to the phytosterol, polycosanols ND phenolic compounds may regulate the absorption excretion, as well as the cholesterol synthesis.[83]

5.4.1.6 Gut Microbiota

The human gut is populated by a collection of bacteria species that develop vital immune and metabolic functions, having an effect on the host's health and nutritional status. The functional advantages of phenolic components present in food on human health may arise from the direct activity of bioactive components, which are not absorbed but modify the microbial environment.[84] Recent studies suggest that phenolic compounds and their metabolites improved gut health maintenance by modulating the microbial balance through stimulating the growth of useful bacteria as well as preventing pathogenic bacteria which exert prebiotic-like outcomes.[85] Among these phenolics, tannins play a pivotal role due to their abundant nature and because, despite their low-absorption ability, they undergo metabolism by the colon microbiota.[86] Sorghum also possesses dietary fibers and resistant starch, which can alter the microbial community of the gut.[87]

5.4.1.7 Satiety, Blood Glucose, and Metabolism

A positive association between sorghum and satiety has also been reported, showing that red, white, and brown/tannin sorghum varieties resulted in higher satiety than a whole-wheat biscuit. Higher glucose insulinotropic peptide and postprandial glucagon-like peptide-1 contents were evidence of this, as was the contribution of sorghum endosperm to improved glucose metabolism. This group of researchers also reported a positive association with weight loss, insulin, fasting glucose, and cholesterol, when consumed by flaked biscuits made from red sorghum varieties.[88] Food type and the extent of processing it has undergone may contribute to the effectiveness of insulin and blood glucose attenuation, which are observed after consuming sorghum-derived products. Indeed, in the case of sorghum, which is consumed in traditional Indian foods such as porridge, roti, flatbread, and dhokla; and in the Western diet in items such as biscuits, cakes, muffins, and pasta, a reduction in blood glucose has been shown.[70] This implies that sorghum's array of nutrients is active, even after processing of the grains. The presence of starch, and its digestibility may cause this desirable glucose response. Indeed, previous *in vitro* research has shown a decrease in starch digestibility of bread made from sorghum,[89] and an inverse relationship between pasta's sorghum content and starch digestibility.[90] The levels of resistant and slowly digestible starch were greater in muffins made from sorghum than those made from wheat, and this may likely be a contributing factor to the reduction in blood glucose and insulin responses.[91]

5.4.1.8 Sorghum's Ability to Reduce the Glycemic Index

Glycemic response reduction and hypoglycemic effect due to a reduction in the α-amylase activity have been associated with sorghum.[70] Reduction in sorghum's glycemic response is not dependent on its phenolic profile, but it is a result of sorghum endosperm. Several mechanisms have been reported to be responsible for this property, but may be ascribed to sorghum proteins' ability to cross-link as well as to interact with the starch during food processing (hydrothermal). In their study, Cisse *et al.*[92] observed that millet foods and traditional sorghum possessed a gastric emptying rate about twice as long as that of non-traditional foods. This indicates that millet and sorghum have the potential to contribute to a glycemic response reduction, as well as to improve satiety, in contrast with other starchy foods. Gulati *et al.*[93] highlighted that the contributing factor to this gastric emptying process on pearl wheat and sorghum needs to be investigated. They suggested a partial relationship of this phenomenon with the disposition of some proteins in millet and sorghum to extensive cross-linking during processing using hydrothermal process to be less digestible.

5.4.2 Antioxidant Potential

Sorghum polyphenols are responsible for protective action against diseases and insects, thereby acting as antioxidants.[94] The action of antioxidants commonly leads to a beneficial effect on plants' physiological functions.[95] Sorghum is abundant in phenolic compounds like phenolic acids (*p*-coumaric and ferulic acids), flavones, flavonoids, 3-deoxyanthocyanidins, and polymeric compounds such as hydrolyzed and condensed tannins.[96] Due to genetic diversity, some sorghum varieties containing high phenolic and tannin content display high antioxidant activity.[97] These compounds are important antioxidants and are present in higher amounts than antioxidants found in grapes, strawberries, and blueberries.[94] The antioxidant activity may assist in the radical reaction disruption through electrons or hydrogen atoms transfer to radicals, promoting the formation of more stable compounds, and lowering the adverse events of several diseases that are associated with the action of free radicals, such as atherosclerosis, cataracts, cancer, inflammatory bowel diseases and rheumatoid arthritis.[98] In addition to its ability to reduce free radicals, sorghum also has antiviral, antibacterial, and antifungal abilities.[99] The varying potential ability of sorghum has led to research into the isolation and characterization of its phenolic compounds to ascertain the sources of compounds exhibiting unique characteristics. Awika *et al.*[94] reported that the bran of sumac sorghum and high tannins of sorghum possess a higher value of oxygen radical scavenging capacity than common fruits. Hence, these types of sorghum exhibit good antioxidant properties, and have the potential to be used in the nutraceutical and functional food industries. Dykes *et al.*[95] analyzed sorghum's antioxidant ability for condensed tannins, total polyphenols, anthocyanin, and flavan-4-ols, taking into account that the phenol content is affected by cultivars. They concluded that the dominant sorghum B_1, B_2 genes exhibited the highest antioxidant activity, especially the S gene. This antioxidant activity was mainly sourced from condensed tannins, which have shown a higher *in vitro* activity than other phenolic compounds.[100] The sorghum from red/purple plants exhibited a higher antioxidant activity than those with thin pericarp, confirming that this majorly comes from the sorghum pericarp, which is usually abundant in polyphenols.[101] For centuries, tannin-rich sorghum has been consumed as food by humans (*e.g.* alcoholic beverages, bread, and porridges) in Asia and Africa, and several varieties are favored for certain types of products. Rudiger[102] reported that bread high in tannins with a natural brown color, an entire grain taste, and high dietary and antioxidant levels can be comparable to commercial bread. Sorghum with tannins has antioxidant potential in stored and precooked meat patties,[103] and can be utilized to retard oxidative damage due to high energy irradiation.[104] Hence, tannin-rich sorghum needs to be regarded as an abundant source of dietary fiber, natural color, and antioxidants. Furthermore, phenols – as well as having antioxidant activity – are highest in sorghum varieties with a dark red or black, thick pericarp, a secondary red/purple plant color, as well as in sorghum varieties possessing a pigmented testa, having a spreader gene (S). This discovery provides beneficial guidelines for production of sorghum with the highest antioxidant levels, which can be used to create healthy foods.[8]

5.4.3 Antimicrobial Activity

Sorghum's antimicrobial activity has been associated with its phenolic content. Sorghum phenolic extracts consist of a combination of plant-based compounds and a very powerful natural antimicrobial replacement.[105] Dykes *et al.*[95] correlated the high phenolic compound concentration with greater antimicrobial activity. A report showed that the phenolic extracts from a sorghum variety exhibited minimum inhibitory concentrations for *Bacillus subtilis*, *Streptococcus aureus*, *Escherichia coli*, *Klebsiella pneumonia* and *Salmonella typhimurium* of 250, 250, 250, 500, and 500 μg mL, respectively.[106] Furthermore, the antimicrobial activity of phenolics isolated from soy flour was demonstrated by Kil *et al.*[106] against *Listeria monocytogenes, Enterococcus faecalis*, *Bacillus cereus*, *Listeria innocua* and *Staphylococcus aureus*.

5.5 Recent Updates in Pigmented Sorghum Utility in the Food Industry

5.5.1 Traditional Processing of Food Products

The traditional processing of pigmented sorghum in food products generally involves soaking, fermentation, and cooking methods like steaming, boiling in water and roasting. Table 5.2 presents some common traditional sorghum-based products and their production regions.[10] Fermented tea was prepared from the red hybrid sorghum cultivar through four steps of soaking, fermentation, steaming, and roasting, with active cultures of *Lactobacillus plantarum* subsp. *Argentoratensis* and *Saccharomyces cerevisiae*.[107] Reductions in the condensed tannin, total phenolic, and total flavonoid contents were observed during the soaking, steaming, and fermentation stages, while increases were observed with roasting. There were also increases in esters during soaking, steaming, and fermentation, and decreases during roasting. Overall, the processing affected ketones, carboxylic acids, alkanes, and alcohols, changed the red sorghum grain volatile profile, and *Saccharomyces cerevisiae* presented the higher possibility for producing sorghum grain tea. A longer fermentation period led to increased bioactive contents of fermented dough (*Ajin*) from high-tannin (*Wad Ahmed*) and low-tannin (*Tabat*) sorghum cultivars during the preparation of Sudanese fermented flatbread (*kisra*). Baking further improved the antioxidant activity, and *kisra* prepared from *Wad Ahmed* cultivar displayed lower total flavonoid and phenol contents and antioxidant activity compared with that prepared from the *Tabat* cultivar.[108]

The fermentation parameters for *ting*, a fermented product used in producing Southern African gruel (*motogo*) or porridge (*bogobe*) from dark brown whole sorghum, were optimized.[109] Optimal fermentation conditions of time (72 h) and temperature (28 °C) reduced the pH value to 5.9, with high levels of total phenols (46.1 mg GAE (gallic acid equivalent) per g), titratable acidity (1.1 g kg^{-1}), tannin (14.1 mg CE (catechin equivalent) per g), total flavonoid (40.9 mg CE per g), antioxidant activity (3.7 μM TE (trolox equivalent) per g), and comparable levels of quercetin (0.61 μg g^{-1}), gallic acid (0.75 μg g^{-1}) and catechin (12.4 μg g^{-1}), with respect to *ting* samples from refined grains. The nutritionally balanced fermented Southern African non-alcoholic beverage (*mahewu*) was developed from compositing sorghum with different

Table 5.2 Some traditional sorghum-based products and regions of production. Reproduced from ref. 10 with permission from Elsevier, Copyright 2021.

Product type	Characteristics	Region of production
Bread		
Sourdough	Fermentation of dough with naturally occurring yeast and *lactobacilli*	African countries
Kisra	Gluten-free fermented flatbread with high biological activity and nutritional value	Algeria, Chad, Saudi Arabia, Sudan
Khamir	Gluten-free yeasted flat bread	Arabian countries
Porridge		
Ogi-baba	Probiotic, gluten-free, smooth, creamy free-flowing thin porridge with variable bioactive and nutritional content	West African countries
Fermented beverage		
Pito	Light yellow to dark brown and slightly sweet to very sour alcoholic drink obtained through lactic acid bacteria and yeast fermentation; good source of folic acid, thiamine, nicotinic acid and riboflavin	Ghana, Nigeria and Togo
Bouza	Light-yellow, thick and sour alcoholic drink and good source of bioactive compounds, amino acids, vitamin C with high biological activity	Egypt, Turkey and some Eastern European countries
Kunun-zaki	Non-alcoholic drink with high biological activity and bioactive compounds, characterized by cheap preparation and easy accessibility	Northern part of Nigeria
Mahewu	Non-alcoholic sour drink with high biological activity and good source of bioactive compounds prepared with *Lactococcus lactis*	African and some Arabian Gulf countries
Sorghum grain tea	Non-alcoholic tea with high biological activity and good source of bioactive compounds	China

proportions of bambara groundnut (BMG), and it was shown to significantly reduce *mahewu* phytates content by more than 90%.[110] BMG supplementation enhanced the contents of essential amino acids like phenylalanine, histidine, threonine, isoleucine, methionine, lysine, and leucine. In particular, an 80 : 20 sorghum:BMG blend presented a lower final pH value after 24 h when compared with that for 100% sorghum (4.3 *vs.* 5.1, respectively) with significantly higher titratable acidity (0.7%), protein (7.7%), and fiber (7.9%) contents, in comparison with values of 0.4, 6.1, and 5.7% for 100% sorghum, respectively. Similarly, a comparison between them constituted *kunun-zaki*, a non-alcoholic fermented sorghum beverage from Northern Nigeria, and fresh samples were prepared. No significant differences in ash, protein, carbohydrate, and fiber contents were evidenced but, on the other hand, significant variation in pH value was revealed (3.61 *vs.* 2.03 for the reconstituted and fresh *kunun-zaki*, samples, respectively), suggesting higher hydrogen ion concentration in the fresh samples.[111]

5.5.2 Challenges of Sorghum as a Modern Food Ingredient

One of the major challenges facing sorghum as a modern food ingredient is its nutrient bioavailability after processing. Although sorghum contains a rich array of phytochemicals, its consumption in the wet-cooked form is often accompanied by weight loss, probably caused by its low protein availability, due to the proteins cross-linking.[112] Low levels of essential amino acids such as isoleucine, methionine, and lysine upon cooking, in addition to anti-nutritional factors such as trypsin inhibitors, phytic acid and tannins, inhibit the nutrients' bioavailability in the sorghum. Processing techniques like wet-cooking and dry heating reduce sorghum protein's digestibility to a large extent, leading to consumption of high-energy cereals such as wheat and rice.[112,113] Other processing methods including fermentation, extrusion, malting, and dry roasting may also influence sorghum's protein digestibility, but to a lesser degree.[113] Thus, an insight into appropriate methods for achieving improved nutritional benefits of sorghum grains by increasing carbohydrate and protein content, while decreasing fat content through processing, was presented.[112] The results showed a 58% decrease in anti-nutritional factors like trypsin inhibitors, a 28% decrease in total phenolics, a 30% decrease in tannins and a 6-fold increase in antioxidant activities upon fermentation. The predicted protein efficiency and essential amino acid to total amino acid ratios were also high during fermentation, and the biological value of sorghum amino acids increased by 2.9-fold when fermented grains were steamed prior to flaking.[112]

The potential of thermal procedures to modify sorghum compounds' bioaccessibility and enhance their concentration, while reducing anti-nutritional tannins and improving antioxidant capacity has also been reported.[114] Nixtamalization (alkaline cooking) proved to be more effective for reducing tannins (74.3%), even though cooking was more effective for increasing phenolic compounds in red sorghum. Both thermal treatments increased intestinal absorption and gastric bioaccessibility of flavonoids and phenolic compounds, with gallic acid being the most bioaccessible phenolic compound at 6359 $\mu g\ g^{-1}$. Likewise, extruding sorghum flour prior to incorporation was effective for enhancing the sorghum-wheat composite dough quality.[115] The incorporation of gluten-free sorghum is often seen as having a negative impact on the wheat dough quality during bread production. Extruding at 110–160 °C, 10–18% feed moisture, and incorporating 10% of the extruded flour effectively increased dough development time and sorghum-wheat composite water absorption, but reduced dough stability through starch granules and partial gelatinization. The viscoelastic solid-like structures from sorghum extrusion, in contrast to the viscoelastic liquid-like form of the native sorghum-wheat composite, indicated the potential of dough containing extruded sorghum flour for producing high-quality bread from less compact non-uniform dough structures, with reduced viscosity and elasticity.

Information on the nutrient bioavailability challenges associated with sorghum is limited to some sorghum genotypes, and mostly to laboratory studies. Therefore, scaled-up studies encompassing a large number of sorghum cultivars, their derivatives and food products are necessary for providing all-inclusive data regarding sorghum protein digestibility. Such a database may be crucial for the effective selection and breeding for broad applications in food systems and other industrial uses. However, scaled-up studies from laboratory-scale and traditional mills to industrial (wheat) milling system with numerous passages are challenging. Most industrial milling systems are developed for rye and wheat milling, and separating the milled sorghum into different fractions requires a process with several breaks, reduction, and purification passages. Nevertheless, a pilot scale roller and stone mill designed for wheat was successfully adapted to produce physically and chemically distinct bran and flour fractions from sorghum kernels. The results showed that the fractions with increased bran material contained less starch and higher amounts of protein ash, fat, total phenolic content, and dietary fiber, and presented higher solubility and water absorption indices.[116] Similar successes with conventional wheat pasta facilities were also reported for the industrial production of gluten-free sorghum pasta.[117] Together, these works presented promising solutions to the lack of processing technologies specific to sorghum grain.[116,117]

5.5.3 Potential Use of Pigmented Sorghum in Modern Food Systems

Several studies have begun to present irrefutable opportunities for diversifying and valorising pigmented sorghum as a potential ingredient in food systems, through high-temperature short-time extrusion or flaking process to produce either crunchy or crispy snacks and cereals. The majority of these products are the results of replacing sorghum in part or whole with wheat, or incorporating it as an ingredient. For example, red, brown, and white sorghum have great potential for producing flaked breakfast cereal.[118] The *in vitro* mineral availability in produced breakfast cereals produced was higher than in the raw flour, due to the concomitant reductions in tannin and phytate levels after processing; but were lower when compared with flaked wheat breakfast cereals. In another study, polyphenols from pigmented sorghum flours were evaluated as possible starch digestibility modulators and were seen as good candidates for functional foods formulation.[119] The study showed that raw pigmented flours were characterized by higher kafirin and tannin content than white flours, and the phenolic profile alterations during *in vitro* digestion indicated that enzyme activities were modulated by flavanol and anthocyanin equivalents. In addition, the *in vitro* starch digestibility of the pigmented flours after cooking presented lower rapidly digestible starch amounts (19.4–30.4 g 100 g dry matter) and starch hydrolysis index (55–70), in addition to higher slowly digestible starch (34.1–40.8 g 100 g dry matter) and resistant starch (4.2–21.4 g 100 g dry matter) contents when compared with white sorghum flours.

It has been shown that changing the kneading time and speed and resting the dough makes it possible to produce shortbread biscuits with low technological properties from 100% sorghum.[120] However, 70% replacement of wheat flour with sorghum flour presented better acceptability with significantly higher intakes of total energy (454.1 kcal), dietary fibers (2.15 g) and carbohydrates (58.51 g). Cookies of acceptable chemical and physical composition and comparable to cookies made from 100% wheat flour could also be produced from sorghum-wheat flour composite.[121] Substituting wheat flour with 5–50% sorghum flour significantly increased crude fiber, moisture, fat, ash, and protein contents, but decreased starch and sugar contents. The calorific value of the cookies reduced from 489 to 421 kcal 100 g, suggesting the promising potential of sorghum-wheat blend cookies for alleviating the heavy demands of importing wheat by developing countries. Whole-sorghum and whole-wheat breakfast cereals were compared in a study, showing that the former had better sensory acceptance (70.6%) when compared with wheat breakfast cereals (41.18%). The contents of 3-deoxyanthocyanidin, total phenolic compounds, and antioxidant activity were 100, 98.2, and 87.9% higher in sorghum cereals, while vitamin E content was 78.6% higher in wheat cereals.[122] The addition of 30% red sorghum flour to Chinese steam bread formulations significantly increased the polyphenolic content and antioxidant activity, while compositing red sorghum with the cream-colored cowpea remarkably incremented the total flavonoid and radical scavenging activities of the sorghum-cowpea composite porridge, with promising potential for alleviating radical induced oxidative stress in comparison with the maize-soybean composite porridge.[123,124]

5.5.4 Pigmented Sorghum for the Development of Gluten-free Foods

Sorghum is considered suitable for consumption by celiacs, since it does not contain gluten proteins, and therefore is a recommended ingredient for gluten-free foods with low glycemic index.[125,126] Accordingly, several studies have begun to show that pigmented sorghum has potential for the development of gluten-free foods. For instance, gluten-free pasta produced from brown and white sorghum flours demonstrated improved nutritional value and resulted in promising alternatives to commercially available pasta.[125] The sorghum flour-based pasta displayed high protein levels (≈170 g kg^{-1}), polyphenols (2.6 g GA kg^{-1}), dietary fibers (≈80 g kg^{-1}), and antioxidant activity, in addition to low starch *in vitro* digestion, moderate glycemic index and a high level of protein hydrolysis (76%) when compared with other gluten-free pasta. Bio-accessible polyphenol contents were 2.4- and 2.9-fold higher, respectively, for brown and white sorghum gluten-free pasta, when compared with pasta from corn and rice. Furthermore, brown sorghum flour was demonstrated to be superior to other genotypes and classified as a low glycemic index ingredient for preparing gluten-free bread with enhanced postprandial glycemic responses in healthy adult male subjects, due to brown sorghum's significantly higher fiber content and antioxidant activity.[127]

Proteins, soluble starch and carbohydrates were observed as the drivers and key elements in determining the pleasant taste of gluten-free bread produced from brown and red sorghum flours.[128] The negative effects of fibers, tannins, and phenolics on the acceptability of the bread were also highlighted. While brown sorghum displayed higher condensed tannin (609.9 mg per 100 g) and total phenolic (1493 mg per 100 g) contents and attended to the appeal of polyphenolic-enriched foods, red sorghum presented lower condensed tannin (21.4 mg per 100 g) and total phenolic (218 mg per 100 g) contents with higher acceptance. Results from a pilot-scale study showed that gluten-free sorghum pasta could be produced on an industrial scale using conventional wheat pasta facilities, with better cooking properties with respect to that produced from the laboratory and commercially available pasta from rice.[117] In this study, it was reported that distinct ingredients influenced the quality of gluten-free pasta prepared from brown and white sorghum flours differently, ranging from increases in firmness and chewiness, material loss reduction during cooking, luminosity improvements to the swelling index minimization. The variations were also dependent on the flour type, considering that white sorghum presented better acceptability than brown sorghum.

In some instances, it is necessary to modify flour's properties to achieve improved results. In this vein, the milling, pearling, and sifting processes were employed for modifying brown and white sorghum flour composition and properties, such as the dimension of their particles, since reduced particle size leads to improved pasting properties.[126] Smaller particle-sized flours with improved pasting ability thus led to production of highly viscous and dense batters, with small homogenously distributed air bubbles (Figure 5.8). Irrespective of the utilized sorghum variety, small particle-sized flours produced gluten-free sponge cakes with low firmness and high volume (Figure 5.9). However, only flour with 90% of the particle population below 211.5 μm in diameter could produce sponge cake with acceptable quality devoid of collapsed crumb.[126]

5.5.5 Pharmaceutical and Other Applications of Pigmented Sorghum

Research on pigmented sorghum is increasingly enhancing their usability beyond the traditional markets, to encompass pharmaceuticals and animal feed production. The biological activities of pigmented sorghum and its derivatives for protecting and modulating chronic diseases are linked with low glycemic index, lipid, dietary fiber and bioactive compound contents.[10] Epidemiological studies show that consumption of pasta containing whole red or white sorghum flour can enhance the plasma

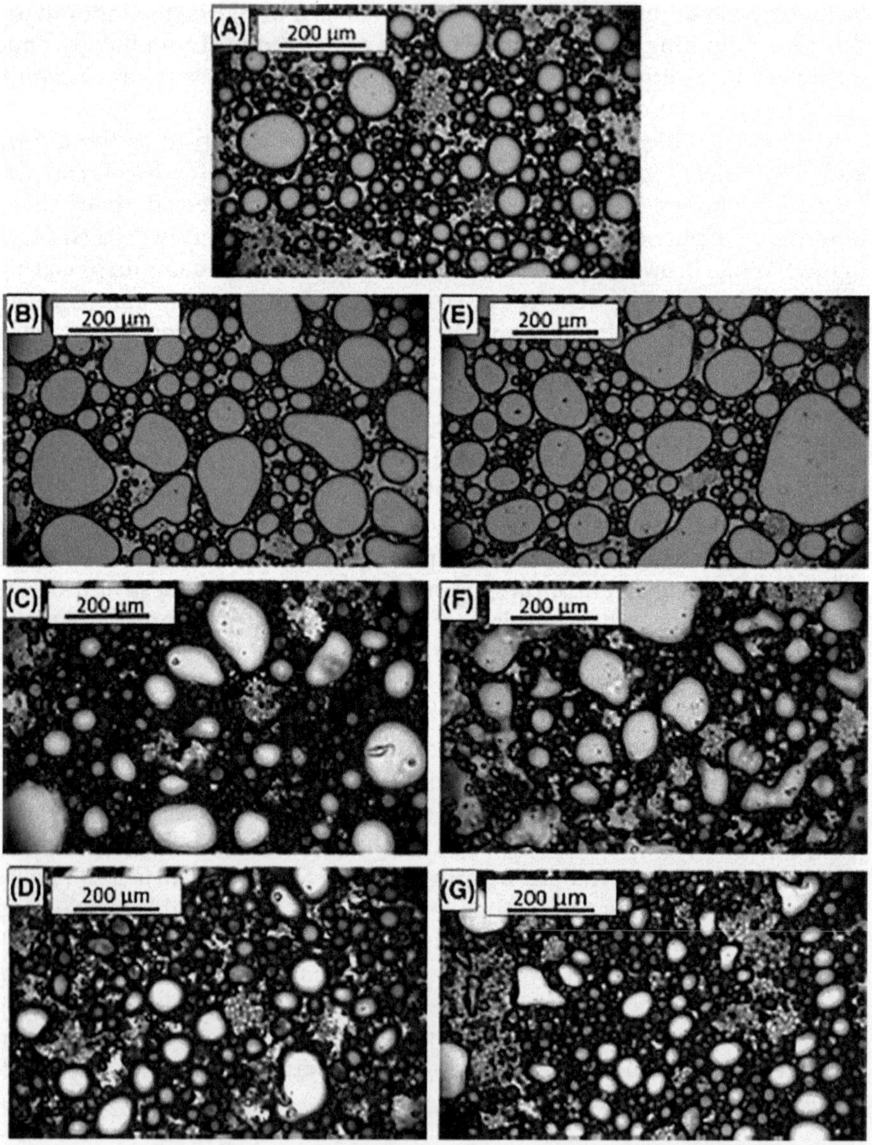

Figure 5.8 Micrographs of batter during gluten-free sponge cake preparation from: (A) pearled white sorghum flour; (B) high speed milled pearled white sorghum flour; (C) 80 µm sieved pearled white sorghum flour; (D) pearled brown sorghum flour; (E) high-speed milled pearled brown sorghum flour; (F) 80 µm sieved pearled brown sorghum flour; (G) whole white sorghum flour. Reproduced from ref. 126 with permission from Springer Nature, Copyright 2021.

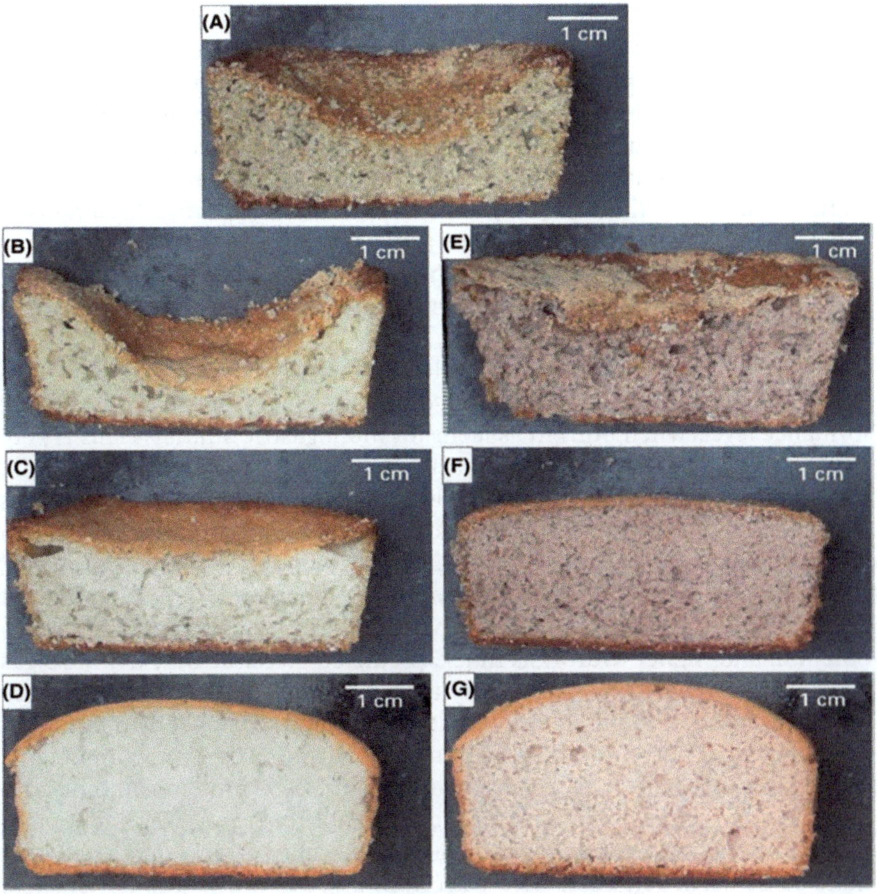

Figure 5.9 Gluten-free sponge cake prepared from: (A) pearled white sorghum flour; (B) high-speed milled pearled white sorghum flour; (C) 80 μm sieved pearled white sorghum flour; (D) pearled brown sorghum flour; (E) high-speed milled pearled brown sorghum flour; (F) 80 μm sieved pearled brown sorghum flour; (G) whole white sorghum flour. Reproduced from ref. 126 with permission from Springer Nature, Copyright 2021.

polyphenols, antioxidant status and markers of oxidative stress in the healthy subjects.[129] Significant increases in hematocrit and hemoglobin levels were detected in anemic women when sorghum extract was added to routine hematinics; compared with the use of hematinics individually.[130] The use of hematinics alone only produced a significant increase in hemoglobin content, and changes in creatinine, liver enzyme, and urea were within the normal range.

Moreover, sorghum's anticancer action was demonstrated. In particular, 3-deoxyanthocyanidins (concentration 200 μM) extracted from black sorghum were able to reduce the viability of human hepatocarcinoma and human leukemia cells by 50 and 90%, respectively, suggesting its cytotoxicity against human cancer cells.[131] Sorghum is a key source of 3-deoxy-anthocyanidins as phytoalexins and seed pigments for countering the microbial attack. The consumption of phenolic compounds from toasted brown sorghum flour by male hypertensive rats was associated with lower superoxide dismutase activity, reduced alanine aminotransferase concentration, improved lipid profile, and increased lipid profile high-density lipoprotein.[132] These experimental findings were ascribed to the elimination of reactive oxygen species, making brown sorghum flour a promising ingredient for the development of sorghum-based functional foods and pharmaceutical products.

Another important industrial application of pigmented sorghum is in animal feed production, even if the sorghum grain is linked with low performance levels when utilized as feeds for monogastric animals. This is generally attributed to the poor amino acid profile and low solubility of kafirin, which inhibit starch enzymatic hydrolysis, and tannins that form complexes with endogenous or digestive enzymes and food proteins to cause low protein and starch digestibility.[133] In this vein, the combination of soaking, grinding, rehydration, and ensiling was employed for reducing the tannin content of red sorghum grains, and these treatments led to increased starch digestion potential and higher hydrolysis index, predicted glycemic index, *in vitro* gas production rate and digestion of dry matter during simulated gastrointestinal tract digestion in pig. Another study concluded that specific non-tannin phenolics were harmful rather than innocuous grain sorghum components in the perspective of chicken-meat production.[134] The protein solubility was positively correlated with apparent metabolizable energy (AME) and metabolizable energy to gross energy ratio (ME : GE), while certain conjugated phenolic acids negatively correlated with AME and ME : GE in broiler chicken. The birds were fed unprocessed sorghum-casein mash diets containing 130 g kg^{-1} casein and 809 g kg^{-1} sorghum from six hammer-milled red sorghum varieties to compare protein digestibility coefficients and nutrient utilization.[134]

5.6 Conclusion and Perspectives

Pigmented sorghum and its products have recently gained admiration as a natural functional food, globally, due to its unique phytochemical profile and bioactive diversity. Pigmented sorghum is rich in all the essential nutrient bioactive molecules such as peptides, polyphenols, flavoinds, tannins, carotenoids, anthocyanins, and other bioactive colorants. These components provide a powerful bioactive role, including the capability to reduce the glycemic index, improve lipid and glucose metabolism, and insulin sensitivity, anticancer, antimicrobial, antioxidant, antitumor, antidiabetic, anti-inflammatory, antiobesity, and cardioprotective activities aimed at providing health benefits for humans against numerous chronic diseases, in comparison with other common cereals. Recent investigations regarding pigmented sorghum's exclusive functional profile, characterized, for example, by the presence of the 3-deoxyanthocyanins and proanthocyanidins, have led to the utilization of these bioactive constituents in food systems to enhance the end products' qualities and health potentials, as well as in the pharmacological industry. Research on pigmented sorghum is increasingly enhancing its usability beyond the traditional markets, and such utilization includes pharmaceuticals and animal feed production. The both available *in vivo* and *in vitro* data recommend that pigmented sorghum's colorants profile mainly impresses the specific physiological responses relevant to disease prevention.

Our data stated that several sorghum applications have been accomplished in the current food market, *e.g.* novel baked good products, cereals breakfast, fermented foods, beverages, and gluten-free foods. One of the major challenges facing sorghum as a modern food ingredient is its nutrient bioavailability after processing. Thus, an insight into approaches for achieving sorghum improved nutritional and techno-functional levels was addressed *i.e.* biotechnology and fermentation. Therefore, recent updates about the innovations associated with the need to tackle the technically challenging aspects of pigmented sorghum as a novel ingredient in the new food markets could be beneficial to researchers involved in the food sector and plant-based food industries. Further studies concerning improving sorghum grain components' functionality and value are needed. However, pigmented sorghum has diverse biological activities; therefore, further biological investigations, including clinical trials, are recommended to explore the its health-promoting effects. According to the above insights, pigmented sorghum and its products can be consumed all around the world as a novel functional food.

References

1. J. R. N. Taylor, in *Sorghum and Millets*, Elsevier, 2019, pp. 1–21.
2. A. Dabija, M. E. Ciocan, A. Chetrariu and G. G. Codin, *Appl. Sci.*, 2021, **11**, 3139.
3. C. R. Abah, C. N. Ishiwu, J. E. Obiegbuna and A. A. Oladejo, *Eur. J. Nutr. Food Saf.*, 2020, 101–111.
4. J. M. Awika, in *Gluten-Free Ancient Grains: Cereals, Pseudocereals, and Legumes: Sustainable, Nutritious, and Health-Promoting Foods for the 21st Century*, Elsevier, 2017, pp. 21–54.
5. FAOSTAT, Sorghum production. Food and agriculture organization of the united nations, https://www.fao.org/faostat/en/#data/QCL/visualize, accessed 21 January 2022.
6. T. F. Teferra and J. M. Awika, *Cereal Foods World*, 2019, **64**, 1–8.
7. A. L. Girard and J. M. Awika, *J. Cereal Sci.*, 2018, **84**, 112–124.
8. Y. Xiong, P. Zhang, R. D. Warner and Z. Fang, *Compr. Rev. Food Sci. Food Saf.*, 2019, **18**, 2025–2046.
9. T. Mehany, I. Khalifa, H. Barakat, S. A. Althwab, Y. M. Alharbi and S. El-Sohaimy, *Food Biosci.*, 2021, **40**, 100891.

10. A. K. Rashwan, H. A. Yones, N. Karim, E. M. Taha and W. Chen, *Trends Food Sci. Technol.*, 2021, **110**, 168–182.
11. O. A. Adebo, *Nutrients*, 2020, **12**, 1111.
12. J. Taylor and J. R. N. Taylor, *J. Agric. Food Chem.*, 2011, **59**, 2386–2392.
13. A. Khoddami, V. Messina, K. Vadabalija Venkata, A. Farahnaky, C. L. Blanchard and T. H. Roberts, *Crit. Rev. Food Sci. Nutr.*, 2021, 1–17.
14. B. Olawoye and S. O. Gbadamosi, *SN Appl. Sci.*, 2020, **2**, 607.
15. P. Shawrang, A. A. Sadeghi, M. Behgar, H. Zareshahi and G. Shahhoseini, *Food Chem.*, 2011, **125**, 376–379.
16. J. R. N. Taylor and M. N. Emmambux, *Cereal Chem.*, 2010, **87**, 263–271.
17. Y. Wu, L. Yuan, X. Guo, D. R. Holding and J. Messing, *Nat. Commun.*, 2013, **4**, 2217.
18. É. Moraes, A. V. A. Queiroz, R. E. Shaffert, N. M. Costa, J. D. Nelson, S. M. Ribeiro and H. S. Martino, *Food Chem.*, 2012, **134**, 1549–1555.
19. V. Kamath, S. Niketh, A. Chandrashekar and P. S. Rajini, *Food Chem.*, 2007, **100**, 306–311.
20. U. Shah, D. Dwivedi, M. Hackett, H. Al-Salami, R. P. Utikar, C. Blanchard, A. Gani, M. R. Rowles and S. K. Johnson, *Sci. Rep.*, 2021, **11**, 15204.
21. L. I. Ezeogu, K. G. Duodu, M. N. Emmambux and J. R. N. Taylor, *Cereal Chem.*, 2008, **85**, 397–402.
22. T. Kumar, I. Dweikat, S. Sato, Z. Ge, N. Nersesian, H. Chen, T. Elthon, S. Bean, B. P. Ioerger, M. Tilley and T. Clemente, *Plant Biotechnol. J.*, 2012, **10**, 533–544.
23. A. E.-M. M. Afify, H. S. El-Beltagi, S. M. Abd El-Salam and A. A. Omran, *Not. Bot. Horti Agrobot. Cluj-Napoca*, 2012, **40**, 86–92.
24. Z. Hadbaoui, A. Djeridane, M. Yousfi, M. Saidi and B. Nadjemi, *Med. J. Nutrition Metab.*, 2010, **3**, 215–220.
25. P. A. Akin, B. Sezer, S. R. Bean, K. Peiris, M. Tilley, H. Apaydin and I. H. Boyaci, *J. Sci. Food Agric.*, 2021, **101**, 1076–1084.
26. S. O. Serna-Saldivar and J. Espinosa-Ramírez, in *Sorghum and Millets*, ed. J. R. N. Taylor and K. G. Duodu, AACC International Press, 2nd edn, 2019, pp. 85–129.
27. C. V. Ratnavathi and V. V. Komala, in *Sorghum Biochemistry*, ed. C. V Ratnavathi, J. V Patil and U. D. Chavan, Academic Press, 2016, pp. 1–61.
28. S. Mehmood, I. Orhan, Z. Ahsan, S. Aslan and M. Gulfraz, *Food Chem.*, 2008, **109**, 855–859.
29. C. S. Silva, V. A. V. Queiroz, M. L. F. Simeone, C. C. Guimaraes, R. Schaffert, J. A. S. Rodrigues and R. A. Miguel, in *Congresso Nacional de Milho e Sorgo*, Águas de Lindóia. Diversidade e inovações na era dos transgênicos: resumos expandidos. Campinas: Instituto Agronômico, Associação Brasileira de Milho e Sorgo, Sete Lagoas, vol. 29, 2012.
30. J. Kruger, J. R. Taylor, X. Du, F. F. De Moura, B. Lönnerdal and A. Oelofse, *Food Chem.*, 2013, **141**, 1019–1025.

31. USDA, Food Data Central United States Department of Agriculture, https://fdc.nal.usda.gov/fdc-app.html#/food-details/169716/nutrients, accessed 21 January 2022.
32. A. L. Girard, S. R. Bean, M. Tilley, S. L. Adrianos and J. M. Awika, *Food Chem.*, 2018, **245**, 1154–1162.
33. J. M. Awika, D. J. Rose and S. Simsek, *Food Funct.*, 2018, **9**, 1389–1409.
34. V. G. Kamath, A. Chandrashekar and P. S. Rajini, *J. Cereal Sci.*, 2004, **40**, 283–288.
35. L. Yang, K. F. Allred, B. Geera, C. D. Allred and J. M. Awika, *Nutr. Cancer*, 2012, **64**, 419–427.
36. C. Chiremba, J. R. N. Taylor, L. W. Rooney and T. Beta, *Food Chem.*, 2012, **134**, 81–88.
37. A. E. M. M. R. Afify, H. S. El-Beltagi, S. M. Abd El-Salam and A. A. Omran, *Asian Pac. J. Trop. Biomed.*, 2012, **2**, 203–209.
38. E. S. M. Abdel-Aal, T. M. Choo, S. Dhillon and I. Rabalski, *Cereal Chem.*, 2012, **89**, 198–204.
39. A. S. Hole, I. Rud, S. Grimmer, S. Sigl, J. Narvhus and S. Sahlstrøm, *J. Agric. Food Chem.*, 2012, **60**, 6369–6375.
40. D. N'Dri, T. Mazzeo, M. Zaupa, R. Ferracane, V. Fogliano and N. Pellegrini, *J. Sci. Food Agric.*, 2013, **93**, 29–36.
41. A. Przybylska-Balcerek, J. Frankowski and K. Stuper-Szablewska, *Eur. Food Res. Technol.*, 2019, **245**, 1075–1080.
42. J. M. Awika and L. W. Rooney, *Phytochemistry*, 2004, **65**, 1199–1221.
43. J. M. Awika, C. M. McDonough and L. W. Rooney, *J. Agric. Food Chem.*, 2005, **53**, 6230–6234.
44. L. Dykes, L. M. Seitz, W. L. Rooney and L. W. Rooney, *Food Chem.*, 2009, **116**(1), 313–317.
45. L. Dykes and L. W. Rooney, *J. Cereal Sci.*, 2006, **44**, 236–251.
46. V. Taleon, L. Dykes, W. L. Rooney and L. W. Rooney, *J. Cereal Sci.*, 2012, **56**, 470–475.
47. C. H. Shih, S. O. Siu, R. Ng, E. Wong, L. C. M. Chiu, I. K. Chu and C. Lo, *J. Agric. Food Chem.*, 2007, **55**, 254–259.
48. D. H. Rhodes, L. Hoffmann, W. L. Rooney, P. Ramu, G. P. Morris and S. Kresovich, *J. Agric. Food Chem.*, 2014, **62**, 10916–10927.
49. C. Petti, R. Kushwaha, M. Tateno, A. E. Harman-Ware, M. Crocker, J. Awika and S. Debolt, *J. Agric. Food Chem.*, 2014, **62**, 1227–1232.
50. A. P. P. Kayodé, M. J. R. Nout, A. R. Linnemann, J. D. Hounhouigan, E. Berghofer and S. Siebenhandl-Ehn, *J. Agric. Food Chem.*, 2011, **59**, 1178–1184.
51. B. Geera, L. O. Ojwang and J. M. Awika, *J. Food Sci.*, 2012, **77**, 566–572.
52. A. Khalil, R. Baltenweck-Guyot, R. Ocampo-Torres and P. Albrecht, *Phytochem. Lett.*, 2010, **3**, 93–95.
53. M. T. Escribano-Bailón, C. Santos-Buelga and J. C. Rivas-Gonzalo, *J. Chromatogr. A*, 2004, **1054**, 129–141.
54. L. Dykes, G. C. Peterson, W. L. Rooney and L. W. Rooney, *Food Chem.*, 2011, **128**, 173–179.

55. T. yang Wang, Q. Li and K. shun Bi, *Asian J. Pharm. Sci.*, 2018, **13**, 12–23.
56. L. Yang, K. F. Allred, L. Dykes, C. D. Allred and J. M. Awika, *Food Funct.*, 2015, **6**, 749–755.
57. R. Zamora-Ros, V. Knaze, J. A. Rothwell, B. Hémon, K. Overvad, A. Tjønneland, C. Kyrø, G. Fagherazzi, C. Boutron-ruault, M. Touillaud, V. Katzke, T. Kühn, J. Förster, A. Trichopoulou, E. Valanou and E. Peppa, *Eur. J. Nutr.*, 2018, **55**, 1359–1375.
58. M. Martínez, M. J. Motilva, M. C. López de las Hazas, M. P. Romero, K. Vaculova and I. A. Ludwig, *Food Chem.*, 2018, **245**, 61–70.
59. R. C. Kaufman, T. J. Herald, S. R. Bean, J. D. Wilson and M. R. Tuinstra, *J. Sci. Food Agric.*, 2013, **93**, 1233–1241.
60. Y. Wu, X. Li, W. Xiang, C. Zhu, Z. Lin, Y. Wu, J. Li, S. Pandravada, D. D. Ridder, G. Bai, M. L. Wang, H. N. Trick, S. R. Beane, M. R. Tuinstra, T. T. Tesso and J. Yu, *Proc. Natl. Acad. Sci. U. S. A.*, 2012, **109**, 10281–10286.
61. J. M. Awika, L. Dykes, L. Gu, L. W. Rooney and R. L. Prior, *J. Agric. Food Chem.*, 2003, **51**, 5516–5521.
62. N. L. Mkandawire, R. C. Kaufman, S. R. Bean, C. L. Weller, D. S. Jackson and D. J. Rose, *J. Agric. Food Chem.*, 2013, **61**, 4448–4454.
63. A. E. Hagerman, K. M. Riedl, G. A. Jones, K. N. Sovik, N. T. Ritchard, P. W. Hartzfeld and T. L. Riechel, *J. Agric. Food Chem.*, 1998, **46**, 1887–1892.
64. G. R. Beecher, *Pharm. Biol.*, 2004, **42**, 2–20.
65. A. Floegel, D. O. Kim, S. J. Chung, W. O. Song, M. L. Fernandez, R. S. Bruno, S. I. Koo and O. K. Chun, *Int. J. Food Sci. Nutr.*, 2010, **61**, 600–623.
66. P. Van Hung, *Crit. Rev. Food Sci. Nutr.*, 2016, **56**, 25–35.
67. R. A. Ortíz Cruz, J. L. Cárdenas López, G. A. González Aguilar, H. Astiazarán García, S. Gorinstein, R. Canett Romero and M. Robles Sánchez, *Biomed Res. Int.*, 2015, **2015**, 164725.
68. S. D. Sharma, S. M. Meeran and S. K. Katiyar, *Mol. Cancer Ther.*, 2007, **6**, 995–1005.
69. R. Zbasnik, T. Carr, C. Weller, K. T. Hwang, L. Wang, S. Cuppett and V. Schlegel, *J. Agric. Food Chem.*, 2009, **57**, 10435–10441.
70. T. G. Simnadis, L. C. Tapsell and E. J. Beck, *Nutr. Rev.*, 2016, **74**, 690–707.
71. C. Isaacson, *Med. Hypotheses*, 2005, **64**, 658–660.
72. J.-M. Hwang, K.-C. Choi, S.-J. Bang, Y.-O. Son, B.-T. Kim, D.-H. Kim, G. J. Choi, D. H. Kim, X. Shi and J.-C. Lee, *Food Sci. Biotechnol.*, 2013, **22**, 265–272.
73. J. H. Park, P. Darvin, E. J. Lim, Y. H. Joung, D. Y. Hong, E. U. Park, S. H. Park, S. K. Choi, E.-S. Moon and B. W. Cho, *PLoS One*, 2012, **7**, 40531.
74. S. Lee, Y. Park, M. Y. Zuidema, M. Hannink and C. Zhang, *World J. Cardiol.*, 2011, **3**, 18.
75. É. A. Moraes, D. I. G. Natal, V. A. V Queiroz, R. E. Schaffert, P. R. Cecon, S. O. Paula, L. Anjos Benjamim, S. M. R. Ribeiro and H. S. D. Martino, *Food Res. Int.*, 2012, **49**, 553–559.
76. P. S. Devi, M. S. Kumar and S. M. Das, *Int. J. Breast Cancer*, 2011, **2011**, 891481.

77. N. M. Ali, A. H. El Tinay, O. E. Abd Elmoneim, O. Salih and N. Yousif, *Food Chem.*, 2009, **114**, 646–648.
78. F. Barros, J. Awika and L. W. Rooney, *J. Sci. Food Agric.*, 2014, **94**, 1212–1217.
79. G. S. Hotamisligil, *Nature*, 2006, **444**, 860–867.
80. N. Carvalho Teixeira, V. A. V Queiroz, M. C. Rocha, A. C. P. Amorim, T. O. Soares, M. A. M. Monteiro, C. B. Menezes, R. E. Schaffert, M. A. V. T. Garcia and R. G. Junqueira, *Food Chem.*, 2016, **197**, 291–296.
81. R.-L. Shen, W.-L. Zhang, J.-L. Dong, G.-X. Ren and M. Chen, *Food Agric. Immunol.*, 2015, **26**, 120–130.
82. S. W. Pelpolage, K. Han, H. Koaze, T. Hamamoto, M. Hoshizawa and M. Fukushima, *J. Food Nutr. Res.*, 2019, **58**(2), 135–145.
83. I. Martínez, G. Wallace, C. Zhang, R. Legge, A. K. Benson, T. P. Carr, E. N. Moriyama and J. Walter, *Appl. Environ. Microbiol.*, 2009, **75**, 4175–4184.
84. I. Fernandes, A. Faria, C. Calhau, V. Freitas and N. Mateus, *J. Funct. Foods*, 2014, **7**, 54–66.
85. F. Cardona, C. Andrés-Lacueva, S. Tulipani, F. J. Tinahones and M. I. Queipo-Ortuño, *J. Nutr. Biochem.*, 2013, **24**, 1415–1422.
86. T. Requena, M. Monagas, M. Pozo-Bayón, P. J. Martín-Álvarez, B. Bartolomé, R. Del Campo, M. Ávila, M. C. Martínez-Cuesta, C. Peláez and M. Moreno-Arribas, *Trends Food Sci. Technol.*, 2010, **21**, 332–344.
87. I. Martínez, J. Kim, P. R. Duffy, V. L. Schlegel and J. Walter, *PLoS One*, 2010, **5**, 15046.
88. A. Stefoska-Needham, E. J. Beck, S. K. Johnson and L. C. Tapsell, *Food Rev. Int.*, 2015, **31**, 401–437.
89. A. Yousif, D. Nhepera and S. Johnson, *Food Chem.*, 2012, **134**, 880–887.
90. I. Khan, A. Yousif, S. K. Johnson and S. Gamlath, *Food Res. Int.*, 2013, **54**, 578–586.
91. N. M. Poquette, X. Gu and S. O. Lee, *Food Funct.*, 2014, **5**, 894–899.
92. F. Cisse, D. P. Erickson, A. M. Hayes, A. R. Opekun, B. L. Nichols and B. R. Hamaker, *Nutrients*, 2018, **10**, 124.
93. P. Gulati, A. Li, D. Holding, D. Santra, Y. Zhang and D. J. Rose, *J. Agric. Food Chem.*, 2017, **65**, 1952–1959.
94. J. M. Awika, L. W. Rooney, X. Wu, R. L. Prior and L. Cisneros-Zevallos, *J. Agric. Food Chem.*, 2003, **51**, 6657–6662.
95. L. Dykes, L. W. Rooney, R. D. Waniska and W. L. Rooney, *J. Agric. Food Chem.*, 2005, **53**, 6813–6818.
96. R. Liu, C. Xu, X. Cong, T. Wu, Y. Song and M. Zhang, *Food Chem.*, 2017, **221**, 2010–2017.
97. T. J. Herald, P. Gadgil and M. Tilley, *J. Sci. Food Agric.*, 2012, **92**, 2326–2331.
98. A. J. Parr and G. P. Bolwell, *J. Sci. Food Agric.*, 2000, **80**, 985–1012.
99. J. B. Harborne and C. A. Williams, *Phytochemistry*, 2000, **55**, 481–504.
100. R. Amarowicz, M. Karamac and F. Shahidi, *J. Food Lipids*, 2003, **10**, 1–10.
101. N. Rezaee, W. B. Fernando, E. Hone, H. R. Sohrabi, S. K. Johnson, S. Gunzburg and R. N. Martins, *Front. Aging Neurosci.*, 2021, **13**, 729949.

102. C. R. Rudiger, *Doctoral Dissertation, The Formulation of a Nutraceutical Bread Mix Using Sorghum, Barley, and Flaxseed*, Texas A&M University. 2003.
103. B. E. Jeschke, *MSc Thesis, Chemical, Colour and Sensory Attributes of Sorghum Bran Enriched Beef Patties in A High Oxygen Environment*, Texas A&M University. 2004.
104. C. McDonough, J. Awika, N. Turner, L. Xu and L. Rooney, in *AACC Annual Meeting Abstracts*, 2004.
105. M. C. Villalobos, M. J. Serradilla, A. Martín, E. Ordiales, S. Ruiz-Moyano and M. G. Córdoba, *J. Sci. Food Agric.*, 2016, **96**, 2116–2124.
106. H. Y. Kil, E. S. Seong, B. K. Ghimire, I.-M. Chung, S. S. Kwon, E. J. Goh, K. Heo, M. J. Kim, J. D. Lim and D. Lee, *Food Chem.*, 2009, **115**, 1234–1239.
107. H. Sun, H. Wang, P. Zhang, S. Ajlouni and Z. Fang, *Cereal Chem.*, 2020, **97**, 612–625.
108. M. Zaroug, I. E. Orhan, F. S. Senol and S. Yagi, *Food Chem.*, 2014, **156**, 110–116.
109. O. A. Adebo, P. B. Njobeh, J. A. Adebiyi and E. Kayitesi, *Food Biosci.*, 2018, **25**, 118–127.
110. X. W. Qaku, A. Adetunji and B. C. Dlamini, *J. Food Sci.*, 2020, **85**, 1661–1667.
111. J. C. Ndulaka, N. Obasi and G. C. Omeire, *Niger. Food J.*, 2014, **32**, 66–72.
112. D. Mohapatra, A. S. Patel, A. Kar, S. S. Deshpande and M. K. Tripathi, *Food Chem.*, 2019, **271**, 129–135.
113. D. K. Weerasooriya, S. R. Bean, Y. Nugusu, B. P. Ioerger and T. T. Tesso, *PLoS One*, 2018, **13**, e0203005.
114. I. Luzardo-Ocampo, A. K. Ramírez-Jiménez, A. H. Cabrera-Ramírez, N. Rodríguez-Castillo, R. Campos-Vega, G. Loarca-Piña and M. Gaytán-Martínez, *Food Chem.*, 2020, **309**, 125684.
115. M. Jafari, A. Koocheki and E. Milani, *J. Cereal Sci.*, 2017, **77**, 49–57.
116. R. Rumler, D. Bender, S. Speranza, J. Frauenlob, L. Gamper, J. Hoek, H. Jager and R. Schonlechner, *Foods*, 2021, **10**, 870.
117. P. M. Palavecino, M. C. Bustos, M. B. H. Alabí, M. S. Nicolazzi, M. C. Penci and P. D. Ribotta, *J. Food Sci.*, 2017, **82**, 2085–2093.
118. G. Wu, J. Ashton, A. Simic, Z. Fang and S. K. Johnson, *Food Res. Int.*, 2018, **103**, 509–514.
119. G. Rocchetti, G. Giuberti, M. Busconi, A. Marocco, M. Trevisan and L. Lucini, *Food Chem.*, 2020, **312**, 126077.
120. R. D. Darman, M. B. Sidoine and V. W. Lendzemo, *Food Sci. Nutr.*, 2020, **8**, 3181–3189.
121. S. A. O. Adeyeye, *Cogent Food Agric.*, 2016, **2**, 1245059.
122. P. C. Anunciacao, L. de Morais Cardoso, J. V. P. Gomes, C. M. D. Lucia, C. W. P. Carvalho, M. C. Galdeano, V. A. V. Queiroz, R. C. G. Alfenas, H. S. D. Martino and H. M. Pinheiro-Saint Ana, *Food Chem.*, 2017, **221**, 984–989.
123. F. Apea-Bah, A. Minnaar, M. J. Bester and K. G. Duodu, *Food Chem.*, 2014, **157**, 157–166.
124. G. Wu, Y. Shen, Y. Qi, H. Zhang, L. Wang, H. Qian, X. Qi, Y. Li and S. K. Johnson, *LWT--Food Sci. Technol.*, 2018, **91**, 77–83.

125. P. M. Palavecino, P. D. Ribotta, A. E. León and M. C. Bustos, *J. Sci. Food Agric.*, 2019, **99**, 1351–1357.
126. M. I. Curti, M. Belorio, P. M. Palavecino, J. M. Camiña, P. D. Ribotta and M. Gómez, *J. Food Sci. Technol.*, 2021, 1–12.
127. L. R. dos Reis Gallo, C. E. G. Reis, M. A. Mendonça, V. S. N. da Silva, M. T. B. Pacheco and R. B. A. Botelho, *Foods*, 2021, **10**, 2256.
128. L. d. L. de Oliveira, G. T. de Oliveira, E. R. de Alencar, V. A. V. Queiroz and L. F. de Alencar Figueiredo, *LWT--Food Sci. Technol.*, 2022, **153**, 112407.
129. I. Khan, A. M. Yousif, S. K. Johnson and S. Gamlath, *Clin. Nutr.*, 2015, **34**, 415–421.
130. A. O. Tayo, A. O. Dosunmu, I. O. Akinola, A. Adewunmi, O. A. Oloyede, A. A. Akinbami, B. I. Osikomaiya and S. B. L. Makanjuola, *Nutrition*, 2017, **33**, 113–117.
131. C.-H. Shih, S.-O. Siu, R. Ng, E. Wong, L. C. M. Chiu, I. K. Chu and C. Lo, *J. Agric. Food Chem.*, 2007, **55**, 254–259.
132. T. L. Silva, U. V. Lacerda, S. L. P. da Matta, V. A. V. Queiroz, P. C. Stringheta, H. S. D. Martino and F. A. R. de Barros, *J. Food Sci.*, 2020, **85**, 2236–2244.
133. S. Brambillasca, M. Fernandez-Garcia, M. Aguerre, J. L. Repetto and C. Cajarville, *J. Cereal Sci.*, 2019, **89**, 102817.
134. A. Khoddami, H. H. Truong, S. Y. Liu, T. H. Roberts and P. H. Selle, *Anim. Feed Sci. Technol.*, 2015, **210**, 190–199.

CHAPTER 6

Pigmented Pseudocereals: Chemistry, Functionality, and Technological Aspects in Food Systems

TAHA MEHANY(0000-0001-7960-7558)*[a,b], AHMED TAHA[c,d], BABATUNDE OLAWOYE[e], SAMEH A. KORMA[f,g], OYEKEMI OLABISI POPOOLA[e], OKON JOHNSON ESUA[h,g] AND MUHAMMAD FAISAL MANZOOR[i,g]

[a]Food Technology Department, Arid Lands Cultivation Research Institute, City of Scientific Research and Technological Applications, 21934 Alexandria, Egypt; [b]Department of Chemistry, University of La Rioja, C/ Madre de Dios 51, 26006 Logroño, La Rioja, Spain; [c]State Research Institute, Center for Physical Sciences and Technology, Saulėtekio al. 3, Vilnius, Lithuania; [d]Department of Food Science, Faculty of Agriculture (Saba Basha), Alexandria University, Alexandria 21531, Egypt; [e]Department of Food Science and Technology, Faculty of Engineering and Technology, First Technical University, Ibadan, Nigeria; [f]Department of Food Science, Faculty of Agriculture, Zagazig University, Zagazig 44519, Egypt; [g]School of Food Science and Engineering, South China University of Technology, Guangzhou 510641, China; [h]Department of Agricultural and Food Engineering, University of Uyo, Uyo 520101, Nigeria; [i]Guangdong Provincial Key Laboratory of Intelligent Food Manufacturing, Foshan University, Foshan 528011, China
*E-mail: tmehany@srtacity.sci.eg

Food Chemistry, Function and Analysis No. 38
Pigmented Cereals and Millets: Bioactive Profile and Food Applications
Edited by Sneh Punia Bangar, Sajid Maqsood and Anil Kumar Siroha

Published by the Royal Society of Chemistry, www.rsc.org

6.1 Introduction

The rapid increase in the global population, food insecurity, and people living with degenerative diseases had led to greater demand for functional foods. Functional or nutraceutical food provides nutritional value and can prevent some health conditions.[1] These food products can be derived from pigmented pseudocereals (PPs). Pseudocereals (Figure 6.1) are edible grains belonging to species of *dicotyledonous* that are similar to the real cereals (monocotyledonous of the *Poaceae* family) in terms of their physical appearance and high starch ratio.[2] Pseudocereals are categorized as superlative foods in the 21st century.[3] PPs are a promising future food crop due to their ability to adapt to climate change, from tropical to moderate climates; and their high genetic variability. The most important species of pseudocereals are buckwheat (*Fagopyrum* sp.), quinoa (*Chenopodium quinoa* Willd), and amaranth (*Amaranthus* sp.). Amaranth and quinoa belong to the family Chenopodiaceae that are native to South America in the Andean region. While buckwheat belongs to the Polygonaceae family; genus *Fagopyrum*, and includes three cultivated species: (a) Tartary buckwheat (*F. tataricum*); (b) common buckwheat (*F. esculentum*); and (c) tall buckwheat (*F. cymosum*), which originated in Western and Central China. Pseudograins are a special trend in the food industry sector because they are a gluten free (GF) and also have an ideal nutritional and functional profile.[4] Furthermore, many recent investigations have focused on the health benefits of pseudocereals, showing that these grains are suitable sources for the development of functional food. Table 6.1 shows the comparable nutritional composition of PPs, compared with that of wheat and rice.

Figure 6.1 Image of major staple pseudocereal grains: (a) amaranth (*Amaranthus cruentus*); (b) quinoa (*Chenopodium quinoa*); and (c) buckwheat (*Fagopyrurn esculentum*). Reproduced from ref. 5 with permission from Elsevier, Copyright 2008.

Table 6.1 A proximate nutritional composition of PPs compared with wheat and rice (g per 100 g).

PPs & cereals	Protein	Lipid	Ash	Fiber	Carbohydrates	Ref.
Amaranth	13.1–21.0	5.6–10	3.2	6.6	48–69	21–23
Buckwheat	13.3–15.6	1–3	3.5–3.9	27.3	59–70	22and24–26
Quinoa	10–18	4.5–8.8	3.2	10.5	54.1–64.2	25and27
Wheat	14.4	2.3	1.9	2.9	64	28
Rice	7.5	2.4	4.7	10.2	72.2	28

Table 6.2 Bioactive compounds illustrations of PPs.

PPs	Phenolic compounds	Flavonoids	Ref.
Amaranth	Anthocyanins, rutin, isoquercetin, nicotiflorine, and quercetin	Protocatechuic acid, *p*-hydroxybenzoic acid, syringic acid, gallic, *p*-coumaric, and vanillic acid	29
Buckwheat	Ferulic acid, protocatechuic acid, caffeic acid, gallic acid, *p*-hydroxyl benzoic acid, *o*-coumaric acid, syringic acid, sinapic acid, vanillic acid, and *p*-coumaric acid	Rutin, quercetin-3 rutinosid, isovitex-invitexin, homoori-entin, and orientin	29 and 30
Quinoa	Sesamolin, 7-hydroxymatairesinol, lariciresinol, syringaresinol, ses-amin, episesamin, episesaminol, sesaminol secoisolariciresinol-ses-quilignan, sesamol, 7-hydroxyse-coisolariciresinol, sesamolinol, ferulic acid-4-glycoside, caffeic acid, protocatechuic acid, cin-namic acid, gallic acid, vanillic acid, ferulic acid, *p*-coumaric acid, and *p*-hydroxybenzoic acid	Genistin, isorhamne-tin glycitein/bioch-anin, myricetin, A, daidzein, Kaemp-ferol6″-*O*-acetyl-daidzin, and quercetin	31 and 32

Pigmented amaranth (PA) is an Andean crop grown mainly for its leafy vegetable, but whose grain had been reported to contain a high level and quality of protein compared with some legumes. Aside from protein quality, its high polyunsaturated fatty acids and dietary fiber content allow the seeds to be utilized as a functional food or food ingredient. Another component that distinguishes PA from other cereals and pseudocereal is the bioactive compounds (Table 6.2), especially phenolic compounds, squalene, sterols, and anthocyanins. Anthocyanins that give the pigmented grain amaranth its color are often present in the outermost layer or the cell wall of the grain. It is a water-soluble compound that gives the grain amaranth its color, which ranges from blue, black, red, orange, blue, and purple. PA differs in color depending on the mutual ratio and content of anthocyanins.[6] In this regard, the blue color is mainly due to the derivatives of delphinidin commonly present in the aleuronic layer of the grain. The purple coloration is due to the derivatives of cyanidin, found typically in the pericarp layer. Aside from the pigmentation of the amaranth grain, anthocyanins also contribute to the functionality of the grain, such as its antioxidant and anti-hypercholesterol activities.

In pigmented buckwheat (PB), the two most important species are *Fagopyrum tataricum* Gaertner and *Fagopyrum esculentum* Moench, referred to as Tartary and common buckwheat, respectively.[7] There are different main pigments in buckwheat grains like anthocyanins and photosynthetic pigments, *e.g.* carotenoids and chlorophyll a and b.[8] The embryo of buckwheat's seed is situated in the endosperm's center, and it contains 2 cotyledons. The pericarp contains a structure with hard fibers tightly surrounding the endosperm, embryo, and seed coat. The immature fruits are green or red, and the hull darkens as ripening occurs. In the hypocotyledon of buckwheat seedlings, more than 3 red pigments have been identified; *i.e.* cyanidin and 2 others, assumed to be cyanidin glycosides. Several components such as dietary fiber, vitamins, D-chiro-inositol, minerals, polyphenols, phytosterols, carotenoids, essential amino acids, and phytosterols, are responsible for the various health benefits of buckwheat.[9] These health benefits include anti-cancer, anti-hypertensive, hypoglycemic, antioxidant, and anti-inflammatory properties.[10] PB is also considered a GF food product. Methods such as solubilization–precipitation have been employed in the extraction of the protein of buckwheat, while its polysaccharides have been extracted by washing, drying, and grinding, by using hot water and acid, in addition to alkali.[9] The seeds of buckwheat are processed into flour, while a large number of processed foods such as yoghurt, pasta, dark sauce, noodles, alcoholic beverages, tea, vinegar, and bakery foods have been derived from PB.[11] Figure 6.2 presents the production share of buckwheat worldwide between 1994 and 2020 as an average, according to FAOSTAT.[12]

Pigmented quinoa (PQ) is known as a superfood, and 'the mother of grains' in the Inca language. PQ is an edible seed, that is stress-tolerant, and provided survival, nutrition, and medication for the Andean people, thousands

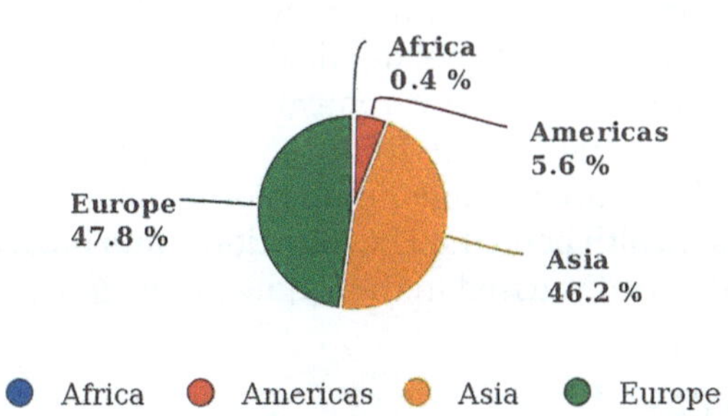

Figure 6.2 Production share of buckwheat worldwide according to FAOSTAT (2022).[12]

of years ago.[13] The color of quinoa seed varies, and the quinoa seed coat contains the pigments that decide the color of the seed (white, purple, red, brown, *etc.*) depending on the occurrence of betacyanin pigments.[14] Recently, PQ has received much attention due to its unique food components, high protein content, bioactive ingredients, mineral and vitamin content, low glycemic index, lactose-free property, and high nutritional and functional value. It has been considered a desirable food in both developing and developed countries, where there is a need to introduce novel and functional foods. Moreover, PQ is a suitable protein source for vegans because it is GF, and an alternative grain source for people who are suffering from gluten allergies or celiac disease.[15] PQ also prevents type 2 diabetes due to its low glycemic index, so it can replace common grains in a diabetic diet.[16] Plant-based proteins commonly have less nutritional value than animal proteins. They are less digestible and are deficient in one or more essential amino acids.[17] Thanks to its high protein content and functional properties, PQ offers a great chance to develop plant-based food substitutes. FAO spotted that PQ is closer to the ideal protein balance than any other plant or cereals, being at least equal to casein quality.[18] Hence, PQ has become known as a complete food, because of its protein content, and particularly its essential amino acid richness.[19] The attractive nutritional quality of PQ supports its utilization in different food products in many studies.[20] This chapter provides a comprehensive overview of the nutritional and functional characteristics of the main PPs, such as amaranth, buckwheat, and quinoa, paying particular attention to the health benefits of the bioactive pigments. The recent applications of PPs in the food industry and their techno-functional properties are also discussed in this chapter.

6.2 Pigmented Amaranth

6.2.1 Chemical Composition of Amaranth and Its Nutritional Value

Like other pigmented pseudocereals, grain amaranth is a good source of protein, carbohydrate, fats, dietary fiber, minerals, and bioactive compounds. These chemical compounds differ in amaranth with respect to species, varieties, climate conditions, and geographical locations. Generally, grain amaranth is composed of: 74–77.5% carbohydrate, 10.68–16% protein, 7–8.3% fats, and 3–3.7% minerals. Aside from the nutritional composition, grain amaranths contain an appreciable number of bioactive compounds, which convey additional health benefits. The nutritional and bioactive components in amaranth grain are discussed in the following sections.[1]

6.2.1.1 Protein

Grain amaranth has a protein content of 16–18%, which is significantly higher than some common cereals, such as wheat (13.2%), rice (7.0%), or maize (9.2%). The protein content of grain amaranth, however, varies among species. Mota *et al.*[32] reported an average of 13.5% for *Amaranthus caudatus*, 15.7% for *A. cruentus*, and 15.5% for *A. hypochondriacus*. In their research, Olawoye *et al.*[33] reported an average of 16.3% for the protein content of *A. virides*. Aside from its high protein content, it also contains a high proportion of essential amino acids (especially tryptophan and lysine), which compare favorably with the quality of animal protein. Grain amaranth protein contains predominantly globulins and albumin with low or no detection of alpha-gliadin and prolamins, making it suitable for treating celiac disease (an auto-immune disease caused by the ingestion of gluten).

6.2.1.2 Carbohydrates, Starch, and Dietary Fiber

Amaranth grain carbohydrate exists in the form of starch and dietary fiber. Although the starch content in grain amaranth is lower than in other cereal grains, its composition (61–69%) by weight makes it the most abundant component in the grain. Amaranth starch comprises amylose (an essentially linear polymer of glucose linked together by alpha-1-4-glycosidic linkage) and amylopectin (a highly branched type of starch with alpha-1-4-glycosidic linkage at the main chain coupled with a short-chain alpha-1-6-glycosidic linkage branches). The amylopectin is the most abundant component of amaranth starch, with about 1700 amylopectin on average and ranging from 90–98%. The amylose content of amaranth starch varies from 2–10%, which is dependent on the species, varieties, and genotypes. The variation in amylose content with respect to species and genotype influenced some properties of the amaranth starch, such as starch digestibility, pasting, and thermal properties. A study on the effect of different amaranth species on the amylose content of the starch revealed a lower peak viscosity, setback, and breakdown viscosity in *A. hypochondriacus* compared with *A. caudatus*.[1] The amylose content of amaranth starch is lower than that of other cereal and pseudocereal starch; Kong *et al.*[34] reported an amylose content of 8.8% for quinoa starch. The higher amylopectin content of the starch makes it highly digestible. The granules of amaranth starch are uniform in size and are 1–3 μm in diameter. Granule size, coupled with high levels of readily digestible starch, while leaching the crystalline structure of the starch during heating contributes to the high glycemic index of grain amaranth. Other carbohydrates found in grain amaranth, aside from starch, are small amount of sugars. Olawoye *et al.*[35] reported 2.1% monosaccharide, 2.4% disaccharides, and 3.8% pentosans in *A. virides,* while 0.39% raffinose, and 0.15% stachyose were detected in *A. cruentus.*[36] The indigestibility of dietary fiber in the gastrointestinal tract makes it an important health-promoting food component. The whole amaranth seed contains around 8–20% dietary fiber (56–67% of which is insoluble and 33–44% soluble). Compared with other pseudocereal grains, the dietary fiber found in grain amaranth is lower when compared to that of quinoa and buckwheat. Insoluble fiber of 7.91 g per 100 g and soluble fibre of 5.66 g per 100 g in *A. caudatus*.[37] Dietary fiber in amaranth is composed mainly of pectic polysaccharides, as well as xyloglucans in varying amounts. Hence, the incorporation of grain amaranth in the diet of cardiovascular disease patients can help to alleviate the disease.

6.2.1.3 Lipids

Generally, amaranth grain contains more lipids than other pseudocereal and cereal grains. Different concentrations of lipids have been reported in *Amaranthus* spp., ranging from 3.24–8.60%.[38,39] However, Olawoye *et al.*[35] reported a similar lipid content of 5.50–7.65% in *A. virides* species after being subjected to different processing conditions. *Amaranthus* spp. Grain has saturated and unsaturated fatty acids, with the latter being the predominant fatty acids (75%). Among the unsaturated fatty acids, linoleic is the most dominant (33.0–55.9%), followed by oleic acid, palmitic acid, and stearic acid in concentrations of 18.7–38.9%, 14.04–26%, and 3.11–4.47%, respectively. Linolenic acid is, however, present in grain amaranth in very low amounts (0.2–1.97%). The high concentration of polyunsaturated fatty acids in grain amaranth makes it suitable for the management of cardiovascular diseases. Other lipid components found in grain amaranth are significant, and varying proportions include fat-soluble vitamins (especially a-tocopherol), sterols, triglycerides, phospholipids, and glycolipids.[35]

6.2.1.4 Vitamins and Minerals

In comparison with other cereals like wheat and maize, pigmented grain amaranths contain higher mineral content, especially calcium, in which it is five times higher than in wheat and quinoa. Other macro-minerals such as Na, K, and P are found in reasonable concentrations in grain amaranth, while Mn, Zn, and Zn appear in *Amaranthus* spp. at a moderate level. In their research, Palombini *et al.*[40] reported a high concentration of Fe (29.4 mg per 100 g) in *Amaranthus* spp. Grain, a value that is 3–5 times higher than that obtained for wheat grain. Other essential micro-minerals found in pigmented grain amaranth in detectable amounts include Cr (6.8–14.4 µg per 100 g) and Mo (59.6 µg per 100 g). The high concentration of Ca and Fe found in pigmented grain amaranth makes it an essential pseudocereal in the diets of children and pregnant women.

In addition to minerals, pigmented grain amaranth is also a rich source of vitamins. In their study, Gamel *et al.*[36] reported that the vitamins present in *Amaranthus caudatus* L. was ascorbic acid, with a value of 12.5 mg kg^{-1}, a riboflavin level of 2.4 mg kg^{-1}, and niacin level of 12.5 mg kg^{-1}. Olawoye *et al.*[35] reported a folic acid content of 82 µg per 100 g in grain amaranth, a higher value than the value reported for wheat grain (43 µg per 100 g). Important fat-soluble vitamins such as tocopherols and tocotrienols had been reported in pigmented grain amaranth in major and minor concentrations. Tocopherol and tocotrienols act as free-radical scavengers, as well as antioxidants, which make them active against hypercholesterolemia.

6.2.2 Bioactive Compounds and Pigments of Pigmented Amaranth

To guard against pests, insects, and microbial invasion, a secondary metabolite that serves as a defense mechanism is synthesized by plants. These secondary metabolites, which are found in the cell wall of seeds, are bioactive and help in fighting against degenerative diseases. Grain amaranth had been ranked among PPs with bioactive compounds, including phenolic compounds, squalene, phytosterols, terpenes, and anthocyanin. Polyphenols have health benefits for the treatment of various types of diseases. Polyphenols and mainly natural plant extract of flavonoids have recently been useful in various food applications and pharmaceutical industries because of the biological activities of these bioactive molecules, principally antimicrobial activities, antioxidant potential, and anti-inflammatory characteristics.[41]

Phenolic compounds have one or more hydroxyl groups attached to a benzene ring. They include, for example, flavonoids, phenolic acids, coumarins, tannins, and alkyl-resorcinols. Phenolic compounds present in pigmented amaranth grain differ in species, varieties, genotype, and origin. A study carried out by Olawoye *et al.*[42] on the antioxidant properties of methanol extracted from Nigerian *Amaranthus virides* grain revealed a total phenolic content of 0.73 μg g^{-1} and a tannin content of 1.5 mg per 100 g. Lopez *et al.*[43] reported that the total phenolic compound (TPC) of *A. hypochondriacus* extracted with methanol, hexane, and ethanol ranged from 16–25 mg GAE per 100 g. This resulted in differences in the TPC value due to the polarity and solubility of the solvents. A different research group had identified different bound and unbound phenolic acids in PA. Alvarez-Jubete *et al.*[44] found abundant protocatechuic acid (13.4 μmol per 100 g). In *A. cruentus*, gallic acid (400 mg kg^{-1}) was the predominant phenolic acid, followed by *para*-hydroxybenzoic acid (14.63 mg kg^{-1}). Rutin and nicotiflorin had also been identified in the flour of various species of PA. Rutin was detected at 99 mg kg^{-1} in *A. hybridus*, 70 mg kg^{-1} in *A. hypochondriacus*, 7 mg kg^{-1} in *A. tricolor* and 11 mg kg^{-1} in *A. retroflexus*. Flavanones, isovitexin and vitexin were found in grain amaranth in a small amount. PA had been ranked among plants as the most important source of phytosterol and squalene. Grain amaranth contains about 8% squalene, which plays an important role in the biosynthesis of coenzyme Q10; an important part of the cell membrane. Anthocyanins are glycosides of polyhydroxy derivatives of 2-phenylbenzopyrlium salts, and are colored pigments that are responsible for blue, orange, brown, purple and red colors in most plants. Pigmented grain amaranths derive their color from anthocyanin; the pigment varies according to variety and genotype. Anthocyanin in grain amaranth has been detected at a moderate level of 90.83–103.6 mg per 100 g.[45]

6.2.3 Functional and Health Benefits of Pigmented Amaranth

6.2.3.1 *The Health Benefits of Pigmented Amaranth in Pharmacological and Clinical Studies*

The hypocholesterolemic, hypolipidemic, antitumor, effect on liver function, antidiabetic, and antihypertensive properties of amaranth have been studied.[46] Squalene and several amaranth components such as amino acids, oils, proteins, sterols, or fiber are responsible for amaranth's hypocholesterolemic effect. Several studies have also been performed to justify this effect on several products of amaranth such as the seed, oil, flour, and grain.[47] The oil extracted from amaranth played a crucial role in its hypocholesterolemic effect.[48] Using defatted amaranth, Mendonça *et al.*[49] reported that some other components, such as proteins could be responsible for amaranth's hypocholesterolemic effect. The pathology, seed, oil, and flour of amaranth play an important role in the hypolipidemic effect of amaranth. *In vitro* assays have been performed on the antidiabetic and antihypertensive activities of PA, with potential being shown for these two biological effects. Proteins present in amaranth have the capability of producing antihypertensive effects, as reported by some authors. For instance, Tovar-Pérez *et al.*[50] studied these effects with peptides from amaranth. Ethyl acetate and methanol extracts from the seeds of amaranth have been shown to have an appreciable activity against diabetes.[51] Moreover, Kim *et al.*[52] reported that amaranth grain increased serum insulin and decreased blood glucose level. The protein of amaranth possesses antitumor properties through a supposed mechanism of action.[53]

6.2.3.2 *Antioxidant Potential*

Several researchers have studied the antioxidant effect of amaranth fractions. Klimczak *et al.*[54] reported the *in vitro* antioxidant assay of 2 species of amaranth, *A. panicalatus* and *A. caudatus*. The content of the phenolic compounds using the Folin–Ciocalteu approach was estimated to be 56.22 mg per 100 g in *A. panicalatus* and 39.17 mg per 100 g in *A. caudatus*. Also, the quantity of squalene, phenolic compounds, oil, and antioxidant properties of 2 varieties of *A. caudatus* has been examined.[51] In an *in vitro* assay by Paśko *et al.*,[55] there was high antioxidant activity in the sprouts and seeds of amaranth. In linoleate and beta-carotene models, an appreciable activity of antioxidants was shown by Klimczak *et al.*[54] The protein in amaranth possesses antioxidant properties, and this increases after digestion.[56]

6.2.3.3 *Antimicrobial Activity*

Studies have indicated the antimicrobial activity of PA. The leaf extract of *A. caudatus* has been reported to contain varying active pharmacological compounds that possess antimicrobial activity.[57] Ac-AMPand Ac-AMP2 were two

(2) antimicrobial peptides isolated from *A. caudutus* seeds, and demonstrated activity against several strains of bacteria. The two (2) peptides also demonstrated activity against fungi, while the extract of *A. caudatus* also inhibits *E. coli* cell invasion into the uroepithelial cells.[58]

6.2.4 Food Applications of Pigmented Amaranth

6.2.4.1 Preprocessing of Pigmented Amaranth for Antinutritional Factor Removal

The reduction of the levels of antinutrients in amaranth grain before consumption can enhance nutrient bioavailability.[23] Cooking, germination, extrusion, fermentation, dehulling, irradiation, and ultrasound wave technology have been used to remove phytic acid in cereal grains. A higher reduction of phytate occurs during autoclaving (pressure cooking) in comparison with atmospheric cooking.[59] Extrusion exposes cereals to high shear forces and temperatures, reducing phytic acid. Soaking before extrusion may enhance the activity of the phytase present in amaranth grain, resulting in a reduction in phytic acid levels.[60] Dry heat treatment (roasting and popping) and moist heat treatment (boiling and slurring) have been applied to amaranth grains to reduce the phytate content. Boiling and slurring reduced the phytate content significantly, while popping and roasting resulted in a slight reduction.[23] These heat treatments also reduced the oxalates and tannins.

6.2.4.2 Traditional Processing of Pigmented Amaranth in Food Products

Traditional food applications such as cooking, roasting, flaking, extrusion, popping, and fermentation are applied to amaranth. Whole seeds of amaranth after cooking can be eaten or popped using short dry heat. These processes (cooking and popping) are the oldest consumption forms that have been applied traditionally. The process of popping amaranth enhances palatability due to the nutty flavor and soft texture it results in. It also compliments sweet products like granolas, cakes, mueslis or cookies. In recent times, amaranth that is popped and extruded is utilized as a major ingredient in breakfast cereals, crunchy bars, and snacks.[61] Milling (dry or wet) and fractionation are essential steps that influence the further development of amaranth products. Dry milling aims to produce whole wheat flour or to separate grains by techniques that are physically fractionated. These techniques include grinding, sieving, sifting, *etc.* the grain into parts, which involves separating the perisperm from the outer part (bran and embryo). Chemical components (starch, dietary fiber, oils, or proteins) of kernels are separated using wet milling. Steps involved in wet milling include amaranth grain cleaning, soaking (inside an aqueous solution), pulverization, filtration using different screens of decreasing sizes of mesh, and separation of the starch by the process of centrifugation. Techniques such as milling and fractionation (pilot scale) have been applied to amaranth to produce protein-rich, as well as starch-rich fractions of flour. For example, Berghofer *et al.*[62] utilized a plan-sifter and roller mill, while Tosi *et al.*[63] utilized the process of differential milling and sieving, in addition to pneumatic separation. Santra *et al.*[64] reported that the seed morphology of amaranth is a factor influencing its processing, as a result of its embryo surrounding the perisperm (starchy endosperm), which is rich in starch, and the size of its tiny seeds.

6.2.4.3 Current Applications of Pigmented Amaranth for the Development of Gluten-free Foods

The deficiency of gluten in amaranth makes it suitable for producing gluten-free food products. Several studies involving gluten-free products – such as biscuits, cakes, cookies, beverages, pasta, and bread – have been carried out.[65] Gluten-free bakery products have been produced with flour derived from amaranth, as a result of the increasing prevalence of gluten sensitivity and celiac disease. Amaranth is usually blended with other cereals containing gluten, to enhance its nutritional value. As with other pseudocereals, amaranth is superior nutritionally compared with many other gluten-free food materials. The market for the growth of gluten-free foods is a driving force in use and increased amaranth production. Cookies made from amaranth have been reported to possess excellent sensory and texture characteristics.[66]

6.2.4.4 The Potential Use of Pigmented Amaranth in Food Systems

In the application of food, amaranth and its isolated fractions offer distinct properties, as well as technological functionality. The concentrates, isolates, and proteins of amaranth have excellent foaming, stabilizing, emulsifying, and solubility properties.[67] Hence, it has high potential for use as a functional food. The high protein content and the bioactive colorant compounds of PA make it a primary option as a food supplement.[68] The starch of amaranth is typical of low content of amylose, small granular size, good gelling and thickening characteristics, excellent freeze/thaw stability, in addition to a high water-binding capacity.[69] These functional characteristics make PA a good option for use in food preparations such as pastes and custards, as well as salad. The flour of amaranth is utilized as thickeners in stews, gravy and soups. Fats in mayonnaise and salad cream have been partially replaced with starches of *A. viridis.*[70] Amaranth sprouts are utilized in salads, while flour from amaranth may replace full-fat flours, due to their better flavor, and may also be marketed as low-calorie or low-fat products. Amaranth flour is used to prepare muffins, pancakes, cakes, bread, pasta, biscuits, and extruded products, among others. Amaranth flour blended with flour from wheat is utilized in preparing unleavened flatbread called 'chapati' and tortilla in India and Latin America, respectively. The grains of amaranth are commonly utilized in India in a type of candy called 'laddoos'. The vegetables of amaranth are eaten as delicacies or as staples in many regions of the world, such as Northern India.

6.3 Pigmented Buckwheat

6.3.1 Chemical Composition, Nutritional, and Functional Value of Pigmented Buckwheat

Similarities have been observed between the chemical compositions of PB's chemical and cereals like wheat. Starch makes up its main component, with protein following, dietary fibers, lipids, and ash (approximately 70%, 12%, 10%, 3% and 2.5%, respectively).[71] Tartary buckwheat consists of a range of nutrients such as bioactive proteins and carbohydrates, vitamins, polyphenols, carotenoids, minerals, and phytosterols.

6.3.1.1 Protein and Amino Acids of Pigmented Buckwheat

All the essential amino acids are contained in PB. Similarities have been observed in the composition of protein and amino acids of Tartary and common buckwheat.[72] There is a balance in the protein regarding the composition of essential amino acids. Joshi *et al.*[73] observed the protein content to be 5.7–14.2%, with 8S globulin and 13S globulins being the major protein components. Prolamins, glutelins, and 2S globulins constitute the minor protein fractions for Tartary and common buckwheat.[74]

6.3.1.2 Carbohydrates, Starch, and Dietary Fiber of Pigmented Buckwheat

Starch makes up the main constituent of the seeds of buckwheat, amounting to about 54.5–54.7% of the total dry matter.[75] They have a high amylose content of approximately 18.3–47%, with Tartary buckwheat having a polygonal starch granule with a range of size of approximately 2–15 μm. The amylopectin possesses unit chains (DP > 100) that are extra-long with the granules possessing A-type polymorph. A considerable amount of resistant starch, which ranges from 27–33.5%, exists in Tartary, as well as common buckwheat.[74] The structural properties, as well as the applications of the starches of buckwheat, have been recently studied. Tartary buckwheat starch has been observed to have structural and properties diversity, with little difference in the structure and properties of starch between Tartary and common buckwheat.[10] The fiber of buckwheat seeds, which is water-soluble, consists mainly of xyloglucans, arabinoglucans arabinogalactans, and pectins. Other components of the fiber include hemicellulose, cellulose, and lignin.[76] Similarities in dietary fiber have been observed in Tartary and common buckwheat, when a genotype was analyzed for each. As analyzed by

Bonafaccia *et al.*,[77] 26%, 24%, and 0.54% were observed for the total dietary fiber, insoluble, and soluble fibers, respectively, for a genotype of Tartary buckwheat. In another variety of Tartary buckwheat, fiber content of 8.4% (8.2% insoluble; 0.2% soluble) was observed.[78] A typical dietary fiber category, resistant starch, is considered to have distinct health benefits, including lowering obesity and colon cancer risk. A range of values between 13.1% and 22.5% was observed for the resistant starch contents in 21 genotypes of Tartary buckwheat. These values were lower in comparison with values obtained for common buckwheat.

6.3.1.3 Lipids

The embryo constitutes most lipids in buckwheat, and lipids play a vital physiological role. Joshi *et al.*[73] observed the lipid content of buckwheat as 0.75% to 7.5%, which is more than that of other cereal types. The endosperm consists of less than 0.4% oil content, while the embryo has an average oil content of 6.5%.[79] Similarity was observed in the lipid content of 18 genotypes of common buckwheat and Tartary buckwheat (approximately 1–5%).[72] Palmitic acid, arachidic, oleic, eicosenoic, linoleic, and stearic acid constitutes the fatty acids of Tartary buckwheat.[80] The major ones include oleic and linoleic acid, with relative content of 53.8% and 27.9%, respectively. Similarities in the fatty acid value were observed between common and Tartary buckwheat, with common buckwheat abundant in unsaturated fatty acids.

6.3.1.4 Vitamins and Minerals

Different vitamins, such as Vitamin B1, B2, B6, C, and E are contained in Tartary buckwheat.[80] Majorly, vitamin E in the seeds takes the form of γ-tocopherol and is followed by δ and α-tocopherol, with values of 117.8 $\mu g\ g^{-1}$, 7.3 $\mu g\ g^{-1}$ and 2.1 $\mu g\ g^{-1}$, respectively. The bran is the location of the majority of minerals present in buckwheat. Bonafaccia *et al.*[77] evaluated the mineral content of Tartary and common buckwheat and observed a higher mineral level in Tartary compared with common buckwheat. Minerals that include potassium, magnesium, and phosphorus are considered the major minerals contained in pseudocereals. The bran houses the majority of the minerals in Tartary buckwheat. Zinc, magnesium, copper, potassium, and iron were observed in 123 genotypes of Tartary buckwheat. The pericarp in common buckwheat had a lower amount of iron, zinc, nitrogen, cobalt, and arsenic than Tartary buckwheat in a study compared to other genotypes.[77]

6.3.1.5 Bioactive Compounds and Pigments of Pigmented Buckwheat

6.3.1.5.1 Polyphenols. Polyphenol was the most studied phytochemical in Tartary buckwheat. It is a well-known fact that polyphenol's high intake lowers the risk of diseases like cancer and cardiovascular diseases, in addition to neurodegeneration.[81] Polyphenols' active substances have potential in treating various types of human diseases.[41]

6.3.1.5.2 Phenolic Compounds. Among the PPs, buckwheat is an excellent polyphenolic source. The phenolic profile of buckwheat includes alkylphenol, phenolic acids, and tyrosol.[82] A higher concentration of phenolic compounds (soluble) is possessed by genotypes of Tartary buckwheat in comparison with common buckwheat. Protocatechuic, vanillic, hydroxybenzoic, caffeic, gallic, ferulic, syringic, chlorogenic, and *p*-coumaric acids, are being identified as the phenolic acids in Tartary buckwheat. These acids are found in milled fractions (coarse bran, hull, fine bran, and light flour) by several means, including nuclear magnetic resonance, high-performance liquid chromatography, and mass spectrometry. Phenolic acids, which are caffeic, vanillic, and protocatechualdehyde acid, are the predominant soluble fraction of common buckwheat, while salicylic acid is predominant in Tartary buckwheat.[73] A higher concentration of phenolic is in buckwheat's bran, which is usually in a form that is free and is also, to a lesser extent, available in bound form.[30] Caffeic, *p*-hydroxybenzoic, chlorogenic, and protocatechuic are the main phenolic acids in the bran of buckwheat, where they are found in abundance. Kim *et al.*[83] recounted that, to increase the content of buckwheat's phenolic compound, sprouting can be employed, such as the increase in the chlorogenic acid, which extends to the other kinds of phenolic acids.

6.3.1.5.3 Flavonoids. Anthocyanins, flavones, and flavonols are the types of flavonoids present in PB. The availability of the important flavonoids, orientin, rutin, homoorientin, quercetin isovitexin, vitexin, and quercitrin in buckwheat are responsible for its nutraceutical properties.[84] The relative abundance, bioactivities and unique composition of flavonoids in Tartary buckwheat has drawn attention in recent years. In a Chinese study, 21 genotypes of Tartary, and 18 genotypes of common buckwheat were compared in terms of their total flavonoid content. The results showed that the flour of Tartary buckwheat had a total flavonoid content ten (10) times greater than that of common buckwheat. In comparison to common buckwheat, the groats of Tartary buckwheat have been shown to have a higher rutin concentration.[84]

6.3.1.5.4 Anthocyanins. Four anthocyanins (including cyanidin 3-*O*-rutinoside, cyanidin 3-*O*-galactopyranosyl-rhamoside, cyanidin 3-*O*-glucoside in addition to cyanidin 3-*O*-galactoside) were identified in the sprouts of common buckwheat.[85] Among 19 genotypes of Tartary buckwheat, a high level of anthocyanin was identified in the genotype Hokkai T10. Breeding or genetic modification may be used in buckwheat seed processing. Tartary buckwheat leaves and sprouts were identified using high-performance liquid chromatography.[85]

6.3.1.5.5 Flavonols and Flavones. Flavonols constitute the main polyphenols in Tartary buckwheat. Flavonols such as catechin, and epicatechin were discovered in the root, sprout, and stem, as well as in the leaves of Tartary buckwheat.[85] At the same time, isoretin, orientin, isovitexin, and vitexin were types of flavones identified in the seeds, germinated seeds, and sprouts of Tartary, and common buckwheat, using HPLC.[83] Other flavonols identified in several PB components (*i.e.* sprouts, leaves, seeds, and flowers) are rutin, kaempferol, isoquercitrin, quercetin-3-*O*-galactosyl-rhamnoside, kaempferol-3-*O*-β-D-glucoside, among others. Rutin (quercetin-3-rutinoside) from the stem, fruits, leaves, and flowers of buckwheat, is utilized to treat disorders in the vascular system.

6.3.2 Functional and Health Benefits of Pigmented Buckwheat

6.3.2.1 The Health Benefits of Pigmented Buckwheat in Pharmacological and Clinical Studies

The antitumor and immunomodulatory effects of polysaccharides (natural) are the major significant biological activities. Several health benefits of PB bioactive components have been reported. Zhu[71] reported the bio-functions of a bioactive compound present in PB (anthraquinones) as anti-cancer, anti-bacterial, and liver protective. Resveratrol, a stilbenoid, has shown a great range of activities, including cardio-protection, anti-cancer, neuro-protection and anti-inflammation.[86] Anti-proliferative activities have been shown in the extract of PB rich in phototoxic compound fagopyrin in addition to several essential bio-activities.[71] The anti-hypertensive activity of PB has been reported. High dietary consumption of carotenoids shows an association with reduced occurrence of several types of diseases, including cardiovascular diseases, UV-induced damage to the skin, some kinds of cancer, muscular degeneration and cataracts.[87] β-Carotene and lutein constitute the major carotenoids available in Tartary buckwheat sprouts. Gu *et al.*[88] showed that buckwheat contains polysaccharide, which helps prevent diabetes by lowering blood sugar through the inhibition of α-glucosidase. Reports by Zhou *et al.*[74] also indicate the reduction in pancreatic lipase catalytic activity of buckwheat polysaccharide by altering its conformation. The polysaccharide of buckwheat may inhibit pancreatic lipase, which could help in obesity management.

6.3.2.2 Antioxidant Activity

Polysaccharide possesses antioxidant activity, as discovered through *in vitro* assays. This is attributed to its several bioactive compounds.[89] Studies have shown that extracted materials of different parts and end products like noodles from Tartary buckwheat have antioxidant attributes. For this reason, the extract can serve as a kind of preservative. The ethanol extract from the sprouts of Tartary buckwheat contained highly efficient antioxidant properties in reducing oxidative stress in human hepatoma compared with ethanol extracts of common buckwheat. Buckwheat polysaccharide's antioxidant mechanism using *in vitro* assay is accredited to an enhancement in antioxidant activity of enzymes, protecting biomolecules from free radical-induced damage.[90] *In vivo* assays employed in determining the activities of antioxidants of polysaccharides of buckwheat are not much described.

6.3.2.3 Antimicrobial Activity

The phenolic compounds like isoquercetin, quercetin, and extracts from the bran of Tartary buckwheat have been demonstrated to act against *Staphylococcus aureus*, and *S. epidermis* as bacterial activity, as well as *Propionibacterium acnes*. The bran of Tartary buckwheat may be utilized in cosmetics to prevent and relieve acne.[91]

6.3.3 Food Applications of Pigmented Buckwheat

6.3.3.1 Preprocessing of Pigmented Buckwheat for Removal of Antinutritional Factors

Various processing methods have been employed to limit antinutritional factors in novel and traditional foods. These factors include trypsin, tannins, oxalate, phytic acid, saponins, and trypsin inhibitors, which enhance cereal grains' eating quality and safety. Techniques such as boiling, high hydrostatic pressure, soaking, dehulling, cooking, germination, extrusion, radiation, and fermentation, using phytase and ultrasound waves, as well as storage, have been employed to reduce phytic acids in foods.[92] Xie *et al.*[90] reported that boiling, steaming, and Se-enrichment improved edibility and eliminated the antinutrients (trypsin, polyphenols, amylase activity in addition to phytic acid) in Tartary buckwheat. A method to reduce antinutritional factors in buckwheat includes microwave processing, which is advantageous because of its speed, safety, consistency, and environmental friendliness.[92] Pre-treatment, such as soaking, also reduces buckwheat's antinutritional factors, through hydrolysis and leaching.

6.3.3.2 Traditional Processing of Pigmented Buckwheat in Food Products

Primarily, buckwheat processing involves cleaning of the matured grains, dehulling, and milling.[93] Dehulling is carried out to separate the groats of buckwheat from its hulls by abrasion or impact of the seeds against steel or stones, followed by screen or air separation of the hulls and groats. Milling is usually carried out on buckwheat to remove the hull, after which a second milling is done to remove the middlings. This process aims to produce buckwheat flour that is light brown in color and white flour when further milling is done. Buckwheat flour is used to produce pancakes, noodles, and baked goods. It is incorporated into gluten flour, such as wheat flour or semolina. Cracked groats of buckwheat can also be sold as grits, while groats that are roasted are sold as 'kasha', a popular food consumed in Russia as well as Eastern Europe. Reports on the applications of the polysaccharide and fiber fractions of buckwheat have been examined.[94]

6.3.3.3 Current Buckwheat Applications of Pigmented Buckwheat for the Development of Gluten-free Foods

Pseudocereals as gluten-free grains are becoming prevalent in the diets of humans, as they have exceptional nutritional as well as nutraceutical importance. They have become known as alternatives in the formulation of gluten-free foods, as they constitute an essential source of several bioactive substances. Buckwheat forms the main ingredient in their daily diet for people suffering from celiac disease. Buckwheat is commonly utilized in flour production, which is employed in product formulations such as gluten-free bread.[95] Other gluten-free foods developed using buckwheat include pancakes, snacks, noodles, cookies, crackers, muffins, pasta, beverages, and beer. Its incorporation into several food formulations has also been established. Dehulling is usually done when buckwheat is processed as an ingredient in gluten-free foods.[96]

6.3.3.4 Potential Use of Pigmented Buckwheat in Food Systems

Modern foods made with PB include vermicelli (dried noodles from buckwheat), instant noodles, spaghetti, macaroni, cake, pastry, bread, vinegar, biscuits, breakfast cereals, and beverages (*e.g.* buckwheat beer, buckwheat tea, and liquor). Tender shoots of buckwheat are utilized as vegetables, which can also be processed as canned vegetables. India and China are known for producing honey from buckwheat. In Russia, Poland and parts of Eastern Europe, groat porridge, meals, and soups from buckwheat flour are favorite indigenous foods. In China, it serves as a medicinal herb and a staple nutritional crop cultivated for a long while. Foods such as wantuo, hele, porridge, griddle cakes, pancakes, and jelly noodles are popular foods in China due to their special flavour, different processing methods, excellent nutritional properties, long history, and role as functional foods in the maintenance of human health. Soba and mook are noodles and jelly-like foods derived from buckwheat in Japan and Korea, respectively. The fiber and polysaccharide fractions of buckwheat have been utilized in various applications, such as bread making. Ballester-Sánchez *et al.*[94] reported that the hemicellulose of the hull of buckwheat was used in the production of bread. Buckwheat functions as a major potential food ingredient in functional and clinical food systems. Several chemical modifications, such as acetylation, carboxymethylation, sulfation, selenization, and phosphorylation have been carried out on buckwheat polysaccharides to improve their functionality.[9] Wang *et al.*[97] chemically modified the fiber of buckwheat using carboxymethylation, hydroxypropyla, and cross-linking. Some countries in Asia recognize the functionality of products from buckwheat in the maintenance of human health. Alkaline extracted from the polysaccharides of Tartary buckwheat was used to coat tilapia fillets.[98] The polysaccharide gelation enhanced the physicochemical, senso, and microbiological properties of the fillet, thereby enhancing its shelf life. The nutritional value of buckwheat enhances its potential for nutraceutical and pharmaceutical possibilities.

6.4 Pigmented Quinoa

6.4.1 Chemical Composition, Nutritional, and Functional Value of Pigmented Quinoa

The characteristics of the seed components are the factors that make PQ attractive to the production system. PQ (Figure 6.3) is a starchy material with a high carbohydrate content that mostly includes starch and a small amount of sugar. PQ has a high nutritional value, mainly due to its high content of good-quality protein. It also has dietary fiber, B-complex vitamins, vitamin C and E and wide minerals like sodium, copper, phosphorus, zinc, manganese, potassium, iron, magnesium, and calcium. Compared with true grains such as rice, wheat, maize, and barley, PQ contains higher amounts of protein and essential amino acids, *e.g.* methionine and lysine, with fatty acids comparable to the lipid composition of soybean.[99] The National Academy of Sciences recommended PQ among 23 promising foodstuffs for investigations, pointing to its potential to improve nutrition and quality of life in developing countries.[100] Cereals play a major role in human nutrition by fulfilling about 50% of the daily protein and energy requirement. The most vital foods in human and animal diets globally are wheat, rice, corn, barley, rye, oat, and sorghum. Abdellatif[14] compared the nutritional contents of common cereals

Figure 6.3 Pigmented quinoa seeds with different colors. Source: Personal archive of Taha Mehany.

(rice, barley, wheat, corn, rye, and sorghum) with PQ and demonstrated the richness of these nutrients in PQ compared with the true cereals, which contain higher protein, lipid, and ash concentrations.

6.4.1.1 Protein and Amino Acids

Compared with the true cereals, the total protein content of PQ (12–23%) is higher than that of rice (7.5%), corn (13.4%), and barley (11.0%), and is comparable to that of wheat grains (15.4%).[101] Proteins contribute to building and maintaining tissues, forming hormones, enzymes, and antibodies, providing energy, and regulating metabolism. Amino acids are a source of sulfur compounds, which are needed by our bodies. Globulins (11S) and albumins (2S) are the main protein subunits from PQ at approximately 37% and 35%, respectively, whereas the prolamins are low in content. 2S has a molecular mass of 8–9 kDa and contains a heterogeneous group of polypeptides. PQ amino acid composition showed high histidine, cysteine, and arginine content, which have vital nutritional aspects, particularly for children. 2S contributes to the more significant level of sulfur amino acids, showing the potential of dietary supplements.[100] 11S, also called chenopodin, is an oligomeric protein with a molecular mass of 320 kDa. It is composed of an acid and a basic polypeptide with molecular masses of 30–40 kDa and 20–25 kDa, respectively. 11S contains a high amount of glycine, arginine, leucine, serine, asparagine–aspartic acid, and glutamine–glutamic acid, but it has a low amount of sulfur amino acids.[100] Elsohaimy *et al.*[102] reported that quinoa protein isolate bands with 55 KDa correspond to globulin, while 31–33 kDa match to hemoprotein. PQ has a low gluten content, due to the low levels of glutamines and prolamines, and this makes it suitable for celiac disease.[103] Prolamins in wheat and barley cause digestive problems for those suffering from celiac disease. To treat this disease, patients have to avoid gluten in their daily diet. Gluten allergy is one of the major nutrition issues worldwide.[104] The number of essential amino acids reflects the nutritional quality of proteins. If only one of these amino acids is limiting, the others will not be fully absorbed, causing loss of dietary protein and less growth. PQ protein supplies more than 150% of children's requirements and more than 200% of adults. The essential acids for humans are valine, histidine, threonine, phenylalanine, isoleucine, methionine, tryptophan, lysine, and leucine. PQ protein can supply around 338% of the lysine, 331% of the threonine, 320% of the phenylalanine and tyrosine, 228% of the tryptophan, 274% of the isoleucine, 212% of the methionine and cysteine, 180% of the histidine, and 323% of the valine suggested in protein sources for adult diet.[105]

6.4.1.2 Carbohydrates, Starch, and Dietary Fiber

Carbohydrate is the main constituent of PQ; starch and fiber are the significant components. Starch is the highest carbohydrate constituent of quinoa grain and is present in 52–69%. Total dietary fiber (TDF) is the plant-derived indigestible portion of food and has two main components in terms of soluble fiber content (SDF), and insoluble dietary fiber (ISDF). SDF dissolves in water and is readily fermented into gases and physiologically active substances in the colon, with prebiotic properties. ISDF does not dissolve in water, it provides bulking mass and is metabolically inert, or has prebiotic properties and metabolically ferments in the large intestine.[106] TDF in PQ is close to the real grain content (7–9.7%). SDF is ranges from 1.3% to 6.1%. PQ contains 3% of simple sugars, normally maltose, followed by D-galactose and D-ribose, while it contains low glucose and fructose ratio.[100] PQ is an outstanding source of fiber, containing about 2.6–10%, of which about 22% is SDF, and 78% is ISDF. More fiber-rich whole grain consumption is related to a lower risk of type 2 diabetes[107] and cardiovascular disease.[108]

6.4.1.3 Lipids

PQ is an alternative oilseed crop, due to its lipid quantity and quality. PQ has a fat content between 2.0% and 9.5%, which is higher than corn (4.7%) and other common cereals. PQ lipids are rich in essential fats like α-linolenic (omega 3), oleic (omega 9), and linoleic (omega 6), and also contain high levels of natural antioxidants, such as α- and γ-tocopherol. A comparison of the fatty acids in PQ oil to those in soybeans and corn showed that they have similar levels of linoleic, oleic, and linolenic. The unsaturated fatty acids (UFAs) constituted (87.2–87.8%) of the total fats present in PQ oil, similar to soy oil structure. UFAs consist of oleic (19.7–29.5%), linoleic (49.0–56.4%), and linolenic (8.7–11.7%). The major saturated fatty acid found in PQ fatty acids was palmitic, which constituted about 10% of the total fats. UFAs are essential for reducing the risk of cancer, cardiovascular, inflammatory, and autoimmune diseases.[109]

6.4.1.4 Vitamins and Minerals

PQ is rich in vitamins that are essential for human health. According to their solubility, they are divided into two groups: water-soluble vitamins like B-complex and vitamin C, and fat-soluble vitamins such as K, D, E, and A. 100 g of PQ contains 16.4 mg of vitamin C, 0.4 mg of thiamine, 0.20 mg of vitamin B6, 78.1 mg of folic acid, and 0.61 mg of pantothenic acid. B1 in PQ is lower than in barley and oats, but B2, B6, and folic acid are higher than in PQ than in most common cereals, *e.g.* wheat, rice, barley, and corn. PQ is an excellent source of vitamin E, in which it is higher than in wheat, and B2 in PQ contributes 80% of children's daily requirements and 40% of adults'.[110] PQ also is rich in vitamins A, B2, and E.[111]

Unlike carbohydrates, proteins, and lipids, minerals are inorganic matter that cannot be produced by living beings. They have very vital functions in the body. The most well known are iron, potassium, sulfur, calcium, phosphorus, sodium, magnesium, zinc, copper, selenium, and chromium. These vital micronutrients must be consumed through a balanced diet in order to achieve the daily requirement.[110] PQ ash content is 3.4%, which is more than that of wheat (1.8%), and rice (0.5%). PQ contains a high amount of minerals. The amounts of iron and calcium are higher than in common cereals.[111] In PQ, the mineral content is: calcium (27.5–148.7), copper (0.2–5.1), iron (1.4–16.8), magnesium (26.0–502.0), phosphorus (140.0–530.0), potassium (7.5–1200.0), and zinc (2.8–4.8) mg per 100 g, respectively. There is variation in the mineral content in PQ, due to the seeds being from different genotypes and varying soil types with different mineral compositions.[110]

6.4.1.5 Bioactive Compounds and Pigments

PQ has exceptional bioactive substances due to its individual phytochemical constituents that have complete biological activities and several applications for human health. The major groups of secondary metabolites in PQ are triterpenoids such as phenolics, saponins, and phytosterols.[13,14] In most developed countries and in developing countries, heart disease are the major reason for death. Foods that include PPs and whole cereals may contribute to cardiovascular protection. PQ is often an important source of natural antioxidants, bioactive colorants, and other healthy micronutrients, and plays a significant part in coronary heart diseases defense.[112]

6.4.1.5.1 Phenolic Compounds. PQ is one of the best sources of bioactive polyphenols, which prevent oxidative stress and have the potential to change the antioxidant status of the cell. The main phenolics found in PQ are vanillic acid, gallic acid, caffeic acid, *p*-hydroxybenzoic acid, *p*-coumaric acid, cinnamic, syringic acid, and ferulic acid.[113] Yawadio Nsimba *et al.*[114] reported PQ extract from Bolivia and Japan; they found that the Japanese one had the highest phenolic content with high antioxidant activity, whereas the Bolivian sample offered the lowest content. The difference in the TPC between both PQ ecotypes is due to genetic factors or different environmental conditions. Polyphenols are classified into numerous subclasses *e.g.* flavonoids, anthocyanins, flavonol glycosides, and isoflavones.[115]

6.4.1.5.2 Flavonoids. There is a wide diversity of flavonoids in PQ; for example, anthocyanins, flavones, flavanones, flavonols, isoflavonoids, and flavanols have many health benefits, especially their contribution to chronic disease prevention. The phenolic profile of flavonol compounds in PQ is: Flavonols spinacetin 3-*O*-glucosyl-(1-6)-glucoside, 3-methoxysinensetin, kaempferol 3-*O*-rhamnosyl-rhamnosyl glucoside, kaempferol 3-*O*-xylosyl-rutinoside, kaempferol 3-*O*-xylosyl-glucoside, methylgalangin, kaempferol 3-*O*-rutinoside, kaempferol 3-*O*-glucoside, quercetin 3-*O*-galactoside 7-*O*-rhamnoside, quercetin 3-*O*-rhamnosylgalactoside, kaempferol 3-*O*-sophoroside, quercetin 3-orutinoside, quercetin 4′-*O*-glucoside, myricetin 3-*O*-rhamnoside, quercetin 3-*O*-arabinoside, quercetin 3-*O*-galactoside/

glucoside, quercetin 3-*O*-glucosyl-xyloside, patuletin 3-*O*-glucosyl-(1-6)-[apiosyl(1-2)]-glucoside,quercetin3-*O*-rhamnoside,kaempferol3-*O*-(2″-rhamnosylgalactoside) 7-*O*-rhamnoside, kaempferol 3-*O*-galactoside 7-*O*-rhamnoside, quercetin 3-*O*-xylosyl-rutinoside, quercetin/morin, kaempferol, quercetin 3-*O*-glucuronide, isorhamnetin, kaempferol 3,7-*O*-diglucoside, kaempferol 3-*O*-galactoside, kaempferol 3-*O*-glucosyl-rhamnosyl-glucoside, and quercetin 3-*O*-rhamnosyl-rhamnosyl-glucoside.[116]

6.4.1.5.3 Phytosterols and Phytoecdysteroids. Phytosterols are belonged to the family of triterpenoids, which are similar to humans' and animals' cholesterol structure, while phytoecdysteroids are known as secondary plant metabolites that defend the plant against insects and nematodes attacking. Martínez-Villaluenga *et al.*[116] reported that phytosterol content was (63.7 mg per 100 g) of β-sitosterol, (15.6 mg per 100 g) of campesterol, and (3.2 mg per 100 g) of stigmasterol. Sterols in PQ are higher than in corn, pumpkin, and barley. These sterols have different biological effects, such as anti-inflammatory, antioxidant, and anticarcinogenic benefits.[100] The total phytoecdysteroid content of PQ varies from (138–570 $\mu g\ g^{-1}$). Different phytoecdysteroids have been extracted from PQ, such as 20-hydroxyecdysone (62–90%) of total phytoecdysteroids. Makisterone A, 24-*epi*-makisterone A, and 24 (28)-dehydromakisterone A were also found in PQ.[13]

6.4.1.5.4 Saponins. Saponin is a glycosylated secondary metabolite natural compound used as a killer repellent by plants. The key structures of sapogenins known in PQ are phytolaccagenic acid, oleanolic acid hederagenin, and 30-*O*-methyl-espergulagenate.[117] The quantity of saponins of PQ is minor compared with that found in legumes.[100] Saponin content varies from 0.1–5% and causes a bitter after taste, which could cause gastric irritation; quinoa grain is classified according to its free saponin content, as either bitter quinoa when the content is >0.11%, or sweet when it is <0.11%.[118] Although some are considered toxic, other saponins could also have positive health benefits, acting as a fungicide, insecticide, and as an antibiotic. In the pharmaceutical industry, saponins are used as precursors to synthesize steroids, hormones, contraceptives, and anti-inflammatories. Additionally, saponin is known as a foaming agent in water and is used in soaps, toothpaste, detergents, shampoos, and hair lotions due to its emulsifying and surfactant properties.[117] Similarly, PQ saponins have a broad biological impact, providing anti-cancer, antifungal, antiviral, hypoglycemic, hypocholesterolemic, antithrombotic, and anti-inflammatory protection.[110] To consume quinoa seeds in food products, they must be free from the bitter taste caused mainly by saponins. Abdellatif[14] found an efficient method to remove saponins from raw quinoa seeds by vigorously washing the whole seeds with cold water, then soaking the washed seeds in 1% $NaHCO_3$ solution for about 20 min. After that, the soaked seeds were rinsed with water until no foam appeared, indicating removal the saponins from the quinoa

Figure 6.4 (a) Raw quinoa seeds; (b) quinoa seeds with saponins; (c) the rinsed quinoa seeds without saponins. Source: Personal archive of Taha Mehany.

seeds hull. Figure 6.4 shows the raw quinoa seeds with saponins, and the rinsed quinoa seeds free from saponins.

6.4.2 Functional and Health Benefits of Pigmented Quinoa

6.4.2.1 The Health Benefits of Pigmented Quinoa in Pharmacological and Clinical Studies

PQ is an excellent functional food, which reduces the risk of various diseases. Its functional properties may be associated with the presence of many constituents such as minerals, fiber, essential fatty acids, vitamins, pigments, natural antioxidants, and phytohormones, which contribute to human nutrition. These ingredients give PQ great advantages over other plants in terms of human health maintenance. The high nutritional value, therapeutic characteristics, and GF features of PQ may benefit high-risk consumers, such as the elderly, children, people with dyslipidemia, lactose-intolerant people, osteoporosis-prone women, and those with diabetes, obesity, anemia, or celiac disease.[110] PQ functional-based products like flakes, cereal, and pasta have several health benefits.[2] Many studies regarding PQ consumption have revealed that numerous foods and nutraceutical products have been developed. Several clinical studies were performed to assess the effects of quinoa consumption on human health. Ruales *et al.*,[119] reported that 100 g of quinoa consumption added to baby foods twice a day for 15 days significantly increased the plasma insulin-like growth factor levels, compared with the control. This research concluded that foods were necessary to provide protein and other vital nutrients, which are the main factors in avoiding malnutrition amongst children. Farinazzi-Machado *et al.*,[120] showed that daily consumption of PQ candies by 22 students aged between 18 and 45 years for 1 month significantly reduced their total cholesterol, triglyceride, and low-density lipoproteins (LDL) levels. Furthermore, four clinical investigations have documented that PQ supplementation applies positive effects on cardiovascular, metabolic, and gastrointestinal human health.[13] In another study that aimed at examining the *in vivo* effects of PQ consumption with celiac disease in adults, 19 celiac patients consumed 50 g of quinoa daily for 6 weeks as part of a gluten-free diet. This study revealed that celiac disease patients safely tolerated daily consumption of 50 g of PQ.[121]

6.4.2.2 Antioxidant Potential

PQ is an excellent matrix of natural antioxidants.[122] Antioxidants can play a vital role in free radical inhibition and oxidation, specifically for cell membrane protection. PQ antioxidants showed success in neural functions, reducing the risk of degenerative diseases related to oxidative stress like cancer, cardiovascular disease, and osteoporosis. In addition, when antioxidants are added to food, it decreases the rancidity, retards the formation of toxic products, maintains nutritional quality, and increases the product's shelf life.[110] Hence, Hirose *et al.*[123] stated that PQ extract of quinoa farmed in Japan have phenolic compounds, mainly quercetin glycosides in large amounts, which are responsible for their high antioxidant capacity. In another study by Paśko *et al.*,[124] the effects of a rat's diet supplemented with PQ on oxidative stress and plasma was studied. The findings demonstrated that PQ could act as protecting agents for rats by reducing lipid peroxidation in the plasma and increasing antioxidant power.

6.4.2.3 Antimicrobial Activity

Different PQ varieties originating from Chile were analyzed for their antimicrobial activity through disk diffusion assay. PQ extracts showed antimicrobial activity, and results were around 8.5–15.0 mm inhibition zone for *S. aureus* and 8.3–14.8 mm for *E. coli.*[125] PQ can also produce a range of byproducts for food, cosmetic, and pharmaceutical applications. PQ saponins can also be used in the pharmaceutical industry because they can encourage changes in intestinal permeability, which can be helpful for the absorption of specific drugs and in hypocholesterolemia. PQ saponin can be used as an antifungal and antibiotic, amongst other pharmacological compounds.[19]

6.4.3 Food Application of Pigmented Quinoa

6.4.3.1 *Preprocessing for Removal of Antinutritional Factors*

To provide improved sensorial attributes of PQ-based foods, saponins that give a bitter after taste are regularly removed or reduced from the grains. There are numerous approaches for saponin removal, without any significant change in the nutritional value of PQ. For example, by a wet technique using vigorous cold water washing or in alkaline water solution, or through dry methods like extrusion, roasting, heat treatment, mechanical abrasion, or a combination of these approaches.[126] The wet method is the most used in the removal of the antinutritional factors from quinoa, but this method is non-economic, and has environmental impacts due to the use of large amounts of water and the need for drying these washed grains, also, the water becomes contaminated with wasted saponins.[127] In addition, on a commercial scale, the dry technique of abrasive peeling is utilized to remove saponin. A disadvantage of this process is nutrient loss, *i.e.* protein, minerals, and vitamins. So using the wet method jointly with polishing is highly recommended, giving PQ a successfully reduced saponins ratio.[128]

6.4.3.2 *Traditional Processing of Food Products*

Many studies have been provided to observe the effect of quinoa seeds' applications in food systems as functional food products, because of their nutritional value, their functional properties and their ability to enhance the nutritional qualities of other grains.[129] PQ has been used in a wide variety of food, from whole seed, soups, salads, sweets, beverages, porridges, sauces, breakfast grains, rice, and infant cereal-based food, and fermented or hot beverages formulation. PQ is germinated or malted for beer manufacture.[105,130]

6.4.3.3 *Applications for Pigmented Quinoa in the Development of Gluten-free Foods*

Many investigations have been made into GF foods from PQ. The final products are also assessed for their nutrient content, including protein, fat, dietary fiber, and minerals. PQ-based GF foods have good health benefits for human nutrition.[2] PQ flour can be combined with maize flour or wheat in bread (10–13%), biscuits (60%), noodles and pasta (30–40%).[130] Gluten-free bread blends of quinoa, buckwheat, rice flour, and potato starch were recognized, and the functional and rheological properties of the bread were evaluated. The quinoa flour ratio didn't affect specific volume, baking loss, or protein content. Nonetheless, 25% of PQ-based bread presented higher sensory qualities and a softer texture profile.[131] Recent trends in GF baked goods using PQ are increasingly spreading to the formulation of novel foods, for example, for the development of cookies with PQ that result in a high nutritional value, sensory, functional, and technological qualities in the end products.[116]

6.4.3.4 The Potential Use of Pigmented Quinoa in Food Systems

Many studies have investigated the potential of PQ in food applications. For example, Pineli *et al.*[132] developed a novel quinoa beverage emulsion and exposed its benefits for diabetic individuals, caused by their lower glycemic index properties. Similarly, Kaur and Tanwar[133] prepared quinoa-based milk from raw, soaked, germinated, and malted quinoa grains and examined the antidiabetic, *in vitro* antihypertensive potential antioxidant activity. The results showed that the malted quinoa beverage amongst all beverages showed high protein content, TPC, and antioxidant power. Numerous attempts have also been made to develop a novel functional food based on PQ flour, to supplement other cereals with additional nutritional benefits.[134] Other studies developed a novel flatbread made from wheat flour with PQ flour. The rheological characteristics of the enhanced flatbread such as appearance, texture, specific volume, odor, and color were assessed. They showed an exceptional quality, with high nutrient content in terms of amino acids, minerals, and protein in the developed products, compared with the control.[135] Moreover, a study of bread making from quinoa showed that supplementing corn and rice flour with up to 10% of quinoa bran notably enhanced, not only the bread volume by 7%, but also the protein by 17% in the final product compared with control.[30] Another study examined wheat flour supplementation with quinoa flour up to 30% to produce pan bread. The results revealed that the protein content was higher in the pan bread (up to 66.40%) than in the control, and that essential amino acids were increased. Additionally, calcium, iron, manganese, phosphorus, potassium, magnesium, and zinc were significantly amplified. The produced pan bread with 30% of PQ flour didn't affect the sensorial scores and technological aspects of the final pan bread product.[136]

6.5 Techno-functional Properties of Pigmented Pseudocereal (Amaranth, Buckwheat, and Quinoa) and Its Molecules in Food Systems

6.5.1 Solubility and Absorption

Solubility is the ability of solids to disperse in aqueous solution. It is influenced by the level of interaction among starch chains within the crystalline and amorphous regions of starch granules, constituent amino acids, the hydrophilic–hydrophobic balance of proteins and environmental factors of iconic strength, pH, and temperature.[102,137,138] High solubility of Tartary buckwheat starch (TBS) was observed at high nitrogen levels, attributed to greater water affinity of small granules when compared with large granules at high nitrogen levels.[138] Under low ionic strength conditions, amaranth protein isolates (API) displayed higher solubility at pH of 2.0 compared with pH of 6.3, and solubility of quinoa protein isolates (QPI) reached a maximum of 75.21% at alkaline pH of 10.[102,139,140]

The processing affects water–protein interactions and depolymerization of starch to influence water-binding properties. It was shown that the formation of amylose–lipid complexes in TBS during high hydrostatic pressure treatment inhibited the mobility of soluble amylose molecules to reduce solubility.[137,141] Also, flour composition was revealed as the most significant factor influencing water absorption capacity (WAC) in a study where WAC increased, and water absorption index (WAI) decreased with the increasing particle size of quinoa flour (QF), as larger-sized fractions presented higher protein and starch contents and increased surface area for enhanced interaction and absorption of water.[142]

6.5.2 Foaming Properties

Foaming properties are essential in food systems for whipping and aeration, and are primarily associated with the formation of continuous, cohesive films of proteins around air bubbles in the foam.[141,142] The ability of QPI to create foam has been demonstrated, where foaming capacity (FCY) and foaming stability (FSY) increased with increasing protein concentration and ranged from 58.37–78.62% at time zero; and from 34.83–68.39% after 60 min, respectively.[102] In another study, it was revealed that higher protein concentration and solubility of the pseudograin protein correlated with higher FCY and FSY, with improved values for the spray-dried protein, indicating that a greater number of air bubbles might be trapped in the interfacial film formed around the air bubbles by proteins.[141] However, it was reported that the foam formed by medium fraction QF disappeared after 60 min, and this was attributed to its high lipid content.[139] While high protein concentration increases FSY from the formation of the strong interfacial film, high FSY demonstrates the enhanced surface activity of soluble native proteins in the continuous water phase.[141,142]

6.5.3 Emulsifying Properties

Emulsions are stabilized by membrane matrices that offer electrostatic and steric repulsion on the oil droplet surface for preventing coalescence.[102,142] Emulsions have been applied extensively in the cosmetics, food systems, pharmacological, and agrochemical industries.[143] Emulsifying stability index (ESI) and emulsifying activity index (EAI) of QF from different varieties were 39–100% and 15–41%, respectively, and the high values were attributed to high protein content.[144] QPI showed increasing EAI and decreasing ESI with increasing protein concentration at values of 1.24–3.38 m^2 g^{-1} and 30.37–46.34 min, respectively.[102] Likewise, reduced protein content resulted in reduced emulsion index of Pickering emulsions prepared from amaranth protein (AP) and quinoa protein (QP), with AP exhibiting superior emulsifying ability due to more angular shape and smaller particle size.[145] Different fractions of QF presented similar EAI of 25–62.5% at different temperatures, with whole grain and fine fractions displaying higher values in comparison with coarse fractions, suggesting the reliability of EAI on particle size.[142] Processing conditions also influence ESI and EAI, with decreases in QPI during ultrasound treatment in comparison with native QP, which is attributed to induced aggregation of protein molecules that decreased solubility and surface hydrophobicity.[146]

6.5.4 Gelling Properties

Gel formation is influenced by covalent and non-covalent interactions that can affect protein and starch structure. Gelatinization temperature of amaranth flour (AF) and QF were higher when compared with starch isolated from the seeds, attributed to hydrophobic and hydrophilic interactions/compounds and properties of dietary fiber and protein.[147] Coarse and fine fractions of QF with high protein and starch content required low concentration to form gel, compared with medium fraction and whole grain flour, which is attributed to competition for water between starch gelatinization and protein gelation.[142] Zhang *et al.*[138] suggested that variations in starch granule size and amylose content at various nitrogen levels of TBS may have restricted the hydration of amorphous areas to increase gelatinization temperature. Gelation ability is also influenced by pH, where QPI extracted at low pH of 3–4.5 presented semi-solid gels with dense microstructure compared with higher pH extraction.[139,148] Gamma irradiation and prolonged ultrasound treatments reduced the gelatinization temperature and gelling capacity of TBS and QF, respectively, due to the reduction in overall crystallinity and destruction of amylopectin chains from partial depolymerization.[149,150] In contrast, ultrasound treatment unfolded the compact structure of QP nanoparticles to present shear-thinning behaviour and the formation of rigid gel-like network structure, while complete gelatinization of quinoa starch to form a strong gel network due to amylopectin crystallites breakdown and amorphous state transformation occurred during high-pressure treatment at 600 MPa.[151,152]

6.5.5 Flavor Binding

The enhancement of flavor retention and flavor binding correlates directly with water holding capacity (WHC) and oil absorption capacity (OAC), and improves mouthfeel.[139,144] QPI showed OAC of 1.88 mL g^{-1}, which was comparable with the values for soy protein (2.10 mL g^{-1}) and wheat protein (1.58 mL g^{-1}).[102,144] The WHC of QF from different varieties, with values ranging from 1.44–1.80 g water per g flour, and fractions during dry fractionation with values ranging from 2.29–2.56 g water per g sample did not present significant differences.[142,144] However, there was a significant reduction in the oil binding capacity (OBC) of QPI subjected to ultrasound treatment compared with native QP, even though ultrasonic oxidation of QPI increased OBC with increasing treatment time. This study revealed that ultrasonic oxidation induced QP conformation that allowed entrapment of oil molecules to increase OBC.[146]

6.6 Conclusions and Future Perspectives

Pigmented pseudocereals are considered natural functional foods due to their nutritional values and various industrial applications, including as bioactive pigments. Certainly, pigmentation makes them a valuable source of bioactive compounds with health-promoting benefits. Consumers' demand for PPs has been increasing, due to their role in preventing several diseases, including celiac disease, obesity, cancer, cardiovascular disease, diabetes, and chronic inflammation. Moreover, PPs have high antioxidant activity, boost the immune system, and reduce serum cholesterol levels. This study offers a comprehensive overview of chemistry, the nutritional value, functional benefits, applications, and techno-functional properties of the main PPs, *e.g.* amaranth, buckwheat, and quinoa. Current data reveal that when staple foods such as pigmented amaranth, buckwheat, and quinoa is integrated into a balanced food pattern, it contributes significant and high-quality protein, dietary fiber, polyunsaturated fatty acids, essential minerals (iron, calcium, magnesium, zinc and manganese), vitamins, and bioactive compounds, serving as an antioxidant that helps to prevent degenerative diseases. PPs natural extracts also act as antimicrobial agents. Our findings show that modern food products from PPs can include noodles, spaghetti, macaroni, cake, pastry, muffins, bread, vinegar, gluten-free foods, biscuits, breakfast cereals, plant-based milk, and other beverages. Published studies in both human and animal models revealed the decreasing effect of these pigmented pseudograins on total and low-density lipoprotein cholesterol and triglycerides, making it a strategy in reducing cardiovascular disease. Incorporating bakery products with these PPs flour did not dramatically affect the final products' technological and sensory aspects, which encouraged the food technologists and nutrition scientists to use PPs flour to produce different functional bakery products to overcome the generated diseases. Due to the excellent techno-functional properties of PP flour and its molecules, they have various applications, including as supplemented cereals on various levels, as plant-based beverages, as food additives, and as a gel-forming agent, foam enhancer, and emulsifier.

Globally, the importance of PPs is well known. In comparison with other cereal grains like wheat, maize, rice, and sorghum, there are fewer studies on the breeding quality of PPs, which makes it a crucial topic due to its difficult quality evaluation establishment. Even though there are subjective traits of its palatability as well as its acceptability, an index that is measurable in terms of objective traits which relate to palatability in addition to acceptability is essential in the evaluation of its quality for breeding. In addition, PP sensory evaluation development is as essential as developing expert panels. A powerful tool that has been reported in PPs is genomic selection. This technology would be relevant in both enhancements of agronomic characteristics, as well as quality characteristics of PPs. Future focus should be on research opportunities to further understand the functionality and utilization of fiber polysaccharides from PPs. The development of the processing techniques of PPs is still required by using improved varieties, so these high-quality foods can be spread. For the improvement of PP flavor, an objective evaluation method is essential. As an important famine crop, the growing period of PPs, which is short, is an encouraging characteristic during periods of famine. A high flavonoid content or antinutritional factors that cause a bitter taste in these crops affects consumers' acceptability. The development of cultivars with a less bitter taste or no bitter taste should be carried out. Other parts of PP plants, such as their leaves and flowers, can as well be used in the development of novel products. Moreover, cooperation and investigations between food scientists, agronomists, agriculturists, biotechnologists, economists, local, and international associations is needed to make underutilized PP crop improvement a real accomplishment in the future. Hence, the application of GF bakeries from PPs is required in the industry to the formulation of novel foods especially for people with gluten allergies. Instead of cereal biofortification with PPs' functional ingredients, it is an appropriate time to explore these vital plants, that are functional and nutritionally rich, to fight malnutrition and famine worldwide, especially in developing countries.

References

1. B. Olawoye and S. O. Gbadamosi, *SN Appl. Sci.*, 2020, **2**, 607.
2. L. Alvarez-Jubete, E. K. Arendt and E. Gallagher, *Trends Food Sci. Technol.*, 2010, **21**, 106–113.
3. T. B. Pirzadah and B. Malik, *J. Agric. Food Res.*, 2020, **2**, 100052.
4. C. Martínez-Villaluenga, E. Peñas and B. Hernández-Ledesma, *Food Chem. Toxicol.*, 2020, **137**, 111178.
5. R. Schoenlechner, S. Siebenhandl and E. Berghofer, in *Gluten-free Cereal Products and Beverages*, Elsevier, 2008, pp. 149–190, VI.
6. S. Iametti, P. A. A. Parizad, F. Bonomi and M. Marengo, in *Encyclopedia of Food Security and Sustainability*, 2018, pp. 261–270.
7. F. Zhu, *Carbohydr. Polym.*, 2020, **248**, 116819.
8. A. Gaberščik, M. Vončina, T. Trošt, M. Germ and L. Olof Björn, *J. Photochem. Photobiol., B*, 2002, **66**, 30–36.

9. X. Ji, L. Han, F. Liu, S. Yin, Q. Peng and M. Wang, *Int. J. Biol. Macromol.*, 2019, **127**, 204–209.
10. F. Zhu, *Trends Food Sci. Technol.*, 2016, **49**, 121–135.
11. Y. Z. Cai, H. Corke, D. Wang and W. D. Li, in *Encyclopedia of Food Grains*, ed. C. Wrigley, H. Corke, K. Seetharaman and J. Faubion, Academic Press, Oxford, 2015, 2nd edn, pp. 307–315.
12. FAOSTAT, *Production Share of Buckwheat by Region*, 2022, https://www.fao.org/faostat/en/#data/QCL/visualize.
13. B. L. Graf, P. Rojas-Silva, L. E. Rojo, J. Delatorre-Herrera, M. E. Baldeón and I. Raskin, *Compr. Rev. Food Sci. Food Saf.*, 2015, **14**, 431–445.
14. T. M. H. Abdellatif, Evaluation of nutritional and functional properties of quinoa seeds flour (*Chenopodium quinoa* Willd) and its use in some bakery products *M.Sc. Thesis*, Alexandria University, 2017, 1–109.
15. O. E. Mäkinen, V. Wanhalinna, E. Zannini and E. K. Arendt, *Crit. Rev. Food Sci. Nutr.*, 2016, **56**, 339–349.
16. K. R. Nanduri, A. Hirich, M. Salehi, S. Saadat and S. E. Jacobsen, *Quinoa: A New Crop for Harsh Environments*, Springer, Cham, 2019.
17. M. Friedman, *J. Agric. Food Chem.*, 1996, **44**, 6–29.
18. M. Tapia, *Mt. Res. Dev.*, 2000, **20**(3), 220–225.
19. V. Sharma, S. Chandra, P. Dwivedi and M. Parturkar, *Int. J. Adv. Res.*, 2015, **3**, 725–736.
20. E. Zannini, S. Jeske, K. Lynch and E. K. Arendt, *Int. J. Food Microbiol.*, 2018, **268**, 19–26.
21. S. Grobelnik Mlakar, M. Turinek, M. Jakop, M. Bavec and F. Bavec, *Agricultura*, 2009, **2**, 43–53.
22. S. W. Souci, W. Fachmann and H. Kraut, *Food Composition and Nutrition Tables*, Medpharm GmbH Scientific Publishers, 2000.
23. J. W. Njoki, D. N. Sila and A. N. Onyango, *Food Sci. Qual. Manage.*, 2014, **25**, 10–17.
24. H. Y. Wang and Y. Li, *Food Sci.*, 2004, **25**, 388–391.
25. A. Bertazzo, S. Comai, I. Brunato, M. Zancato and C. V. L. Costa, *Food Chem.*, 2011, **124**, 93–96.
26. G. Bonafaccia, M. Marocchini and I. Kreft, *Food Chem.*, 2003, **80**, 9–15.
27. K. H. Wright, O. A. Pike, D. J. Fairbanks and C. S. Huber, *J. Food Sci.*, 2002, **67**, 1383–1385.
28. S. O. S. Saldivar, in *Encyclopedia of Food and Health*, Elsevier Inc., 2015, pp. 703–711.
29. P. Paśko, H. Bartoń, P. Zagrodzki, S. Gorinstein, M. Fołta and Z. Zachwieja, *Food Chem.*, 2009, **115**, 994–998.
30. X. D. Guo, C. S. Wu, Y. J. Ma, J. Parry, Y. Y. Xu, H. Liu and M. Wang, *Int. Food Res. J.*, 2012, **49**, 53–59.
31. L. Alvarez-Jubete, H. Wijngaard, E. K. Arendt and E. Gallagher, *Food Chem.*, 2010, **119**, 770–778.
32. C. Mota, M. Santos, R. Mauro, N. Samman, A. S. Matos, D. Torres and I. Castanheira, *Food Chem.*, 2016, **193**, 55–61.

33. B. Olawoye and S. O. Gbadamosi, *J. Food Process. Preserv.*, 2020, **44**, e14643.
34. X. Kong, S. Kasapis, J. Bao and H. Corke, *J. Sci. Food Agric.*, 2012, **92**, 1800–1807.
35. B. Olawoye, O. Kadiri and T. D. Oluwajuyitan, in *Handbook of Cereals, Pulses, Roots, and Tubers*, CRC Press, 2021, pp. 221–234.
36. T. H. Gamel, J. P. Linssen, A. S. Mesallam, A. A. Damir and L. A. Shekib, *J. Sci. Food Agric.*, 2006, **86**, 1095–1102.
37. C. Collar and A. Angioloni, *J. Cereal Sci.*, 2014, **59**, 145–154.
38. C. P. Awasthi, A. Kumar, N. Singh and R. Thakur, *Indian J. Agric. Biochem.*, 2011, **24**, 141–144.
39. S. Kaur, N. Singh and J. C. Rana, *Food Chem.*, 2010, **123**, 1227–1234.
40. S. V. Palombini, T. Claus, S. A. Maruyama, A. K. Gohara, A. H. P. Souza, N. E. de Souza, J. V. Visentainer, S. T. M. Gomes and M. Matsushita, *Food Sci. Technol.*, 2013, **33**, 339–344.
41. T. Mehany, I. Khalifa, H. Barakat, S. A. Althwab, Y. M. Alharbi and S. El-Sohaimy, *Food Biosci.*, 2021, **40**, 100891.
42. B. Olawoye and O. Kadiri, *Ann. J. Food Sci. Technol.*, 2016, **17**, 114–123.
43. D. N. Lopez, M. Galante, G. Raimundo, D. Spelzini and V. Boeris, *Food Res. Int.*, 2019, **116**, 419–429.
44. L. Alvarez-Jubete, H. Wijngaard, E. K. Arendt and E. Gallagher, *Food Chem.*, 2010, **119**, 770–778.
45. V. R. Lopez, G. S. Razzeto, M. S. Gimenez and N. L. Escudero, *Plant Foods Hum. Nutr.*, 2011, **66**, 157–162.
46. Z. Chmelik, M. Snejdrlova and M. Vrablik, *Nutr. Res.*, 2019, **72**, 36–45.
47. M. B. Lado, J. Burini, G. Rinaldi, M. C. Anon and V. A. Tironi, *Plant Foods Hum. Nutr.*, 2015, **70**, 371–379.
48. A. A. Qureshi, J. W. Lehmann and D. M. Peterson, *J. Nutr.*, 1996, **126**, 1972–1978.
49. S. Mendonça, P. H. Saldiva, R. J. Cruz and J. A. G. Arêas, *Food Chem.*, 2009, **116**, 738–742.
50. E. G. Tovar-Pérez, I. Guerrero-Legarreta, A. Farrés-González and J. Soriano-Santos, *Food Chem.*, 2009, **116**, 437–444.
51. F. Conforti, G. Statti, M. R. Loizzo, G. Sacchetti, F. Poli and F. Menichini, *Biol. Pharm. Bull.*, 2005, **28**, 1098–1102.
52. H. K. Kim, M. J. Kim, H. Y. Cho, E. K. Kim and D. H. Shin, *Cell Biochem. Funct.*, 2006, **24**, 195–199.
53. A. Chaturvedi, G. Sarojini and N. L. Devi, *Plant Foods Hum. Nutr.*, 1993, **44**, 63–70.
54. I. Klimczak, M. Malecka and B. Pacholek, *Nahrung*, 2002, **46**, 184–186.
55. P. Paśko, H. Barton, M. Folta and J. Gwizdz, *Rocz. Panstw. Zakl. Hig.*, 2007, **58**, 35–40.
56. M. C. Orsini Delgado, V. A. Tironi and M. C. Añón, *LWT–Food Sci. Technol.*, 2011, **44**, 1752–1760.
57. A. O. Ojokoh and R. A. O. Gabriel, *Afr. J. Biotechnol.*, 2010, **9**, 3175–3177.

58. S. Mohanty, S. Zambrana, S. Dieulouard, W. Kamolvit, V. Nilsén, E. Gonzales, C. G. Östenson and A. Brauner, *J. Ethnopharmacol.*, 2018, **220**, 155–158.
59. C. F. Chau and P. C. K. Cheung, *J. Agric. Food Chem.*, 1997, **45**, 4773–4776.
60. M. Albarracín, R. J. González and S. R. Drago, *LWT–Food Sci. Technol.*, 2013, **53**, 76–80.
61. J. M. Ramos Diaz, S. Kirjoranta, S. Tenitz, P. A. Penttilä, R. Serimaa, A. M. Lampi and K. Jouppila, *J. Cereal Sci.*, 2013, **58**, 59–67.
62. E. Berghofer and R. Schoenlechner, Grain amaranth, in *Pseudocereals and Less Common Cereals*, ed. Belton, P. S., and Taylor, J. R. N., Springer-Verlag, Berlin, 2002.
63. E. A. Tosi, E. Ré, H. Lucero and R. MasciarelliFood, *Sci. Technol. Int.*, 2000, **6**, 433–438.
64. D. K. Santra and R. Schoenlechner, in *Sustainable Protein Sources*, ed. S. R. Nadathur, J. P. D. Wanasundara and L. Scanlin, Academic Press, San Diego, 2017, pp. 257–264.
65. A. M. C. de la Barca, M. E. Rojas-Martínez, A. R. Islas-Rubio and F. Cabrera-Chávez, *Plant Foods Hum. Nutr.*, 2010, **65**, 241–246.
66. G. E. Inglett, D. Chen and S. X. Liu, *LWT–Food Sci. Technol.*,, 2015, **63**, 214–220.
67. A. Fidantsi and G. Doxastakis, *Colloids Surf., B*, 2001, **21**, 119–124.
68. M. O. Jimoh, A. J. Afolayan and F. B. Lewu, *Sci. Rep.*, 2019, **9**, 1–14.
69. E. Wilhelm, T. Aberle, W. Burchard and R. Landers, *Biomacromolecules*, 2002, **3**, 17–26.
70. S. Singh, V. Singh, R. A. Salas-Perez, M. v. Bagavathiannan, A. Lawton-Rauh and N. Roma-Burgos, *Pest Manage. Sci.*, 2019, **75**, 1131–1139.
71. F. Zhu, *Food Chem.*, 2016, **203**, 231–245.
72. P. Qin, Q. Wang, F. Shan, Z. Hou and G. Ren, *Int. J. Food Sci.*, 2010, **45**, 951–958.
73. D. C. Joshi, G. v. Chaudhari, S. Sood, L. Kant, A. Pattanayak, K. Zhang, Y. Fan, D. Janovská, V. Meglič and M. Zhou, *Planta*, 2019, **250**, 783–801.
74. Y. Zhou, Y. Jiang, R. Shi, Z. Chen, Z. Li, Y. Wei and X. Zhou, *J. Sci. Food Agric.*, 2020, **100**, 1220–1229.
75. R. Repo-Carrasco-Valencia and J. V. Arana, *Pseudocereals: Chemistry and Technology*, Wiley, 2016, pp. 49–70.
76. J. Ruan, Y. Zhou, J. Yan, M. Zhou, S. H. Woo, W. Weng, J. Cheng and K. Zhang, *Food Rev. Int.*, 2020, 1–15.
77. G. Bonafaccia, L. Gambelli, N. Fabjan and I. Kreft, *Food Chem.*, 2003, **83**, 1–5.
78. N. Yang, Y. M. Li, K. Zhang, R. Jiao, K. Y. Ma, R. Zhang, G. Ren and Z. Y. Chen, *J. Funct. Foods*, 2014, **6**, 311–318.
79. D. G. Dorrell, *J. Am. Oil Chem. Soc.*, 1971, **48**, 693–696.
80. X. Zhou, Y. Lu, Y. Zhou, Y. Xiao and L. He, in *2015 2nd International Workshop on Materials Engineering and Computer Sciences*, Atlantis Press, October 2015, pp. 110-115.

81. D. Angelino, M. Cossu, A. Marti, M. Zanoletti, L. Chiavaroli, F. Brighenti, D. del Rio and D. Martini, *Food Funct.*, 2017, **8**, 2368–2393.
82. G. Rocchetti, L. Lucini, J. M. L. Rodriguez, F. J. Barba and G. Giuberti, *Food Chem.*, 2019, **271**, 157–164.
83. S. J. Kim, I. S. M. Zaidul, T. Suzuki, Y. Mukasa, N. Hashimoto, S. Takigawa, T. Noda, C. Matsuura-Endo and H. Yamauchi, *Food Chem.*, 2008, **110**, 814–820.
84. D. Zielińska, M. Turemko, J. Kwiatkowski and H. Zieliński, *Molecules*, 2012, **17**, 9668–9682.
85. S. J. Kim, T. Maeda, M. Z. I. Sarker, S. Takigawa, C. Matsuura-Endo, H. Yamauchi, Y. Mukasa, K. Saito, N. Hashimoto, T. Noda, T. Saito and T. Suzuki, *J. Agric. Food Chem.*, 2007, **55**, 6314–6318.
86. L. Němcová, J. Zima, J. Barek and D. Janovská, *Food Chem.*, 2011, **126**, 374–378.
87. P. A. Tuan, A. A. Thwe, J. K. Kim, Y. B. Kim, S. Lee and S. U. Park, *Food Chem.*, 2013, **141**, 3803–3812.
88. S. S. Gu, H. Q. Sun, X. L. Zhang, F. N. Huang, L. C. Pan and Z. Y. Zhu, *Int. J. Biol. Macromol.*, 2021, **166**, 730–740.
89. X. He, X. Wang, J. Fang, Y. Chang, N. Ning, H. Guo, L. Huang, X. Huang and Z. Zhao, *Int. J. Biol. Macromol.*, 2017, **101**, 910–921.
90. J. H. Xie, F. Zhang, Z. J. Wang, M. Y. Shen, S. P. Nie and M. Y. Xie, *Carbohydr. Polym.*, 2015, **133**, 596–604.
91. L. Wang, X. Yang, P. Qin, F. Shan and G. Ren, *Ind. Crops Prod.*, 2013, **49**, 312–317.
92. R. Suhag, A. Dhiman, G. Deswal, D. Thakur, V. S. Sharanagat, K. Kumar and V. Kumar, *LWT–Food Sci. Technol.*, 2021, **150**, 111960.
93. G. Mazza, Storage, processing, and quality aspects of buckwheat seed, *New Crops*, Wiley, New York, 1993.
94. J. Ballester-Sánchez, M. T. Fernández-Espinar and C. M. Haros, *Food Hydrocolloids*, 2020, **101**, 105513.
95. X. N. Guo, S. Yang and K. X. Zhu, *Food Chem.*, 2019, **283**, 52–58.
96. K. Dziedzic, A. Szwengiel, D. Górecka, M. Rudzińska, J. Korczak and J. Walkowiak, *J. Cereal Sci.*, 2016, **69**, 25–31.
97. F. Wang, H. Zhang, W. Jin and L. Li, *J. Sci. Food Agric.*, 2018, **98**, 2880–2888.
98. K. Y. Lee, K. H. Park, J. B. Ahn and H. G. Lee, *Food Sci. Biotechnol.*, 2014, **23**, 1469–1475.
99. S. Sohaimy, S. Mohamed, M. Shehata, T. Mehany and M. Zaitoun, *Annu. Res. Rev. Biol.*, 2018, **22**, 1–11.
100. L. E. A. James, *Adv. Food Nutr. Res.*, 2009, **58**, 1–31.
101. L. E. Abugoch, N. Romero, C. A. Tapia, J. Silva and M. Rivera, *J. Agric. Food Chem.*, 2008, **56**, 4745–4750.
102. S. A. Elsohaimy, T. M. Refaay and M. A. M. Zaytoun, *Ann. Agric. Sci.*, 2015, **60**, 297–305.
103. I. Arneja, B. Tanwar and A. Chauhan, *Pak. J. Nutr.*, 2015, **14**, 1034–1040.

104. V. D. Capriles and J. A. G. Arêas, *Compr. Rev. Food Sci. Food Saf.*, 2014, **13**, 871–890.
105. B. Valcárcel-Yamani and S. C. D. S. Lannes, *Food Public Health*, 2012, **2**, 265–275.
106. M. I. González Martín, G. Wells Moncada, S. Fischer and O. Escuredo, *J. Sci. Food Agric.*, 2014, **94**, 876–881.
107. K. C. Maki and A. K. Phillips, *J. Nutr.*, 2015, **145**, 159S–163S.
108. Y. Wu, Y. Qian, Y. Pan, P. Li, J. Yang, X. Ye and G. Xu, *Clin. Nutr.*, 2015, **34**, 603–611.
109. H. Ando, Y. C. Chen, H. Tang, M. Shimizu, K. Watanabe and T. Mitsunaga, *Food Sci. Technol. Res.*, 2002, **8**, 80–84.
110. A. Vega-Gálvez, M. Miranda, J. Vergara, E. Uribe, L. Puente and E. A. Martínez, *J. Sci. Food Agric.*, 2010, **90**, 2541–2547.
111. R. Repo-Carrasco, C. Espinoza and S. E. Jacobsen, *Food Rev. Int.*, 2003, **19**, 179–189.
112. S. Gorinstein, A. Lojek, M. Číž, E. Pawelzik, E. Delgado-Licon, O. J. Medina, M. Moreno, I. A. Salas and I. Goshev, *Int. J. Food Sci.*, 2008, **43**, 629–637.
113. P. Paśko, M. Sajewicz, S. Gorinstein and Z. Zachwieja, *Acta Chromatogr.*, 2008, **20**, 661–672.
114. R. Yawadio Nsimba, H. Kikuzaki and Y. Konishi, *Food Chem.*, 2008, **106**, 760–766.
115. R. Tsao, *Nutrients*, 2010, **2**, 1231–1246.
116. C. Martínez-Villaluenga, E. Peñas and B. Hernández-Ledesma, *Food Chem. Toxicol.*, 2020, 137.
117. I. Dini, O. Schettino, T. Simioli and A. Dini, *J. Agric. Food Chem.*, 2001, **49**, 741–746.
118. A. M. M. Filho, M. R. Pirozi, J. T. D. S. Borges, H. M. Pinheiro Sant'Ana, J. B. P. Chaves and J. S. D. R. Coimbra, *Crit. Rev. Food Sci. Nutr.*, 2017, **57**, 1618–1630.
119. J. Ruales, Y. de Grijalva, P. Lopez-Jaramillo and B. M. Nair, *Int. J. Food Sci. Nutr.*, 2002, **53**, 143–154.
120. F. M. V. Farinazzi-Machado, S. M. Barbalho, M. Oshiiwa, R. Goulart and O. Pessan Junior, *Food Sci. Technol.*, 2012, **32**, 239–244.
121. V. F. Zevallos, L. I. Herencia, F. Chang, S. Donnelly, H. J. Ellis and P. J. Ciclitira, *Am. J. Gastroenterol.*, 2014, **109**, 270–278.
122. M. N. Laus, A. Gagliardi, M. Soccio, Z. Flagella and D. Pastore, *J. Food Sci.*, 2012, **77**, C1150–C1155.
123. Y. Hirose, T. Fujita, T. Ishii and N. Ueno, *Food Chem.*, 2010, **119**, 1300–1306.
124. P. Paśko, P. Zagrodzki, H. Bartoń, J. Chłopicka and S. Gorinstein, *Plant Foods Hum. Nutr.*, 2010, **65**, 333–338.
125. M. Miranda, J. Delatorre-Herrera, A. Vega-Gálvez, E. Jorquera, I. Quispe-Fuentes and E. A. Martínez, *Agric. Sci.*, 2014, **05**, 1015–1024.

126. I. Dini, O. Schettino, T. Simioli and A. Dini, *J. Agric. Food Chem.*, 2001, **49**, 741–746.
127. C. Wahli, *Quinua: hacia su cultivo comercial*, ING. CHRISTIAN WAHLI, 1990.
128. J. T. Borges, R. C. Bonomo, C. D. Paula, L. C. Oliveira and C. Márcia, *Temas Agrarios*, 2010, **15**, 9–23.
129. S. H. Park, T. Maeda and N. Morita, *J. Appl. Glycosci.*, 2005, **52**, 337–343.
130. S. A. Valencia-Chamorro, Quinoa, in *Encyclopedia of Food Science and Nutrition*, Academic Press, Amsterdam, 2003.
131. G. M. Turkut, H. Cakmak, S. Kumcuoglu and S. Tavman, *J. Cereal Sci.*, 2016, **69**, 174–181.
132. L. L. O. Pineli, R. B. A. Botelho, R. P. Zandonadi, J. L. Solorzano, G. T. de Oliveira, C. E. G. Reis and D. da S. Teixeira, *LWT–Food Sci. Technol.*, 2015, **63**, 1261–1267.
133. I. Kaur and B. Tanwar, *Rom. J. Diabetes Nutr. Metab. Dis.*, 2016, **23**, 215–225.
134. C. M. Rosell, G. Cortez and R. Repo-Carrasco, *Cereal Chem.*, 2009, **86**, 386–392.
135. S. A. El-Sohaimy, M. G. Shehata, T. Mehany and M. A. Zeitoun, *Int. J. Food Sci.*, 2019, 4686727.
136. S. A. El-Sohaimy, M. G. Shehata, T. A. Djapparovec, T. Mehany, M. A. Zeitoun and A. M. Zeitoun, *J. Food Process. Preserv.*, 2021, **45**, 1–10.
137. H. Liu, X. Guo, Y. Li, H. Li, H. Fan and M. Wang, *J. Food Eng.*, 2016, **189**, 64–71.
138. W. Zhang, Q. Yang, M. Xia, W. Bai, P. Wang, X. Gao, J. Li, B. Feng and J. Gao, *Int. J. Biol. Macromol.*, 2019, **129**, 799–808.
139. D. N. López, R. Ingrassia, P. Busti, J. Wagner, V. Boeris and D. Spelzini, *LWT–Food Sci. Technol.*, 2018, **97**, 523–529.
140. S. E. Suarez and M. C. Añón, *Food Hydrocolloids*, 2018, **74**, 115–123.
141. Y. P. Timilsena, R. Adhikari, C. J. Barrow and B. Adhikari, *Food Chem.*, 2016, **212**, 648–656.
142. Á. G. Solaesa, M. Villanueva, A. J. Vela and F. Ronda, *Food Hydrocolloids*, 2020, **105**, 105770.
143. A. Taha, E. Ahmed, A. Ismaiel, M. Ashokkumar, X. Xu, S. Pan and H. Hu, *Trends Food Sci. Technol.*, 2020, **105**, 363–377.
144. M. Pellegrini, R. Lucas-Gonzales, A. Ricci, J. Fontecha, J. Fernández-López, J. A. Pérez-Álvarez and M. Viuda-Martos, *Ind. Crops Prod.*, 2018, 38–46.
145. A. Kierulf, J. Whaley, W. Liu, M. Enayati, C. Tan, M. Perez-Herrera, Z. You and A. Abbaspourrad, *Food Chem.*, 2015, **315**, 126246.
146. H. Cao, R. Sun, J. Shi, M. Li, X. Guan, J. Liu, K. Huang and Y. Zhang, *Ultrason. Sonochem.*, 2021, **77**, 105685.
147. S. Srichuwong, D. Curti, S. Austin, R. King, L. Lamothe and H. Gloria-Hernandez, *Food Chem.*, 2017, **233**, 1–10.

148. G. A. Ruiz, W. Xiao, M. van Boekel, M. Minor and M. Stieger, *Food Chem.*, 2016, **209**, 203–210.
149. M. Z. Dar, K. Deepika, K. Jan, T. L. Swer, P. Kumar, R. Verma, K. Verma, K. S. Prakash, S. Jan and K. Bashir, *Int. J. Biol. Macromol.*, 2018, **108**, 1348–1356.
150. F. Zhu, *Trends Food Sci. Technol.*, 2021, **110**, 155–167.
151. X. Zhang, Z. Zuo, W. Ma, P. Yu, T. Li and L. Wang, *Food Hydrocolloids*, 2021, **117**, 106748.
152. J. Ahmed, L. Thomas, Y. A. Arfat and A. Joseph, *Carbohydr. Polym.*, 2018, **197**, 649–657.

CHAPTER 7

Pigmented Millets: Nutritional Quality and Potential Benefits for Human Health

ANUPREET KAUR SOBTI[a], VASUDHA BANSAL*[a], NITYA SHARMA*[b], PRIYANKA PRASAD[b], J. K. SAHU[b], SUKIRTI JOSHI[b,c] AND DIVYANG SOLANKI[d]

[a]Department of Foods and Nutrition, Govt Home Science College, Sector 10, Chandigarh, India; [b]Centre for Rural Development and Technology, Indian Institute of Technology Delhi, Hauz Khas, New Delhi, India; [c]Soft Solids Research Group, Department of Mechanical Engineering, Imperial College London, London, South Kensington, UK; [d]Department of Food Science, School of Agriculture and Food Sciences, The University of Queensland, Australia

*E-mail: vasu22bansal@gmail.com, nitya.sharma64@gmail.com

7.1 Pigmented Millets and Their Classification

Cereals and cereal variants are consumed as part of most people's diets,[1] supplying a significant share of dietary calories and nutrients in developed and developing nations. They are made up of around 75% carbohydrates (mostly starches) and about 6–15% protein and, moreover, they provide more than half of the world's total energy.[2] The relevance of cereals and cereal products is further exemplified by the evidence-based notion that cereal production,

Food Chemistry, Function and Analysis No. 38
Pigmented Cereals and Millets: Bioactive Profile and Food Applications
Edited by Sneh Punia Bangar, Sajid Maqsood and Anil Kumar Siroha

Published by the Royal Society of Chemistry, www.rsc.org

which amounts to around 2600 million tonnes annually, is the most important factor in world food security.[3]

Coloured cereal grains have a high concentration of phenolics and flavonoids, such as anthocyanins, which are found in various areas of the grain, and are responsible for the colour. Several studies have linked anthocyanins and other phenolic compounds to antioxidant and anti-inflammatory properties. By altering NF-κB-mediated pathways, these drugs and their derivatives can inhibit inflammatory responses by acting on natural cell signalling pathways.[4] Pancreatic-amylase and brush-border-glucosidase enzyme activity has also been demonstrated to be inhibited by polyphenols and anthocyanins. Intestinal-glucosidase inhibition reduces carbohydrate digestion and absorption, resulting in lower glucose levels post-meal (post prandial glycemia), which is regarded as an effective method of controlling type 2 diabetes.

Millets are amongst the most prominent cereal crops on the planet, and are derived from the French word *mille*, which means 'thousand', suggesting that a fistful of millet comprises countless grains. Moreover, one-third of the population globally consume millets. In terms of global agricultural production, it is the sixth most significant cereal crop.[5] *Jowar* (sorghum), *Sama* (little millet), *Ragi* (finger millet), *Korra* (foxtail millet), and *Variga* (proso millet) are all millets. These grains have been farmed for more than a thousand years and are utilised worldwide; in the Middle Ages, Romans and Gauls preferred millets over wheat porridges. China, Greece, India, Africa, and Egypt contribute the entirety of the global commercial millet crop. The major millet group includes sorghum (*Sorghum bicolor* L.) and pearl millet (*Pennisetum typhoides* L.), while finger millet or ragi (*Eleusine coracana* L. Gaertn), barnyard millet (*Echinochloa frumentacea* L.), foxtail of Italian millet (*Setaria italica* L.), kodo millet (*Paspalum scrobiculatum* L.), little millet (*Panicum sumatrens* L.), proso millet (*Panicum miliaceum* L.), and brown-top millet (*Brachiaria ramosa* L. Stapf; *Panicum ramosum* L.) are classified as small millet. Small millets are superior in terms of nutritional value to rice and wheat, because they are high in dietary fibre, proteins, minerals, and vitamins, and so are classified further as nutri-cereals.[6] In India, these crops were farmed in 36.90 million hectares from 1965–1966, yielding 16.4 million tonnes of grain, but in 2017–2018, the cultivation area was reduced to 14.25 million ha, yielding 16.4 million tonnes, a 61.4% contraction. Even though the area under cultivation was reduced in 2017–18 as compared to 1965–1966, the overall production has increased.

Millet consumption has decreased by 0.9% globally, but is anticipated to increase between 2019 and 2024. India, China, and Niger are the world's top millet producers, accounting for greater than 55% of across-the-board output. India has been the leading millet producer globally, accounting for 41.04% of the global market. However, its output in Africa has risen considerably in recent years. Millet production globally was expected to be 27.8 million tonnes. Millet's prominence as a food staple has declined in India over the previous two decades due to a variety of causes, including increased earnings, urbanization, and government policy. More than half of millet production is now going to alternative applications, rather than being consumed as a staple. African countries, primarily Niger, Burkina Faso, Mali, Nigeria, and Sudan, consume more than 40% of the world's millet. Millets are widely consumed in poor countries such as in India and the countries of Africa, where food and nutrition security are key concerns. The crop's capacity to tolerate drought allows it to be grown in regions with harsh conditions of environment. India is the leading millet producer in the world. Millet seed development and production are the only public and private investments. Millets' international prices are highly volatile, mostly driven by supply volumes, and are sometimes unconnected to the prices of other important grains like maize, sorghum, or barley.[7]

Millets can withstand adverse conditions like drought, and some wilder species may even thrive in flooded areas and swampy areas. These have a lower glycemic index, contain gluten-free protein, and are high in minerals, B vitamins, and antioxidants (calcium, iron, copper, magnesium, and so forth). These exceptional characteristics make them nutrient-dense and climate-resilient crops. These can be used not only as a source of income for farmers, but also to improve the overall health and well-being of the community. Scientific interventions can overcome existing restrictions, such as anti-nutritional components and low sensory acceptance of millet-based products.

Millets are divided into two categories based on their grain size: major millets and minor millets. Pseudo millets are so named because they are not members of the Poaceae botanical family, which includes 'true' grains. However, they are nutritionally alike to 'true' grains and are utilised in similar ways (Figure 7.1).

7.1.1 Major Millets

7.1.1.1 Finger Millet

Finger millet (*Eleusine coracana* L.) is a significant millet grain that is widely grown in India and countries in Africa, and is a staple food for a considerable proportion of the population in both areas. In India, it stands sixth in terms of production behind wheat, rice, maize, sorghum, and bajra.[8] It is a naked caryopsis (Figure 7.1) with a brick red seed coat that is commonly used as a whole meal in the production of traditional meals like *roti* (unleavened breads or pancakes), *mudde* (dumpling), and *ambali* (thin porridge). Finger millet is an excellent subsistence grain. Its little seeds can be stored for a long time with very little damage from insects, and little loss of viability. The grain of finger millet can be used in a variety of ways. For diabetics and the elderly, the ground flour is converted into porridge or bread, producing nourishing and wholesome foods.[9] Millets have a nutritional potential comparable to that of common cereals such as rice, wheat, barley, or bajra in terms of protein, carbohydrate, and calories. The nutrient content of finger millet grain is given in Table 7.1.

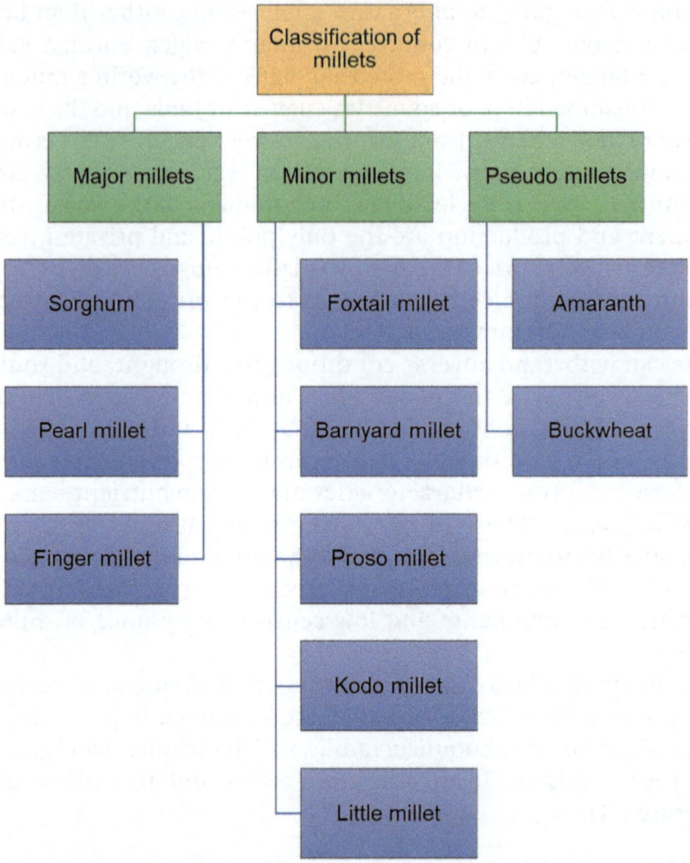

Figure 7.1 Classification of millets according to their grain size.

To improve the nutritional qualities of cereal grains, various food processing procedures are applied. Fermentation is one of the oldest methods of food preservation, and it has long been employed in the processing of cereals like sorghum and millets. Because of their distinct flavors and high nutritious content, fermented foods have gotten much attention worldwide. A small number of studies have found that spontaneous fermentation improves the nutritional value of cereal grains by increasing protein content, digestibility, and lysine content. Furthermore, spontaneous fermentation increases the bioavailability of micronutrients such as iron, zinc, and calcium, while also creating antimicrobial compounds that inhibit harmful bacteria development and the formation of bacterial toxin. It also boosts cereals' phenolic components and antioxidant activity. According to the existing literature, spontaneous fermentation has a very limited impact on FM flours' bioactive components, antioxidants, and physicochemical properties. The effect of this procedure on the bioactive components and antioxidant activity of pearl millet, on the other hand, has been thoroughly investigated.[10]

Table 7.1 Millets and their proximate analysis and dietary fibre.[a]

						Proximate analysis and dietary fibre (per 100 g edible portion)				
S. no.	Name of the millet	Scientific name	Vernacular names	Place of origin	Colours	Energy (kJ)	Carbohydrates (g)	Protein (g)	Fat (g)	Dietary fibre (g)
1	Finger millet	*Eleusine coracana* (L.) Gaertn.	Ragi, mandua, nagli, kapai, marua, nachni, African bird's foot, rapoko, hunsa, wimbi, bulo, telebun, koracan, kurakkan	East African highlands	White to brown	1342 ± 10	66.82 ± 0.73	7.16 ± 0.63	1.92 ± 0.14	11.18 ± 1.14
2	Pearl millet	*Pennisetum glaucum* (L.) R. Br (= *P. americium* (L))	Bajra, cattail, bulrush, candlestick, sanyo, munga, seno	West African Savannah	Creamy, white to gray and purple	1456 ± 18	61.78 ± 0.85	10.96 ± 0.26	5.43 ± 0.64	11.49 ± 0.62
3	Sorghum	*Sorghum bicolor* (L.) Moench ssp. *bicolor*	Great millet, jowar, cholam, jola, jonna, durra, Egyptian millet, feterita, Guinea corn, jwari, juwar, milo, shallu, gaoliang, kaoliang, kafir corn, dura, dari, mtama, solam	African Savannahs	Tan (nonpigmented), red, or purple	1398 ± 13	67.68 ± 1.03	9.97 ± 0.43	1.73 ± 0.31	10.22 ± 0.49
4	Foxtail millet	*Setaria italica* (L.) P. Beauv ssp. *italica*	Kauni, kangni, korra, tenai, rala, Italian, German, Hungarian, Siberian, navane, thanahal	China	Pale yellow to orange, red, brown, or black	331[b]	60.09[b]	12.30[b]	4.30[b]	—

(Continued)

Table 7.1 (*continued*)

S. no.	Name of the millet	Scientific name	Vernacular names	Place of origin	Colours	Proximate analysis and dietary fibre (per 100 g edible portion) Energy (kJ)	Carbohydrates (g)	Protein (g)	Fat (g)	Dietary fibre (g)
5	Kodo millet	*Paspalum scrobiculatum* L.	Varagu, bastard, ditch, naraka, water couch, Indian paspalum, creeping paspalum, amu	India	Blackish brown to dark brown	1388 ± 10	66.19 ± 1.19	8.92 ± 1.09	2.55 ± 0.13	6.39 ± 0.60
6	Barnyard millet	*Echinochloa crus-galli* (L.) P. Beauv. (syn. *E. esculenta* (A. Braun)	Japanese, jhingora, kudraivali, oodalu, sanwa, sawan, Korean, kweichou	Japan	Straw white, grey, little grey to dull white	307[b]	65.55[b]	06.20[b]	2.20[b]	—
7	Little millet	*Panicum sumatrense* Roth. ex Roem. & Schult. Subsp. *sumatrense* (syn. *P. miliare* auct. pl.)	Little millet, kutki, samalu, same, samai, blue panic, heen meneri	India	Grey to straw white	1449 ± 19	65.55 ± 1.29	8.92 ± 1.09	2.55 ± 0.13	6.39 ± 0.60
8	Proso millet	*Panicum miliaceum* L. ssp. *miliaceum*	Cheena, panivaragu, variga, baragu common, hog, broom, samai, Russian, panic, mahameneri	China	White cream, yellow, orange, red, brown to black	341[b]	70.04[b]	12.50[b]	1.10[b]	—

[a]Indian Food Composition Table, NIN.[29]
[b]Nutritive Value of Indian Foods, NIN.[30]

Finger millet grain bioactive substances have been studied more and more over the years. Several studies on grains have identified specific physiologically active chemicals that are also nutritionally significant. These molecules that are concentrated around plant nutrients including dietary fibre (DF), minerals, and polyphenols, and they have been discovered to have potential nutraceutical effects due to their involvement in a variety of biological systems. DF is a multi-component plant-based mixture that modulates the rate and degree of blood glucose that increases after consuming a carbohydrate or carbohydrate-like food. Non-starch polysaccharides, non-glucan oligosaccharides, resistant starches, certain polyols, and modified starches are examples of primary plant dietary fibres.

Finger millet's polyphenols and dietary fibre levels are credited with boosting public awareness of the grain's nutritional and health benefits. The polyphenol content of finger millet grain is higher than that of rice, wheat, barley, and maize. These phenolic compounds are not deposited in the grain, but rather in the outer aleurone layer, testa, and pericarp of fruit, which make up most of the bran portion, where they exist as free, soluble conjugates, and insoluble bound forms.[11] In research from McKeown *et al.*,[12] many epidemiological studies have emphasised the importance of consumption of whole grains and their variants to prevent the risk of CVDs, type II diabetes, and a range of other metabolic disorders. Because millets are usually cooked whole, the dietary fibre, minerals, phenolics, and vitamins concentrated in the grain's outer layer or seed coat become part of the dish and provide nutritional and health advantages.[13] If the millet is processed to remove the seed coat materials, as is done in millet malting and milling, it could be used as an additive in various foods. Finger millet is a gluten-free grain that is ideal for celiac disease patients (gluten-allergic).

A brief nutritional profile of finger millet:

- The content of calcium in finger millet is the highest (300–350 mg per 100 g).[11]
- The mineral content of ragi is the highest.[14]
- It has a comparatively lower protein (6–8% protein) and fat content (1.5–2%) and contains mainly unsaturated fats.[14]
- The sulphur-rich amino acid composition of finger millet proteins makes them stand out.[14,15]
- The grains are well known for their usage as weaning foods due to their outstanding malting qualities.[9]
- It has a great deal of antioxidant properties.[15]

7.1.1.2 Pearl Millet

The most extensively produced millet crop is pearl millet (*Pennisetum glaucum*). The states of Rajasthan, Gujarat, Haryana, and Madhya Pradesh are the top producers of this grain, which is largely cultivated and consumed in India. Bajra, as it is known locally, is a major source of macro- and micronutrients for millions of the poorest people in the areas where it is grown. It is also a key food staple in Africa and Asia's semi-arid and arid regions, and it is well suited to drought and low-fertility sandy acid soil. The crop is the fourth most tropical food cereal on the planet.[16] Pearl millet's nutritional profile reveals that it is a rich source of energy and protein.[17] Compared with corn protein, it has a 40% higher lysine and methionine content, and a 30% higher threonine content.[18] According to nutritional research, the metabolic energy of pearl millet for non-ruminant animals is similar to that of maize. Like other cereals, pearl millet is related to nutritional inhibitors such as enzyme inhibitors, phytic acid, and tannins. These variables reduce protein and starch digestibility and mineral bioavailability, lowering the grain's nutritional value.

The nutrient composition of pearl millet is given in Table 7.1. Pearl millets have been shown to improve insulin sensitivity and lower triglyceride levels. Pearl millet is an excellent diabetes remedy. It digests slowly and releases glucose into the blood at a slower rate than other foods due to its high fibre content. This efficiently aids people with diabetes to keep a steady blood sugar level for an extended length of time. Due to its high iron content (8 mg per 100 g) and high zinc content (3.1 mg per 100 g), millet is used to treat iron deficiency anaemia.[19] Stomach ulcers can be treated with pearl millet. Pearl millet is one of the few foods that alkalises the stomach and helps to prevent or lessen the effects of stomach ulcers.

Compared with other cereals, pearl millet is considered a nutri-cereal because of its high grain Fe and Zn content, metabolizable energy, high antioxidant and polyphenol content, and high proportion of slowly digested starches, dietary fibres, and good essential amino acid profile. Higher genetic diversity in Fe and Zn content in grain has aided in the mapping and mining of QTLs, alleles, and genes involved in micronutrient metabolism. Over the last decade, pearl millet has developed efficient genetic and genomic resources, making this possible. Bi-parental RIL mapping populations, association mapping panels, chromosomal segment substitution lines (CSSLs), and TILLING populations are examples of genetic stocks. On the genomics side, significant progress has been made in developing genomic markers, such as the development of SSR marker repositories. Following that, a genome-wide SNP repository based on next-generation sequencing was created. A substantial influence was provided by the 1000 genomes re-sequencing project. A high-quality reference genome was created by re-sequencing the world diversity panel and mapping population parents and hybrid parental lines.[20]

A brief nutritional profile of pearl millet:

- The lignin and phytonutrients in pearl millet act as strong antioxidants, thus preventing heart-related diseases.[19]
- Pearl millet has high protein, as well as high fat content (Table 7.1).
- It has a dietary fibre content of 11.5%. Food transit time in the intestine is lengthened. As a result, the risk of inflammatory bowel disease is reduced.

- Pearl millet has a greater niacin concentration than any other cereal.[14]
- Folate, magnesium, iron, copper, zinc, and vitamins E and B-complex are also present.[19]
- When compared with other millets, it has a high energy level (Table 7.1).
- It also contains calcium and unsaturated fats, both of which are beneficial for health.[14]

7.1.1.3 Sorghum

Sorghum is an annual plant that emerged as a wild crop in African savannas about 10 000 years ago and was domesticated in Ethiopia around 5000 years ago. Since then, sorghum cultivation has spread all over the world. Sorghum is the fifth most important cultivated cereal in the world, following rice and wheat, and is widely utilised as a food staple in Africa and Asia, due to its drought resistance.[21]

Sorghum's principal bioactive constituents are phenolic compounds found in all types of this cereal. Sorghum contains almost all phenolic classes, but phenolic acids, tannins, and flavonoids are the most abundant. The capacity to analyse the bioaccessibility of phenolic compounds in sorghum, and the dietary factors that can affect it is critical for determining their functional capacity in humans and, in the long run, implementing therapeutic measures for health professionals. The nature, quantity, and distribution of individual oligomers and polymers in sorghum tannins varies. Type I (no detectable quantities), type II (tannins extractable only in acidified methanol), and type III (tannins extractable in both methanol and acidified methanol) are the three types. Almost all of the tannins in sorghum are oligomers or polymers of catechins (flavan-3-ols and/or flavan-3,4-diols) that have been condensed.

Many flavonoids found in sorghum have been extracted and identified. In sorghum, the anthocyanins are the most studied flavonoids. This group of chemicals is responsible for the blues, purples, and reds found in plants. Cyanidin, delphinidin, malvidin, pelargonidin, petunidin, and peonidin are the six most frequent anthocyanidins. Sorghum anthocyanins are distinct from other anthocyanins in that they lack the hydroxyl group in the 3-position of the C-ring, giving them the name 3 deoxyanthocyanins. Compared with conventional anthocyanins, this unique property boosts their stability at high pH. Flavan-4-ol compounds, such as luteoforol and apiforol, are formed from flavanones (*i.e.* naringenin and eriodictyol) in red pericarp sorghums, and may be precursors of sorghum anthocyanidins. Flavan-4-ols may play a key role in mould resistance. According to various studies, there is a link between flavan-4-ol content and mould resistance in sorghums. Sorghum contains substantial amounts of three types of flavonoids: anthocyanins, flavones, and flavanones. Sorghum anthocyanins are 3-deoxyanthocyanidins, which account for up to 79% of the flavonoids in the grain. Because they lack a hydroxyl group at position C-3, they have a distinct structure and are hence more stable than other anthocyanins. Sorghum's major anthocyanin 3-deoxyanthocyanidins are non-methoxylated forms (luteolinidin and apigeninidin). Small amounts of methoxylated 3-deoxyanthocyanidins (5-methoxy-luteolinidin, 7-methoxy-luteolinidin, and 7-methoxy-apigeninidin) as well as methoxylated 3-deoxyanthocyanins (5-methoxy-luteolinidin 7-glucoside, 7-methoxy-luteolinidin 5-glucoside, and 7-methoxy-luteolinidin 5-glu (apigeninidin 5-glucoside and luteolinidin 5-glucoside) glucoside). The proportion of 3-deoxyanthocyanidins in sorghum is related to its colour and antioxidative properties. Total 3-deoxyanthocyanidins (5.4 to 6.1 mg g^{-1}) are 3 to 4 times higher in pericarp and black testa varieties than in red and brown variants (1.6 to 2.8 mg g^{-1}).[22]

Grain sorghum possesses qualities that make it ideal for consumption by people who suffer from chronic illnesses. Each component in sorghum has a distinct nutritional value that may be useful in preventing and controlling lifestyle illnesses and disorders. Because of agronomic factors (water availability, soil fertility, temperatures, and environmental variables during grain development) and genotype, sorghum's protein content and composition differ from cultivar to cultivar. Sorghum proteins are mostly found in the endosperm (80%), germ (16%), and pericarp (6%). The primary components in sorghum are kaffirin or prolamins and glutelins, which are found in the endosperm, whereas the germ contains a lot of albumins and globulins. Seeds of sorghum, lupin, and millet are significant crops that contain important health-promoting components. Lupin variants have higher protein, oleic acid, amino acid, and mineral content than millet phenol, anthocyanin, and sorghum seeds. Lupin seed protein content was greater than millet and sorghum seed protein content. In all of the samples, glutamic acid was the most prevalent amino acid. In general, the amino acid content of lupin seeds was found to be higher than that of millet and sorghum seeds. The primary fatty acids found in seed samples were palmitic, stearic, oleic, linoleic, and linolenic acids. The main sugar in lupin seeds is stachyose, while the main sugars in millet and sorghum samples are fructose and glucose.[23]

A brief nutritional profile of sorghum:

- Prolamin (kaffirin), a major component of sorghum protein, has the unique property of lowering digestibility when cooked, which may be beneficial to some dietary categories.
- Cooked sorghum proteins are much less digestible than other grain proteins, which could be beneficial to some dietary groups.
- Protein, fibre, thiamine, riboflavin, folic acid, and carotene are all abundant.
- It contains plenty of minerals like potassium, phosphorus, calcium, iron, zinc, and sodium.

7.1.2 Minor Millets

Food consumption has shifted, and now health-conscious people with increased spending power are looking for nutritious foods. The consumption of foods containing nutraceuticals provides health and well-being advantages by reducing the risks of chronic illnesses such as cardiovascular disease, cancer, and obesity. Nutraceuticals contain 'antioxidants', which are micronutrients that can counteract the effects of free radicals, which are potentially damaging chemicals produced during various natural processes in the body and linked to cell and tissue ageing. Small millets have a significant amount of phenolics, a powerful antioxidant, and as a result, small millets and their value-added products are in high demand among consumers. Furthermore, finger millet-based diets showed significantly lower plasma glucose levels due to the high fibre content, compared with rice and wheat. Small millets contain bound polyphenols, which inhibit the action of digestive enzymes. Some polyphenolic substances, like flavonoids, phenolic acids, and proanthocyanidins, are known to forage free radicals, thus preventing many diseases and morbid conditions. Millets, which are nutritionally dense, have significant potential to provide health advantages when ingested as a staple cereal or as a value-added product, and hence these little grains actually deserve to be called nutri-cereals.

All vital components, such as carbohydrates, dietary fibres, proteins, lipids, vitamins, essential elements, and antioxidants, are found in minor millets. Nutraceutical research is becoming more popular as people become more mindful of their health and nutrition. Scientific studies have demonstrated the value of dietary fibres, antioxidants, and phytochemicals, hence there is a continuing hunt for new sources of nutraceuticals. Millets have higher protein levels and a more balanced amino acid profile than grains like wheat and rice, making them nutritionally equal or perhaps superior (good source of methionine, cystine, and lysine). Millets are a basic food items that contain not only macronutrients such as protein, carbohydrates, and fats, but also a variety of vitamins and minerals. Malnutrition and numerous health problems such as obesity, diabetes, cardiovascular diseases, cancer, celiac disease, and others are most prevalent in developing nations due to insufficient nutrition. This is primarily due to a lack of awareness and information among the general public about the various types of food available, particularly small millets. Minor millets are highly significant natural resources in this context, since they contain all essential elements and can be utilised as functional food. Small millets are a staple part of their diet for low-income people in several areas of India. The biochemical components of these millets explain why they are regarded as functional foods and they are useful in treating diabetes, tumours, atherosclerosis, and have a microbial action.

7.1.2.1 Foxtail Millet

The annual grass plant foxtail millet (*Setaria italica*) produces seeds with health-promoting characteristics due to its unique protein composition, which contains a considerable concentration of important AA. Setarins, proline-rich, alcohol-soluble proteins (prolamin) that make up around 60% of the total protein in mature foxtail seeds have fewer disulphide cross-linked proteins than other cereals and millets.[24]

Protein fractionation techniques are a useful tool for learning about the nature of foxtail proteins and how they might be used in agriculture, food, pharmaceuticals, and bio-based technologies. The constituents, formation, and nutritional profile of the protein concentrate can be influenced by different processing procedures. Furthermore, foxtail protein or its hydrolysate has demonstrated a number of bioactive properties that could be investigated further to treat chronic disorders in humans. Furthermore, foxtail millet might be considered a strong choice for replacing animal protein sources due to its inexpensive cost and outstanding functional qualities of flour and protein concentrate. Furthermore, there is a lot of promise for producing low-cost, protein-rich functional food products that can aid with chronic disease prevention and management.

A brief nutritional profile of foxtail millet:

- It contains a high amount of carbohydrates.
- In comparison to rice, it contains twice as much protein.
- Copper and iron are among the minerals found in it.
- It has a sweet nutty flavor.
- It is one of the most digestible and hypoallergenic grains.

7.1.2.2 Kodo Millet

Kodo millet, a drought-resistant plant that originated in Africa and was domesticated in India a thousand years ago, is one of the world's oldest cereals. Arid and semi-arid parts of African and Asian countries grow this millet crop. Kodo millet is primarily farmed in India's Deccan region, but it is also grown in the Himalayan foothills.

Dietary fibre and minerals such as iron, as well as antioxidants, are abundant in kodo millet. Kodo millet has a lower phosphorus level than any other millet and has a significantly stronger antioxidant potential than any other millet or major cereal. The mineral and fibre content are affected by processing, such as parboiling and debranning, although anti-nutritional elements such as phytate are reduced. To improve the nutritional value, palatability, and functionality of certain traditional Indian foods, kodo flour has been used alone or in combination with other cereal and legume flours. Kodo millets can be utilised in both traditional and creative dishes. Whether unprocessed or processed, the grain can be boiled whole, decorticated, and milled to flour using conventional or industrial methods. Kodo millet flour is pounded into flour and used to create pudding in India. It is also cooked as rice in tribal areas, and the flour is used to make a variety of dishes. It is also cooked like rice in Africa. Cattle, goats, pigs, lambs, and poultry can all benefit from it as animal fodder. *P. scrobiculatum* grows well on hillside slopes in Hawaii, where other grasses do not thrive.[25]

A brief nutritional profile of kodo millet:

- It has a significant protein level (11%) and low-fat content (4.2%), as well as a high fibre content (14.3%).
- B vitamins, including niacin, pyridoxin, folic acid, and minerals like calcium, iron, potassium, magnesium, and zinc, are abundant in kodo millet.
- It includes a lot of lecithin and is great for boosting the nervous system's strength.

7.1.2.3 Barnyard Millet

Barnyard millet has long been used to replace rice in the Indian Himalayan region. The grains are processed (dehulled and cooked) and consumed in the same way as rice. *Mad-ira ki kheer* (barnyard millet porridge) is a traditional sweet delicacy in Uttarakhand (India). Barnyard millet is used in classic South Indian dishes such as *idli, dosa*, and *chakli*. Biscuits, sweets, noodles, rusk, ready mix, popped products, and other specialty foods have all been developed with millet, but large-scale production requires industrial contribution to commercialise these products on a wider scale. Because of its high iron content, this millet is ideal for commercial diets for diabetics, newborns, and pregnant women. Despite its nutritional advantages, however, the lack of ready-to-use processed products has hampered the use and acceptance of barnyard millet. To suit the demands of today's consumers, millet-based food items in the form of ready-to-use products and functional foods must be developed. Value addition to minor millets provides consumers with more variety, convenience, and high-quality food, but it is also critical for the resurgence of barnyard millet cultivation. In the Himalayan region, barnyard millet is a significant source of feed. The leaves of barnyard millet are broad, and the plant grows quickly, producing a large amount of fodder. Barnyard fodder is delicious and can be made into hay or silage.[26]

A brief nutritional profile of barnyard millet:

- It is the optimal source to get crude fibre and iron.
- Other useful elements found in its grains include gamma amino butyric acid (GABA) and beta-glucan, which act as antioxidants and help lower blood lipid levels.

7.1.2.4 Little Millet

Little millet is another fiber rich coarse cereal, less fibrous than barnyard millet. According to a study, some types of kodo millet and little millet have 37–38% dietary fibre: the highest among cereal grains. Fibre promotes healthy digestion and satisfies hunger rapidly. Little millet, like foxtail and barnyard millet, is high in iron. It contains a lot of fat, mostly polyunsaturated fatty acids (PUFAs), which are good for you. The flavonoids in tiny millet are antioxidants that have a variety of roles in the body's immune system. Little millet has a weak amino acid makeup despite its high protein content. Antioxidants found in little millet, such as polyphenols, phenolic compounds, tannins, and flavonoids, are not directly related to nourishing the body. However, they do play a role in promoting health by assisting in diseases like diabetes, cardiovascular disease, cataract, cancer, inflammation, and gastrointestinal issues. Polyphenols are the most common type of phytochemical found in plant-based meals, and they have been connected to a variety of health advantages. They are regarded as 'life span essential' because of their significance in preserving health in the later stages of life.[27]

The soluble- and insoluble-bound phenolic extracts of different millet types are rich sources of phenolic chemicals. Antioxidant, metal chelating, and reducing properties are all present in them. Germination, steaming, and roasting effects on the nutraceutical and free radical scavenging characteristics of little millet were also examined. The total phenolic, flavonoid, and tannin contents of processed little millet increased significantly compared with the original sample, demonstrating that processing improves the nutraceutical and antioxidant capabilities of the grain.

Millet grains are higher in polyphenol and anti-oxidant content than millet flour and flakes, according to a number of studies. Theis research serves as a springboard for future investigations on the effects of polyphenol-rich millet products on glycemic response.

A brief nutritional profile of little millet:

- It is tinier than other millets.
- It has considerable iron content.
- It has significant antioxidant activities.
- It contains about 38% of dietary fibre.

7.1.2.5 Proso Millet

Because of its health benefits, including its strong nutritional potential, low GI, and gluten-free status, proso-millet has been drawing attention. The starch characteristics are determined by the content and distribution of brained chains of amylose and amylopectin. The waxy and non-waxy cultivars have drastically different starch structures and morphologies. According to the evidence, many kinds had amylose content ranging from 0.75–28.3%, while the Hongmeizi variety had 32.3%. There is a link between amylose content and cooking quality and thermal and pasting qualities. The smallest to largest starch granules ranged in size and form from 0.3 to 17 μm and spherical to polygonal, respectively.[28] Because of their excellent resistance to swelling after heat treatment, non-waxy starch types of proso millet are commonly utilised in food processing. Bakery items such as gluten-free bread, porridge, pasta, ready-to-eat breakfast cereals, infant foods, and beverages are just a few of the food applications of proso millet. As a result, it is a quality alternative to existing starch and provides insight to various food processing companies.

A brief nutritional profile of proso millet:

- It has the highest concentration of proteins (12.5%).
- The specific features of proso millet contribute to its health benefits. It contains a lot of carbohydrates and fatty acids.
- It is a less expensive source of manganese than other traditional sources such as nuts and spices.
- It has a substantial calcium content, which is necessary for bone formation and maintenance.
- It lowers cholesterol levels and lowers the risk of CVDs.

7.2 Composition and Distribution of Carotenoids and Anthocyanin in Pigmented Millets

Plants, insects, and microorganisms are examples of biological sources of natural pigments. Despite the fact that these sources are exclusive, food coloured with other food has always been preferred, primarily using extracts or pigments from plants, such as beetroots, berries, or tomatoes for red colour; saffron or marigold for yellow colour; annatto for orange colour; and leafy vegetables and plants for green colour. Microorganisms and microalgae that produce colour or pigment are abundant in nature. Melanins, violacein, carotenoids, quinines, flavins, monascins, phycocyanin, and other pigments/dyes derived from natural sources are some of the most prevalent pigments/dyes.[31] Figure 7.2 shows a detailed taxonomy of natural food pigments.

7.2.1 Carotenoids

Plant carotenoids are lipid-soluble pigments found embedded in the membranes of chloroplasts and chromoplasts in the colours red, orange, and yellow. Chlorophyll in photosynthetic tissues masks their colour, but these pigments contribute to the rich colours of many flowers and fruits, as well as the carrot root, in late stages of plant growth. Carotenoids are structural components of the photosynthetic antenna and reaction centre complexes that safeguard photosynthetic organisms from potentially hazardous photooxidative reactions.[32] Carotenoids with provitamin A activity are necessary components in diet, and many carotenoids have been shown to have anti-cancer properties. Carotenoids are dietary pigments converted into other substances and offer colour to various natural food products, including millets.

Plant carotenoids perform critical functions in photosynthesis as accessory pigments, photooxidation protection, and structural determinants in plastid pigment–protein complexes. The function of these pigments is mainly influenced by whether or not the tissue is photosynthetic. Photoprotection against hazardous oxygen species is the most critical function of photosynthetic tissues. Carotenoids contribute to the colour of flowers and fruits in

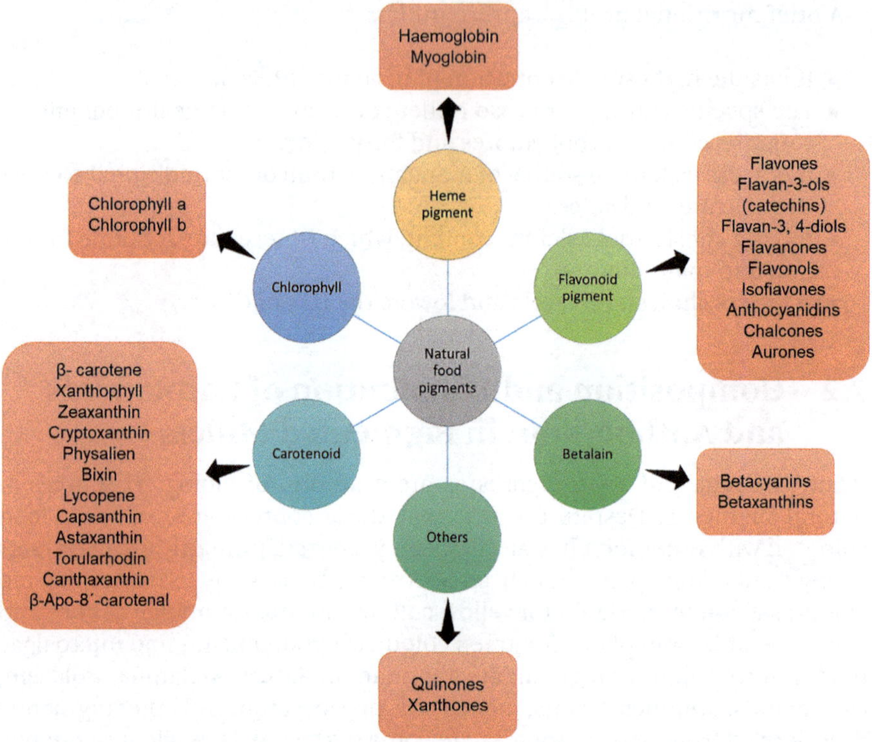

Figure 7.2 Detailed taxonomy of natural food pigments.

non-photosynthetic tissues. Anticarcinogenic and antioxidant properties are also found in carotenoids. For example, the provitamin A carotenoid β-carotene is an important nutritional element for chronic disease prevention, and it has been proven that carotenoids can lower blood levels of bad cholesterol and triglycerides. As a result, various carotenoids have been used as colourants, nutritional supplements, and nutraceuticals in the food, cosmetics, and pharmaceutical industries.

So far, numerous types of carotenoids have been detected in different cereals using various methods, with carotene, cryptoxanthin, *trans* or *cis*-lutein, and zeaxanthin being the most common. In wheat, the carotenoid lutein has the highest concentration, followed by zeaxanthin and cryptoxanthin. The majority of the total yellow pigments in proso millet were lutein (5–15 mg kg^{-1}) and zeaxanthin (16–17 mg kg^{-1}), according to results for carotenoid concentration of several proso millet cultivars.[26] In line with these findings, researchers discovered that, in comparison to lutein and zeaxanthin, carotene is a modest component of carotenoids in pearl, foxtail, and finger millets, averaging 0.076, 0.028, and 91 0.032 mg kg^{-1}, respectively.[33]

In research from Kim *et al.*,[34] HPLC was used to measure the amounts of five distinct carotenoids in millet (*Panicum miliaceum* L.) cultivars from China and Korea: lutein, zeaxanthin, -carotene, 9-*cis*-carotene, and 13-*cis*-carotene. The carotenoid concentration of cultivars from the two locations differed significantly, with the Chinese cultivars having a larger content than the Korean cultivar. Lutein levels were highest in all cultivars, however, Joongback (China) had 34%, 33%, and 16% greater levels than Joongjukyumi (China), Hwangguem (Korea), and Joongjuk (China), respectively. Joongback also had 58%, 55%, and 37% greater zeaxanthin concentrations than Joongjukyumi, Hwangguem, and Joongjuk, respectively, and a 33% higher carotene concentration than Hwangguem (Korea). The levels of the other two carotenoids (9-*cis*-carotene and 13-*cis*-carotene) were comparable in all 4 cultivars (Figure 7.3). The results suggest that the millet cultivar Joongback from China could potentially be used as a source of carotenoids in the human diet.

A high content of lutein and zeaxanthin is useful for the manufacture of functional cereal food due to the bioactive characteristic of carotenoids. Sorghum (0.003–0.17 mg kg^{-1}), barley (0.86 mg kg^{-1}), oat (0.17–0.23 mg kg^{-1}), and maize (2.23 mg kg^{-1}) were all found to have all-*trans* lutein in numerous

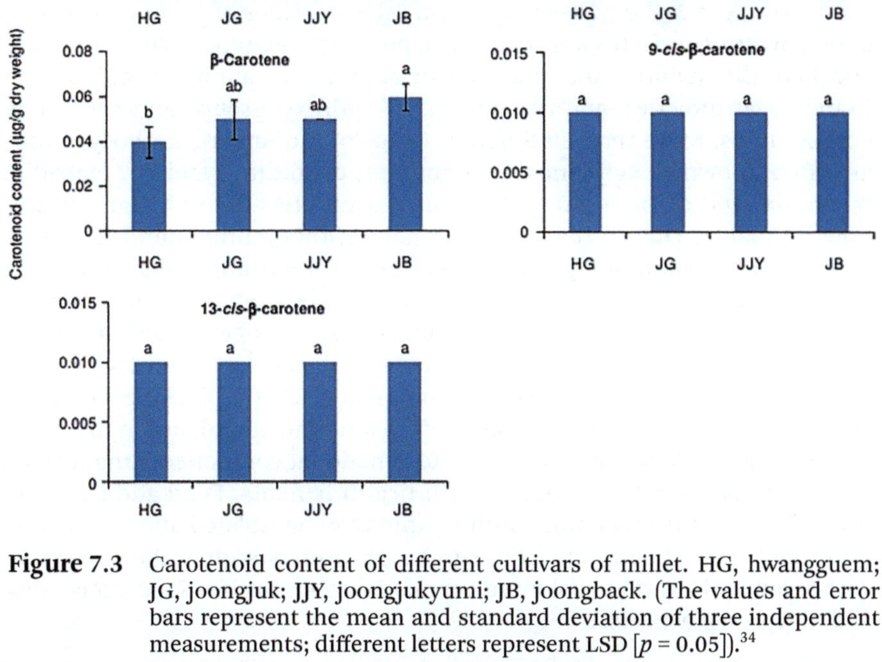

Figure 7.3 Carotenoid content of different cultivars of millet. HG, hwangguem; JG, joongjuk; JJY, joongjukyumi; JB, joongback. (The values and error bars represent the mean and standard deviation of three independent measurements; different letters represent LSD [$p = 0.05$]).[34]

studies. On the other hand, the current findings revealed a much higher amount of all-*trans*-lutein (6.18–13.99 mg per kg DW) than the previously listed other commonly consumed grains. This high level of all-*trans*-lutein is comparable to carotenoid-rich cereals like einkorn wheat (7.41 mg kg^{-1}) and high-carotene maize (11.03 mg kg^{-1}).

When compared with other millet types in the domain of small millets, it was discovered that carotenoid content varied between varieties. According to research, Xanthophyll (lutein) content ranged from 4.9 to 15.1 mg kg^{-1}, whereas zeaxanthin concentration ranged from 16.0 to 16.8 mg kg^{-1}. Foxtail millet is a possible option for the development of functional cereal foods due to its high level of carotenoids (lutein and zeaxanthin).[35,36]

One of the most significant cereals in the Chinese diet is foxtail millet. Researchers have been studying its antioxidant properties in recent years. The significant quantity of natural antioxidants detected in the samples, such as phenolic compounds, phenolic acids, and carotenoids, may be connected with foxtail millet's outstanding peroxyl radical scavenging capacity and cellular antioxidative activity. Furthermore, human breast and liver cancer cells *in vitro* were suppressed by foxtail millet extracts. When taken as a whole, foxtail millet is a beneficial cereal that has shown potential in preventing and treating cardiovascular, geriatric, and cancer disorders.

7.2.2 Anthocyanins

Anthocyanins are a family of vividly coloured pigment compounds that give many plants their hues of red, orange, brown, purple, and blue. Anthocyanins are flavylium or 2-phenylbenzopyrylium salt glycosides of polymethoxy and/or polyhydroxy derivatives. The number, position, and type of sugar moieties attached, the quantity and character of aromatic or aliphatic acids linked to the sugar moieties, and the number of hydroxyl groups all vary among anthocyanins. More than 30 different forms of monomeric anthocyanidins are often known as aglycones. Pelargonidin, peonidin, cyanidin, malvidin, petunidin, and delphinidin are the only six varieties found in various portions of plants. One of the most prevalent anthocyanins found in nature appears to be cyanidin 3-glucoside. Because of their appealing nutritional profile, anthocyanins have garnered considerable attention. Antioxidation, anti-cancer, retinal protection, hypolipidaemia, anti-ageing, and improved gut health are among the many purported health benefits of anthocyanins.[37]

Cereals that are naturally coloured comprise black, purple, blue, pink, red, and brown varieties.[38] Anthocyanins can be found in abundance in these pigmented grains. They can be turned into functional components that can be used to create new food products with health benefits. Food additives and natural colourants made from anthocyanins can be isolated and have a long shelf life (*e.g.* anthocyanins from sorghum). Understanding the content of anthocyanins in various coloured cereals is important for food processing and nutritional aspects.

Escribano-Bailón *et al.*[37] investigated the presence of anthocyanins in a variety of coloured cereals from over a decade ago. The composition of anthocyanins in coloured grains (wheat, maize, rice, and sorghum) and their extraction and identification procedures were evaluated. Concerning the non-phytoalexin anthocyanidins, cyanidin-3-dimalonyl glucoside has been found in sorghum, however, the glycosides of cyanidin and pelargonidin have not been identified. Though it has been discovered that red light and UV-A stimulate their synthesis, the formulation of these pigments is independent of the plant's prospective fungal infection. Varietal variances, pericarp colour, and soil and climatic circumstances could all contribute to differences in anthocyanin levels in sorghum cultivars. Compared with other sorghum cultivars, a study[38] found that sorghum flour from the JS-263 cultivar had a good source of anthocyanins (16 g per g). Apigeninidin levels were lower in all other sorghum cultivars (3-deoxyanthocyanin). In millet varieties, none of the anthocyanins were identified. Sorghum cultivar PC-1 (179.76 mg per 100 g) has a greater concentration of condensed tannins. Condensed tannins were discovered to be more plentiful in MB-87 (174.76 mg per 100 g) than in other millet cultivars. Antioxidant levels in sorghum and millet are higher than in other cereals. They could be good for human health if mixed with wheat flour and used to replace a portion of wheat, preferably 30%. Cereal brans can be separated as a functional element, and their qualities can be altered physically or biologically. The findings suggested that bran's structural changes might be made without affecting its nutritional value. As a result, this functional element could be improved for value addition and end use.

No anthocyanins were found in millet types, which could be attributable to millet's genetic makeup, while apigeninidin- and luteolinidin-type chemicals were found in finger millet using only an ethyl acetate extract (Table 7.2).[36] Millet's anthocyanin profile is unknown, and more research is needed.

Prevention against obesity, diabetes, cardiovascular disease, and cancer has all been associated with whole grain cereal consumption.[40]

Table 7.2 Anthocyanin content of pigmented cereal grains.[39]

Sample	Amount (μg g^{-1})
Blue barley	4
Maize	
Pink	93
Red	558
Purple	965
Blue	225
Black rice	2283
Black sorghum	944
Wheat	
Blue	106–153
Purple	13–139

The high dietary fibre content is responsible for some of these health benefits. Phenolic chemicals, in general, have been shown to improve these health outcomes. The exact health-favourable effect of each component has not yet been fully defined due to the complex nature of these effects, and despite attempts to examine the contributable role of different chemicals.[41]

7.3 Biological Activities From Pigmented Millets

Apart from natural small molecules, the increased frequency of lifestyle-related disorders and corresponding knowledge of natural therapies to treat these ailments has pushed the investigation of biologically active components. Through *in vitro* and animal research, several substances from various sources in diet have shown compelling evidence as health-improving elements against a variety of issues, including diabetes, obesity, inflammation, CVD, and cancer.

Exploratory studies on naturally produced antimicrobial/viral agents, such as proteins and their chemically synthesised analogues that may have clinical value are necessitated by the emergence of drug-resistant strains of bacteria and viruses. In this regard, Camargo Filho *et al.*[42] discovered the existence of antiviral peptides in jower. The 2 kDa peptide from jower inhibited the replication of bovine herpes virus (BHV), herpes simplex virus type 1 (HSV-1) and poliovirus type 1 with a weak activity. Furthermore, the peptides were demonstrated to have a preventive effect against secondary infection and inhibit the spread of HSV-1 infection *in vitro*. Furthermore, antibacterial activity was observed in proteins/peptides isolated from finger millet, barnyard millet, and proso millet against *Pseudomonas aeruginosa* (MTCC 424) and *Salmonella entrica* (MTCC 739). Similarly, three peptides produced from foxtail millet (FFMp4, FFMp6, and FFMp10) showed high antibacterial action against *E. coli* ATCC 8099. These studies show that millet-derived peptides have the potential to be effective antimicrobials against viruses, bacteria, and fungi.[43]

Melanogenesis is a multi-step process governed by a variety of processes. Microphthalmia-associated transcriptional factor (MITF) is required for the development and differentiation of melanocytes, and it is induced by α-MSH. TRP-1 and TRP-2 are two melanogenesis-related proteins that trigger tyrosinase, and MITF operates on their promoter regions. The identification of chemicals against mushroom tyrosinase and B16 melanoma cells resulted from research into the melanogenesis inhibitory activity of barnyard millet grain extracts (BME). The structure related to the inhibitory actions was also investigated using molecular docking simulations. As a result, BME can be utilised to prevent melanogenesis by acting as a natural source of depigmentation.[44]

Millets have indeed been proven to contain significant levels of antioxidants in the form of phenolics and tannin, which have been shown to effectively scavenge free radicals *in vitro* and *in vivo*. By participating in a single electron transfer process, these antioxidant phytochemicals display radical-quenching properties. Nonetheless, there is a wealth of information about millet peptides and protein hydrolysates that have antioxidant capabilities in various oxidative reaction systems. Furthermore, many peptidic amino acid residues capable of transferring electrons to free radicals at physiological pH contribute to their antioxidant effects as observed in commercial varieties of red sorghum.[45]

The findings of a study by Senevirathne *et al.*[46] showed that some millet and sorghum varieties in Sri Lanka had a variety of health benefits in relation to diabetes and its consequences. Inhibition of α-amylase (a major enzyme involved in starch digestion), inhibition of both early and middle-stage protein glycation, and reversal of both early- and middle-stage glycation products were among the biological activities reported. When compared with the other millet samples evaluated, the pigmented samples, such as sweet sorghum, oshadha, and rawana, showed the strongest antiamylase, antiglycation, and glycation-reversing properties. In 11 widely available red sorghums, diverse chemical compositions of soluble substances and insoluble phenolic substances, including four phenolic acid derivatives, three flavonoids, and one proanthocyanidin, were discovered for the first time in sorghum. Four main 3-deoxyanthocyanidins were measured in sorghum as a class specific to phenolic chemicals. According to Pearson's correlation analyses, 3 deoxyanthocyanins may contribute significantly to the antioxidant capabilities of sorghums, particularly the soluble phenolic fractions. *In vitro* anti-proliferation and anti-inflammatory actions were also demonstrated in the soluble phenolic fraction of B13, which might be attributable to its comparatively high TPC, TFC, and TPAC values, 3- deoxyanthocyanin concentration, and free radical scavenging capacities.[47]

For the first time, cellular antioxidant and antiproliferative activities of proso millet were discovered in another investigation. The antioxidant activity and phenolic acid content of various proso millet types are described. The bound fraction accounted for approximately 65 percent of the total phenolic content in the proso millet types studied. Proso millet also contains phytochemicals such as ferulic acid, chlorogenic acid, syringic acid, caffeic acid, and *p*-coumaric acid, all of which have been linked to improved human health.[48]

A study was conducted to see how various biological processing treatments affected the bio-techno-functional features of foxtail millet (*Setaria italica* L.). Anti-nutritional factors, *in vitro* starch and protein digestibility, bioactive constituents, and associated antioxidant potential were measured after grains were exposed to soaking, germination, fermentation, and the aforementioned techniques. Considerable variation (p 0.05) in anti-nutritional factors, *in vitro* starch and protein digestibility, bioactive constituents, and associated antioxidant potential was observed.[49]

A summary of biological activities from pigmented millets have been summarised in Table 7.3.

Table 7.3 Biological activities from pigmented millets.

S. no	Millet type	Biological activity	Biological effect	References
1	Barnyard millet	Melanogenesis inhibitory effect	The extracts of barnyard millet grain (BME) fostered the isolation of compounds against mushroom tyrosinase and B16 melanoma cells	Seo *et al.*, (2012)[44]
2	Proso millet, white finger millet, kodo millet, foxtail millet, and finger millet, sweet sorghum	Antiamylase, antiglucosidase, and antiglycation activities of millets and sorghum	Millets and sorghum showed significant dose-dependent anti-amylase early-stage antiglycation middle-stage antiglycation, early-stage glycation reversing, and middle-stage glycation reversing activities	Senevirathne *et al.*, (2021)[46]
3	Proso millet	Phytochemical and antiproliferative	The cellular antioxidant activity (CAA) of the extract was assessed using the HepG2 model against MDA human breast cancer and HepG2 human liver cancer cells	Zhang *et al.*, (2014)[33]
4	Proso millet	Secondary metabolite production on millet	Millet showcased a potential substrate over rice, due to higher lovastatin and lower citrinin production	Marič *et al.*, (2019)[50]
5	Proso millet	Coloured grains showed higher phenolic acid and flavonoid contents	Functional activity of millet grains was found to be more	Li *et al.*, (2021)[47]
6	Commercial variety of red sorghum (red sorghums (B11, B12, B13, B14, J124, J127, J138, J140, J142, J152, and J158)	Proliferation inhibition, antioxidant activity	High content of phenolics, flavanoid, proanthocynadin added to radical scavenging activities	Li *et al.*, (2021)[47]
7	Pearl millet	Graphine oxide showed the potential results for growth enhancement of pearl millet	Higher growth, yield and antioxidant content was found	Mahmoud *et al.*, (2021)[51]

8	Foxtail millet	Bioprocessing of millet resulted in improved bio-functionality of millets	Soaking, fermentation, germination showed improved bioactive constituents and antioxidant potential	Sharma *et al.*, (2022)[49]
9	Proso millet	Detected new functional ingredients as potent antioxidants	Huge and diverse phenolics resulted to the antioxidant properties of proso millet	Yuan *et al.*, (2022)[48]
10	Proso millet	Enhanced bio-functionality of millet flour	Combination of germination and fermentation tremendously increased the biological properties of protein digestibility	Sharma *et al.*, (2022)[49]
11	Pearl millet	Low glycemic activity	*In vitro* oro-gastro intestinal simulation observed pearl millet of low IGP (63.71%) relative to rice (65.89%).	Mondal *et al.*, (2022)[52]

7.4 Conclusion

The need of the hour is to assist people in the understanding and appreciation of food and the introduction of millets as a nutrient-dense food that can meet the nutritional needs of the population worldwide, as well as to create strategies to consume millets appropriately in order to mark a decline in cases of malnourishment and co-morbidities. All millet foods have significant health benefits, thanks to their high fibre content, which aids in metabolic disorders such as type II diabetes, obesity, cardiovascular diseases (CVDs), and others, as well as their high protein content, which aids in child growth and development. Calcium content, on the other hand, aids in bone development in both children and the elderly, high iron content, which aids in management of anaemia, and gluten-free properties helps patients with gluten allergy. The waxy coats of millet include cardio-protective substances such as phytosterols and policosanols. If these millets are made into flour without being dehulled, they can provide a number of advantages. Millets retain their antioxidant properties, *i.e.* protecting body cells from free radicals, if used without being dehulled. Biscuits and confectionary, beverages, weaning foods, and beer are some of the emerging industrial raw materials. Grits, flour, and meals made from cereals including millet, sorghum, and corn are increasingly widely available. Sorghum, maize, and wheat composites are used to manufacture soft biscuits and cookies, while cakes and non-wheat breads have been a focus of increased sci-tech inquiry, with promising outcomes. The composition and distribution of pigments in these millets critically determine their biological activities. Carotenoids have anticarcinogenic and antioxidant effects. For instance, the provitamin A carotenoid, carotene, is a key component for chronic disease management, and carotenoids have been shown to lower harmful cholesterol and triglyceride levels in the blood. As a result, carotenoids have recently been used in the food, cosmetics, and pharmaceutical industries as pigmenting agents, nutritional supplements, and nutraceuticals. Millets have been proven to contain significant amounts of antioxidants, which have been shown to help reduce oxidative stress *in vitro* and *in vivo*. The emergence of drug-resistant bacterial and viral strains have necessitated exploratory research on naturally occurring antimicrobial/viral agents, such as proteins and their synthetic chemical counterparts, that could have clinical utility. Therefore, millets are an actual magic ingredient in the human diet and must be popularised to a great extent for sustainable health and wellness.

References

1. B. McKevith, *Nutr. Bull.*, 2004, **29**(2), 111.
2. B. Margetts, *Public Health Nutr.*, 2003, **6**(4), 323.
3. FAO Cereal Supply and Demand Brief, World Food Situation, Food and Agriculture Organization of the United Nations, 2021, accessed January 2022, https://www.fao.org/worldfoodsituation/csdb/en/.
4. P. A. Parizad, J. Capraro, A. Scarafoni, *et al.*, *Plant Foods Hum. Nutr.*, 2019, **74**(1), 128.

5. K. Ambati and K. V. Sucharitha, *Int. J. Recent Sci. Res.*, 2019, **10**(7), 33943.
6. B. Dayakar Rao, K. Bhaskarachary, G. D. Arlene Christina, G. Sudha Devi, A. T. Vilas, and A. Tonapi, *Nutritional and Health Benefits of Millets*, ICAR-Indian Institute of Millets Research (IIMR), Hyderabad, Indian, 2017, p. 112.
7. Markets R., Global Millet Markets 2019–2024 – India Dominates Global Production & Africa Dominates Global Consumption, 2019, accessed-January 2022, https://www.prnewswire.com/news-releases/global-millet-markets-2019-2024---india-dominates-global-production--africa-dominates-global-consumption-300888582.html.
8. P. B. Devi, R. Vijayabharathi, S. Sathyabama, N. G. Malleshi and V. B. Priyadarisini, *J. Food Sci. Technol.*, 2014, **51**(6), 1021.
9. M. M. Dida, and K. M. Devos Finger millet, in *Cereals and Millets*, Springer, Berlin, Heidelberg, 2006, p. 333.
10. S. S. Purewal, R. K. Salar, M. S. Bhatti, K. S. Sandhu, S. K. Singh and P. Kaur, *J. Food Meas. Charact.*, 2020, **14**(1), 150.
11. D. Chandra, S. Chandra and A. K. Sharma, *Food Sci. Hum. Wellness*, 2016, **5**(3), 149.
12. N. M. McKeown, J. B. Meigs, S. Liu, P. W. Wilson and P. F. Jacques, *Am. J. Clin. Nutr.*, 2002, **76**(2), 390.
13. U. Antony, G. Sripriya and T. S. Chandra, *J. Agric. Food Chem.*, 1996, **44**(9), 2616.
14. J. R. Taylor, and M. N. Emmambux Gluten-free foods and beverages from millets, in *Gluten-free Cereal Products and Beverages*, Academic Press, 2008, p. 119.
15. J. A. Adebiyi, A. O. Obadina, A. F. Mulaba-Bafubiandi, O. A. Adebo and E. Kayitesi, *J. Cereal Sci.*, 2016, **70**, 132.
16. R. K. Salar, S. S. Purewal and M. S. Bhatti, *Resour.-Effic. Technol.*, 2016, **2**(3), 148.
17. M. A. Gassem and M. A. Osman, *J. Saudi Soc. Agric. Sci.*, 2008, **7**, 1.
18. G. W. Burton, A. T. Wallace and K. O. Rachie, *Crop Sci.*, 1972, **12**(2), 187.
19. S. Malik, *Int. J. Adv. Res. Innovative Ideas Educ.*, 2015, **1**(3), 414.
20. R. K. Srivastava, C. T. Satyavathi, M. D. Mahendrakar, R. B. Singh, S. Kumar, M. Govindaraj and I. A. Ghazi, *Front. Genet.*, 2021, 12.
21. M. Suguna, C. Aruna, C. Deepika, C. V. Ratnavathi and V. A. Tonapi, *J. Cereal Sci.*, 2021, **100**, 103226.
22. L. de Morais Cardoso, S. S. Pinheiro, H. S. D. Martino and H. M. Pinheiro-Sant'Ana, *Crit. Rev. Food Sci. Nutr.*, 2017, **57**(2), 372.
23. F. Al Juhaimi, Ş. Şimşek, K. Ghafoor, E. E. Babiker, M. M. Özcan, I. A. M. Ahmed, *et al.*, *J. Oleo Sci.*, 2019, **68**(11), 1063–1071.
24. N. Sachdev, S. Goomer and L. R. Singh, *J. Sci. Food Agric.*, 2021, **101**, 831.
25. M. Johnson, S. Deshpande, M. Vetriventhan, H. D. Upadhyaya and J. G. Wallace, *Plant Genome*, 2019, **12**(3), 190021.
26. S. Sood, R. K. Khulbe, A. K. Gupta, P. K. Agrawal, H. D. Upadhyaya and J. C. Bhatt, *Plant Breed.*, 2015, **134**(2), 135.
27. Medindia, Little Millet - The Immune Boosting Indigenous Grain, 2022, [online] available at: <www.medindia.net/dietandnutrition/little-millet-the-immune-boosting-indigenous-grain.htm>, accessed 4 March 2022.
28. S. P. Bangar, A. O. Ashogbon, S. B. Dhull, R. Thirumdas, M. Kumar, M. Hasan, V. Chaudhary and S. Pathem, *Int. J. Biol. Macromol.*, 2021, **190**, 960.

29. T. Longvah, R. Ananthan, K. Bhaskarachary and K. Venaiah, *Indian Food Composition Table*, National Institute of Nutrition, (Indian Council of Medical Research), Department of Health Research, Ministry of Health & Family Welfare, Government of India, Hyderabad-500 007, Telangana State, INDIA, 2017, http://www.ifct2017.com/frame.php?page=home.
30. C. Gopalan, B. V. Rama Sastri, and S. C. Balasubramanian, *Nutritive Value of Indian Foods*, ed. B. S. N. Rao, Y. G. Deosthale, and K. C. Pant, National Institute of Nutrition, ICMR, Hyderabad, 2010.
31. R. Dikshit, and P. Tallapragada, Comparative study of natural and artificial flavoring agents and dyes, in *Natural and Artificial Flavoring Agents and Food Dyes*, Academic Press, 2018, p. 83.
32. G. E. Bartley and P. A. Scolnik, Plant carotenoids: pigments for photoprotection, visual attraction, and human health, *Plant Cell*, 1995, **7**(7), 1027.
33. L. Zhang, R. Liu and W. Niu, *PLoS One*, 2014, **9**(8), e104058.
34. Y. B. Kim, J. K. Kim, M. R. Uddin, C. H. Park, H. H. Kim, E. Chung, *et al.*, *Asian J. Chem.*, 2014, **26**(2), 464.
35. R. Shen, S. Yang, G. Zhao, Q. Shen and X. Diao, *J. Cereal Sci.*, 2015, **61**, 86.
36. F. Zhu, *Food Res. Int.*, 2018, **109**, 232.
37. M. T. Escribano-Bailón, C. Santos-Buelga and J. C. Rivas-Gonzalo, *J. Chromatogr. A*, 2004, **1054**, 129.
38. F. Ahmad, I. Pasha, M. Saeed and M. Asgher, *Int. J. Food Prop.*, 2018, **21**(1), 1586.
39. L. Dykes and L. W. Rooney, *Cereal Foods World*, 2007, **52**(3), 105.
40. A. Gani, S. M. Wani, F. A. Masoodi and G. Hameed, *J. Food Process. Technol.*, 2012, **3**(3), 146.
41. A. Fardet, *Nutr. Res. Rev.*, 2010, **23**, 65.
42. I. Camargo Filho, D. A. G. Cortez, T. Ueda-Nakamura, C. V. Nakamura and B. P. Dias Filho, *Phytomedicine*, 2008, **15**(3), 202.
43. M. Thapliyal, A. Bisht and A. Singh, *Int. J. Curr. Pharm. Res.*, 2016, **8**(4), 24.
44. W. D. Seo, J. Y. Kim, K. C. Jang, S. I. Han, J. E. Ra, S. H. Oh, *et al.*, *J. Korean Soc. Appl. Biol. Chem.*, 2012, **55**(5), 579.
45. H. Agrawal, R. Joshi and M. Gupta, *LWT–Food Sci. Technol.*, 2017, **84**, 608.
46. I. G. N. H. Senevirathne, W. K. S. M. Abeysekera, W. P. K. M. Abeysekera, N. Y. Jayanath, S. P. Galbada Arachchige and D. C. M. S. I. Wijewardana, *J. Evidence-Based Complementary Altern. Med.*, 2021, 1–10.
47. Y. Li, M. Li, J. Liu, W. Zheng, Y. Zhang, T. Xu, ... and L. Yu, *J. Agric. Food Chem.*, 2021, **69**(32), 9407.
48. Y. Yuan, J. Xiang, B. Zheng, J. Sun, D. Luo, P. Li and J. Fan, *LWT–Food Sci. Technol.*, 2022, **154**, 112611.
49. R. Sharma and S. Sharma, *Food Chem.*, 2022, **368**, 130815.
50. A. Marič, M. Skočaj, M. Likar, K. Sepčić, I. K. Cigić, M. Grundner and A. Gregori, *J. Food Sci. Technol.*, 2019, **56**(7), 3364.
51. N. E. Mahmoud and R. M. Abdelhameed, *Plant Stress*, 2021, **2**, 100025.
52. D. Mondal, M. Awana, S. Aggarwal, D. Das, B. Thomas, S. P. Singh, ... and V. Krishnan, *Food Hydrocolloids*, 2022, 107481.

CHAPTER 8

Pigmented and Non-pigmented Cereals: Comparative Study of Properties

KHONGSAK SRIKAEO*[a]

[a]Faculty of Food and Agricultural Technology, Pibulsongkram Rajabhat University, Muang Phitsanulok 65000, Thailand
*E-mail: khongsak@live.psru.ac.th

8.1 Introduction

Cereals are plants that belong to the monocot Gramineae family. They include: common wheat (*Triticum aestivum* L.), durum wheat (*Triticum durum* Desf.), maize (*Zea mays* L.), oats (*Avena sativa* L.), rye (*Secale cereale* L.), rice (*Oryza sativa* L.), barley (*Hordeum vulgare* L.), sorghum (*Sorghum bicolor* L. Moench), pearl millet (*Pennisetum glaucum* L.), finger millet (*Eleusine coracana* L.), teff (*Eragrostis tef*), and triticale (x *Triticosecale Wittmack*). Pseudo-cereals such as quinoa (*Chenopodium quinoa* Willd), amaranth (*Amaranthus* sp.), buckwheat (*Fagopyrum esculentum* Moench), and chia (*Salvia hispanica* L.) do not belong to the Gramineae family. However, due to their nutrient-rich, gluten-free properties,[1] they can be used in a similar manner to cereal grains. Cereals are essential components of human nutrition on a daily basis. They can be converted into functional ingredients and used to create novel foods with potential health benefits.

Food Chemistry, Function and Analysis No. 38
Pigmented Cereals and Millets: Bioactive Profile and Food Applications
Edited by Sneh Punia Bangar, Sajid Maqsood and Anil Kumar Siroha

Published by the Royal Society of Chemistry, www.rsc.org

Pigmented or coloured cereal grains have gained much attention recently due to their attractive nutritional values. Their colours range from black to purple, blue, pink, red, and brown. The major types of pigments responsible for the colours, as well as the health benefits of the cereals are flavonoids, carotenoids and some phenolic compounds.[2,3] Anthocyanins, an important class of flavonoids, are concentrated in the aleurone or pericarp layer and produce blue, purple, or a combination of these colours. Flavonoids, such as yellow *C*-glycosides of flavones, flavonols, flavanonols, proanthocyanidins, and reddish-coloured phlobaphenes, are concentrated in the outer layer of grains, whereas carotenoids, which contribute to the yellow colour of the grain, are concentrated in the endosperm.[4]

The colour range of the majority of grain cereals follows N. I. Vavilov's 1920 formulation of 'The Law of Homologous Series in Hereditary Variability'. It states that a similar series of genetic variability usually characterizes species and genera that are genetically closely related. A similar range of grain colours, therefore, exists in all cereals.[5] Variation in pigments as well as phenolic content was also observed in cereal species grown in different geographical locations and when subjected to variable levels of salinity, water, and temperature.[6]

8.2 Wheat

The significance of pigmented wheat is that it contains a higher concentration of tannins, polyphenols, carotenoids, and anthocyanins, which are responsible for the grains' various colours. In wheat grains, anthocyanins can be expressed in either the pericarp (*e.g.* purple pericarp) or the aleurone layer (*e.g.* blue aleurone). Anthocyanin in wheat results in blue aleurone and purple pericarp grains. These anthocyanins are controlled by the Ba and Pp genes. Carotenoids are responsible for the wheat grain's yellow endosperm, controlled by the Psy genes.[7] Figure 8. 1 shows the wheat kernels with accumulated anthocyanin. The genes responsible for purple pericarp (Pp, Figure 8.1b) and blue aleurone (Ba, Figure 8.1c) can be pyramidized by sophisticated crossbreeds to breed wheat grains with increased anthocyanin content (Pp and Ba, Figure 8.1d).

Wheat grain is purple due to the anthocyanin found in the pericarp. The principal anthocyanidin-3-*O*-glycosides found in purple wheat grains are cyanidin 3-glucoside, cyanidin 3-rutinoside, and cyanidin 3-(6″-succinylglucoside).[9] Contrasts in anthocyanin synthesis are responsible for the colour difference between purple and blue grains. Delphinidin 3-rutinoside and delphinidin 3-glucoside are the primary anthocyanins found in blue aleurone wheat.[9,10]

Pigmented wheat possesses a plethora of health benefits and has been shown to be beneficial in preventing and combating a variety of chronic diseases, including cancer, cardiovascular disease (CVD), diabetes, inflammation, obesity, and the effects of aging.[7] Pigmented wheat's health benefits are attributed to its high anthocyanin and phenolic content, as well as its

Figure 8.1 Wheat grain colour due to accumulation of anthocyanins in different grain layers: (a) white (no anthocyanins); (b) purple pericarp (Pp); (c) blue aleurone (Ba); and (d) deep purple (Ba + Pp). Reproduced from ref. 8 with permission from *Plant Methods*, Copyright 2018.

high antioxidant activity. Apart from phytochemical compounds, pigmented wheat is a good source of macro- and micronutrients, which are necessary for the human body to function normally.[3,11,12]

8.2.1 Nutritional Composition

A whole grain of wheat is composed of three layers: aleurone, germ, and bran. Wheat grains are high in macronutrients such as carbohydrates, fats, and proteins; micronutrients such as vitamins and minerals; phytochemicals, dietary fibres, and a variety of bioactive compounds.[3]

For macronutrients, a balance of protein and amino acid is critical in determining the nutritional quality of wheat. Starch, present in the endosperm, is the major carbohydrate. Common wheat contains 2.15% ash, 12.93% protein, 3.33% oil, 61.85% starch, and 2.29% cellulose,[13] which are considerably lower than pigmented wheat, both qualitatively and quantitatively. Compared with common wheat, approximately 11.74% to 18.17% higher protein content, 7.31% to 18.13% higher essential amino acids, and 8.88% to 18.91% higher total amino acids were recorded in pigmented wheat.[12] Purple wheat bran powder contains 28 g per 100 g of protein content, 4.3 g per 100 g of ash content and a small amount of sugars, the majority of which is glucose. Additionally, arabinoxylan and β-glucan gums are also found in wheat bran powder.[9] Although removing wheat bran affected the macronutrients present in the bran and research found a 0.98%, 10.45%, 3.02%, 71.24%, and 0.24% reduction in the concentration of ash, protein, oil, starch, and cellulose content, respectively.[13] Debranned durum wheat was indicated to have a 60%, 18%, and 90% reduction in the concentration of ash, protein and cellulose, respectively, when compared with whole durum wheat.[3,13] Pigmented wheat produces less flour than common wheat due to its low volume, weight, and plumpness, which reduces the endosperm content. Because fat and dietary fibre are located in the embryo and seed coat, they do not differ significantly between pigmented and common wheat.[12,14]

In terms of micronutrients, pigmented wheat is a rich source of B-group vitamins especially thiamine (B1), riboflavin (B2), niacin (B3), pyridoxine (B6), folate (B9), and vitamin E (tocopherol/tocotrienol). Additionally, it contains a trace amount of provitamin A (beta-carotene), vitamin D (calciferol), and vitamin K. (phylloquinone).[15] Zinc, iron and magnesium content in pigmented wheat is about 108.54% to 142.68%, 8.57% to 42.86%, and 5.31% to 40.63% higher than common wheat, respectively. Green wheat contains the highest Zn, while blue wheat contains the highest calcium content.[12,16] It should be noted that pigmented wheat does not always contain more micronutrients than common wheat, especially under varying planting conditions.

8.2.2 Bioactive Compounds

Phenolic acids, flavonoids, and anthocyanins are the most abundant bioactive compounds found in pigmented wheat grains. Most of these compounds exist in the entire grain as glycosides linked to various sugar moieties or as various complexes linked to natural acids, amines, lipids, starches, and various phenols.[17] Additionally, other significant phytochemicals found in each wheat variety include alkylresorcinols, folate, sterols, and tocols.[2]

8.2.2.1 Phenolic Acids

Phenolic acids are a major component of grain crop metabolites, occurring in free, conjugated, or insoluble bound structures.[18] Approximately 95% of grain phenolic compounds are ester- or ether-linked to the cell wall polysaccharides and act as an intra- and intermolecular interface to shape systems. They are identified as dietary fibre-phenolic compounds.[19]

Individually, the phenolic acids in pigmented and unpigmented wheat grains are classified as hydroxy-benzoic acids and hydroxycinnamic acids based on their C_1–C_6 and C_3–C_6 skeletons. Subordinates of the hydroxybenzoic acids include *p*-hydroxybenzoic, vanillic, syringic, and gallic acids, while the hydroxycinnamic acids include *p*-coumaric, caffeic, ferulic, and sinapic acids. These mixtures are available bound and as esters and glycosides. Cinnamic acids have been classified as bioactive components of the diet due to their association with basic cell wall components.[20] Ferulic acid is the major hydroxycinnamic acid. The various derivates occur in trace amounts. Black wheat contains the most phenolic compounds, flavonoids, and antioxidant activity, while light purple wheat contains the least.[21] Black wheat contained 659.8 g g^{-1} gallic acid equivalents and 319.3 g g^{-1} rutin equivalents of total phenolics and flavonoids. Additionally, black wheat has the highest total anthocyanin content, ranging from 128 to 198 mg kg^{-1}, while yellow wheat has the lowest total anthocyanin content, starting from 0.925 mg kg^{-1}. Blue wheat contains a total phenolic content of 6.8 g kg^{-1}, while white wheat contains 0.50 g kg^{-1}. Purple wheat contained a significant amount of flavonoids, ranging from 21.59 to 102.95 mg per 100 g. White wheat contained the fewest flavonoids (9.6 mg per 100 g).[2] In summary, current research indicates that dark-coloured wheat contains higher levels of flavonoid, phenolic, and antioxidant compounds than light-coloured wheat.

8.2.2.2 Flavonoids

Flavonoids are significant bioactive mixtures or polyphenols that have a high antioxidant capacity and a variety of medical benefits. They are generally found in coloured fruits grown from the ground.[22] The highest concentration of flavonoids is found in the bran layer and germ of wheat grain.[23] Chemically, the overall structure of flavonoids is a 15-carbon skeleton, composed of two phenyl rings (A and B) and a heterocyclic ring (C). It can be classified into six distinct subtypes based on a typical C_6–C_3–C_6 carbon skeleton, which include flavones (*e.g.* apigenin, luteolin, chrysoeriol, and their glycosides), isoflavones, chalcones, flavonols (*e.g.* kaempferol, quercetin, myricetin, and their glycosides), flavanones (*e.g.* naringenin, eriodictyol, and their glycosides), and anthocyanidins.[24]

The flavonoids found in whole wheat are primarily flavones, flavone *C*-glycosides, and anthocyanins. Total flavonoids in various wheat varieties ranged from 20.14 mg per 100 g catechin equivalent to 67.69 mg per 100 g catechin equivalent.[25] Additionally, the proportion of free flavonoid to bound flavonoid in wheat varied according to variety. The results indicated that the composition and content of flavonoids varied significantly by wheat variety. Antioxidant agent properties of wheat were fundamentally related to the total substance of flavonoids; however, no discernible relationship between the ratio of free to bound flavonoid and antioxidant activity was discovered. Additionally, the flavonoid content varies among whole wheat varieties. According to the literature, bran or germ parts contained flavonoid substances ranging from 214.8 to 272.9 mg of CE per 100 g, while endosperm contained between 17.4 and 23.2 mg of CE per 100 g.[26] Recently, it has been found that flavonoids in wheat grains generally occur in their *C*-glycoside structures.[27] Tricin is a prevailing flavone pigment in wheat grain.[28] Apigenin-C glycosides were separated and identified from wheat flour.[29] Apigenin-6-*C*-glucoside-8-*C*-arabinoside largely exists in the wheat germ.[30] From wheat sprouts, a few *C*-monoglycosides, including isoorientin and isoscoparin, and an unusual flavonoid, were identified.[31] There were 72 flavone *C*-glycosyl derivatives isolated from wheat germ, with several of them serving as significant marker mixtures for separating whole wheat and refined wheat food products.[27] In summary, the flavonoids found in wheat-based ingredients and nourishments vary according to the variety and part of wheat. Tricin (5,7,4-trihydroxy 3,5-dimethoxyflavone) was identified as the flavone pigment with the highest concentration in wheat.[28] Isoflavones were also detected in the free phenolic components of some wheat varieties.[32] Purple wheat has the highest total flavonoid content, ranging from 21.59 to 102.95 mg per 100 g catechin equivalents.[33] Different varieties of wheat, such as yellow, red, and normal wheat, contained 13.44, 10.72, and 9.6 mg per 100 g of cyanide, respectively.[2]

8.2.2.3 *Anthocyanins*

Anthocyanins are a class of water-soluble pigments known as flavonoids. They are responsible for the vibrant and alluring orange, red, purple, and blue hues found in the majority of organic products. Anthocyanins are based on anthocyanidin (aglycone), with saccharide residues bound at different hydroxylated positions. Individual anthocyanins differ in the amount of hydroxy- or methoxy-groups attached to aglycones and, more specifically, sugars *via O*-glycosidic bonds at various positions, as well as in the aliphatic or aromatic acids attached to the sugars. Six anthocyanidins: cyanidin, delphinidin, malvidin, pelargonidin, peonidin, and petunidin are commonly found in pigmented wheat and are responsible for the grain's purple, blue, or red colour. Grain anthocyanins are abundant in the outer layers of wheat. Black grained wheat, blue grained wheat, and wheat with a purple pericarp all contain a high concentration of total anthocyanins,[7] and some purple wheat contains a lot higher anthocyanin levels than blue wheat.[9] A total of 22 diverse anthocyanins from blue, 23 from purple, and 26 from black wheat lines were characterized.[7] Four principal anthocyanins, delphinidin-3-glucoside (45%), cyanidin-3-glucoside (28%), delphinidin-3-rutinoside (22%), and cyanidin-3-rutinoside (2%) were identified in blue wheat. Major anthocyanins in blue aleurone were delphinidin-3-rutinoside (33.44 mg kg^{-1}), delphinidin-3-glucoside (13.68 mg kg^{-1}) and malvidin-3-glucoside (12.04 mg kg^{-1}), followed by cyanidin-3-rutinoside (8.42 mg kg^{-1}), cyanidin-3-glucoside (3.07 mg kg^{-1}), peonidin-3-arabinoside (2.22 mg kg^{-1}), peonidin-3-galactoside (1.94 mg kg^{-1}), and peonidin-3-glucoside (0.88 mg kg^{-1}).[34] Cyanidin-3-glucoside, cyanidin-3-galactoside, and malvidin-3-glucoside are the most abundant anthocyanins in purple grains. There were twelve different anthocyanin glycosides identified: delphinidin-3-galactoside, delphinidin-3-arabinoside, cyanidin-3-arabinoside, petunidin-3-glucoside, cyanidin-3-galactoside, cyanidin-3-glucoside, pelargonidin-3-glucoside, peonidin-3-glucoside, malvidin-3-glucoside, pelargonidin-3-galactoside, pelargonidin-3-arabinoside and peonidin-3-arabinoside. The total anthocyanin content of purple wheat was 11.81 mg kg^{-1}, with cyanidin-3-glucoside (10.34 mg kg^{-1}), peonidin-3-galactoside (0.58 mg kg^{-1}), and malvidin-3-glucoside (0.48 mg kg^{-1}) being the most abundant.[7,34]

Further research is being conducted to determine the anthocyanin glycosides present in coloured grained wheat. New anthocyanin glycosides identified in blue, purple, and black grained wheat lines with better return were accounted for cyanidin-3-(2G xylosylrutinoside), cyanidin-3 (300,-600 dimalonyl-glucoside),cyanidin-3 rutinoside, 3′glucoside, cyanidin-3 (600-feruloyl-glucoside) 5 glucoside, delphinidin-3 sambubioside, delphinidin-3 (600 malonyl-glucoside), delphinidin-3 caffeoyl glucoside, malvidin-3-(600-*p*-caffeoylglucoside), malvidin-3 rutinoside-5 glucoside, pelargonidin-3 (600 malonyl glucoside), pelargonidin-3-rutinoside, peonidin-3, 5di-glucoside,peonidin-3-rutinoside-5-glucoside,andpetunidin-3-rutinoside-5-glucoside.[7,35] Anthocyanin levels are influenced by genotype and by climatic and environmental conditions.[10]

8.2.2.4 *Pro-anthocyanidins and Phlobaphenes*

In addition, the colourless polyphenols flavan-3-ols, flavan-4-ols, and flavan-3,4-diols may be converted to coloured phlobaphenes or other proanthocyanidin oxidized subsidiaries *via* the flavonoid biosynthesis pathway. Commercial wheat grain contain trace amounts of catechins and di-, tri-, and colourless oligomeric proanthocyanidins soluble in aqueous acetone. Attempts to distinguish these compounds in wheat previously appeared to fail due to interference from methoxyhydroquinone glycosides. Oligomeric proanthocyanidins typically contain prodelphinidin and a small amount of procyanidin, whereas dimeric proanthocyanidins may also contain some propelargonidin units. Procyanidin B3 (catechin-[4α-8]-catechin) and prodelphinidin B3 (gallocatechin-[4α-8]-catechin) were separated. During depolymerization under oxidative conditions, oligomeric proanthocyanidins can yield dull flavan-3,4-diols and then shaded anthocyanidins.[36] By oxidation, flavan-3-ols and proanthocyanidins provide ascension to brown subordinates. Phlobaphenes (deoxy flavonoids) with a reddish brown hue framed by oxidation and polymerization of flavan-4-ols in red grained wheat coats and coleoptiles.[2]

8.2.2.5 *Flavones, Flavanols and Flavanones*

Wheat grain contains flavonoids that appear colourless, white, or yellowish, but turn yellow when concentrated in alkaline solutions. Wheat flavonoid content was reported to range between 14.7 and 39.7 mg per 100 g, with a mean of 25.2 mg per 100 g, with the mean flavonoid content of white, red, and black wheat being 24.1, 29.0, and 36.1 mg per 100 g, respectively.[37] Rutin is not detected following alkaline and acid hydrolysis, most likely due to the hydrolysis of the *O*-glycosidic bond. Due to their anti-inflammatory, antimicrobial, and anticancer properties, flavones have gained increased attention.[2]

8.2.2.6 Coumarins, Stilbenes and Lignans

It has been reported that grains of certain wheat varieties contain a variety of coumarins, including the compound coumarin and its hydroxylated derivatives.[38] Coumarins, as benzopyrones mixtures, may exhibit a variety of bioactivities, including anticoagulant, estrogenic, dermal photosensitizing, antimicrobial, vasodilatory, molluscacidal, antihelmintic, calming and attractive, pain-relieving, and physiological state movement. Coumarin was clearly discernible within the free phenoplast division of an ancient durum wheat genotype.[32]

There are reports on stilbenes in cereals. In some investigated durum wheat genotypes, researchers discovered 2-stilbenes, two-fold glycosylated pinosylvin, and glycosylated pinosylvin, as well as 2-isomers of glycosylated pinosylvin that occurred within the bond phenolics of some ancient wheat genotypes.[32] Lignans in wheat grain are gathering of di-phenolic exacerbates. The concentration of a significant polymer in wheat bran, secoisolariciresinol di-glycoside, increased from 42.7 to 82.9 $\mu g\ g^{-1}$.[39]

8.2.2.7 Carotenoids

Carotenoids are also known as tetraterpenoids and are another collection of polyphenolic compounds found in pigmented wheat and commonly found in fruits and vegetables.[40] Wheat contains a-carotene, b-carotene, b-cryptoxanthin, lutein, lutein dilinooleate, and zeaxanthin. The carotenoids antheraxanthin, taraxanthin (lutein-5, 6-epoxide), triticoxanthin, and flavoxanthin are present in trace amounts in various pigmented wheat grain carotenoids. Carotenoids are also present in blue and purple grain wheat, where their content is comparable to that of carotenoids in conventional wheat. Purple varieties, on the whole, exhibit more significant levels than do blue. The processing of purple Ethiopian wheat (*T. aethiopicum*) and blue aleurone wheat (*T. aestivum*) revealed that only 42% of the total grain carotenoids were recovered in the flour portion of purple wheat, whereas the flour portion of blue wheat contained more than twice the amount in the bran, indicating a higher concentration in the wheat endosperm.[41] As for pigmented wheat, lutein is the most plentiful among any remaining carotenoids,[4] followed by zeaxanthin, antheraxanthin, α-carotene, and β-carotene, while β-cryptoxanthin is a minor part or not recognized. A carotenoid is present in einkorn wheat, at a concentration of 1.63–4.90 mg kg^{-1}, in emmer wheat at a concentration of 4.73–13.64 mg kg^{-1}, in durum wheat at a concentration of 2.69–8.38 mg kg^{-1}, in spelt wheat at a concentration of 1.62–2.98 mg kg^{-1}, and in bread wheat at a concentration of 1.40–4.90 mg kg^{-1}.[42] Correlation of carotenoids scattering within cultivars' grains revealed that the starchy endosperm, aleurone layer, and germ contained 0.557, 0.425, and 2.157 mg kg^{-1} lutein, respectively. While the aleurone and germ contained 0.776 and 3.094 mg kg^{-1} zeaxanthin, respectively, the starchy endosperm lacked any. In purple wheat, higher lutein levels were detected in the aleurone (0.534 mg kg^{-1}) and lower levels in the germ (1.714 mg kg^{-1}), while higher zeaxanthin levels (1.217 mg kg^{-1} in aleurone and 3.233 mg kg^{-1} in a germ) appeared differently when compared with mean assessments of non-pigmented wheat grains.[43]

8.2.2.8 Tocols

Tocopherols (vitamin E) are composed of a chromanol ring and a C16 phytol side chain. They are classified into two types according to whether the side chain is saturated (tocopherols) or contains three double bonds at carbons 3, 7, and 11 (tocotrienols). These exist in eight structures (α, β, γ, and δ in both tocopherols and tocotrienols), which shift in spots of methyl groups on the chromanol ring.[44] Tocols (especially unsaturated tocotrienols) have malignant growth avoidance specialist properties which may be liable for their clinical preferences. Tocol content was reported to be 32.68 mg kg^{-1} in blue-grained spring wheat and 30.33 mg kg^{-1} in purple-grained wheat, which was equivalent to white-grained standard spring wheat (32.00 mg kg^{-1}) and red-grained standard winter wheat (31.01 mg kg^{-1} dry weight), respectively.[4] Purple-grained winter wheat varieties exhibited significantly lower tocol levels (26.56 mg kg^{-1}) than blue-grained winter wheat varieties (24.64 mg kg^{-1}). Additionally, yellow-grained spring wheat TA 4024 (21.89 mg kg^{-1}) and yellow-grained winter wheat varieties (18.51 mg kg^{-1}) had low tocol levels.[45]

8.3 Rice

Rice is a staple food for over half of the world's population, as indicated by the FAO World Rice Production Report. Rice grains are consumed in their polished form or as white rice in several countries as part of the meal; however, some Asian countries also consume pigmented rice cultivars.[46] Pigmented rice or coloured rice, distinguished by its red, black, or purple aleurone layer, contains rich phytochemicals of high nutritional quality.[46–48] Rice grains differ in size and shape of kernel, varying from tiny to big in size, short to long in length, and round to bold in shape, as influenced by cultivars and genetic backgrounds. Rice growing conditions and environments could also play an important role in controlling the rice properties.

Due to the diverse set of phenolic compounds found in pigmented rice, such as phenolic acids, flavonoids, and proanthocyanidins, pigmented rice has garnered considerable attention in recent years.[49] The distribution of phenolic compounds in various rice fractions increases in the order of endosperm < whole-grain < embryo < bran.[50] The majority of nutrients in rice grain are concentrated in the outer aleurone layer and embryo, while the endosperm is primarily composed of starch. Dehulling and milling remove the majority of micronutrients, fatty acids, antioxidants, and fibre. As a result, diets that are excessively reliant on white rice risk nutritional deficiencies.[51,52] Rice breeding has long been focused on increasing crop productivity, although some attention has been paid to grain size, shape, and amylose content.[53,54] The nutritional quality of grain produced by certain pigmented rice varieties has been demonstrated to be superior to grain produced by conventional, modern non-pigmented rice varieties, owing largely to their more efficient accumulation of bioactive compounds.[46,55,56] Consumer interest in health-promoting food products is creating a sizable market for more nutritionally valuable rice, benefiting many people who rely on rice as a staple food, while also benefiting producers.[57] As a result, a number of major rice research programs are focusing on nutritional quality, which include increased micronutrient and antioxidant content, as well as a decrease in the grains' glycaemic index.[48] The properties of pigmented rice, with a special focus on nutritional composition as well as their bioactive compounds, in comparison with non-pigmented rice are discussed here.

8.3.1 Nutritional Composition

As mentioned earlier, most of the nutrients found in rice grain accumulate in the outer aleurone layer and embryo. However, rice is commonly consumed after polishing or whitening and the polished grain is called white rice, which comprises mostly starch. The removal of the outer bran layer during rice milling results in a loss of most micronutrients, dietary fibre, and bioactive components.[48] In brown rice the process removes only the outermost layer of the rice kernel, the hull, which has the least detrimental effect on its nutritional value. When brown rice is completely milled and polished, it loses 67% of its vitamin B3, 80% of its vitamin B1, 90% of its vitamin B6, half of its manganese, half of its phosphorus, 60% of its iron, and all of its dietary fibre and essential fatty acids. White rice that has been fully milled and polished is required to be enriched with vitamins B1, B3, and iron.[58] Nutritionally, brown rice is superior to white rice and an improved source of vitamins, minerals, fibre, γ-oryzanol, and phytochemical components, which can support human health.[59] Consequently, regardless of colour, whole brown rice obtained from pigmented and non-pigmented rice varieties are rich in nutrients and phytochemicals. It is believed that the synergy between the three anatomical fractions of rice grains (bran layer, germ, and endosperm) results in greater health benefits than any single component.[60] Nonetheless, the presence of bran and germ in brown rice is inextricably linked to the following intrinsic and extrinsic factors that contribute to whole brown rice's low consumer acceptance. These include the perceived hard texture and chewing palatability of brown rice due to the high-fibre content, the low water absorption characteristics and prolonged cooking time, the increased susceptibility to oxidation and off-flavor formation when compared to white rice, the presence of anti-nutritional factors (*e.g.* phytic acid and tannin), the habitual preference for refined products, and the unfamiliarity of average consumers with the health benefits of brown rice.[61] Industrial milling usually removes about 7–12% of the bran fraction, improving the sensory properties and storage stability. Pigmented rice varieties, after being completely milled (more than 12% degree of milling) could visually look white, similar to polished non-pigmented rice. However, the endosperm colour may be slightly different in its translucence and opacity.[62] The properties of brown rice and its health benefits are not covered here; some excellent reviews on brown rice properties are available.[58,59,63]

Generally, as discussed earlier, pigmented varieties are healthier than non-pigmented ones because of their greater antioxidative and free radical scavenging properties that provide and promote human health by reducing the concentration of radical oxygen species and free radicals. Red and black paddy varieties contain approximately 38% more protein, 18% more crude fibre, and are higher in lysine, vitamin B1, and other minerals than conventional varieties. Apart from providing high-quality protein, fibre, and vitamins, pigmented rice varieties provide functional benefits due to their antioxidant compounds, which inhibit the formation of reactive cell-damaging free radicals.[55]

Well-milled rice obtained from pigmented and non-pigmented rice varieties should contain similar macro- and micronutrients as they only comprise starchy endosperm. The outer aleurone layer and embryo with denser phytonutrients are removed during the milling process. Starches from white, red and black pigmented rice were extracted and analysed for their composition. It was found that starches presented statistical differences for some components according to the rice variety; however, the major components were the carbohydrates, which are represented mainly by starch. Even with the differences between rice varieties, the starches present the same granules of polygonal and angular formats.[64]

Waxy pigmented and non-pigmented rice varieties in Cambodia were analysed for their compositions. They found that the mean values of most physical and chemical properties of waxy pigmented rice were slightly higher than those of non-pigmented rice, except for starch.[65] In addition, nutrient composition and glycaemic potential of pigmented and non-pigmented Indian rice cultivars, namely Kattuyanam (brown), Kavuni (red), Kavuni (black), and Karudan samba (white) were studied and found to have immense nutrient potentials. Kattuyanam had a comparatively higher content of protein and amylose, followed by red and black Kavuni. Amylose content (27.28–30.03%) correlated inversely with the glycaemic index and glycaemic load. Kattuyanam, an indigenous pigmented rice variety, with 30.03% amylose and 9.33% total dietary fibre had a low glycaemic index of 47.19 and glycaemic load of 8.80, whereas Karudan samba (white rice) recorded a high glycaemic index of 69.74 and glycaemic load of 13.84.[66] Two pigmented and two non-pigmented aromatic rice varieties were compared and it was found that pigmented rice had higher protein and fat content. Non-pigmented rice varieties had a high amylose content and a high pasting profile, whereas pigmented rice varieties had the opposite characteristics.[67]

8.3.2 Bioactive Compounds

Rice is made up of phytochemicals and nutrients that contain bioactive compounds, including flavonoid (especially anthocyanin and proanthocyanidin), carotenoids (such as α-, β-carotene, lutein, and lycopene), phenolic compounds (such as caffeic acid, ferulic acid *etc.*), phytosterols (such as β-sitosterol, stigmasterol, and capesterol), the vitamin E isoforms (α-, γ-, δ-tocotrienols and tocopherols), γ-oryzanol, coumaric acid, phytic acid, tricin *etc.*[46,68] Higher levels of the majority of the compounds are present in the outer layer of rice grain. However, modern technology such as the high-efficiency vector system for transgene stacking could be used to engineer anthocyanin biosynthesis in rice endosperm, providing a new rice variety with purple endosperm.[69]

8.3.2.1 Phenolic Acids

Pigmented rice contains phenolics or polyphenols, which include the hydroxybenzoic acid skeleton C_6–C_1 derivatives (gallic, *p*-hydroxybenzoic, protocatechuic, syringic acids, and vanillic) and the hydroxycinnamic acid skeleton C_6–C_3 derivatives (caffeic, chlorogenic, cinnamic, ferulic, sinapic acids, and *p*-coumaric).

Ferulic acid, protocatechuic acid, *p*-coumaric acid, synaptic acid, and vanillic acid were identified as the major phenolic acids in both pigmented and non-pigmented rice.[70] A total of 29 phenolic acids in rice were identified, among them ferulic, sinapic, and *p*-coumaric acids were predominant phenolic compounds.[68] The soluble form of polyphenols comprised approximately 38–60% of the total polyphenol content in light brown rice grains and approximately 81% in pigmented grains, particularly the black and red pericarp.[71,72] As mentioned earlier, because phenolics are concentrated on the grain's outer surface, polishing significantly reduces their concentration. Around 70–90% of the phenolic acids in rice bran are found in the pericarp of light brown rice grains.[73]

The majority of phenolic acids in rice are also insoluble phenolic acids. Insoluble phenolics account for the majority of polyphenols and play a significant role in the structural formation of the rice grain cell wall, as evidenced by whole grains of Thai rice.[67,74]

8.3.2.2 Flavonoids

Numerous flavonoids have been identified in diverse types of rice. These flavonoid compounds have been classified into several classes, including flavanols (flavan3-ols), flavanones, flavanonols, flavones, flavonols, and isoflavones, which are typically found as *O*- or *C*-glycosides. Tricin is one of the seven flavonoids commonly found in rice. It is primarily found in the bran, which contains 77% of all seven flavonoids (131.5 mg per 100 g). Other flavonoids include luteolin (14%), apigenin (6%), quercetin (3%), isorhamnetin

(1%), myricetin (<1%), and kaempferol (<1%).[68] Similar to other cereals, seeds of black rice had higher flavonoid content (500 mg QE per 100 g) than red rice (400 mg QE per 100 g) and total flavonoid content values were lowest in the seeds of white rice (260–270 mg QE per 100 g).[75] Pigmented and brown rice had higher phenolics and flavonoids than non-pigmented rice.[76]

8.3.2.3 Anthocyanin and Proanthocyanidin

Anthocyanin and proanthocyanidin, as well as apigenin, catechin, epicatechin, and tricin, are bioactive pigments found in a wide range of rice varieties. Rice pigments containing bioactive compounds are found in the aleurone layer. They come in a variety of colours, including black, brown, dark brown, red, and dark purple.[77]

Anthocyanins are the primary functional components in most pigmented types.[68,78] There are about 18 anthocyanins that have been recognized in rice, but only four among them have been quantified: (1) cyanidin-3-*O*-glucoside; (2) peonidin-3-*O*-glucoside; (3) cyanidin-3-*O*-rutinoside; and (4) cyanidin-3-*O*-galactoside. The following is the order of the most abundant quantified anthocyanins in rice: cyanidin-3-*O*-glucoside (51–84%), peonidin-3-*O*-glucoside (6–16%), cyanidin-3-*O*-rutinoside (3–5%) and cyanidin-3-*O*-galactoside (1–2%) of the total anthocyanin content.[68] The mean values of all these four quantified anthocyanins found in the rice bran of pigmented rice are about four times higher than in whole rice grain, depending on the colour of rice bran and a fraction of the rice.[79,80] Cyanidin-3-*O*-glucoside is the primary component of the pigmented rice bran, with concentrations ranging from 130–2640 mg per 100 g. Peonidin-3-*O*-glucoside (11.4–534.1 mg per 100 g) is also prevalent, while cyanidin-3-*O*-galactoside and cyanidin-3-*O*-rutinoside are present in trace amounts.[68] However, malvidin and petunidin-3-*O*-glucoside have seldom been identified as important rice anthocyanins. In a study of different Japanese rice cultivars, it was discovered that petunidin-3-*O*-glucoside was the predominant anthocyanin in the Chinakuromai rice cultivar, accounting for about half of the total anthocyanin content. Moreover, malvidin was found at 17–28% of the total anthocyanin content in the Asamurasaki cultivar.[80]

Proanthocyanidins are synthesized in a wide range of rice varieties, with anthocyanidins serving as critical intermediates. These bioactive pigmented compounds are responsible for rice's red and purple hues. Proanthocyanidins are primarily composed of flavon 3-ol units (such as 3-*O*-gallates, catechin, epicatechin, and epigallates).[30,81] The husks of pigmented rice varieties were found to be abundant in catechin and epicatechin, containing 1.57–3.16 mg per 100 g and 0.57–6.14 mg per 100 g, respectively.[68] The amount of oligomers and polymers in the extractable fraction samples of rice varied by 4.3 times the amount of total proanthocyanidins. However, the fraction concentration was highly correlated with bioactive compounds (such as flavonoids, total phenolics, and total proanthocyanidins) and with antiradical capacity.[82]

8.3.2.4 Phytic Acid

Phytic acid is a naturally occurring bioactive compound that is also known as phytate phosphorus, inositol hexaphosphate (IP6), and myoinositol-1.2.3.4.5.6-hexakisphosphate. Phytic acid is a nutrient antagonist found in cereals and pulses. It is well established that it impairs mineral bioavailability and inhibits the starch-digesting enzyme α-amylase (which requires calcium to function) in the human gut.[83] They are found in whole rice grain and bran as the most abundant form of phosphorus, which is 65–73% of the total content of phosphorus.[84,85] In rice bran, phytic acid is a negative molecule capable of binding proteins and starch. As a result, it aids in reducing absorption and increases faecal bulk.[86]

8.3.2.5 Tocols

As mentioned, in wheat, tocols are a group of fat-soluble vitamins generally termed vitamin E. They are the major group of antioxidants. Tocols are classified into eight types. Two groups *viz.* tocotrienols and tocopherols; with four in the tocotrienols group and four in the tocopherols group, are naturally occurring in cereals.[87–89] There are some important tocols that have been reported in rice and rice bran.

Tocopherols (α-tocopherol and γ-tocopherol) are regarded as the most important natural antioxidants.[90] There are many reports that show the presence of α-tocopherol and γ-tocopherol in rice and rice bran.[68,91] Rice whole grain contains about 2.40–43.81 mg kg^{-1} of α-tocopherol.[68] In addition, α-tocopherol was the main tocol found in two Taiwanese rice varieties.[92] A report on Southern US rice varieties found that α-tocotrienol content was the highest, followed by γ-tocotrienol, α-tocopherol, and γ-tocopherol.[93] Rice bran oil comprises up to 70% of tocotrienols.[94] Two novel tocotrienols have been isolated in heated rice bran and characterized as didesmethyl and desmethyl tocotrienols.[95] The vitamin E distribution in rice is ranked in the following increasing order: husk < endosperm < whole grain < bran, depending on the rice pigment.[81,96]

Tocotrienols are vitamin E isoforms traditionally regarded as the most powerful natural antioxidants.[90] In rice whole grain, the most abundant tocotrienol is γ-tocotrienol, which is reported to be 8.60 mg kg^{-1} to 46.95 mg kg^{-1}.[68] The study recorded an average of 830 μg g^{-1} of total tocotrienol content in 109 samples of Japanese rice bran. Toctrienol-rich rice bran (1350–1430 μg T3 per g) was found in the cultivars, Joushuu, Kouchi-Akamai, and Wataribune.[97] In addition to known α-, β-, γ- and δ-tocotrienols, recent research successfully isolated two new tocotrienols from the stabilized and heated rice bran: (1) desmethyl tocotrienol [3, 4-dihydro-2-methyl-2-(4,8,12-trimethyltrideca-3′(*E*),7′(*E*),11′-trienyl)-2*H*-1-benzopyran-6-ol]; and (2) didesmethy tocotrienol [3, 4-dihydro-2-(4,8,12-trimethyltrideca-3′(*E*),7′(*E*),11′-trienyl)-2*H*-1-benzopyran-6-ol].[95]

8.3.2.6 Gamma(γ)-oryzanol (Steryl Ferulate)

The phytochemical component γ-oryzanol in rice is a mixture of different steryl ferulates. They result from esterification of the hydroxyl group of sterols and/or triterpene alcohols with the carboxylic acid group of ferulic acid.[88,98,99]

The distribution of γ-oryzanol in various parts of rice is as follow: 3174.2–3176.4 mg kg^{-1} (bran) > 413.3–473.3 mg kg^{-1} (whole grain) > 102.4–323.2 mg kg^{-1} (husk) > 49.1–231.8 mg kg^{-1} (endosperm).[96,99,100] The bran of rice was clearly stated as predominantly sources of γ-oryzanol and presumed as the first single compound of rice-bran oil.[100] Later, γ-oryzanol was described as a mixture of different substances. Those substances were ferulic acid, phytosterol ferulates, and sterols. The phytosterol ferulates compound results from the esterification of methyl sterols to ferulic acid. Different components from the mixture of steryl ferulates that compose γ-oryzanol were reported in different parts of non-pigmented rice varieties. The mean values of those components were 58.9 mg kg^{-1} (endosperm), 3067.1 mg kg^{-1} (bran), and 288.6 mg kg^{-1} (whole grain).[96,99,100] Whole grain rice contained 18–63 mg per 100 g γ-oryzanol on a dry weight basis.[101] While rice bran has been reported to contain 185–421 mg per 100 g γ-oryzanol.[102] These concentrations depend upon the various agricultural and processing factors. During the processing steps, a large amount of oryzanol is lost. Bran collected immediately following the milling break contained the highest concentration of oryzanol.[103] In rice bran oil processing, degumming and dewaxing of crude RBO removed only 1.1–5.9% of oryzanol, while the alkali treatment removed 93.0–94.6% of oryzanol from the original crude oil.[104]

8.4 Maize

Maize, also known as corn, is one of the major cereal crops in the world. It is considered a staple food in some areas of Asia, Africa, and Latin America, where it is usually consumed as corn kernels or corn on the cob.[105] The colours of corn kernels are white, yellow, red, and purple. White corn varieties are commonly used as a substitute staple for rice, while the yellow corn varieties are utilized in animal feed manufacturing and other industrial uses.[106]

Purple maize, which is typical of the Andean Region of South America, where it has been used for centuries to prepare traditional food and beverages, is considered a good source of natural colourants, being rich in anthocyanins.[107] It is well known that the maize plant genotype is not the only factor determining corn pigmentation, as different environmental factors, altitude, temperature, and abiotic stress, in addition to the maturation stage, could highly affect pigment accumulation in maize.[108] Cultivation practices (*e.g.* potassium fertilization and seeding density) were reported to have less impact.[109] Post-harvest and some processes such as germination can also affect polyphenol concentration, especially anthocyanins, as demonstrated by the decrease of their amount in purple corn seed, except for peonidin malonylated forms.[110]

Pigment production is possible in almost every plant tissue in maize. One ear of maize can illustrate diversity in pigment production. The corn colour is mainly due to the presence of a huge number of secondary metabolites, such as phenolic acids, carotenoids, and flavonoids. The different expression of these pigments imparts to corn tissues different colours, from yellow-orange to dark purple-blue, as well as ivory, and cream colours. Pigments are mostly concentrated in the thick pericarp or in the aleurone layers of kernels, and also in corn cobs.[111–113]

Rather than referring to maize varieties according to their pigment location (*e.g.* 'aleurone-pigmented' or 'pericarp-pigmented' corn), it has been suggested that coloured maize varieties are typically classified according to their colour. For example, blue corn refers to corn that contains anthocyanin in the aleurone tissue, whereas purple corn refers to corn that contains anthocyanin in the pericarp. These names can be confusing; while aleurone-pigmented varieties are frequently blue, they can also be pink, purple, black, or even green in the presence of carotenoids and anthocyanins. Corn with anthocyanin pigmented in the pericarp is more consistently purple in colour, but it can range from dark red to purple to black. Red corn may refer to lines with brick-red to orange phlobaphene pigments in the pericarp or to aleurone lines with a high pelargonidin content. Commercial varieties of glass gem maize contain mixtures of carotenoids and anthocyanin in aleurone, which produce the majority of the visible spectrum's colours. Purple, pink, red, and blue kernels almost certainly contain anthocyanins; orange and yellow kernels almost certainly contain carotenoids; and red–orange and green kernels almost certainly contain a combination of both.[114]

8.4.1 Nutritional Composition

Chemical composition, thermal and pasting properties of whole grain flour from purple maize varieties in Argentina were determined, compared to popcorn and white corn, which were used as controls. The composition of purple maize did not show significant differences with controls, but amylose content was significantly lower. High variability in pasting and thermal properties of flour was observed between genotypes.[115] Similar results were reported; in brief, the chemical compositions and some physicochemical properties of pigmented and non-pigmented maize varieties are similar.[116]

Similar to other pigmented cereals, pigmented corn varieties are said to have a high nutritional value due to their high content of essential nutrients and antioxidants such as anthocyanins, carotenoids, phenolic acids, and flavonoids.[117] Anthocyanins in pigmented maize were reported to modify nutritional *in vitro* starch fraction, contributed to resistant starch formation and, consequently, low starch digestibility.[118]

Flours from three different pigmented maize varieties were evaluated under cooking conditions and compared to common yellow maize. In general, raw pigmented maize flours contained significantly more protein and less starch than commercial common yellow maize flour (11.4 *vs.* 8.3 g per 100 g and 68.8 *vs.* 71.8 g per 100 g, respectively). Additionally, a numerically greater total dietary fibre content was discovered when pigmented maize flours were compared with normal yellow maize flour. On the other hand, similar crude lipid and ash concentrations were determined, averaging 4.4 g per 100 g and 1.3 g per 100 g, respectively.[119] Maize flours are rich in protein, starch and dietary fibre, as a function of the individual maize varieties. Greater values of proximate compositions were reported for blue maize when compared with commercial yellow maize flours.[120]

Pigments in coloured maize have the potential to promote different physiological effects, especially the formation of resistant starch. When compared with cooked pigmented maize flours, it has been reported that flours from common yellow maize contained the most rapidly digestible starch and the least resistant starch. The higher resistant starch and slowly digestible starch contents of cooked pigmented maize flours and the lower rapidly digestible starch content of cooked common yellow maize flour may be attributed to the molecular structure of selected pigmented maize starches and possible anthocyanins-starch interactions that may inhibit/retard enzyme activity during *in vitro* digestion.[121] It has been reported that pigmented maize starch may have different molecular properties (crystallinity) than commercial maize starch, resulting in increased resistance to enzyme hydrolysis.[122] Additionally, phenolic compounds have been associated with modified starch digestion in the past, with positive correlations between slowly digestible and resistant starch fractions.[118] This effect could be explained by the inhibition of digestive enzymes *via* binding to their active sites[123] and the formation of strong non-covalent phenolic/starch interactions during cooking, resulting in insoluble networks with limited availability for amylolytic attack.[124]

8.4.2 Bioactive Compounds

Pigmented grain maize contains phenolic compounds, carotenoid, and anthocyanin, which are compounds associated with nutraceutical properties and health promotion. Therefore, such grain is considered a functional food with high antioxidant properties and preventive functions against cancer, diabetes, obesity, and neurodegenerative disorders.[125]

8.4.2.1 Phenolic Acids

The main phenolic acids presented in the 10 selected Philippine pigmented maize samples were *p*-coumaric, ferulic, and gallic acids, which showed accession variations. The contents of *p*-coumaric, ferulic, and gallic acids ranged from 44.24 to 96.32 μg g^{-1}, 1.79 to 4.93 μg g^{-1} and 9.84 to 25.74 μg g^{-1}, respectively.[126] Several studies confirmed the presence of *p*-coumaric acid in light (2.76 μg g^{-1}) and dark blue (5.12 μg g^{-1}) maize,[127] as well as in yellow corn (230.1 μg g^{-1}) and red corn (319.90 μg g^{-1}).[128] In addition, ferulic acid in different coloured maize such as in light (2.8 μg g^{-1}), dark blue (5.1 μg g^{-1}), purple (55.2 μg g^{-1}), yellow 4543.2 μg g^{-1}.[13,128,129] The majority of these phenolic acids can be found in the aleurone layer, germ, and pericarp while a trace amount is present in the endosperm of the corn.[126]

8.4.2.2 Anthocyanins

Anthocyanins are a class of water-soluble flavonoids found in the pericarp, the kernel's outermost layer, the aleurone, the endosperm's peripheral layer, or both structures.[125] The difference between blue and purple maize varieties is that purple maize produces the most anthocyanins in the kernel's pericarp. Typically, blue maize varieties produce anthocyanins in the aleurone.[130] Anthocyanins were determined in whole kernels of four different pigmented maize genotypes (yellow, red, blue, and purple). They were not detected in yellow maize kernels, but their contents ranged from 38.9 g g^{-1} (red maize) to 780 g g^{-1} (purple maize); blue maize contained an intermediate amount (215.6 g g^{-1}); the value was significantly different for the three genotypes.[125] Similarly, three types of maize with different colours were reported to contain anthocyanins ranging from 0.09 to 276.11 mg CGE per 100 g at maturity[131] and 2.50 to 696.07 mg CGE per kg.[127] In brief, the dark red maize had the highest concentration of total anthocyanins.[125]

8.4.2.3 Carotenoids

Carotenoids are naturally occurring pigments ranging from yellow to orange to red; over 600 carotenoids have been identified in nature. Carotenoids, which give maize grain its colour, are also the pigments that give egg yolks their colour. Lutein, zeaxanthin, β-cryptoxanthin, α and β-carotene are all found in yellow maize grains.[132,133] Maize hybrids play a major role in controlling the carotenoid content, but the most abundant substances are lutein and zeaxanthin.[134] Total carotenoid contents in white, purple, orange and maroon maize were reported to be 52.23, 172.45, 1610.93 and 930.36 μg per 100 g, respectively.[135]

8.4.2.4 Tocols

Tocols are present in pigmented maize varieties. While purple kernels contained more total phenolics, flavonoids, and proanthocyanidin, yellow maize contained more carotenoids, and red maize contained the most total tocols.[125]

Tocopherols and tocotrienols were measured in white, purple, orange, and maroon maize. Total tocopherols ranged from not detected (purple maize) to 192.72 μg per 100 g (maroon maize). While tocotrienols ranged from 40.33 μg per 100 g (purple maize) to 231.39 μg per 100 g (maroon maize).[135]

Vitamin E in maize is concentrated in the germ, which contains about 45–50% of oil that is used for cooking purposes and is obtained from wet milling process.[136] The oil contains approximately 14% saturated fatty acids, 30% monounsaturated fatty acids, and 56% polyunsaturated fatty acids. The refined maize oil contains linoleic acid 54–60%, oleic acid 25–31%, palmitic acid 11–13%, stearic acid 2–3%, and linolenic acid 1%.[137] The two main forms of vitamin E present in our diet are alpha (α) and gamma (γ) tocopherols. Maize oil is among the rich sources of these tocopherols, especially γ-tocopherol and their reported concentration was 21.3 and 94.1 mg per 100 g, respectively.[138,139]

8.5 Other Cereals

Apart from wheat, rice, and maize, other pigmented cereals also exhibited high levels of phytochemicals including anthocyanin, polyphenols, and flavonoids. This may contribute to health-promoting activities and also add to dietary diversity.

8.5.1 Barley

Barley is one of these grains, because it matures faster than wheat, adapts well to harsh environments of salinity stress and water scarcity, and is suitable for growing in drought-prone areas, making it an important part of sustainable agriculture. While hulled barley is the most widely grown type and is used in malting and brewing, some hulless barley is also grown in smaller quantities. Barley is structurally and visually similar to wheat, but contains more bioactive components, gaining widespread acceptance due to the reported health benefits associated with dietary fibre, beta-glucan, antioxidants, and phenolic components.[140,141]

There has been documented genetic variation in the colour (*e.g.* black, blue, and purple) and anthocyanin content of barley kernels.[142] The average anthocyanin content of kernels from seven barley groups (127 genotypes) with varying colour characteristics ranged between 60 and 350 g g^{-1}.[143] Like the other cereals, most of the anthocyanins are concentrated in the bran of the kernels.[144] For example, the anthocyanin content in the bran of one barley genotype amounted to 1.66 g kg^{-1}.[145] A range of anthocyanin compounds has been identified in barley kernels from various studies.[143–147] They included delphinidin 3-glucoside, cyanidin 3-glucoside, delphinidin 3-malonylglucoside, cyanidin 3-malonylglucoside, cyanidin 3-(6″-succinyl) glucoside, and peonidin 3-(6″-succinyl) glucoside.[144,145,147] The composition of these anthocyanins varies significantly between barley genotypes. For example, delphinidin 3-glucoside and cyanidin 3-glucoside contained 15–168 g g^{-1} and 9–240 g g^{-1}, respectively (average values of seven groups differing in colour features).[143]

8.5.2 Sorghum

There is a large diversity in the colour of sorghum kernels. Similar phytochemicals found in most cereals were also found in sorghum.[142] All sorghums contain phenolic acids in abundance, mostly flavonoids. The phenolic acids in sorghum are primarily present in bound form with ferulic acid being dominant (24–47%). In addition, gallic acid is found only in bound form

(12.9–46.0 μg g^{-1}), whereas cinnamic acid is found only in free form (2.0–10.7 μg g^{-1}). One variety with red pericarp and pigmented testa contained cinnamic acid in bound form only (19.7 μg g^{-1}).[148]

Numerous flavonoids from sorghum have been isolated and identified. The major class of flavonoids studied in sorghum is anthocyanins. Sorghum anthocyanins are unique in that they lack a hydroxyl group in the *C*-third ring's position, which increases their stability at high pH compared to other anthocyanins.[149] Sorghum with a pigmented testa have condensed tannins, also known as proanthocyanidins or procyanidins. Condensed tannins are composed of flavan-3-ols with variable length. These tannin sorghums are excellent antioxidants, which slow hydrolysis in foods, produce naturally dark-coloured products and increase the dietary fibre levels of food products. Sorghum has a high concentration of 3-deoxyanthocyanins (*e.g.* luteolinidin and apigenidin) that give stable pigments at high pH.[150]

Recent studies have revealed even greater variation in the anthocyanin content of sorghum kernels by examining additional genetic resources. For instance, 381 sorghum genotypes were analysed and the anthocyanin content varied significantly (0–149 absorbance per mL per g, arbitrary unit).[151] In addition, 13 sorghum genotypes were also analysed and found the contents of 3-deoxyanthocyanins to be up to 680 μg g^{-1}.[152] Apart from the kernels, the coloured leaf sheath has been extensively analysed as a potential source of anthocyanins for food and other applications.[153–155] Mutagenesis-assisted breeding resulted in sorghum genotypes with high levels of 3-deoxyanthocyanidins in the leaves. The leaves of this sorghum mutant contained 10.05 mg luteolinidin equivalents per g anthocyanin.[153]

Individual anthocyanins were identified from leaves and kernels of various sorghum genotypes, and it was discovered that 3-deoxyanthocyanidins are unique anthocyanins found only in sorghum and not in other cereals.[156] Genetic diversity in the anthocyanin composition of sorghum has been recorded. For example, the sorghum kernels (13 genotypes) contained luteolinidin (0–282 μg g^{-1}), apigeninidin (0–166 μg g^{-1}), 5-methoxyluteolinidin (0–154 μg g^{-1}), and 7-methoxyapigeninidin (0–137 μg g^{-1}). Sorghum leaf sheath (6 genotypes) contained apigeninidin (17–46 mg g^{-1}), luteolinidin (0.4–2.4 mg g^{-1}), and malvidin (0.6–1.0 mg g^{-1}).[157]

8.5.3 Millet

Millet is a term describing a number of small-seeded grains in the Paniceae and Chlorideae (tribes) of the family Poaceae. Commonly cultivated millets include pearl (*Pennisetum glaucum*), proso (*Panicum miliaceum*), foxtail (*Setaria italica*), and finger (*Eleucine coracana*) millets.[158]

All millets contain phenolic acids in which ferulic, *p*-coumaric, and cinnamic acids are the major phenolic acids in millets. Anthocyanins in millets vary from purple to dark colour in pericarp or seed coat.[159] The dark pigmented testa is often associated with increased total phenol content, providing antimicrobial and antioxidant properties in finger millet.[160]

Analyses of the anthocyanin components in an ethyl acetate extract of finger millet revealed the presence of apigeninidin- and luteolinidin-type anthocyanins.[161] Finger millet is the only millet reported to contain condensed tannins. Brown finger millets were found to contain tannins (0.12–3.47% catechin equivalents) compared with white finger millets (0.04–0.06%).[150]

Due to yellow pigments, foxtail millet is a bright yellow colour. The total carotenoid content of various millet varieties varied, averaging 1.99, 0.78, 1.73, and 3.66 mg kg^{-1} for finger, little, foxtail, and proso millets, respectively.[162] These values are comparable to wheat's (1.5–2 mg kg^{-1}) and sorghum's (1.8–2.3 mg kg^{-1}) carotenoid content, but significantly less than maize's (18–55 mg kg^{-1}) carotenoid content.[163] While lutein is the most abundant carotenoid in wheat, results for the carotenoid content of various proso millet cultivars indicated that lutein (5–15 mg kg^{-1}) and zeaxanthin (16–17 mg kg^{-1}) accounted for the majority of the yellow pigments in proso millet.[164] Compared with lutein and zeaxanthin, β-carotene, is a minor component of carotenoids in pearl, foxtail, finger millets, which averages 0.076, 0.028, 0.032 mg kg^{-1}, respectively.[165]

8.5.4 Rye

Like other cereal grains, Rye grain contributes significant quantities of energy, protein, selected micronutrients, and non-nutrients to the human diet.[166] Rye grains are oval/wedge-shaped and range in colour from light to dark brown. Genetically, rye grains are related to wheat and barley grains. However, the primary distinction is in the grain's size, nutritional composition, and organoleptic properties.[167]

Studies that focus on pigmented rye grains are limited. Total phenolic compounds, total flavonoids, tocopherols and tocotrienols, and inositol phosphates, reduced and oxidized glutathione contents, and glutathione peroxidase activity of the respective extracts originating from the whole rye grain and its morphological fractions were studied. Similar to other cereals, it was found that the content and properties of antioxidants are rich in pericarp with testa fraction, rather than in the whole grain.[168] One early study examined the anthocyanins in rye's pericarp, coleoptile, and first leaves, and discovered that cyanidin 3-glucoside, peonidin 3-glucoside, and acylated were formed in the pericarp, cyanidin 3-rutinoside was formed in the coleoptile, and cyanidin 3-glucoside was formed in the first leaves.[169] There have been very few reports on pigmented rye. Coloured rye genotypes remain to be developed through genetic means.[142] Recent bioactive profiling of two rye cultivars (black and white) revealed the presence of total phenolic compounds (0.08–2.62 mg GAE per g), total antioxidant capacity (0.9–6.8 mg AAE per g), and condensed tannin content (4.24–9.28 mg CE per 100 g). The presence of catechol (91.1–120.4 mg per 100 g), resorcinol (52–70.3 mg per 100 g), vanillin (1.3–5.5 mg per 100 g), ferulic acid (1.4–1.5 mg per 100 g), quercetin (4.6–4.67 mg per 100 g), and benzoic acid (5.3 mg per 100 g) was determined by HPLC using the best solvent (water). Black rye contained a greater concentration of phenolic compounds in the various extraction mediums than white rye.[170]

8.5.5 Oat

Oat grain is made up of two parts: the hull and the kernel (groat). Depending on environmental and genetic factors, the whole oat is typically 25–35% hull, but naked oat cultivars are also available. The hulls of different varieties vary in colour, with the most common being white, yellow, reddish-brown, and black. In contrast to black rice or wheat, the dark pigment found in the black hull of a variety of *Avena sativa* has been identified as melanin, a mixture of oligomers formed from *p*-coumaric acid.[171]

Oat contains a variety of different phenolic antioxidants, both free and bound. Avenanthramides are soluble amides of cinnamic acids or avenalumic acids (*p*-coumaric, caffeic, ferulic, and sinapic acid) and an anthranilic acid (or a hydroxylated, and/or methoxylated derivative of anthranilic acid). Over 35 avenanthramides have been identified in oats, three of which have been identified as major components.[172] On the other hand, significantly less research has been conducted on oat glyceride esters. Over 24 phenolic esters with antioxidant activity have been isolated, including caffeoyl and/or feruloyl esters of long chain wax alcohols, ω-hydroxy fatty acids, and their monoglycerides.[173] While most studies on oats focus on the soluble phenolics, only a few describe the bound phenolic composition. Not only in the groat, but also in the hull, soluble phenolic compounds have been detected. Only a few studies on the hull composition have been published, but all of them conclude that the groat has a higher antioxidant activity than the hulls.[174–176] Currently, there are no published data on the bound phenolic content of the hull.[177]

While comparative studies on normal and pigmented cereals' phenolic content and antioxidant activity have been conducted, the relationship between antioxidant activity and oat hull colour has received scant attention.[177] Black-hulled oat varieties have been found to exhibit high antioxidant activities as compared to the white varieties[175] In addition, antioxidant properties of red-hulled oat varieties were comparable to other pigmented varieties.[177]

8.6 Conclusion and Future Perspectives

Due to their appealing nutritional values, pigmented cereal grains have gained considerable attention in recent years. Flavonoids, carotenoids, and certain phenolic compounds are the primary pigments responsible for cereals' colours and health benefits. Anthocyanins, an important class of flavonoids, are accumulated in the aleurone or pericarp layer and give a blue, purple or a combination of these colours. Flavonoids are concentrated in the outer layer of grains, whereas carotenoids, which contribute to the yellow colour of the grain, are concentrated in the endosperm. A number of reports have associated anthocyanins and other phenolic compounds to antioxidant and anti-inflammatory activities.

This chapter discusses the properties of pigmented cereals such as wheat, rice, maize, barley, sorghum, millet, rye, and oat. Pigmented cereals, on average, have higher nutritional values than non-pigmented varieties. The importance of pigmented cereals is due to the high concentration of phytonutrients they contain.

Pigmented wheat contains a higher concentration of tannins, polyphenols, carotenoids, and anthocyanins. Anthocyanins are expressed in either the pericarp or the aleurone layer of wheat grains. For pigmented rice, a diverse array of phenolic compounds such as phenolic acids, flavonoids, and proanthocyanidins are present. While most nutrients in rice grain are concentrated in the outer aleurone layer and embryo, the dehulling and milling processes remove the majority of micronutrients, fatty acids, antioxidants, and fibre. Unlike rice, pigment production is possible in almost every plant tissue in maize. Similar to other cereals, pigmented maize varieties are said to have a high nutritional value due to their high content of essential nutrients and antioxidants such as anthocyanins, carotenoids, phenolic acids, and flavonoids. Although varieties of barley, sorghum, millet, rye, and oat have received less attention than other cereals, they contain significant amounts of phytochemicals such as anthocyanin, polyphenols, and flavonoids.

Numerous studies have been conducted recently on pigmented cereals. However, extensive research on the nutritional properties, functional properties, product-making characteristics, and new processing techniques of various pigmented cereals is necessary. The contents and compositions of anthocyanins, carotenoids, and flavonoids should also be further studied and evaluated. Breeding technologies and genetic engineering offer effective tools for developing new pigmented cereal cultivars with higher yields and that are more susceptible to different agro-climatic conditions. Moreover, development of processed products from pigmented cereals is limited due to the unavailability of proper processing technology.

References

1. P. Thakur, K. Kumar and H. S. Dhaliwal, *Food Biosci.*, 2021, **42**, 101170.
2. R. Gupta, M. Meghwal and P. K. Prabhakar, *Trends Food Sci. Technol.*, 2021, **110**, 240.
3. P. Saini, N. Kumar, S. Kumar, P. W. Mwaurah, A. Panghal, A. K. Attkan, V. K. Singh, M. K. Garg and V. K. Singh, *Crit. Rev. Food Sci. Nutr.*, 2021, **61**, 3197.
4. J. Lachman, P. Martinek, Z. Kotíková, M. Orsák and M. Šulc, *J. Cereal Sci.*, 2017, **74**, 145.
5. E.-S. M. Abdel-Aal, J. C. Young and I. Rabalski, *J. Agric. Food Chem.*, 2006, **54**, 4696.
6. S. Rao, L. J. Schwarz, A. B. Santhakumar, K. A. Chinkwo and C. L. Blanchard, *Cereal Chem.*, 2018, **95**, 589.

7. M. Garg, M. Chawla, V. Chunduri, R. Kumar, S. Sharma, N. K. Sharma, N. Kaur, A. Kumar, J. K. Mundey, M. K. Saini and S. P. Singh, *J. Cereal Sci.*, 2016, **71**, 138.
8. S. Böhmdorfer, J. T. Oberlerchner, C. Fuchs, T. Rosenau and H. Grausgruber, *Plant Methods*, 2018, **14**, 29.
9. E.-S. M. Abdel-Aal, P. Hucl and I. Rabalski, *Food Chem.*, 2018, **254**, 13.
10. D. C. Knievel, E.-S. Abdel-Aal, I. Rabalski, T. Nakamura and P. Hucl, *J. Cereal Sci.*, 2009, **50**, 113.
11. O. O. Onipe, A. I. O. Jideani and D. Beswa, *Int. J. Food Sci. Technol.*, 2015, **50**, 2509.
12. S. Tian, Z. Chen and Y. Wei, *J. Cereal Sci.*, 2018, **83**, 63.
13. S. Žilić, A. Serpen, G. Akıllıoğlu, M. Janković and V. Gökmen, *J. Cereal Sci.*, 2012, **56**, 652.
14. D. Giordano, M. Locatelli, F. Travaglia, M. Bordiga, A. Reyneri, J. D. Coïsson and M. Blandino, *Food Chem.*, 2017, **233**, 483.
15. H. S. Balyan, P. K. Gupta, S. Kumar, R. Dhariwal, V. Jaiswal, S. Tyagi, P. Agarwal, V. Gahlaut and S. Kumari, *Plant Breed.*, 2013, **132**, 446.
16. L. Chen, H. Zhang, D. J. McClements, Z. Zhang, R. Zhang, Z. Jin and Y. Tian, *Carbohydr. Polym.*, 2019, **215**, 47.
17. Y. Liu, J. Qiu, Y. Yue, K. Li and G. Ren, *Ther. Clin. Risk Manage.*, 2018, **14**, 247.
18. A. A. M. Andersson, L. Dimberg, P. Åman and R. Landberg, *J. Cereal Sci.*, 2014, **59**, 294.
19. P. Vitaglione, A. Napolitano and V. Fogliano, *Trends Food Sci. Technol.*, 2008, **19**, 451.
20. S.-J. Wang, J. Zeng, B.-K. Yang and Y.-M. Zhong, *Pharm. Biol.*, 2014, **52**, 1150.
21. Y. Li, D. Ma, D. Sun, C. Wang, J. Zhang, Y. Xie and T. Guo, *Crop J.*, 2015, **3**, 328.
22. J. M. Harnly, R. F. Doherty, G. R. Beecher, J. M. Holden, D. B. Haytowitz, S. Bhagwat and S. Gebhardt, *J. Agric. Food Chem.*, 2006, **54**, 9966.
23. L. Dykes and L. W. Rooney, *Cereal Foods World*, 2007, **52**, 105.
24. A. N. Panche, A. D. Diwan and S. R. Chandra, *J. Nutr. Sci.*, 2016, **5**, e47.
25. E. Leoncini, C. Prata, M. Malaguti, I. Marotti, A. Segura-Carretero, P. Catizone, G. Dinelli and S. Hrelia, *PLoS One*, 2012, **7**, e45997.
26. K. K. Adom, M. E. Sorrells and R. H. Liu, *J. Agric. Food Chem.*, 2005, **53**, 2297.
27. P. Geng, J. Sun, M. Zhang, X. Li, J. M. Harnly and P. Chen, *J. Mass Spectrom.*, 2016, **51**, 914.
28. M. Naczk and F. Shahidi, *J. Pharm. Biomed. Anal.*, 2006, **41**, 1523.
29. R. E. Asenstorfer, Y. Wang and D. J. Mares, *J. Cereal Sci.*, 2006, **43**, 108.
30. R. Hirawan, W. Diehl-Jones and T. Beta, *J. Agric. Food Chem.*, 2011, **59**, 12330.
31. B. T. T. Luyen, N. P. Thao, B. H. Tai, J. Y. Lim, H. H. Ki, D. K. Kim, Y. M. Lee and Y. H. Kim, *Arch. Pharmacal Res.*, 2015, **38**, 1011.

32. G. Dinelli, A. Segura Carretero, R. Di Silvestro, I. Marotti, S. Fu, S. Benedettelli, L. Ghiselli and A. Fernández Gutiérrez, *J. Chromatogr. A*, 2009, **1216**, 7229.
33. Q. Liu, Y. Qiu and T. Beta, *J. Agric. Food Chem.*, 2010, **58**, 9235.
34. D. B. M. Ficco, V. De Simone, S. A. Colecchia, I. Pecorella, C. Platani, F. Nigro, F. Finocchiaro, R. Papa and P. De Vita, *J. Agric. Food Chem.*, 2014, **62**, 8686.
35. P. Bartl, A. Albreht, M. Skrt, B. Tremlová, M. Ošťádalová, K. Šmejkal, I. Vovk and N. P. Ulrih, *Int. J. Food Sci. Nutr.*, 2015, **66**, 514.
36. J. A. McCallum and J. R. L. Walker, *Cereal Chem.*, 1990, **67**, 282.
37. D. Ma, D. Sun, Y. Zuo, C. Wang, Y. Zhu and T. Guo, *J. Integr. Agric.*, 2014, **13**, 1258.
38. J. Lachman, J. Dudjak, M. Orsák and V. Pivec, *Plant, Soil Environ.*, 2003, **49**, 1.
39. H. Qu, R. L. Madl, D. J. Takemoto, R. C. Baybutt and W. Wang, *J. Nutr.*, 2005, **135**, 598.
40. M. Rosa Perez-Gregorio and J. Simal-Gandara, *Curr. Pharm. Des.*, 2017, **23**, 2742.
41. S. Siebenhandl, H. Grausgruber, N. Pellegrini, D. Del Rio, V. Fogliano, R. Pernice and E. Berghofer, *J. Agric. Food Chem.*, 2007, **55**, 8541.
42. P. R. Shewry and S. Hey, *J. Cereal Sci.*, 2015, **65**, 236.
43. V. U. Ndolo and T. Beta, *Food Chem.*, 2013, **139**, 663.
44. P. R. Shewry and J. L. Ward, *Food Energy Secur.*, 2012, **1**, 47.
45. L. Granda, A. Rosero, K. Benešová, H. Pluháčková, J. Neuwirthová and R. Cerkal, *J. Food Sci.*, 2018, **83**, 2439.
46. D. Samyor, A. B. Das and S. C. Deka, *Int. J. Food Sci. Technol.*, 2017, **52**, 1073.
47. G.-F. Deng, X.-R. Xu, Y.-J. Guo, E.-Q. Xia, S. Li, S. Wu, F. Chen, W.-H. Ling and H.-B. Li, *J. Funct. Foods*, 2012, **4**, 906.
48. E. G. N. Mbanjo, T. Kretzschmar, H. Jones, N. Ereful, C. Blanchard, L. A. Boyd and N. Sreenivasulu, *Front. Genet.*, 2020, **11**, 229.
49. V. Cheynier, *Phytochem. Rev.*, 2012, **11**, 153.
50. Y. Shao, F. Xu, X. Sun, J. Bao and T. Beta, *J. Cereal Sci.*, 2014, **59**, 211.
51. R. Sarma, H. Swamy and H. E. Shashidhar, *Int. J. Curr. Microbiol. Appl. Sci.*, 2018, **7**, 1887.
52. P. Saneei, B. Larijani and A. Esmaillzadeh, *Public Health Nutr.*, 2017, **20**, 233.
53. Y. Rao, Y. Li and Q. Qian, *Plant Cell Rep.*, 2014, **33**, 551.
54. F. Breseghello and A. S. G. Coelho, *J. Agric. Food Chem.*, 2013, **61**, 8277.
55. F. M. Bhat, S. R. Sommano, C. S. Riar, P. Seesuriyachan, T. Chaiyaso and C. Prom-u-Thai, *Agronomy*, 2020, **10**, 1817.
56. G.-F. Deng, X.-R. Xu, Y. Zhang, D. Li, R.-Y. Gan and H.-B. Li, *Crit. Rev. Food Sci. Nutr.*, 2013, **53**, 296.
57. T. I. Terungwa and M. A. Yuguda, *Int. J. Sustain. Dev. World Policy*, 2014, **3**, 90.

58. P. D. Babu, R. Subhasree, R. Bhakyaraj and R. Vidhyalakshmi, *Magnesium*, 2009, **187**, 67.
59. S. A. Mir, M. A. Shah, S. J. D. Bosco, K. V. Sunooj and S. Farooq, *Cereal Chem.*, 2020, **97**, 895.
60. C. Schaffer-Lequart, U. Lehmann, A. B. B. Ross, O. Roger, A. L. L. Eldridge, E. Ananta, M.-F. Bietry, L. R. R. King, A. V. V. Moroni and S. Srichuwong, *Crit. Rev. Food Sci. Nutr.*, 2017, **57**, 1562.
61. Q. Xia, B. D. Green, Z. Zhu, Y. Li, S. M. T. Gharibzahedi, S. Roohinejad and F. J. Barba, *Crit. Rev. Food Sci. Nutr.*, 2019, **59**, 3349.
62. F. F. Paiva, N. L. Vanier, J. D. J. Berrios, J. Pan, F. de A. Villanova, G. Takeoka and M. C. Elias, *J. Food Compos. Anal.*, 2014, **35**, 10.
63. A. S. M. Saleh, P. Wang, N. Wang, L. Yang and Z. Xiao, *Compr. Rev. Food Sci. Food Saf.*, 2019, **18**, 1070.
64. L. R. da Silva, C. W. Piler de Carvalho, J. I. Velasco and F. M. Fakhouri, *Int. J. Biol. Macromol.*, 2020, **156**, 485.
65. C. Chay, W. A. Hurtada, E. I. Dizon, F. B. Elegado, C. Norng and L. C. Raymundo, *Food Res.*, 2017, **1**, 9.
66. K. Meera, M. Smita, S. Haripriya and S. Sen, *J. Cereal Sci.*, 2019, **87**, 202.
67. S. Saikia, H. Dutta, D. Saikia and C. L. Mahanta, *Food Res. Int.*, 2012, **46**, 334.
68. P. Goufo and H. Trindade, *Food Sci. Nutr.*, 2014, **2**, 75.
69. Q. Zhu, S. Yu, D. Zeng, H. Liu, H. Wang, Z. Yang, X. Xie, R. Shen, J. Tan, H. Li, X. Zhao, Q. Zhang, Y. Chen, J. Guo, L. Chen and Y.-G. Liu, *Mol. Plant*, 2017, **10**, 918.
70. M. Zaupa, L. Calani, D. Del Rio, F. Brighenti and N. Pellegrini, *Food Chem.*, 2015, **187**, 338.
71. N. V. M. de Mira, I. L. Massaretto, C. de S. C. I. Pascual and U. M. L. Marquez, *J. Food Compos. Anal.*, 2009, **22**, 405.
72. K. K. Adom and R. H. Liu, *J. Agric. Food Chem.*, 2002, **50**, 6182.
73. P. Stratil, B. Klejdus and V. Kubáň, *Talanta*, 2007, **71**, 1741.
74. P. Maisuthisakul and L. Changchub, *Int. J. Food Prop.*, 2014, **17**, 855.
75. S. Nayeem, B. Venkidasamy, S. Sundararajan, S. P. Kuppuraj and S. Ramalingam, *J. Food Sci. Technol.*, 2021, **58**, 884.
76. J. Vichapong, M. Sookserm, V. Srijesdaruk, P. Swatsitang and S. Srijaranai, *LWT--Food Sci. Technol.*, 2010, **43**, 1325.
77. D. K. Verma and P. P. Srivastav, *Trends Food Sci. Technol.*, 2020, **97**, 355.
78. S. Rattanachitthawat, P. Suwannalert, S. Riengrojpitak, C. Chaiyasut and S. Pantuwatana, *J. Med. Plants Res.*, 2010, **4**, 796.
79. B. Min, A. M. McClung and M. Chen, *J. Food Sci.*, 2011, **76**, C117.
80. X. Q. Chen, N. Nagao, T. Itani and K. Irifune, *Food Chem.*, 2012, **135**, 2783.
81. F. Finocchiaro, B. Ferrari, A. Gianinetti, C. Dall'Asta, G. Galaverna, F. Scazzina and N. Pellegrini, *Mol. Nutr. Food Res.*, 2007, **51**, 1006.
82. M.-H. Chen, A. M. McClung and C. J. Bergman, *Food Chem.*, 2016, **208**, 279.
83. A. Kumar, C. Sahu, P. A. Panda, M. Biswal, R. P. Sah, M. K. Lal, M. J. Baig, P. Swain, L. Behera, K. Chattopadhyay and S. Sharma, *J. Sci. Food Agric.*, 2020, **100**, 1598.

84. N. Depar, I. Rajpar, N. B. Sial and M. I. Keerio, *J. Basic Appl. Sci.*, 2013, **9**, 26.
85. T. Frank, B. S. Meuleye, A. Miller, Q.-Y. Shu and K.-H. Engel, *J. Agric. Food Chem.*, 2007, **55**, 11011.
86. L. U. Thompson, *Food Res. Int.*, 1993, **26**, 131.
87. G. Panfili, A. Fratianni and M. Irano, *J. Agric. Food Chem.*, 2003, **51**, 3940.
88. S.-H. Huang and L.-T. Ng, *J. Agric. Food Chem.*, 2011, **59**, 11150.
89. S. Fritsche, X. Wang and C. Jung, *Antioxidants*, 2017, **6**, 99.
90. A. A. Qureshi, W. A. Salser, R. Parmar and E. E. Emeson, *J. Nutr.*, 2001, **131**, 2606.
91. G. Perretti, E. Miniati, L. Montanari and P. Fantozzi, *J. Supercrit. Fluids*, 2003, **26**, 63.
92. P.-Y. Lin and H.-M. Lai, *Food Chem.*, 2011, **127**, 86.
93. C. J. Bergman and Z. Xu, *Cereal Chem.*, 2003, **80**, 446.
94. T. C. Raghuram and C. Rukmini, *Indian J. Med. Res.*, 1995, **102**, 241.
95. A. A. Qureshi, H. Mo, L. Packer and D. M. Peterson, *J. Agric. Food Chem.*, 2000, **48**, 3130.
96. T. L. Jeng, Y. J. Shih, P. T. Ho, C. C. Lai, Y. W. Lin, C. S. Wang and J. M. Sung, *J. Sci. Food Agric.*, 2012, **92**, 1468.
97. P. Sookwong, K. Nakagawa, K. Murata, Y. Kojima and T. Miyazawa, *J. Agric. Food Chem.*, 2007, **55**, 461.
98. P. Goufo, J. Pereira, N. Figueiredo, M. B. P. P. Oliveira, C. Carranca, E. A. S. Rosa and H. Trindade, *J. Cereal Sci.*, 2014, **59**, 15.
99. N. B. Tuncel and N. Yılmaz, *Eur. Food Res. Technol.*, 2011, **233**, 577.
100. E. Mandak and L. Nyström, *J. Agric. Food Chem.*, 2012, **60**, 6123.
101. S. J. Britz, P. V. V Prasad, R. A. Moreau, L. H. Allen, D. F. Kremer and K. J. Boote, *J. Agric. Food Chem.*, 2007, **55**, 7559.
102. S. Yu, Z. T. Nehus, T. M. Badger and N. Fang, *J. Agric. Food Chem.*, 2007, **55**, 7308.
103. B. J. Lloyd, T. J. Siebenmorgen and K. W. Beers, *Cereal Chem.*, 2000, **77**, 551.
104. A. G. G. Krishna, S. Khatoon, P. M. Shiela, C. V. Sarmandal, T. N. Indira and A. Mishra, *J. Am. Oil Chem. Soc.*, 2001, **78**, 127.
105. Y. Ai and J. Jane, *Compr. Rev. Food Sci. Food Saf.*, 2016, **15**, 581.
106. P. Ranum, J. P. Peña-Rosas and M. N. Garcia-Casal, *Ann. N. Y. Acad. Sci.*, 2014, **1312**, 105.
107. K. Petroni, R. Pilu and C. Tonelli, *Planta*, 2014, **240**, 901.
108. X. Hu, J. Liu, W. Li, T. Wen, T. Li, X.-B. Guo and R. H. Liu, *J. Cereal Sci.*, 2020, **95**, 103065.
109. P. U. Jing, V. Noriega, S. J. Schwartz and M. M. Giusti, *J. Agric. Food Chem.*, 2007, **55**, 8625.
110. L. M. Paucar-Menacho, C. Martinez-Villaluenga, M. Dueñas, J. Frias and E. Peñas, *LWT--Food Sci. Technol.*, 2017, **76**, 236.
111. F. Lao and M. M. Giusti, *Food Anal. Methods*, 2016, **9**, 1367.
112. M. Paulsmeyer, L. Chatham, T. Becker, M. West, L. West and J. Juvik, *J. Agric. Food Chem.*, 2017, **65**, 4341.

113. R. Colombo, L. Ferron and A. Papetti, *Molecules*, 2021, **26**, 199.
114. L. A. Chatham, M. Paulsmeyer and J. A. Juvik, *Theor. Appl. Genet.*, 2019, **132**, 2927.
115. P. S. Mansilla, M. C. Nazar and G. T. Pérez, *Int. J. Biol. Macromol.*, 2020, **146**, 311.
116. S. Trehan, N. Singh and A. Kaur, *J. Food Sci. Technol.*, 2018, **55**, 2334.
117. D. Del Pozo-Insfran, C. H. Brenes, S. O. S. Saldivar and S. T. Talcott, *Food Res. Int.*, 2006, **39**, 696.
118. G. A. Camelo-Méndez, E. Agama-Acevedo, M. M. Sanchez-Rivera and L. A. Bello-Pérez, *Food Chem.*, 2016, **211**, 281.
119. G. Rocchetti, G. Giuberti, A. Gallo, J. Bernardi, A. Marocco and L. Lucini, *Food Res. Int.*, 2018, **108**, 183.
120. G. A. Camelo-Méndez, E. Agama-Acevedo, J. Tovar and L. A. Bello-Pérez, *J. Cereal Sci.*, 2017, **76**, 179.
121. N. Bordenave, B. R. Hamaker and M. G. Ferruzzi, *Food Funct.*, 2014, **5**, 18.
122. K. Hanhineva, R. Törrönen, I. Bondia-Pons, J. Pekkinen, M. Kolehmainen, H. Mykkänen and K. Poutanen, *Int. J. Mol. Sci.*, 2010, **11**, 1365.
123. F. Barros, J. M. Awika and L. W. Rooney, *J. Agric. Food Chem.*, 2012, **60**, 11609.
124. F. Zhu, *Trends Food Sci. Technol.*, 2015, **43**, 129.
125. S. Suriano, C. Balconi, P. Valoti and R. Redaelli, *LWT--Food Sci. Technol.*, 2021, **144**, 111257.
126. J. A. M. A. Libron, D. E. M. Cardona, J. M. C. Mateo, A. K. M. Beltran, A. P. P. Tuaño and T. P. Laude, *J. Food Compos. Anal.*, 2021, **101**, 103954.
127. S. Žilić, A. Serpen, G. Akıllıoğlu, V. Gökmen and J. Vančetović, *J. Agric. Food Chem.*, 2012, **60**, 1224.
128. A. K. Das and V. Singh, *J. Funct. Foods*, 2015, **13**, 363.
129. F. Ramos-Escudero, A. M. Muñoz, C. Alvarado-Ortíz, A. Alvarado and J. A. Yáñez, *J. Med. Food*, 2012, **15**, 206.
130. Q. Li, P. Somavat, V. Singh, L. Chatham and E. G. de Mejia, *Food Chem.*, 2017, **231**, 332.
131. Q. Hu and J. Xu, *J. Agric. Food Chem.*, 2011, **59**, 2026.
132. M. Alfieri, A. Hidalgo, N. Berardo and R. Redaelli, *J. Cereal Sci.*, 2014, **59**, 181.
133. A. C. Kurilich and J. A. Juvik, *J. Agric. Food Chem.*, 1999, **47**, 1948.
134. F. Karadas, E. Grammenidis, P. F. Surai, T. Acamovic and N. H. C. Sparks, *Br. Poult. Sci.*, 2006, **47**, 561.
135. S. Lakshmi, G. Goudar, M. Singh, H. S. Dhaliwal, P. Sharma and T. Longvah, *J. Food Meas. Charact.*, 2021, **15**, 2774.
136. C. O. Egesel, J. C. Wong, R. J. Lambert and T. R. Rocheford, *Crop Sci.*, 2003, **43**, 818.
137. G. E. Orhun, *Int. J. Food Sci. Nutr. Eng.*, 2013, **3**, 13.
138. C. K. Sen, S. Khanna and S. Roy, *Life Sci.*, 2006, **78**, 2088.
139. T. Rouf Shah, K. Prasad and P. Kumar, *Cogent Food Agric.*, 2016, **2**, 1166995.

140. J. Moza and H. S. Gujral, *Food Chem.*, 2016, **194**, 561.
141. T. Madhujith and F. Shahidi, *J. Agric. Food Chem.*, 2007, **55**, 5018.
142. F. Zhu, *Food Res. Int.*, 2018, **109**, 232.
143. M.-Y. M.-J. Kim, J.-N. Hyun, J.-A. J.-G. Kim, J.-C. Park, M.-Y. M.-J. Kim, J.-A. J.-G. Kim, S.-J. Lee, S.-C. Chun and I.-M. Chung, *J. Agric. Food Chem.*, 2007, **55**, 4802.
144. C. Lee, D. Han, B. Kim, N. Baek and B. Baik, *Int. J. Food Sci. Technol.*, 2013, **48**, 984.
145. S. H. Lee, E. Jeong, S.-S. Paik, J. H. Jeon, S. W. Jung, H.-B. Kim, M. Kim, M.-H. Chun and I.-B. Kim, *Curr. Eye Res.*, 2014, **39**, 79.
146. Y. Yao, W. Sang, M. Zhou and G. Ren, *J. Agric. Food Chem.*, 2010, **58**, 770.
147. I. Diczházi and L. Kursinszki, *Cereal Chem.*, 2014, **91**, 195.
148. D. H. Hahn and L. W. Rooney, *Cereal Chem.*, 1986, **63**, 4.
149. J. M. Awika, L. W. Rooney and R. D. Waniska, *Food Chem.*, 2005, **90**, 293.
150. L. Dykes and L. W. Rooney, *J. Cereal Sci.*, 2006, **44**, 236.
151. D. H. Rhodes, L. Hoffmann Jr, W. L. Rooney, P. Ramu, G. P. Morris and S. Kresovich, *J. Agric. Food Chem.*, 2014, **62**, 10916.
152. L. Dykes, L. M. Seitz, W. L. Rooney and L. W. Rooney, *Food Chem.*, 2009, **116**, 313.
153. C. Petti, R. Kushwaha, M. Tateno, A. E. Harman-Ware, M. Crocker, J. Awika and S. DeBolt, *J. Agric. Food Chem.*, 2014, **62**, 1227.
154. B. Geera, L. O. Ojwang and J. M. Awika, *J. Food Sci.*, 2012, **77**, C566.
155. A. Khalil, R. Baltenweck-Guyot, R. Ocampo-Torres and P. Albrecht, *Phytochem. Lett.*, 2010, **3**, 93.
156. M. T. Escribano-Bailón, C. Santos-Buelga and J. C. Rivas-Gonzalo, *J. Chromatogr. A*, 2004, **1054**, 129.
157. A. P. P. Kayodé, M. J. R. Nout, A. R. Linnemann, J. D. Hounhouigan, E. Berghofer and S. Siebenhandl-Ehn, *J. Agric. Food Chem.*, 2011, **59**, 1178.
158. F. Zhu, *Food Res. Int.*, 2014, **64**, 200.
159. V. G. Renganathan, C. Vanniarajan, N. Senthil, A. Nirmalakumari, A. Karthikeyan, K. Veni and J. Ramalingam, *Plant Breed.*, 2021, **140**, 246.
160. D. Kumari, T. Madhujith and A. Chandrasekara, *Food Sci. Nutr.*, 2017, **5**, 474.
161. M. Siwela, J. R. N. Taylor, W. A. J. de Milliano and K. G. Duodu, *Food Chem.*, 2010, **121**, 443.
162. V. T. Asharani, A. Jayadeep and N. G. Malleshi, *Int. J. Food Prop.*, 2010, **13**, 41.
163. R. Shen, S. Yang, G. Zhao, Q. Shen and X. Diao, *J. Cereal Sci.*, 2015, **61**, 86.
164. L. Zhang, R. Liu and W. Niu, *PLoS One*, 2014, **9**, e104058.
165. B. S. Mamatha, R. K. Sangeetha and V. Baskaran, *Int. J. Food Sci. Technol.*, 2011, **46**, 315.
166. M. S. Edge, J. M. Jones and L. Marquart, *J. Am. Diet. Assoc.*, 2005, **105**, 1856.
167. W. Bushuk, *Cereal Foods World*, 2001, **46**, 70.
168. H. Zieliński, A. Ceglińska and A. Michalska, *Food Chem.*, 2007, **104**, 980.
169. W. Dedio, R. D. Hill and L. E. Evans, *Can. J. Plant Sci.*, 1972, **52**, 977.

170. P. Kaur, K. S. Sandhu, S. P. Bangar, S. S. Purewal, M. Kaur, R. A. Ilyas, M. R. M. Asyraf and M. R. Razman, *Antioxidants*, 2021, **10**, 1214.
171. M. Varga, O. Berkesi, Z. Darula, N. V. May and A. Palágyi, *Phytochemistry*, 2016, **130**, 313.
172. F. W. Collins, *J. Agric. Food Chem.*, 1989, **37**, 60.
173. D. G. H. Daniels and H. F. Martin, *J. Sci. Food Agric.*, 1967, **18**, 589.
174. C. L. Emmons and D. M. Peterson, *Cereal Chem.*, 1999, **76**, 902.
175. S. Bryngelsson, B. Mannerstedt-Fogelfors, A. Kamal-Eldin, R. Andersson and L. H. Dimberg, *J. Sci. Food Agric.*, 2002, **82**, 606.
176. Y. Xing and P. J. White, *J. Am. Oil Chem. Soc.*, 1997, **74**, 303.
177. M. Varga, R. Jójárt, P. Fónad, R. Mihály and A. Palágyi, *Food Chem.*, 2018, **268**, 153.

CHAPTER 9

Pigment in Cereals: Extraction, Quantifications, and Characterization

SIMMI RANJAN KUMAR[a], ANJELINA SUNDARSINGH[b] AND MUHAMMAD BILAL SADIQ*[c]

[a]Department of Biotechnology, Faculty of Science, Mahidol University, Phaya Thai, 10400, Thailand; [b]Department of Food Technology, Ghani Khan Choudhury Institute of Engineering and Technology, Malda, 732141, India; [c]School of Life Sciences, Forman Christian College (A Chartered University), Lahore, 54600, Pakistan
*E-mail: bilalsadiq@fccollege.edu.pk

9.1 Introduction

Cereals are a traditional food for humans, due to the health benefits associated with them.[1] Rice, maize, wheat, sorghum and barley are major cereals that are globally produced in significant quantities. Cereals are naturally enriched with various colored pigments such as red, black, purple, brown, blue and pink.[2] Color is a primary food characteristic which has played a substantial role in food for many years, to enhance food's appearance and quality. Natural pigments, for instance, anthocyanins, chlorophyll, betalains and carotenoids, provide a wide range of color shades to food.[3] Various

Food Chemistry, Function and Analysis No. 38
Pigmented Cereals and Millets: Bioactive Profile and Food Applications
Edited by Sneh Punia Bangar, Sajid Maqsood and Anil Kumar Siroha

Published by the Royal Society of Chemistry, www.rsc.org

phytochemicals are found in cereal grains, some of which significantly impact grain pigmentation.[4] The fundamental structures of anthocyanins are known as anthocyanidins.[5] Anthocyanins are phenolic compounds that give a blue, red, purple or orange color to a range of flowers, fruits, vegetables and cereals. Anthocyanins are well known for their bioactive potential.[6] Antioxidation, anticancerous, retinal protection, hypolipidemia, anti-aging, and enhanced gut health are among the numerous health benefits of anthocyanins.[7] The anthocyanins are normally extracted from their sources using acidified hydroalcoholic solutions and purified by using chromatographic methods. In order to identify the structure of anthocyanins, nuclear magnetic resonance and mass spectrometry are predominantly used.[7] Carotenoids are categorized as yellowish-orange pigments observed in plants that may be classified into hydrocarbon carotenes and their hydroxylated derivatives, known as xanthophylls.[8] Although carotenoids are a minor component of cereals, certain grains have a greater and others a lower amount of carotenoids when compared with vegetables and fruits. Lutein and zeaxanthin are the most plentiful carotenoids found in cereal grains.[9] According to several studies, carotenoid-rich diets have been associated with a low risk of various chronic and degenerative illnesses, cancer, cardiovascular disease and age-related macular degeneration.[8] Flavonoids are responsible for red pigments in cereal grains and have numerous health benefits, such as alleviating many diseases, such as coronary heart disease and cancer.[10] Reverse-phase HPLC and spectrophotometric methods are used for carotenoid and flavonoid separation and quantification.[11] Genetic variables, growth factors, and technical procedures influence the color components of grains and grain products.[4] Growing consumer awareness, along with a focus on the health benefits of specific food components, has increased the popularity of coloured grains and flours.[12] This chapter summarizes cereal pigments, their composition, extraction and quantification.

9.2 Composition of Pigments

Cereals are major sources of proteins, carbohydrates, minerals, vitamins and pigments. While pigments compose a minor portion of cereal, they are the principal coloring compound in cereals. Unlike what we see in fruits and vegetables, whose major composition are phenolic compounds, flavonoids and pigments, a minimal amount of pigments are found in cereals.[13] Cereal crops like wheat, maize, rice, barley, oats, rye and sorghum can have a different range of coloring pigments such as carotenoids, anthocyanins, and photosynthetic pigments like chlorophyll and so on.

9.2.1 Carotenoids

Carotenoids are tetraterpene pigments that are lipophilic in nature and are accountable for the orange, purple, yellow and red color of a wide range of plant and animal surfaces. These pigments are also abundantly available in nature, along with some fungi, bacteria and algae. Eight isoprene units, along with forty carbon skeletons make up the majority of carotenoids. On the other hand, a polyene chain along with nine conjugated double bonds with an end group at both ends denotes their general structures.[13,14]

The advent of carotene dates back to 1906, when Russian-Italian botanist, Zwet, managed to separate carotene, xanthophyll and chlorophyll from green leaves through column chromatography. Further to this, the structure of carotene was annotated by Karrer and Khun in 1928–1930. Over 850 naturally occurring carotene compounds have been reported up to 2018. Carotenes, based on their structure, can be further categorized as alpha and beta carotene, or carotenoid, which is one of the most common forms.[14,15] Other derivatives of carotene include xanthophylls, lutein, xeaxanthin, astaxanthin, cryptoxanthin (Figure 9.1) and several others.

Wheat is one of the major cereals among the cereal crops and is used as a staple in many countries situated in temperate zones. Starch and protein are major macronutrients in wheat, while phytochemicals, including carotenoids, are the main micronutrients. Traditionally, due to consumer preference, wheat-based products like bread, cakes and cookies are made from white flour, which has very low carotenoid composition. On the other hand, species like durum wheat are said to have highest carotenoid composition compared with regular wheat species and are preferred for pasta production. Although carotenoids are tiny components of wheat grains, the high number of wheat-based meals consumed in most societies may make wheat an essential carotenoid source, potentially contributing in the prevention of both chronic diseases and deficiencies.[16]

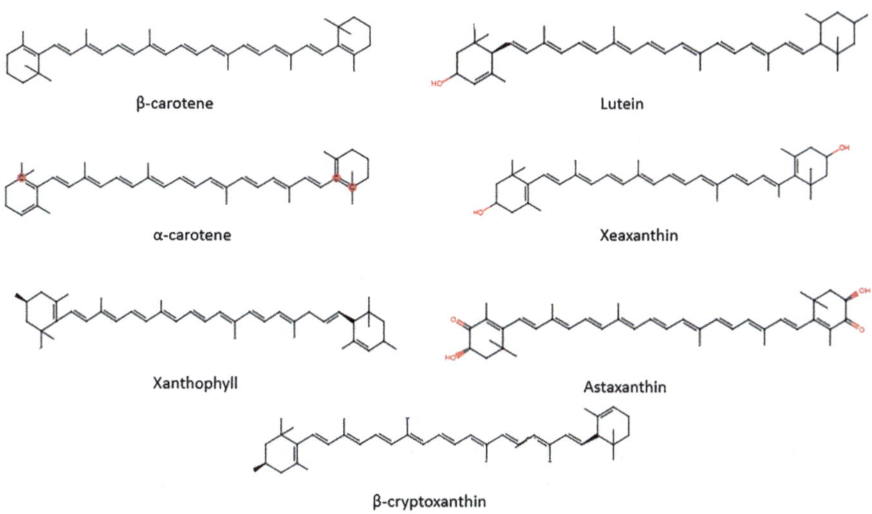

Figure 9.1 Structure of carotene and some of its derivatives. Reproduced from ref. 28, https://doi.org/10.3390/molecules25245785, under the terms of the CC BY 4.0 license https://www.mdpi.com/openaccess.

Among the different species of durum wheat, modern durum wheat varieties have significantly higher carotenoid levels than the old traditional varieties. As stated earlier, total carotenoids account for its other variables: zeaxanthin, astaxanthin, cryptoxanthin, and alpha or beta carotene. Principally, a large number of wheat species are low in carotenoid composition. However, genetically modified wheat varieties are produced to generate higher carotenoids in wheat endosperms. The mean total carotenoid congregation in durum wheat can range from 1.178–3.345 μg g^{-1} to 6.2 ± 0.13 μg g^{-1} in dry weight, giving pasta or semolina its yellow-amber color. Further, the total carotenoid composition of durum wheat can be divided into variables like 0.776 and 3.094 μg g^{-1} zeaxanthin, α- and β-carotene (7.7%) and lutein (86–94%) present in aleurone, germ and across layers of wheat, respectively.[17,18]

The total carotenoid pigment present in maize is significantly higher than in all other cereal crops. In contrast to wheat, where lutein was one of the major pigment, maize has an abundant amount of zeaxanthin and lutein, followed by provitamin A carotenoids, β-carotene, α-carotene and β-cryptoxanthin, which are present at significant levels.[19,20] On the other hand, rice, which is the most copiously found cereal, has the lowest amount of carotenoid pigment. Rice has a limited amount of carotenoids, and most of its derived pigments, such as carotene, lutein and zeaxanthin, are generally separated during white rice processing, since the bran is removed (Table 9.1).[21] This is one of the factors that gave rise to the advent of 'golden rice'. Like rice, which is a staple in many Asian countries, sorghum, a close relative of maize, is considered a staple for over 300 million people in Africa. But, unlike maize, which contains a high amount of carotenoid pigment, sorghum has a meagre concentration, although some species of sorghum like yellow sorghum has provitamin A carotenoids. β-carotene, lutein and zeaxanthin are the major pigments found in sorghum in limited amounts.[22]

9.2.2 Anthocyanins

Anthocyanins are a coloring compound found in various fruits, vegetables, plants and cereals and are accountable for orange, brown, red, blue and purple colours. They naturally occur in the form of methoxylated heterosides and polyhydroxylated from flavylium ion or 2-phenylbenzopyrilium salts. They are also known as aglycons, and occur in various forms, individually or with various other organic groups. According to the research and literature, around 30 anthocyanins have been discovered to date, of which 6 anthocyanins – malvidin, peonidin, delphinidin, cyanidin, petunidin and pelargonidin – are the most commonly found in plants and vegetables[2,7] (Figure 9.2).

Major cereals like rice, wheat, maize, barley and sorghum, and minor cereals like rye, millets and oats can be found in different colors, and this signifies the presence of any pigments such as carotenoids or anthocyanins. The purple, red, black and brown colours of cereals are due to the different

Table 9.1 Carotenoid composition of some selected cereals.[a]

Species	Total carotenoids ($\mu g\ g^{-1}$)	Lutein ($\mu g\ g^{-1}$)	Zeaxanthin ($\mu g\ g^{-1}$)	β-Carotene ($\mu g\ g^{-1}$)	α-Carotene ($\mu g\ g^{-1}$)	β-Cryptoxanthin ($\mu g\ g^{-1}$)	Reference
Durum wheat	1.178–4.416	0.721–3.078	0.088–0.184	0.007–0.033	0.003–0.029	0.002–0.014	52
Bread wheat	1.133–1.135	0.897–1.020	n.d.	n.d.	n.d.	n.d.	53
Maize	16–47	2.93–6.68	3.41–6.21	2.98–8.64	0.28–4.65	0.26–0.74	54
Rice	4.61–5.51	0.16–0.26	0.03–0.17	1.17–2.15	0.70–1.90	0.00–0.17	55
Sorghum	3.82–19.50	0.28–4.05	0.45–6.97	2.23–6.02	n.d.	0.70–3.93	56

[a]n.d.: not determined.

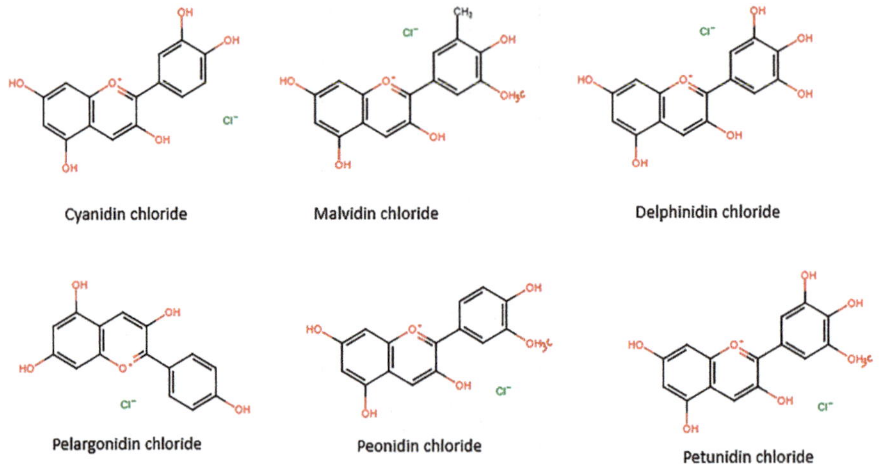

Figure 9.2 Commonly found anthocyanins. Reproduced from ref. 28, https://doi.org/10.3390/molecules25245785, under the terms of the CC BY 4.0 license https://www.mdpi.com/openaccess.

types of anthocyanins present in that particular cereal crop. The composition of anthocyanins in coloured maize, rice, wheat and sorghum, as well as their extraction and identification procedures, have been well researched and documented during the last few decades. Acidified alcohol aqueous solutions are commonly used to isolate anthocyanins, which are then separated using chromatographic methods. A variety of analytical procedures, including high-performance liquid chromatography (HPLC), mass spectrometry (MS), nuclear magnetic resonance (NMR), hydrolysis techniques, and ultraviolet visible spectroscopy (UVV), are used to determine the structure of anthocyanins.[2]

Almost all cereals are rich in anthocyanin, while some of them have major pigmentation, for instance maize, which can be found in different colors (Figure 9.3). Maize of different species can be found in red, purple, black, brown or pink color, with most having a specific part where the color is concentrated. As quoted in a review by Zhu[7] pericarps and aleurone layers are where color pigments are mainly concentrated in purple and blue maize kernels, and this is the distinction between purple and blue maize.[23] In contrast, maize cobs can sometimes be responsible for some anthocyanin pigments.[6,24]

According to a study by Li *et al.*,[23] the total monomeric anthocyanin (MMA) content in purple corn was found to be highest in pericarp, with a measure of 1477.0 ± 84.2 mg C3G equivalent per kg of whole corn, in contrast to other parts of corn like germs, fines, small and large grits that only had about half and less than half what contained in the pericarp. On the other hand, significantly lower total MMA could be extracted from blue maize, which was 76.9 ± 0.6 mg C3G equivalent per kg of whole corn and

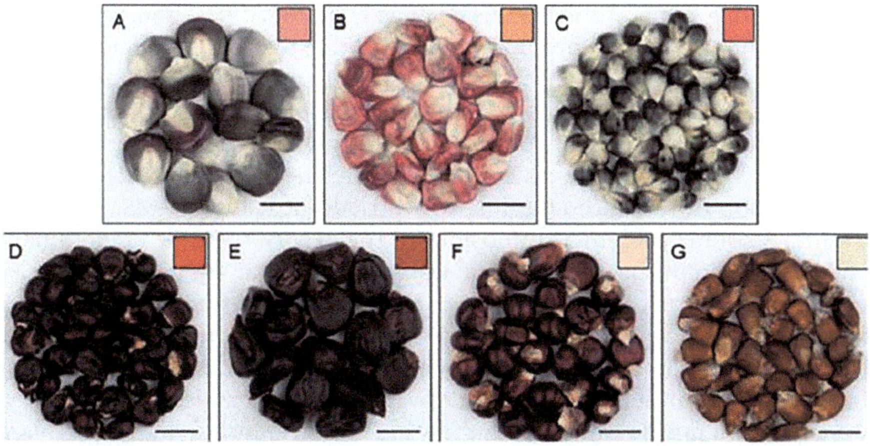

Figure 9.3 Various pigmented maize samples: (A) Blue aleurone; (B) Pink aleurone; (C) Blue without acylated; (D) pericarp without condensed forms; (E) pericarp with condensed forms; (F) phlobaphene; (G) bronze pigment. Reproduced from ref. 24 with permission from American Chemical Society, Copyright 2017.

36.4 ± 7.0 mg C3G equivalent per kg of whole corn in small grits and pericarp, respectively. The contrast between the MMA between purple and blue maize was noticeably large, with a value of 3610.0 ± 34.8 and 257.7 ± 10.9 mg C3G equivalent per kg of whole corn, respectively. Based on studies by Li *et al.*,[23] it was concluded that, depending on the corn cultivar, most pigments were localized in the pericarp or aleurone layer. This showed that, depending on the genotype of the corn, maize processing can yield highly concentrated ANC-containing coproducts.

Blue, purple, red or yellow maize are also commonly found maize varieties. In these varieties, both anthocyanin and carotenoids are predominantly present. Proanthocyanidins, a group of anthocyanins, were found to be major pigment component for blue, purple, yellow and red maize with an average value of 1381 ± 69.8, 1279 ± 56.1, 613 ± 53.7 and 578 ± 36.0 μg g^{-1} DM of catechin, respectively. The difference between blue, purple, yellow and red maize was not only due to genetic varieties but, admittedly, these molecules are produced through structural condensation of flavanols (catechin or epicatechin) with anthocyanins. Polymers of flavan-3-ol subunits (catechins) connected by an interflavan carbon bond makeup proanthocyanidins (condensed tannins) and thus blue and purple maize had the highest concentration of this group of anthocyanins.[25]

Like maize, wheat is another cereal that can be said to have a noticeable amount of anthocyanin in it and usage of anthocyanin-rich products from wheat are mainly produced from blue or purple wheat, which are present in the exterior coating of the kernel, either in the aleurone or pericarp layers. D3G and C3G are the predominant anthocyanins found in purple and blue wheat, among other anthocyanins such as D3R, M3G, P3G and C3G. These anthocyanins are mostly present in their chloride forms, for instance.[26,27]

Among all varieties of wheat, *T. aestivum* (bread wheat), *T. durum* (durum wheat), *T. monococcum* (einkorn) are the most commonly studied for their anthocyanin content. In an attempt to identify the anthocyanins in wheats, a study was carried out on nine different varieties of wheat species and the confirmation of the anthocyanins was carried out using LC–MS. C3G with 19.5–23.8% and cyanidin-3-rutinosid with 10.9–15.9% were the predominant anthocyanin in purple wheat, while D3G (55.0% to 68.9%) and D3R (58.6% to 69.8%) were crucial anthocyanins in blue wheat. Other than the previously mentioned anthocyanins, cyanidin-3-glucoside, cyanidin-3-rutinoside and pelargonidin-3-glucoside (Figure 9.2) were present in trace amounts. Apart from this, some other anthocyanin derivatives, such as malonyl and succinyl were found in purple wheat. Both these are acylated anthocyanin, which are deemed to be a precursor of biosynthesis of all flavonoids.[26] Table 9.1 gives a summary of the concentration of different anthocyanin extracted from wheat flour.[28]

9.3 Extraction of Pigments

The extraction process has been shown to significantly impact recovery and the concentration of targeted compounds. Figure 9.4 depicts the extraction methods for colored pigments from cereals. Microwave, pressurized fluid, supercritical fluid, electric fields, ultrasonic energy and other innovative approaches have been developed, for effective and safe anthocyanin extraction.[29] The use of acidified methanol or ethanol as extractants is one of the most common methods. Anthocyanin extraction with methanol is the most efficient and preferred method, owing to its lower toxicity compared with ethanol.[5] Black sorghum anthocyanins were extracted using 0.1% HCl in MeOH and 70% aqueous acetone.[30] Total anthocyanins were substantially greater in acidified methanol than in aqueous acetone, and acidified

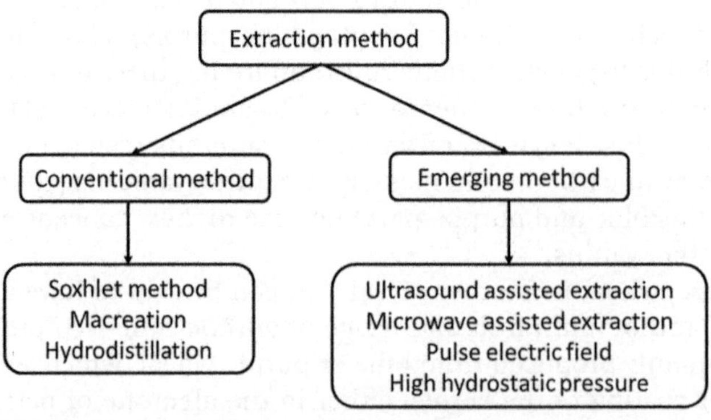

Figure 9.4 Extraction methods for colour pigments from cereals.

methanol extraction was twice as efficient as acetone. The different cereal pigments, their extraction method, solvents and quantity are presented in Table 9.2.

9.3.1 Novel or Emerging Non-conventional Technologies

The limits of traditional technologies encourage the development of innovative techniques. Difficulties in obtaining a high level of purity characterize conventional extraction procedures, the use of expensive solvents, a longer extraction time, the potential for heat-labile chemical degradation and limited extraction selectivity.[31] Innovative techniques have been developed to overcome these limitations. Several unique and developing approaches are now being used for the extraction process. The advantages and disadvantages of the emerging technologies are presented in Table 9.3. The following are the most important new and emerging approaches.

9.3.1.1 Ultrasound-assisted Extraction

Ultrasound-assisted extraction (UAE) has been used to extract bioactive compounds from plant sources with a frequency range of 20 kHz, which is the human hearing threshold.[32,33] The ultrasound's output source is generally a vibrating body that causes the surrounding medium to vibrate and then distributes energy to other nearby particles through the ultrasonic wave. Power, frequency and amplitude are the most important physical parameters in the ultrasound process.[32] Solvent absorption and mass transfer are improved using the ultrasound-assisted technique. This approach uses less solvent; however, it extracts specified compounds well. Ultrasound has been proved to be effective in extracting pigments from various plant resources by having a mechanical impact on the tissue matrix, resulting in cell wall disruption. Proper extraction is important for separating and identifying a compound. UAE has appeared as an efficient technique with prominent compound recovery, specifically from solid materials.[29] Das *et al.*[33] optimized the extraction of anthocyanin using black and purple rice bran utilizing UAE, considering the extraction factors of extraction time, solvent composition, and pH. UAE substantially enhanced the yield of anthocyanin compounds in black and purple rice bran compared with the conventional extraction process.

9.3.1.2 Microwave Assisted Extraction

Microwave assisted extraction (MAE) is an innovative technique that uses microwave energy to extract phytochemicals, and it is a promising alternative to conventional extraction.[31,34] Compared with conventional extraction, some reports have indicated that MAE has benefits, such as shorter processing time, less solvent and energy use, and higher yield.[34] Microwaves have an

Table 9.2 The different cereal pigments and their extraction method, solvents and quantity.

Source	Pigment	Colour	Extraction technique	Solvent	Method	Quantity	Reference
Maize	Anthocyanin	Purple	Conventional	Methanol with 0.01% HCl	HPLC	35.1%	6
				70% Acetone with 0.01% HCl		31.0%	
				Water with 0.01% HCl		32.4%	
				20% Ethanol		28.7%	
				40% Ethanol		31.7%	
				50% Ethanol		32.4%	
				60% Ethanol		32.2%	
				80% Ethanol		28.6%	
				Ethanol With 0.01% HCl		36.0%	
Maize	Anthocyanin	Blue	Conventional	Methanol with 0.001 HCl	HPLC	63.19%	24
		Pink				62.84%	
Maize	Anthocyanin	Blue	MAE	Methanol with 1% HCl	UV-Vis	189–336 μg g^{-1}	42
			Conventional		UPLC–MS	423 μg g^{-1}	
Sorghum	Anthocyanin	Black	MAE	Methanol with 1% HCl	UV-Vis	916–1090 μg g^{-1}	
			Conventional		UPLC–MS	861 μg g^{-1}	
Sorghum	Anthocyanin	Black	Conventional	Methanol with 1% HCl, 70% aqueous acetone	HPLC Spectrophotometric	4.0–9.8 mg luteolinidin equivalents/g	30
Sorghum	Anthocyanin	—	PEF			33.9%	44
Wheat	Anthocyanin	Blue	Conventional	Methanol with 1N HCl	LC-UV-Vis	211.9 μg g^{-1}	56
		Purple				38.0–95.8 μg g^{-1}	
		Red				6.7–7.9 μg g^{-1}	
		White				7.1–7.3 μg g^{-1}	
Barley		Blue				34.6 μg g^{-1}	
Rice		Black				3276.0 μg g^{-1}	
		Red				93.5 μg g^{-1}	
		Wild				27.2 μg g^{-1}	

Rice	Anthocyanin	Purple	Conventional	Ethanol with HCl	Spectrophotometer	21.82 mg cyanidin-3-glucoside/L	33
			UAE			34.86 mg cyanidin-3-glucoside/L	
		Black	Conventional			18.75 mg cyanidin-3-glucoside/L	
			UAE			31.95 mg cyanidin-3-glucoside/L and	
Barley	Carotenoid	Purple	Conventional	Methanol	HPLC Spectrophotometric		
Rice	Anthocyanin	Black	HHP	Methanol with 1% HCl		538.25 mg cyanidin-3-glucoside/100 g	57

Table 9.3 Advantages and disadvantages of non-conventional techniques used for pigment extraction. Reproduced from ref. 35 with permission from Elsevier, Copyright 2021.

Non-conventional techniques	Advantages	Disadvantages	Reference
Ultrasound-assisted extraction	Less time and energy consumption, fast extraction; a small amount of solvent; operated at ambient temperature, high yield and aqueous and non-aqueos solvent can be used	Non-selective,possibility of free radical species formation, with the passage of time, power declines, and scaling up is a challenge	32 and 58
Microwave-assisted extraction	Fast and short extraction time with higher yield, cost-effective, less and no use of solvents, and operational at room temperature	Poor extraction for non-polar or volatile compounds, applicable for a limited amount of samples, can affect the chemical and structure of compounds	34 and 59
Pulsed electric field	Less time and energy consumption, improves extraction and diffusion, non-destructive, suitable for sensitive compounds, operated under room temperature and easy to scale up	Expensive equipment, efficiency of the PEF highly dependable with the amount of electric field strength and electrode gap	60 and 61
High hydrostatic pressure	Energy efficient, nutritional value and quality attributes of the product are retained, reduced thermal exposure, suitable for high moisture-content foods, and eco-friendly	Need preheating step for pressure-assisted thermal processing, higher processing costs and batch operations, and degradation of thermo-labile compounds	62 and 63

electro-magnetic field that ranges from 0.3 to 300 GHz, and is composed of 2 perpendicular magnetic and electric fields. Microwave heating includes the conversion of electromagnetic energy to heat *via* the use of ionic and dipole rotation mechanisms.[35] Microwave irradiation causes thermal stress in cells, leading to cell wall breakage and the release of intracellular compounds. Most of the plant extracts extracted by MAE are dietary polyphenols, including flavonoids, anthocyanins and carotenoids, among other compounds.[34] MAE seems to be the most effective method for increasing extraction efficiency and yield of sorghum anthocyanin (3-deoxyanthocyanin pigment).[36]

9.3.1.3 Pulsed Electric Field

The pulsed electric field (PEF) assisted extraction method is based on the non-thermal cell permeabilization process. PEF implies the utilization of short-duration pulses (μs to ms) of moderate electric voltage (0.5–20 kV cm^{-1}) to material between two electrodes.[35,37] PEF generates a potential difference across the cell membrane called the transmembrane potential difference. Electrical breakdown may occur in the electroporation of the cell membrane when the potential difference reaches a threshold magnitude. This process enhances the permeability of the cytoplasmatic membranes, allowing the intracellular contents to be released more easily.[38] PEF also reduces the amount of energy and organic solvent needed to extract the required compounds. PEF facilitates the extraction of valuable components from fruits, oilseeds, other raw materials and agricultural waste products; the use of PEF as a pretreatment shortens processing time, while lowering power consumption.[39] The anthocyanin content extracted by PEF was found to be 33.9% from sorghum.[38]

9.3.1.4 High Hydrostatic Pressure

High hydrostatic pressure (HHP) processing is a non-thermal technology that includes pressurizing food in flexible pouches by putting them in a container with a pressure of 600 MPa.[29] HHP is considered a good substitute to conventional extraction techniques because of its short time, processing condition, extraction efficiency, and less solvent utilization. HHP primarily impacts non-covalent bonds, which implies that low molecular weight molecules, for instance, many bioactive components (flavonoids, phenolic components, carotenoids and anthocyanin), are less affected, and so this treatment may obtain most of the bioactive compounds.[40,41]

9.4 Quantification and Characterization of Pigments

The pigments from cereals such carotenoids and anthocyanins can be extracted using the methods discussed in the earlier section. Their quantification is commonly performed using HPLC, MS, NMR, hydrolysis techniques, and UVV.[2] These methods are generally used to analyze the structure of anthocyanin and carotenoids, and will be discussed in this section.

9.4.1 High Performance Liquid Chromatography (HPLC)

HPLC is a reliable and fast technology for analyzing organic chemicals in fruits, vegetables, cereals and grain materials. The most widely used techniques for extracting and characterising carotenoids and anthoyanins in food are liquid–liquid and solid–liquid extraction. Because the examination of chemicals like carotenoids and anthocyanins can be complicated, HPLC is thought to be one of the easier methods to use. According to Wald *et al.*,[42] reverse-phase HPLC (RP-HPLC) developed for extraction and characterization of starchy staple foods was comparatively more advantageous. Carotenoids are commonly separated chromatographically using RP-LC. While C18 stationary phases have been widely used in the past, due to their greater selectivity, C30 columns have become increasingly important in carotenoid analysis. They managed to characterize lutein and β carotene from cereals ad some wheat species. Although HPLC can often be used for the quantification of pigment compounds, the extraction process also affects the concentration of the pigments. The solvents used in the extraction process interfere with the chemical structure and change its composition. For instance, anthocyanins can be converted into anthocyanidins during the extraction process.[43]

9.4.2 Liquid Chromatography–Mass Spectrometry (LC -MS)

Due to MS's sensitivity and high specificity compared with other chromatographic detectors, it has consistently been preferable to combine it with chromatographic procedures. Liquid chromatography-mass spectrometry (LC–MS) has become a standard technology since the introduction of electrospray ionization (ESI): a simple and robust interface. It may be used to analyze a wide range of biological chemicals, and the use of tandem MS with stable isotope internal standards allows for the development of extremely sensitive and precise assays, however, some method tuning is required to minimize ion suppression effects.[44] Also, because the technique combines chromatographic separation with sensitive detection of the compound-specific precursor *m*/*z* as well as the distinctive fragment *m*/*z*, hyphenation of LC to MS with various ion sources has proved particularly effective for structural confirmation of anthocyanins. As a result, selective identification and sensitive quantification are possible. Ionization strategies for natural chemicals that are either non-volatile or thermodynamically unstable have recently been developed.[45]

Mass spectrometry is a technique for identifying anthocyanins by detecting the mass of molecular ions in the sample as well as fragments resulting from the rupture of these compounds using higher ionization energy. In certain circumstances, gas chromatography with mass spectrometry has been utilized, however, due to the structural lability of anthocyanins in relation to ionization techniques based on electronic impact, as well as the need to derivatize them, this has proven to be a challenging and costly application.[2]

9.4.3 UV-vis Spectral Analysis

The characterization of anthocyanin using UV-vis spectroscopy was reported in 1939 by Pauling, who observed the structure of flavylium cation. UV-vis spectroscopy can provide information on the structure of various chemicals *via* absorption spectroscopy. Despite the fact that mass spectroscopy is far more helpful than UV-vis spectroscopy, this method is still used for simple pigment characterisation and quantification.[45,46] According to Saha *et al.*,[45] the absorption pattern of anthocyanin shows a distinctive behaviour. The peak is observed around 550 nm, Again, the spectra can differ depending upon acylated and non-acylated anthocyanins.

In the case of carotenoids, the UV-vis spectrophotometric approach is also a useful tool for separating and identifying various carotenoids quantitatively, due to the conjugated double bond, which absorbs in the region of 400–500 nm. But, unlike HPLC and mass spectroscopy, UV-vis spectrometry has more disadvantages and is less sophisticated. However, it is still being used to analyze compounds due to its cost-effectiveness. According to a comparative study between Raman spectroscopy and UV-vis spectroscopy on analysis of carotenoids, β carotene peaks were observed at three distinct points, namely 442 nm, 467 nm, and 500 nm.[47]

An alternative to UV-vis spectrometry, Near Infrared Reflectance Spectroscopy (NIRS) is being currently used for analysis of total phenolic content, antioxidants and pigments such as anthocyanins and carotenoids, but its usage in the characterization and quantification of cereal grains are limited as compared with its usage in fruits and vegetables. Table 9.4 depicts some of the frequently used quantification and characterization methods for cereals.

9.4.4 Nuclear Magnetic Resonance (NMR)

NMR spectroscopy, or nuclear magnetic resonance spectroscopy, is a technique that takes advantage of the magnetic properties of specific nuclei. Antioxidants, phenolic chemicals and colors in food are frequently identified using this non-destructive approach. However, little research has been done on the characterization of pigments in cereal crops, and this is regarded to be an uncharted topic in this field. NMR spectroscopy is being utilized to identify and validate pigments in some grain types, despite its limitations.[48]

Among cereals, components from rice have been mostly explored using various characterization technologies. Rice is coloured in general before milling, and variants of anthocyanin and carotenoid pigments can be found in golden and pigmented rice. As a result, NMR spectroscopy can be utilized

Table 9.4 Characterization and quantification method in some selected cereals. Reproduced from ref. 7 with permission from Elsevier, Copyright 2018.[a]

Cereal type	Quantification and identification methods	Reference
Maize	HPLC–MS, total anthocyanin method UVV	64
	pH differential, HPLC -MS, UVV	65
Rice	HPLC–MS/MS, UVV	66
	HPLC–MS/MS, UVV, pH differential	
Wheat	HPLC–MS, UVV	67
	HPLC–MS, UVV, NMR	68
	HPLC, UVV, pH differential	27
Barley	HPLC, UVV	69
	HPLC, pH differential	70 and 71
	Total anthocyanin, HPLC–MS, UVV	72
Sorghum	HPLC, UVV	73
	UVV, MS, NMR	74
	HPLC, UVV, total anthocyanin	75

[a]HPLC-High Performance Liquid Chromatography, MS-Mass Spectrometry, UVV-Ultra Violet Visual, NMR-Nuclear Magnetic Resonance.

to identify the pigments and metabolites present. More research on identifying specific anthocyanins and carotenoids is still needed.[49]

9.5 Application

Pigments, due to their characteristic colors and sensitivity to various parameters such as pH, can be used in food systems, nutraceuticals, smart packaging and biosensors. Consumption of biologically active components, which are commonly prevalent in dark-coloured fruits, vegetables and cereal grains, has been shown to reduce the risk of several illnesses because of their strong antioxidant activities. As a result of their high concentrations of nutritional fibre and nutrients, bread and bakery goods have risen to prominence in the human diet. However, regular wheat flour baked goods have limited antioxidant potential and must be supplemented with coloured wheat flour to boost their nutritional value. Cereals are a good source of carbohydrates and proteins, and minor micronutrients, including minerals, vitamins, and antioxidants comprising carotenoids. Anthocyanin extraction from fruits and vegetables is more expensive, but coloured grains can be recovered economically. However, research into these pigment molecules' isolation and practical usage from coloured grain is still lacking.[7] Compared with fruit and vegetables, the carotenoid content is lower in cereals utilizing biotechnology; cereals' carotenoid content was enhanced basically in three cereal species: corn, wheat and sorghum. However, a new rice variety, golden rice, was produced at the beginning of the century by engineering grains to generate β-carotene to treat vitamin A deficiency in people with rice-based traditional diets.[13] According to ayurvedic practitioners, coloured rice has numerous health benefits and is effective against ulcers, skin diseases, backache and good for use as weaning food for malnourished babies.[50,51] In terms of genetics, chemistry, and biological activities, Zhu[7] states that there are certain features of anthocyanin from grains that have not been studied enough. As a result, further study is needed to discover, isolate and characterize these compounds to determine their potential impact on human health.

9.6 Conclusion

The existence of pigments such as carotenoids and anthocyanins in various food crops has long been researched, while research on cereal crops has been limited in comparison with other coloured food products. Major cereal crops such as rice and wheat are staples in some parts of the world, whereas minor cereals such as barley and sorghum, which are high in specific pigments, are still neglected. Pigmented cereals are enriched with coloured compounds that exhibit well-documented bioactive potential against various ailments. The utilization of pigmented cereal in the formulation of various food and feed products can enhance food products' nutritional and functional value. Pigmented grains can be used in medications, cuisine, textiles, painting and biotechnology, in addition to their myriad health benefits. Their appropriate application will benefit fields such as biotechnology, where fortification of cereals might improve the health and hunger indices of people in developing nations.

References

1. G. F. Deng, X. R. Xu, Y. J. Guo, E. Q. Xia, S. Li, S. Wu, F. Chen, W.-H. Ling and H. B. Li, *J. Funct. Foods*, 2012, **4**(4), 906.
2. M. T. Escribano-Bailón, C. Santos-Buelga and J. C. Rivas-Gonzalo, *J. Chromatogr. A*, 2004, **1054**(1–2), 129.
3. S. Ghosh, T. Sarkar, A. Das and R. Chakraborty, *LWT--Food Sci. Technol.*, 2022, **153**, 112527.
4. J. Lachman, P. Martinek, Z. Kotíková, M. Orsák and M. Šulc, *J. Cereal Sci.*, 2017, **74**, 145.
5. A. Castañeda-Ovando, M. de Lourdes Pacheco-Hernández, M. E. Páez-Hernández and J. A. Rodríguez, *Food Chem.*, 2009, **113**(4), 859.
6. F. Lao and M. M. Giusti, *J. Cereal Sci.*, 2018, **80**, 87.
7. F. Zhu, Anthocyanins in cereals: Composition and health effects, *Food Res. Int.*, 2018, **109**, 232.
8. E. G. Kean, G. Ejeta, B. R. Hamaker and M. G. Ferruzzi, *J. Agric. Food Chem.*, 2007, **55**(7), 2619.
9. V. U. Ndolo and T. Beta, *Food Chem.*, 2013, **139**(1–4), 663.
10. Z. Liu, Y. Liu, Z. Pu, J. Wang, Y. Zheng, Y. Li and Y. Wei, *Biotechnol. Lett.*, 2013, **35**(11), 1765.
11. M. Naczk and F. Shahidi, *J. Pharm. Biomed. Anal.*, 2006, **41**(5), 1523.
12. G. Rocchetti, L. Lucini, G. Giuberti, S. R. Bhumireddy, R. Mandal, M. Trevisan and D. S. Wishart, *Food Chem.*, 2019, **298**, 125068.

13. E. Mellado-Ortega and D. Hornero-Méndez, *Phytochem. Rev.*, 2015, **14**(6), 873.
14. T. Maoka, *J. Nat. Med.*, 2020, **74**(1), 1.
15. T. Maoka, *Arch. Biochem. Biophys.*, 2009, **483**(2), 191.
16. X. Qin, W. Zhang, J. Dubcovsky and L. Tian, *Plant Mol. Biol.*, 2012, **80**(6), 631.
17. P. Colasuonno, I. Marcotuli, A. Blanco, M. Maccaferri, G. E. Condorelli, R. Tuberosa, R. Parada, A. C. de Camargo, A. R. Schwember and A. Gadaleta, *Front. Plant Sci.*, 2019, **10**, 1347.
18. D. Trono, *Plants*, 2019, **8**(12), 551.
19. E. T. Wurtzel, A. Cuttriss and R. Vallabhaneni, *Front. Plant Sci.*, 2012, **3**, 29.
20. *Maize: Nutrition Dynamics and Novel Uses*, ed. D. P. Chaudhary, S. Kumar and S. Langyan, Springer India, New Delhi, vol. 3, 2014.
21. L. Lamberts and J. A. Delcour, *J. Agric. Food Chem.*, 2008, **56**(24), 11914–11919.
22. S. Oliveira, K. A. Costa, E. Severiano, A. da Silva, M. Dias, G. Oliveira and J. V. Costa, *Agronomy*, 2020, **10**(11), 1714.
23. Q. Li, P. Somavat, V. Singh, L. Chatham and E. G. de Mejia, *Food Chem.*, 2017, **231**, 332.
24. M. Paulsmeyer, L. Chatham, T. Becker, M. West, L. West and J. Juvik, *J. Agric. Food Chem.*, 2017, **65**(21), 4341.
25. S. Suriano, C. Balconi, P. Valoti and R. Redaelli, *LWT--Food Sci. Technol.*, 2021, **144**, 111257.
26. E. S. M. Abdel-Aal, P. Hucl and I. Rabalski, *Food Chem.*, 2018, **254**, 13.
27. D. B. M. Ficco, V. De Simone, S. A. Colecchia, I. Pecorella, C. Platani, F. Nigro, F. Finocchiaro, R. Papa and P. De Vita, *J. Agric. Food Chem.*, 2014, **62**(34), 8686.
28. N. Sharma, V. Tiwari, S. Vats, A. Kumari, V. Chunduri, S. Kaur, P. Kapoor and M. Garg, *Molecules*, 2020, **25**(24), 5785.
29. S. Farooq, M. A. Shah, M. W. Siddiqui, B. N. Dar, S. A. Mir and A. Ali, *J. Food Meas. Charact.*, 2020, **14**(6), 3508.
30. J. M. Awika, L. W. Rooney and R. D. Waniska, *Food Chem.*, 2005, **90**(1–2), 293.
31. N. A. Sagar, S. Pareek, S. Sharma, E. M. Yahia and M. G. Lobo, *Compr. Rev. Food Sci. Food Saf.*, 2018, **17**(3), 512.
32. C. Wen, J. Zhang, H. Zhang, C. S. Dzah, M. Zandile, Y. Duan, H. Ma and X. Luo, *Ultrason. Sonochem.*, 2018, **48**, 538.
33. A. B. Das, V. V. Goud and C. Das, *Ind. Crops Prod.*, 2017, **95**, 332.
34. G. A. Cardoso-Ugarte, M. E. Sosa-Morales, T. Ballard, A. Liceaga and M. F. San Martín-González, *LWT--Food Sci. Technol.*, 2014, **59**(1), 276.
35. N. P. Nirmal, R. Mereddy and S. Maqsood, *Food Chem.*, 2021, **356**, 129611.
36. D. A. Herrman, J. F. Brantsen, S. Ravisankar, K. M. Lee and J. M. Awika, *Food Chem.*, 2020, **333**, 127494.
37. B. Kumari, B. K. Tiwari, M. B. Hossain, N. P. Brunton and D. K. Rai, *Food Bioprocess Technol.*, 2018, **11**(2), 223.

38. U. C. Lohani and K. Muthukumarappan, *Innov. Food Sci. Emerg. Technol.*, 2016, **35**, 29.
39. N. Naliyadhara, A. Kumar, S. Girisa, U. D. Daimary, M. Hegde and A. B. Kunnumakkara, *Trends Food Sci. Technol.*, 2022, 238–255.
40. S. A. Moreira, S. Silva, E. M. Costa, J. A. Saraiva and M. Pintado, *Food Funct.*, 2020, **11**(1), 921.
41. H. Scepankova, M. Martins, L. Estevinho, I. Delgadillo and J. A. Saraiva, *Plant Foods Hum. Nutr.*, 2018, **73**(4), 253.
42. J. P. Wald, D. Nohr and H. K. Biesalski, *J. Food Compost. Anal.*, 2018, **67**, 119–127.
43. F. U. G. Akogou, A. P. P. Kayodé, H. M. W. den Besten, A. R. Linnemann and V. Fogliano, *LWT*, 2018, **90**, 592.
44. J. J. Pitt, *Clin. Biochem. Rev.*, 2009, 19.
45. S. Saha, J. Singh, A. Paul, R. Sarkar, Z. Khan and K. Banerjee, *J. AOAC Int.*, 2020, **103**(1), 23.
46. B. M. Popović, B. Blagojević, R. Ž. Pavlović, N. Mićić, S. Bijelić, B. Bogdanović, A. Mišan, C. M. M. Duarte and A. T. Serra, *Food Chem.*, 2020, **302**, 125373.
47. R. Ullah, S. Khan, A. Shah, H. Ali and M. Bilal, *Laser Phys.*, 2018, **28**(5), 055601.
48. R. S. Parmar and C. Singh, *Biochem. Biophys. Rep.*, 2018, **13**, 22.
49. H. Du, Y. Huo, H. Liu, G. M. Kamal, J. Yang, Y. Zeng, S. Zhao and Y. Liu, *CyTA–J. Food*, 2019, **17**(1), 128.
50. T. S. Rathna Priya, A. R. L. Eliazer Nelson, K. Ravichandran and U. Antony, *J. Ethn. Foods*, 2019, **6**(1), 1.
51. A. M. Digesù, C. Platani, L. Cattivelli, G. Mangini and A. Blanco, *J. Cereal Sci.*, 2009, **50**(2), 210.
52. M. D. Rey, M. C. Calderon, M. J. Rodrigo, L. Zacarías, E. Alos and P. Prieto, *PLoS One*, 2015, **10**(8), e0134598.
53. M. Aluru, Y. Xu, R. Guo, Z. Wang, S. Li, W. White, K. Wang and S. Rodermel, *J. Exp. Bot.*, 2008, **59**(13), 3551.
54. C. Bai, T. Capell, J. Berman, V. Medina, G. Sandmann, P. Christou and C. Zhu, *Plant Biotechnol. J.*, 2016, **14**(1), 195.
55. R. A. Moreau, A. F. Harron, M. J. Powell and J. L. Hoyt, *J. Am. Oil Chem. Soc.*, 2016, **93**(4), 569.
56. E. S. M. Abdel-Aal, J. C. Young and I. Rabalski, *J. Agric. Food Chem.*, 2006, **54**(13), 4696.
57. L. Meng, W. Zhang, X. Zhou, Z. Wu, A. Hui, Y. He, H. Gao and P. Chen, *J. Cereal Sci.*, 2019, **86**, 54.
58. B. E. Esquivel-González, O. M. Rutiaga-Quiñones, N. E. Rocha-Guzmán, L. Medina-Torres, E. D. C. Varela-Santos and L. A. Ochoa-Martínez, *J. Chem. Bio. Phy. Sci. Sec., B*, 2017, **7**(3), 784.
59. A. Singh, M. Ganesapillai, and N. Gnanasundaram, in *IOP Conference Series: Materials Science and Engineering*, IOP Publishing, November 2017, vol. 263, No. 3, p. 032004.

60. J. Azmir, I. S. M. Zaidul, M. M. Rahman, K. M. Sharif, A. Mohamed, F. Sahena, M. H. A. Jahurul, K. Ghafoor, N. A. N. Norulainid and A. K. M. Omar, *J. Food Eng.*, 2013, **117**(4), 426.
61. M. Nowacka, S. Tappi, A. Wiktor, K. Rybak, A. Miszczykowska, J. Czyzewski, K. Drozdzal, D. Witrowa-Rajchert and U. Tylewicz, *Foods*, 2019, **8**(7), 244.
62. S. A. Khan, R. Aslam and H. A. Makroo, *J. Food Proc. Eng.*, 2019, **42**(1), e12896.
63. M. Paciulli, I. G. Medina-Meza, E. Chiavaro and G. V. Barbosa-Cánovas, *LWT--Food Sci. Technol.*, 2016, **68**, 98.
64. S. Žilić, A. Serpen, G. Akıllıoğlu, V. Gökmen and J. Vančetović, *J. Agric. Food Chem.*, 2012, **60**(5), 1224.
65. B. Harakotr, B. Suriharn, M. P. Scott and K. Lertrat, *Euphytica*, 2015, **203**(2), 237.
66. G. Pereira-Caro, G. Cros, T. Yokota and A. Crozier, *J. Agric. Food Chem.*, 2013, **61**(33), 7976.
67. C. E. Tyl and M. Bunzel, *J. Agric. Food Chem.*, 2012, **60**(3), 731.
68. Y. Shao, F. Xu, X. Sun, J. Bao and T. Beta, *J. Cereal Sci.*, 2014, **59**(2), 211.
69. S. L. Yao, Y. Xu, Y. Y. Zhang and Y. H. Lu, *Food Funct.*, 2013, **4**(11), 1602.
70. S. H. Lee, E. Jeong, S. S. Paik, J. H. Jeon, S. W. Jung, H. B. Kim, M.-H. Chun and I. B. Kim, *Curr. Eye Res.*, 2014, **39**(1), 79.
71. C. Lee, D. Han, B. Kim, N. Baek and B. K. Baik, *Int. J. Food Sci. Technol.*, 2013, **48**(5), 984.
72. I. Diczházi and L. Kursinszki, *Cereal Chem.*, 2014, **91**(2), 195.
73. L. Dykes, L. M. Seitz, W. L. Rooney and L. W. Rooney, *Food Chem.*, 2009, **116**(1), 313.
74. A. Khalil, R. Baltenweck-Guyot, R. Ocampo-Torres and P. Albrecht, *Phytochem. Lett.*, 2010, **3**(2), 93.
75. A. P. Kayodé, M. R. Nout, A. R. Linnemann, J. D. Hounhouigan, E. Berghofer and S. Siebenhandl-Ehn, *J. Agric. Food Chem.*, 2011, **59**(4), 1178.

CHAPTER 10

Pigmented Cereals: Key Influencing Factors

SNEHA SWAIN[a], K. R. KARTHIKA PARVATHY[b], BALASUBRAMANIAN PARAMASIVAN[a], ARUN PRASATH*[c] AND R. PANDISELVAM[d]

[a]Department of Biotechnology & Medical Engineering, National Institute of Technology Rourkela, 769008, India; [b]Department of Life Science, National Institute of Technology Rourkela, 769008, India; [c]Department of Food Process Engineering, National Institute of Technology Rourkela, 769008, India; [d]Physiology, Biochemistry, and Post-Harvest Technology Division, ICAR – Central Plantation Crops Research Institute, Kasaragod-671 124, Kerala, India
*E-mail: varun@nitrkl.ac.in

10.1 Introduction

Cereal is any cultivated grass with an edible component, *i.e.*, the grain. They are grown in large quantities due to their small size and ability to provide sufficient energy for humans and animals. Some edible grains belong to other families, typically non-grasses, and are named pseudocereals.[1] Some examples are buckwheat, quinoa, and chia. Alongside provision of energy, cereals are a great source of nutrition. Most of the provided nutrition is carbohydrates, but they also provide proteins, some fats and oils, and vitamins and minerals. Although not all countries consume grain in its primary form, it is a staple food in refined and processed forms; for example, wheat is consumed as chapattis in India and other forms of bread in numerous western

Food Chemistry, Function and Analysis No. 38
Pigmented Cereals and Millets: Bioactive Profile and Food Applications
Edited by Sneh Punia Bangar, Sajid Maqsood and Anil Kumar Siroha

Published by the Royal Society of Chemistry, www.rsc.org

countries. A similar trend is marked in the case of other cereals as well, as, different varieties of rice with different colours are available in various regions, but not all are abundantly consumed.

The history of cereal dates back to early Neolithic villages (around 10700 to 9400 BC, *i.e.*, Stone Age),[2] where they processed the grains to obtain a desired form of cereal for consumption. Domestication of crops like wheat, rye, barley, oats, and flaxseeds were all done during the early Neolithic phase. The first grains were domesticated by primitive humans (around 8000 years ago) and their farming started from communities in the Fertile Crescent Region. The three Neolithic founder crops were emmer wheat, einkorn wheat, and barley, which led to agriculture development. Approximately during the same time, in East Asia, the cultivation of rice (of different kinds) and also millets began.[3] Millets and sorghum cultivation were also initiated during the same time in sub-Saharan West Africa.[4]

The name 'Fertile Crescent'[5] means civilization's dependence on the cultivation of cereals. Many Chinese emperors have stated that agriculture is the foundation of their empire. Cereals can be classified into six major types of grain: millets, adzuki beans, rice, soybean, wheat and barley together, and sesame. The cereal crops mentioned above grow well in moderate weather and gradually abate if the weather changes to a hot environment. Other cereals prefer a hotter climate, which can be grown during hotter weather conditions. This cycle enables us to grow as many as six types of crops, considering that during a very favourable season, two types of crops can be grown on the same piece of land.

A very vital phase in the history of crops/agriculture was the Green Revolution,[6] which occurred during the second half of the 20th century and showed a significant increase in agricultural production. These changes were brought about by high-yield genetically engineered crops and involved the use of fertilizers, pesticides, herbicides, insecticides, *etc.* Although it was the greatest of its kind, but not the first, and it ranked third in the list of agricultural revolutions since the Stone Age. The other two were the Neolithic Revolution[7] and the British Agricultural Revolution.[8] The use of high-class technologies also contributed to the vastness of the Green Revolution. The start of the Green Revolution was in the early 1960s, after mass famine, and the production that took place then has been the greatest achievement in the history of agriculture. An example of the impact on one state of one country can be seen in the following example. The state of Punjab was the first selected to administer the new crops (high-yield) and note the efficiency of Indian land, due to Punjab's reliable water supply and highly fertile plain area. India's Green Revolution began with plant breeding, financing agrochemicals (fertilizers, pesticides, insecticides, *etc.*) and, finally, irrigation development. Statistics state that the yield rose from two tons per hectare in the 1960s to six tons per hectare by the mid-1990s. Since there was a rise in production, the cost reduced from $550 a ton in 1970 to $200 a ton in 2001.[9]

Apart from the function of providing nutrition and energy to humans, cereals have also played other vital roles. To briefly introduce a few, there is the blood cholesterol level lowering activity of cereals due to their fibers' absorption activity on cholesterol. Taking the example of rice, it is a grain with multiple benefits. Rice water has been used for various purposes, including its use to restore skin elasticity, reduce the appearance of aging on the skin, smoothing hair, *etc.*[10] Similarly, we can find a lot of other cereals with exemplary functions that have helped humans in different ways.

Pigments are a vital part of plants. They impart differences between two kinds in the species or even strain levels in both plants and micro-organisms. Considering the variety of pigments in plants, their functions may also vary. As an example, for higher energy, and ease of cultivation, people consume white rice and, for proper nutrition and higher fiber content, people prefer brown rice. Rice has various kinds depending on taste, odour, the height of plants, nutritional content, *etc.*[11] The preference lies in the requirement of any individual. But all of this did not happen in itself. This was brought about by the use of biotechnology. Not just rice, but many other crops have been developed by using new techniques and deeper knowledge of genes, with the use of biotechnology.

This chapter has elaborated on all types of benefits that these developed cereals have provided and what else is held for the future. We also have focused on what components in these pigmented cereals contribute towards the phenotype and functions or benefits that the cereals display. This chapter is inclusive of primary and secondary cereals and their uses from around the world, the factors that influence pigment production in cereals, nutritional and other functions that these cereals have, the development that biotechnology has provided these cereals with, be it genetically, metabolically or other strategies involved and, finally, the challenges that we are currently facing with the development of these crops. To conclude, we will look into some of the future perspectives that may be worked upon soon or have already been taken into our hands.

10.2 Overview of Cereal Production

In February, the FAO raised its prediction for global cereal production in 2021 by 2.1 million tonnes, bringing the total to 2793 million tonnes, an increase of 0.8% year on year.[12] Currently, there are over 50 000 edible plant species, but just 15 contribute to 90% of the world's intake of food energy. Two-thirds of this is made up of three main kinds of cereal: rice, maize, and wheat.[13] Cereal is domesticated grass (monocotyledonous) from the Gramineae family; and is usually cultivated for its small and edible seed. These plants are dioecious, except for maize. Cereals have starch endosperm, an embryo/germ and various bran and husk layers. With maize, wheat and rice, other plants contribute to world nutrition. They are barley, sorghum, millets, oats, and rye.[14] Cereals have gained importance as food plants because of their high yields and nutrition. They contain carbohydrates (70%), protein (10%) and lipids (3%), although this value might vary from crop to crop.

The harvested mass per unit area (yield) has been the measure of greatest importance, to fulfil the target for all crops. World cereal production and yield per hectare have steadily increased over the last forty years. But there is evidence that the increase in genetic yield potential from breeding lags behind the necessary amount needed to maintain food security globally. To date, maize is the only crop to have undergone marketing after being genetically engineered. There is still testing going on in other crops. The shift from other conventional breeding programs to genetic engineering is progressively increasing. Progress in the direction of genetic engineering occurred due to understanding genomics, transcriptomics, proteomics, and phenomics. Genome editing and advances in marker-assisted selection *in vitro* cellular techniques and recombinant DNA technology have played a major role as innovations in plant breeding.[13]

Today, cereals are the major source of vitamin B complex and minerals, proteins and energy in the human diet. They are also considered an important source of dietary fiber and the most calorie-providing and starchy food. Each region has one cereal which is regarded as the staple food. Rice, maize, and wheat are the main 3, contributing about 93% of the total cereal calories. They are the staple foods in Asia, America, and Europe, respectively. Sorghum and millets are widely grown and consumed in Africa and India. Cereals provide approximately 33.8 g of protein, among the estimated total of 73.9 g of protein, according to FAO (Food and Agriculture Organization). But this is just for adults and children to some extent. This quality of cereals does not hold as good for infants. Cereals contributed to nearly 50% of the average *per capita* caloric intake in human diets in 1997. In developing countries, cereals and legumes are usually consumed together, thus their protein intake increases and essential amino acids are covered within the diet. Plant breeders have developed maize, sorghum and barley genotypes with high lysine. This resulted in improved quality of protein and high nutritional value. In different countries around the world, for example, Brazil, Ghana, China, and some African countries, lysine and tryptophan infused maize are grown commercially.[15]

There are two types of cereals based on consumption frequency and amount. They are primary and secondary cereals. The primary cereals are largely consumed, either in their raw form, or after processing. They include wheat, rice and maize. Apart from the above three, barley, sorghum, millets, oats, are rye are also consumed, but to a lesser quantity. Thus, they are considered secondary cereals. Among the major primary pigmented cereals, pigmented rice has a wide array of coloured grains such as yellow, green, brown, red, purple, and black (Table 10.1).[16] Due to the rich nutritional landscape and antioxidant nutraceuticals, pigmented rice varieties have the potential to promote human health by managing various metabolic impairments. However, due to lack of sensitization among humankind on poor awareness, low yield, consumer preferences, and staggered market potential, this still represents only 0.1% of total rice production.

Table 10.1 List of cereals along with their pigmented colour.

Cereals	Pigmented colour
Primary pigmented cereals	
Rice	Black, red, brown and purple
Wheat	White, black, blue, purple and yellow
Maize	Blue, red and purple
Secondary pigmented cereals	
Foxtail millet	Yellow
Little millet	Reddish-brown, light brown, creamish brown and blackish brown
Finger millet	Black
Fonio millet	Black and white
Rye	White, green, purple and brown
Barley	Green, white, yellow, purple, blue and black
Sorghum	Red, yellow, white and black

10.2.1 Primary Cereals

Three major kinds of cereal – rice, wheat, and maize – are considered under the primary category as they contribute majorly towards food, animal feed, and energy all over the world. These are the major contributors with regards to their dietary and subsidiary benefits. Rice is consumed by more than 50% of the world's population, and China and India alone contribute to this amount of rice production and consumption. Although rice is a source of carbohydrates and consists of other nutrients in adequate amounts, there is a very trace amount of vitamins and minerals which decreases even more due to processing. This has been taken up for fortification, and many new technologies have taken up the responsibility of escalating its potential.[17] Wheat and maize, on the other hand, are mainly produced by the United States of America, which accounts for about 68% of the world's supply of food.[18] There are many forms of wheat consumption, and all begin with the conversion of wheat into powder and then making dough for further food preparation. Maize or corn kernels are widely consumed all over the world and in a variety of forms, sometimes ranging from its conversion into a powdered form to consumption in the form of kernels. Due to the large cultivable land area of the countries and cheap manual labour, and high-tech machines, these countries have been able to hold themselves at the positions where they are. The climate required for rice is hot and humid; for wheat it is warm; for maize it is semi-arid temperature, and alternate rainfall or sunny weather is ideal for maize.[19]

As mentioned in the introduction, rice has multiple benefits. Its benefits for skin and hair are numerous. In their native lands, they are majorly consumed for food and energy. Rice has also displayed many medicinal and pharmaceutical uses, not just for these purposes. It starts with controlling high blood pressure, maintaining the body's balance, correcting digestive system issues, calming down skin inflammation, and also helping cure diarrhoea in children.[20] But it holds that any medication is only valid if it is taken/consumed in the proper quantity. Ayurveda and Unani medication systems have proven the pharmaceutical benefits of rice and have been using them to find cures for the above disorders in the body. An example is the consumption of saturated drinks made from Laja rice, which helps in the cure of nausea, burning, thirst, and debility, which people commonly face. Since in this paper we are dealing with pigmented cereals, there are some important benefits of the different varieties of pigmented cereals that should not be overlooked. Rice is present in different colours *i.e.*, white, red, yellow, green, purple, brown, and black, which give different benefits to humans.[21] The traditional varieties of rice, which are white, red, brown, and black, have less fat, and oryzanol in a good amount compared to hybrids, and does not allow the body to build up cholesterol. These rice varieties contribute to a healthy and balanced food habit and display other benefits as well. For example, brown rice is consumed by those who have been trying to lose weight; black rice is rich in antioxidants, thus it is consumed to keep cells defended from attack by free radicals and thus prevents oxidative stress. Red rice is easy to digest and eases digestive issues in certain patients. Aromatic rice, which is mostly of the white rice kind, has a bran layer rich in vitamins and fiber and is thus considered healthier.[22] Other varieties of rice, namely yellow, green, and purple have been genetically modified to produce their colour and respective benefits. Purple rice was developed by the Japanese, with the benefits of antioxidants and other secondary metabolites, which were not naturally prevalent.[23] Yellow rice is also at times regarded as golden rice and has been widely cultivated due to its benefits to eyesight and the immune system. This is developed with added aromatic effect and has been biofortified, which means it has been developed with added nutritional values to produce beta-carotene that is generally not found in rice.[24] The development of green rice was dependent on resolving the long-term challenge of pests and insects, which was the colour of grain attracting them to feed on. This development in the colour of rice contributed to resistance to various diseases, drought resistance, high nutrient efficiency and finally in reducing consumption by pests, thereby having the same effect on the use of chemical fertilizers and water.[25]

Wheat has an outer covering called the wheat bran, which is used to make medicine and is a source of high dietary fiber for preventing colon diseases, including cancer. Cancers such as breast cancer and stomach cancer, haemorrhoids, and hiatal hernia can be prevented by consumption of wheat bran derivatives.[26] Wheat bran is an exemplary source of fiber which helps with dietary obstruction and digestive disorders. A recent study has revealed that wheat bran helps with type 2 diabetes mellitus. But the actual cure is yet to be revealed. The six major kinds of pigmented wheat are white, yellow, red, purple, blue, and black. Like black rice, black wheat also has high free radical scavenging activity against DPPH and high phenolic content. Since this had been genetically engineered, it displayed the high antioxidant activity that could be inducted into the grains for their physical benefits.[27] Blue wheat has a different composition of anthocyanin than purple wheat, which deployed various genetic properties and functional characteristics.[28] This colour of wheat is rich in zinc and other micro-nutrients, which gives it a distinct role in promoting human health. Purple wheat is high in nutrients, fiber, and vital micronutrients compared with other varieties of wheat, and thus could be beneficial for the health of adults and children.[29] Red wheat has high gluten levels and is high in protein, dietary fiber, manganese and selenium, all of which are beneficial and necessary for a functioning metabolism.[30] White wheat has the lowest phenolic and flavonoid content and thus has lower antioxidant properties than blue and other darker varieties. The colour of yellow wheat is due to its carotenoid content; the greater the percentage of carotenoid, the darker the shade in the yellow range.[31]

Maize is one of the more domesticated and evolved plants, and its genetic diversity is concentrated in Mesoamerica, particularly in Mexico, which is the plant's primary source of origin, domestication, and diversification with a significant quantity of 59 maize races.[32] Maize is consumed in a variety of forms. It is consumed majorly as corn starch and also in boiled form. It is a fibrous vegetable (the seeds are considered as cereals) and it has a high carbohydrate and protein content of approximately 66.2% and 11.1%, respectively. Low-income people consume dishes made of corn. Apart from being food, maize is used for medicinal purposes as well. Corn silk is the thread-like yellow part of the corn, *i.e.*, the style part.[33] This is used as a medicine for various purposes like a bladder infection, inflammation of the prostate, the urinary system, bedwetting, and kidney stones. Corn silk has also shown benefits for congestive heart failure, diabetes, high blood pressure fatigue and high cholesterol levels. There are majorly three kinds of corn; white, yellow and bi-colour. When the corn is examined for its quality of taste, it is not the colour that determines taste, rather is the genetic features that are involved. Another variety of corn is flint corn or Indian corn, which seems to be genetically modified, but is a result of open pollination, and the appearance it carries is gem-like. The constituents of yellow maize are pro-vitamin A and non-pro-vitamin A carotenoids, which carry potential benefits for humans. In regions where this is consumed, there has been an improved impact on dietary intake.[4] White maize is chosen for human consumption, whereas yellow is chosen for animal feed; the reason being that yellow contains carotenoids, which, on cooking degrade and cause a strong aroma and flavour. White is devoid of the latter, thus they can be better cooked for human consumption.[34]

10.2.2 Secondary Cereals

Some others are a major part of the cereal industry but not are not as frequently or widely consumed. These are barley, sorghum, millets, oats, and rye, and are in the category of secondary cereals. Barley is the fourth largest grain crop globally, and Russia is currently the largest producer of barley. Following Russia, Germany and France stand in the second and third position.

Sorghum belongs to the *Poaceae* family and is not just food for humans, but also pasture for animals. Its domestication began in Africa and has spread throughout the world. The largest producer of sorghum is the United States of America, with a supply of million metric tons, as of the year 2021/2022. Millets are variable, small-seeded grasses and are widely grown around the globe as both food and fodder. Millets belong to the *Paniceae* family of grasses, and India is ranked the top-most producer of millets with a production of 1.1 crore tons of millet per year, by World Atlas. European Union is the largest producer of oats, which is consumed as either oatmeal and/or rolled oats. People consume oats as a healthy breakfast or meal substitute, but they were originally domesticated for their use as fodder. Oats are a type of cereal from the *Poaceae* family, and, being gluten-free, they are considered the healthiest on earth. Steel-cut oats are the best choice for better control of blood sugar levels. Rye is grown extensively for its use as flour, bread, beer, and use in whiskeys, vodkas, or maybe in crisp bread. They are also used for animal fodder. The largest producer of rye is the European Union, according to 2020 statistics on agriculture and farming.[35]

Apart from being used as food, barley is used in lowering blood sugar levels, blood pressure, and cholesterol levels, and has also proven to aid in weight loss. People with digestion complaints are suggested to consume barley to keep their gastrointestinal tract clean and healthy.[36] Barley exhibits protection against heart diseases and keeps our hearts healthy. There have also been effects of barley on the immune system and lung health, as they aid in preventing asthma. Prevention of cancer and management of diabetes is a couple of other benefits that barley provided along with the provision of good skin care, and help in preventing osteoporosis. The different pigments in barley are white, green, yellow, purple, blue, and black. White barley is the kind whose husk has been removed and polished but is still rich in various beneficial plant compounds, vitamins, and minerals. Its high fiber content has helped in digestive issues and has many other health benefits.[36] Green barley has similarities with black wheat and rice, for an abundance of antioxidants which are vitamin A and C, thereby helping fight free radicals that may damage our cells and tissues. It has been proven to have anticancer effects on cancer types like lymphoma, leukemia, and breast cancer.[37] Black, blue, and yellow barley have similar antioxidant properties, but the differences lie in their phenolic composition, which ultimately brings about differences in free radical scavenging activity.[38] The abundant anthocyanin content in purple barley is a topic of interest to many scientists these days, and its antioxidant properties have led to its increased consumption by humans.[39]

Sorghum is a good source of vitamins and minerals and is rich in antioxidants. It has high dietary fiber content and helps in reducing blood cholesterol, and acts as a natural remedy for diabetes patients.[40] Sorghum helps in weight loss and inhibits tumor growth. It is a staple diet for celiac patients (patients with an immune reaction to gluten). Some other benefits that sorghum provides are improved bone health; it is rich in calcium and magnesium; helps blood circulation and is thus great for enhancing energy levels.

It has been used for helping improve gluten allergies and has provided great results. The different colours that sorghum exhibits are red, yellow, white, and black. Red sorghum is usually used as a livestock feed and displays antioxidant properties that are beneficial to animals and humans. Black sorghum is rich in fiber, antioxidants, and protein. Both black and red sorghum are used in the bakery industry. White sorghum has been used as a complete substitute for corn, for further processed food items.[41] Yellow sorghum has a very thin minimally weathered pericarp, which provides it with the colour it has and has benefitted humans, as it controls blood sugar level, and keeps it within limits, both white and yellow sorghum provide such benefits.[42]

Millets are rich in fiber and antioxidants. This is considered an ideal food for diabetes, as it prevents an irregular spike in blood sugar levels.[43] Along with controlling diabetes, it is good for blood cholesterol levels, relieving menstrual cramps, aiding in sleep, preventing celiac disease, treating coronary artery disease, *etc.* Millets have also been proven to aid in weight loss and reduce the risk of colon cancer. Millets also help lower cholesterol levels and alleviate constipation. Two major supporting compounds present are β-glucan and avenantramides.[44] Avenantramides promote blood vessel dilation and thereby reduce the risk of cardiovascular diseases. It is, in addition, applied topically to regions that are itchy for a soothing feeling. β-Glucan is a soluble fiber that improves glucose metabolism and promotes beneficial bacteria in the guts.

Rye aids in weight reduction, controlling diabetes, fighting cancer, improving digestion, eliminating toxins and, finally, controlling hypercholesterolemia. Along with the above-mentioned benefits, rye also prevents gallstones, artery blockages, cardiovascular diseases and other illnesses.[45] Rye has proven to boost metabolism along with muscle growth, and helps prevent asthma. The different colours that rye displays are green, purple, and white, and sometimes brown. Two genes in the plant control the pigment of green rye. For the purple pigment, cyaniding-3-glucosides is a major compound, along with some cyaniding-3-rutinoside, peonidin-3-rutinoside, and their corresponding acylated derivatives.[46]

10.3 Key Influencing Factors of Pigment Production in Cereals

All cereals have different varieties and these varieties constitute pigments in different compositions. Cell disruption, concentration, and solvent extraction are methods usually taken up to recover the pigments after fermentation. As an example of natural dye production, red koji dates have been used to produce a red dye for the past hundred years. This can further be used to dye cereals manually.[47] Crude pigments are frequently extracted *via* traditional methods, but the emergence of an alternative to the traditional technique, the Green Extraction technique, has benefitted by the use of a lesser amount of solvent and time. Maceration, hydrodistillation and

Soxhlet extraction are traditional processes, which have proven to be simple, economical and straightforward to use. The efficiency of extraction can be intensified and escalated by the use of modern technologies such as negative pressure cavitation, ultra-high pressure, high voltage electrical discharges, pulsed electric fields, *etc.*[48]

For the biosynthesis of carotenoids, genes have been recognized in numerous model crop plants for the function of structural enzymes, catalytic activities, and essential cofactors, which have been partially characterized. Oxygenated derivatives of β-carotene are xanthophylls, lutein and zeaxanthin. By introducing 5,6-epoxy moieties by zeaxanthin epoxidase (ZEP), 3-hydroxyl β-rings of zeaxanthin are further oxygenated. Astaxanthin is not a naturally occurring pigment in most of the higher plants. Over-expression of the gene Psy225 from the daffodil plant, including other varieties of Psy genes, resulted in increased accumulation of Xanthophylls.[49]

Carotenoid synthesis inside a cell begins with the condensation of GGPP (geranylgeranyl pyrophosphate) which is considered a precursor of the methylerythritol 4-phosphate (MEP) pathway.[50] Availability of the precursor influences flux of the entire MEP pathway. The expression pathways of DXR (1-deoxy-D-xylulose 5-phosphate reductoisomerase), DXS (1-deoxyxylulose-5-phosphate synthase) and 1-hydroxy-2-methyl-2-(*E*)-butenyl 4-diphosphate reductase (HDR) genes are involved in the MEP pathway, which is found to influence the carotenoid constituent in maize endosperm positively. Above were the factors that influenced the synthesis of carotenoids. But the degradation of carotenoids also has to be considered, to preserve the viability of pigment in the crop containing the same. There are two basic mechanisms that a degradation pathway follows: specific and non-specific. The specific pathway leads to the degradation of carotenoids into apocarotenoids by the enzyme called dioxygenases. On the other hand, non-specific mechanisms include oxidation, be it enzymatic or non-enzymatic. LOXs (lipoxygenases) trigger enzymatic oxidation and are heme iron-containing enzymes catalyzing the hyper oxidation of polyunsaturated fatty acids. Carotenoid destruction *via* a non-enzymatic pathway is characteristic of conjugate double-bond structure found in the same, which renders electron-rich system at risk of reactions with electrophilic oxidizing agents. Carotenoids showed a decrease in their content in maize when exposed to normal drying than freeze-drying, which ranged 15–45% of the former. While on storing in dark space, the decrease in carotenoid was approximately 24–61% of the former. Overall, temperature and humidity played an important role in the stability and storage of maize and highlighted the genotypic effect that allowed inbred lines with improved stability.[51]

Anthocyanin is a pigment that is water soluble and belongs to the flavonoid family. They are found in the form of glycosides of poly ethoxy and polyhydroxy derivatives of 2-phenylbenzopyrrile salts and consist of aglycone and a carbohydrate residue that may be glucose, xylose, galactose, arabinose, rhamnose or rutinose (Figure 10.1). The ionic nature of the structure of anthocyanin risks it to the influence of pH. This is found in various chemical

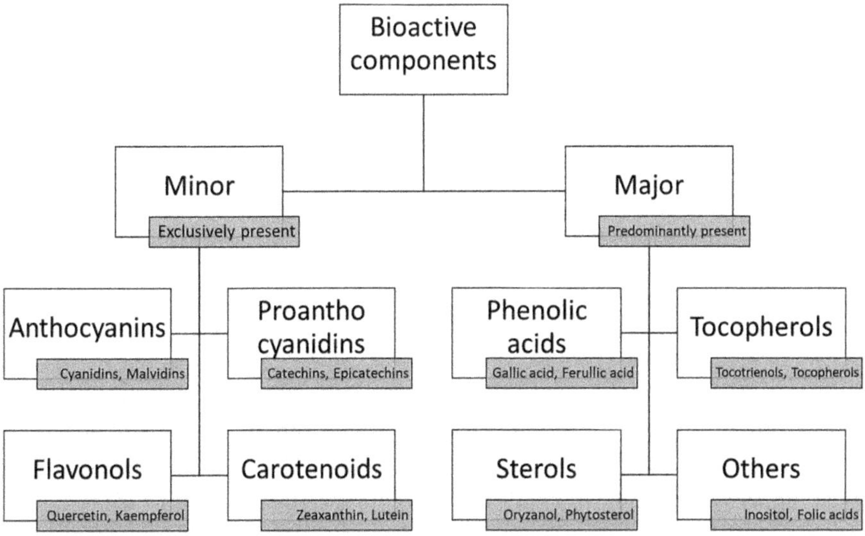

Figure 10.1 Bioactive components present in the pigmented cereals.

forms and is influenced by the pH of the solution. Anthocyanin undergoes multiple mechanisms, for example, nucleophilic attack of water, polymerization, cleavage and glycosylation, which causes the loss of pigments and their degradation. Hence temperature also has effects on the stability of the molecular structure of anthocyanins, and with increasing temperature, these compounds degrade. Browning of these pigments occurs in the presence of oxygen, which is because of their unsaturated structure, and their susceptibility to molecular oxygen. Hence to increase the stability of these compounds, they are stored in a vacuum or nitrogen and argon are good alternatives.[52] The stability of anthocyanin molecules is affected by a variety of factors, including anthocyanin content, temperature, and pH, along with exposure to light, oxygen, enzymes, ions, similar flavonoid and phenolic substances, ascorbic acid, and sulphites. Because of the electron-deficient flavylium cation structure, the anthocyanidin backbone is highly reactive.[53]

Cereals are low in fats, and on average contain about 3.6% of their caloric content. Linoleic acid is majorly found in grains, but linolenic acid can be detected in small quantities, which is influenced by the season of growth and geographical location of harvest. Since the content of PUFAs and lipids in cereals is low, the enhancement techniques available are either fermentation of the cereals in the presence of particular fungi, specific to the final result, or genetically modifying the cereal crop to acquire the gene of interest, which is the PUFA producing gene. One attractive perspective for the enhancement in the content of PUFAs in cereals has its foundation in fermentation methods. Success in the production of microbial PUFAs has led to an interest in developing fungal strains and fermentation processes.[54] The presence of an active biosynthetic apparatus in *Mucorales* fungi enables the production of both PUFAs and carotenoids. The fungi are also considered a potential candidate for the enrichment of the desired bioactive compounds in preferred cereals. Recently, Certik's group selected beneficial fungal strains. The SSF process was optimized to maximize the acquired content of β-carotene and GLA from the cereal-based mass, which was pre-fermented.[55] For the production of PUFA-enriched cereals, regulation of SSF is necessary. This can be carried out by selecting a suitable inoculum and cultivation techniques, providing internal support, taking a close look at moisture and water activity, proper regulation of nutrients and finally modifying the PUFA yield by plant compounds.[56]

The method used to produce fermented cereals is SSF (Solid State Fermentation), in which the growth of micro-organisms is on the moist solid substrate but in the absence of water. The substrate acts as a source of required nutrition and a place for the anchorage of microbes. There are various steps to the development of fungal PUFAs, such as isolation and screening of the microbe, optimization of nutritional and physio-chemical parameters, scale-up of the processes to be carried out without lacunae and finally the regular operation of the bioreactor or fermenter. Although the lipid and fat content in PUFA is not highly appreciable, content in a slightly higher range can be detected in some parts of cereals like rice and wheat bran, peeled barley and oat flakes. In the process of fermentation, various factors play their part, which is fermentation time, nutrient utilization from the substrate, proper light provision, physiological stress parameters and finally, application of various stimulators for the biosynthesis of PUFAs and pigments.[56]

Wheat is a relatively poor insoluble fiber, which is mainly found in the form of water-extractable arabinoxylans.[57] Unlike in barley and oats, water-extractable arabinoxylans in wheat form dietary fiber fractions that are soluble, which are pretty similar to β-galactans and in aqueous environments, it forms a viscous solution.[58] Flavonoids accumulate as C-glycosidic forms in cereals.[59] Zeatin is the second most abundant xanthophyll identified in wheat.[60]

10.4 Nutritional and Functional Properties of Pigmented Cereals

Consumers are looking for natural food colours, such as carotenoids, due to evidence of health problems caused by artificial food colours. The production of carotenoids has medicinal and nutritional benefits. As a result, carotenoids are produced in agricultural systems in a variety of ways through plants.[49] Some of the most important and valuable carotenoids are – carotenoids, astaxanthin, canthaxanthin, lycopene, and lutein, which are being used in a wide range of products.[47] The pro-vitamin A activity of β-carotene is the major reason for carotenoids' nutritional relevance in humans. Carotenoids have been linked to a lower risk of cancer, a lesser incidence of cardiovascular disease, protection of the macula region of the retina, cataract control, and enhanced iron absorption, in addition to their role as vitamin A precursors.[61]

In Asia, rice cultivars with black, red, and purple bran layers that have long been harvested are rich in phytochemicals and nutrients.[62] Rice is made up of two layers: the hull and the caryopsis. The outer layer is the hull, representing around 20% of the total weight of paddy rice and comprises minerals and cellulose. The caryopsis that contains the bran, endosperm, and germ is recovered after the dehulling process. Brown rice is produced from grain that has been dehulled.[63] The pericarp protects the seed coat and gives the rice its colour, resulting in light brown, red, and black rices that are brightly coloured. Black rice is a whole grain, due to its sensory features, high nutritional content, and, most importantly, its good health properties. Black rice has attracted increased attention among the pigmented rice varieties.[64] Black rice's endosperm contains around 75% carbohydrates, mostly starch, and a protein-rich outer layer as sub-aleurone. Other polysaccharides found in small proportions in the outer layer of black rice include dietary fibers such as resistant starch, cellulose, hemicellulose, pectins, and simple sugars like glucose, fructose, and saccharose. The germ, which makes up around 2–3% of the total weight of paddy rice, contains proteins and lipids. These rice varieties were also thought to be effective remedies that could heal various ailments, and they also supply staple food with unique flavors, aroma, and fluffiness.[65] When compared with white rice, black rice has been demonstrated to have positive nutritional and functional properties. This is due to the combination of a diverse array of nutritional and bioactive components in the bran layers and embryo, such as amino acids, functional lipids, dietary fiber, flavonoids specifically anthocyanin and proanthocyanidin, carotenoids like α-β-carotene, lutein and lycopene, vitamins like B complex, A and E, minerals, which includes K, Fe, Zn, Cu, Mg, Mn, and P, phenolic compounds (caffeic acid and ferulic acid), anthocyanins, tocopherols, γ-oryzanols, tocotrienols, phytosterols as β-sitosterol, stigmasterol and capesterol, coumaric acid and phytic acid. Due to proanthocyanidin, purple or black rice bran has been shown to have stronger antioxidant activity than red rice bran.[66–68] The antioxidant activity and scavenging capacity of anthocyanins in pigmented rice reveal their putative role in food compositions. Additionally, investigations on several pigmented and non-pigmented rice varieties will confirm the mechanisms of the compounds' functionality and their key role, such as antioxidant and free radical scavenging, anti-influenza, anti-atherosclerosis, antiobesity, anti-allergic, and antitumor activities against a wide range of degenerative and chronic illnesses (Figure 10.2).[68]

Amongst the cereal grains, maize has the highest antioxidant capacity associated with numerous health benefits that could prevent several chronic degenerative diseases like diabetes, Alzheimer's, arthritis and some types of cancer.[53] Pigmented maize with nutraceutical properties is linked to its high level of anthocyanins, which are produced from secondary metabolites and have positive biological action through its antioxidants in it. These substances are beneficial to one's health since their antioxidant activity inhibits mutagenesis and the proliferation of malignant and anti-inflammatory cells. Furthermore, anthocyanins included in maize grain protect people with type

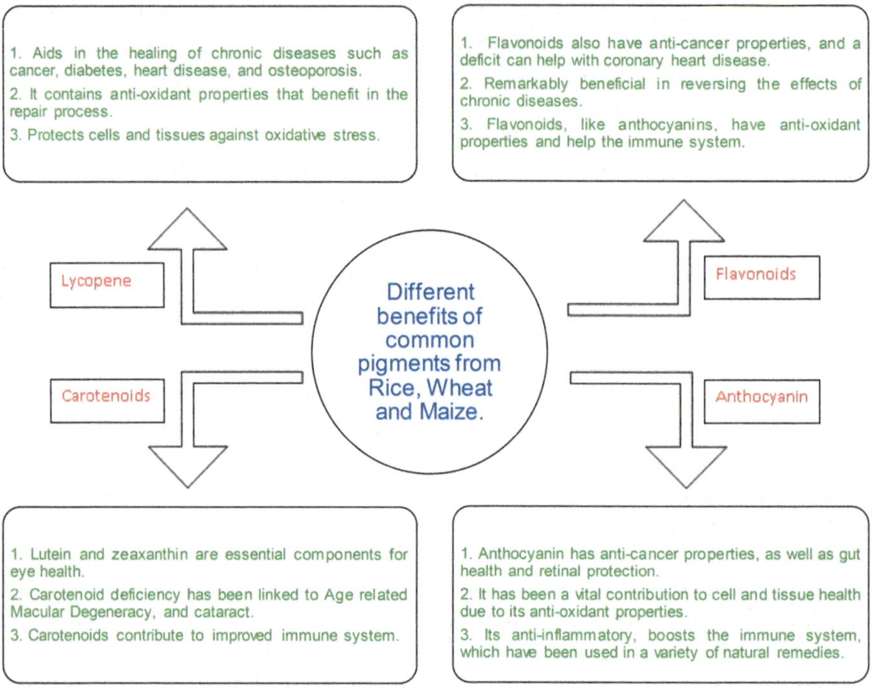

Figure 10.2 Common health benefits of primary crops.

2 diabetes from developing nephropathy. These characteristics, however, are lost when maize is boiled before being consumed. As a result, the cooking method and duration of the thermal treatment should be investigated to discover how these variables affect the anthocyanin characteristics in the human body.[69] For the past severeal decades, people have been suffering from intestinal and extra-intestinal symptoms connected with ingestion of gluten which causes gluten-related disorders (GRD). Instead, pigmented corn varieties contribute to the rapid diffusion of gluten-free functional foods with rich dietary antioxidants, which could counteract the oxidative stress at the intestine in celiac disease patient.[70]

Wheat is the world's most extensively cultivated crop, with more than 758.3 million metric tonnes produced each year. Bread wheat (*Triticum aestivum* L.) accounts for the majority of this, whereas durum wheat accounts for 35–40 million metric tonnes (*Triticum durum* Desf.). The distribution of yellow colour components in whole-grain durum wheat was investigated, and it was determined that, throughout grain milling, a decreased rate of pigment degradation was observed.[71] They reported equal quantities of carotenoids in all of the purple wheat genotype's fractions, with much more anthocyanins in the external layers due to the pigment's location. They also revealed that the exterior layers had 3- to 5-fold more antioxidant activity than similar flours, however, it diminished when particle size from the millstream decreased.[72]

Some researchers studied a significant number of purple/red tetraploid wheat varieties. About bread wheat, they utilized 11 typical anthocyanin standards and found delphinidin (Dp-3-Rut and Dp3-Glc) was the most abundant aglycone in blue wheat, accounting for 68.9% of total anthocyanins, followed by cyanidin (Cy-3-Glu and Cy-3-Rut) (Table 10.2). However, in contrast to

Table 10.2 List of anthocyanins along with their concentration reported in pigmented cereals.

Pigmented cereals	Concentration (mg kg^{-1})	Pigmented cereals	Concentration (mg kg^{-1})
Cyanidin-3-glucoside		**Delphinidin-3-rutinoside**	
Purple corn (whole meal)	1135.4[a]	Blue wheat	49.6[b]
Purple corn	298.9[b]	Blue wheat	33.4[f]
Blue corn (whole meal)	17.4[a]	**Malvidin-3-rutinoside**	
Shaman blue corn	110.1[b]	Blue wheat	2.0[b]
Red corn	284.5[b]	**Malvidin-3-glucoside**	
Purple wheat	4.0[b]	Purple wheat	0.5[c]
Purple wheat	10.3[c]	Blue wheat	12.4[c]
Blue wheat	3.1[c]	Red wheat	0.22[c]
Blue wheat	20.3[b]	**Pelargonidin-3-(6″-malonylglucoside)**	
Red wheat	4.0[c]	Purple corn	51.6[a]
Black rice	2013.0[b]	Blue corn	6.5[a]
Red rice	14.0[b]	**Peonidin-3-glucoside**	
Blue barley	1.2[b]	Black rice	162.1[b]
Purple barley	99.1[d]	Red rice	2.5[b]
Cyanidin 3-rutinoside		Purple wheat	2.1[b]
Black rice	19.9[b]	**Peonidin-3-rutinoside**	
Red rice	1.3[b]	Blue wheat	1.2[b]
Shaman blue corn	1.1[b]	**Peonidin-3-arabinoside**	
Red corn	284.5[b]	Blue wheat	2.2[c]
Blue wheat	16.8[b]	**Peonidin-3-galactoside**	
Cyanidin-3-(6″-succinylglucoside)		Blue wheat	1.9[c]
Purple wheat	0.6[b]	Purple wheat	0.6[c]
Purple wheat	1.1[b]	Red wheat	0.3[c]
Purple wheat	1.2[b]	**Peonidin-3--(6″-malonylglucoside)**	
Cyanidin-3-(3″,6″-dimalonylglucoside)		Purple corn	145.1[a]
Purple corn (whole meal)	398.2[a]	Blue corn	7.4[a]
Blue corn (whole meal)	40.1[a]	Purple wheat	0.6[b]
Shaman blue corn	1.5[b]	Purple wheat	0.9[b]
Black rice	71.8[b]	Purple wheat	0.5[b]
Red rice	1.7[b]	**Petunidin-3-glucoside**	
Cyanidin-3-(6″-malonylglucoside)		Blue wheat	2.2[b]
Purple wheat	1.2[b]	Blue barley	2.9[b]
Delphinidin-3-glucoside		Purple barley	37.0[d]
Purple wheat	65.5[e]	**Petunidin-3-rutinoside**	
Purple wheat	13.7[b]	Blue wheat	4.5[d]
Purple barley	93.0[d]		

[a]Asenstorfer *et al.*[75]
[b]Urias-Lugo *et al.*[76]
[c]Li *et al.*[77]
[d]Bellido and Beta.[78]
[e]Abdel-Aal *et al.*[79]
[f]Garg *et al.*[80]

earlier findings, relatively high quantities of Mv-3-Glc have been detected in all three Sebesta genotypes studied, but the peonidin aglycone is always glycosylated with diverse sugars like glucoside, arabinoside, and galactoside for all the genotypes.[72,73] Anthocyanin has proven to decrease inflammation by modulating various steps of signalling. These inflammatory responses were ina dose-dependent manner.[74]

10.5 Biotechnological Strategies for Pigmented Cereal Production

10.5.1 Genetic Engineering

Biotechnologists must understand all of the parameters which affect pigment production to improve pigment accumulation. Genetic engineering requires genetic transformation technology followed by gene promoters and in addition *cis*-acting regulatory elements with gene sources with known functions (Figure 10.3). Biotechnology has two major strategies for cereals, which are conventional breeding programs where genes on chromosomes control physiological and physical characteristics. Plant breeders benefit greatly from the capacity to monitor the presence or absence of such genes in plants which are identified through their phenotypes. This was enabled through the use of molecular markers, which are unique DNA sequences or fragments connected to the gene or genes. The ability to transfer genes between species is the second major use of biotechnology. This signifies that certain genes can be added to a crop variety in a single step, eliminating the need for all of the back-crossing that is normally necessary, by saving time and effort.

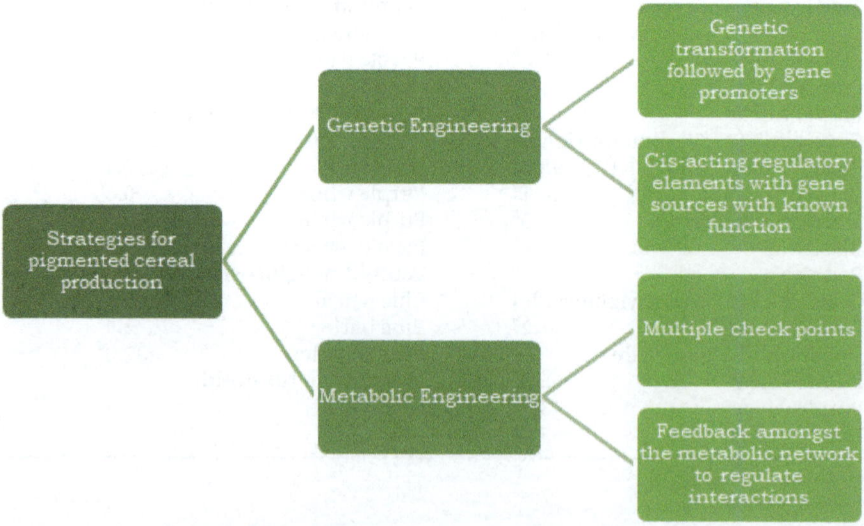

Figure 10.3 Strategies for pigmented cereals production.

Moreover, the genes do not have to come from a sexually compatible species with the crop. Transgenic plants with agronomically valuable genes, in addition to herbicide resistance genes, have been identified. Because insects cause significant damage to crops throughout the world, developing insect-resistant plants has and will remain to be a primary objective. Stem boring insects are prominent maize and rice pests, and tolerance to them has been gained attention through the introduction and expression of modified or synthetic forms of the Bt δ-endotoxin, a natural insecticidal toxin produced by *Bacillus thuringiensis*. There is no common method available for identifying transgenic plants. To identify individuals with single-copy gene insertion, ample molecular analysis, usually incorporating Southern blotting procedures of DNA extracted from individual plants from vast populations of regenerants are required.[14]

Rice cultivars' agronomic and economic properties are biotic and abiotic tolerance, adaptability, eating quality, and a yield that has been vastly increased. Also, numerous genes or alleles implicated in cold, drought, salt stress tolerance, flood, bacterial blight, bacterial leaf streak, rice blast, and brown planthopper resistance, have been discovered and identified in wild rice types.[81] In several crop species, the chemical mutation is applied to improve various plant traits. Previously, Taiwan's Agriculture Research Institute employed sodium azide to generate mutations in two white-pericarp rice lines, SA418 and SA419 which are, red- or purple-pericarp mutants with waxy or non-waxy endosperm traits respectively were studied to understand the expression of the Ra and Rc genes in these mutants. Both spontaneous and chemically induced mutations result in transforming white rice grains to coloured rice grains. Three different types of genes affect the colour of grains (Ra, Rc and Rd). The purple pericarp is controlled by the Ra gene, where purple predominates over white. Brown pericarp is caused by the Rc gene in combination with the absence of the Rd gene. The Rd gene does not create any colour on its own, but when the Rc and Rd genes are crossed, a red pericarp is formed. Red pericarp mutant M-69 with non-waxy endosperm weighed more per 100 grains than white-pericarp SA418, SA419, TNG67, and IR64. TNG67 and IR64 were likewise outperformed by mutant M-69 in terms of grain yield, antioxidants, total phenolics. In addition, the mutant M-69 bran parts are a good natural source of biologically active ingredients that can be used as health-promoting constituents in nutraceutical and functional foods.[82] The interest in rice breeding *via* biotechnology may be revived due to insect resistance traits and high β-carotene varieties of golden rice.[13] Cereals are genetically engineered to enrich their PUFA content.[56] Improvement strategies for the crop are now focussing on specific targetted modification and selection of endogeneous sequences either directly from the plant or by modifying the native form by RNA interference. Mutational breeding, Genetic engineering and marker-assisted selection (MAS) are superior crop improvement methods as they elevate the efficiency and accuracy of pinpoint introgression of value-added traits. It has been shown *via* experiments of gene shuffling that mixing genes from Gene Libraries can help produce novel carotenoids.

Typical microbial production optimization made improvement in the accumulation of novel products.[49,83,84]

Carotenoids, the prime components of wheat's yellow pigment, influence flour colour and impact the grain's nutritional value and its utility in many applications. Free radicals can induce oxidative stress in human cells, and anthocyanins have been shown to scavenge them. For a long time, the abundance of anthocyanins in wheat grain was not factored in genetic improvement unless it was used as a genetic marker for specific purposes to distinguish it from wheat used for human food, or in functional genomics.[85] The colour of flour and semolina is primarily due to carotenoid accumulation in the wheat grain which is a quantitative trait controlled and influenced by environmental factors. Despite this, the final colour of end products is also influenced by losses during grain storage and carotenoid oxidative degradation by enzymes like polyphenol oxidase, lipoxygenase, and peroxidase during processing.[72] Among 14 genotypes of wheat, barley, and tritordeum cultivated in Czechia in 2014 and 2015, five carotenoids, lutein, zeaxanthin, antheraxanthin, α-, and β-carotene, were measured using HPLC-DAD-MS. Yellow-grained tritordeum HT 439 (12.16 g per g DW) had the maximum carotenoid production, second by blue-grained wheat V1-131-15 (7.46 g per g DW) and yellow-grained wheat TA 4024 (7.04 g per g DW). When assessing carotenoid content, blue types had lower levels, whereas purple variants had the same or higher amounts as standard bread wheat. Lutein was the most abundant carotenoid in wheat and tritordeum, while zeaxanthin was the utmost barley.[60]

10.5.2 Metabolic Engineering

Metabolic engineering improves substrate inflow into a pathway through overexpression of a gene coding a rate-limiting enzyme *via* genetic alteration. Currently, metabolic engineers focus on multiple points of control and feedback amongst the metabolic network to empirically regulate the rate controllers' interactions and other factors. Flavonoids (secondary metabolites) are the pigments that give the flower and many other sections of the plant their non-green colour, and they also act as signaling molecules in a plant to microbe and plant to insect interactions. This compound gained attention due to its structural diversity, biological and ecological importance, and health-promoting effects. Their structure consists of a 15-carbon phenylpropanoid core, two six-carbon aromatic rings (designated as rings A and B) connected by a heterocyclic ring with three carbon atoms (designated as ring C), which can be represented as C6–C3–C6.[86] The deamination of phenylalanine to generate cinnamic acid, which is catalyzed by phenylalanine ammonia lyase, is the first step in the production of flavonoids. Cinnamate 4-hydroxylase involves the conversion of cinnamic acid to 4-coumaric acid, which is transformed to 4-coumaroyl-CoA by 4-coumaric acid ligase.[87] The conversion of cinnamic acid to *p*-coumaroyl-CoA is the rate-limiting step, that affects the production of phenolic acids, flavanones, proanthocyanidins, and anthocyanidins.

The majority of the flavonoids found in black and purple rice are anthocyanins, which are responsible for the purple to blue pigmentation. Anthocyanin synthase catalyzes the oxidation of leucoanthocyanidin to cyanidin, pelargonidin, and delphinidin. The compounds cyanidin-3-*O*-glucoside and peonidin-3-*O*-glucoside are the most prevalent, although other compounds such as cyanidin-3,5-diglucoside, cyanidin-3-*O*-(6″-*O*-*p*-coumaroyl)glucoside, pelargonidin-3-*O*-glucoside, peonidin-3-*O*-(6″-*O*-*p*-coumaroyl)glucoside and cyanidin-3-*O*-arabidoside.[86] Anthocyanin reductase converts unstable anthocyanidins into the colourless flavan-3-ols epiafzelechin, epicatechin, and epigallocatechin, which are then glycosylated to produce a diverse range of compounds.[88]

Glyceraldehyde-3-phosphate and pyruvate serve as early substrates in the plastid-localized MEP route, which leads to the synthesis of geranylgeranyl diphosphate (GGPP), the common precursor for the biosynthesis of carotenoids and various other terpenoid substances.[89–91] The condensation of two GGPP molecules by phytoene synthase (PSY) to form 15-*cis*-phytoene is the first step in the carotenoid biosynthesis pathway. Phytoene desaturase (PDS) and ζ-carotene isomerase (ZDS) catalyze two desaturation processes that convert phytoene to lycopene. These enzymes produce poly-*cis* molecules, which are then transformed to all-*trans* compounds by ζ-carotene isomerase (ZISO) and carotene isomerase (CRTISO), and also light-mediated photo-isomerization.[91]

Due to the inadequacy of natural animal and plant sources rich in essential metabolites, supplementation of diet from natural sources is limited. The enhancement of PUFAs and pigments in the diet has been possible.[56] The most important breeding target for all crops is the harvested mass per unit area (yield). But, data has stated that an increase in genetic yield potential is falling behind the necessary limit to stay constant at global food security. Many agencies state their consensus that doubling the crop yield is necessary by 2050. Surveys have concluded that major crops have failed in achieving a 2.4% rate of crop production growth per annum to secure a doubling of agricultural production by 2050.[13] Roughly one-third, approximately 30% of the entire 2 billion tons per year production of cereals is wasted or lost due to several reasons. Overall losses consider all the food parts lost during the entire supply chain. These parts contribute to the high organic load of wastes, during the processing of raw materials.[56,92,93]

10.6 Summary and Future Perspectives

Pigmented cereals and their parts are one of the bioresources for functional foods that can benefit humankind because of their phytochemicals and associated biological activities. The studies of pigmented cereals have been increased in recent past decades, and they revealed that the colouration of blue, purple, pink and red is due to the presence of anthocyanins. Thus the pigments in the major and minor cereals are due to the accumulation of secondary metabolites. Further, a significant quantity of phenolic acids and flavonoids in the whole cereals (kernel, cob, husk, silk) may be different depending on the plant and other geospatial conditions. The differential accumulation of pigments in the cereals is regulated by various proteins and enzymes involved in the biosynthesis of secondary metabolites such as flavonoids and carotenoids. However, each of these compounds has exhibited biological activities (for instance, anthocyanins have anti-cancer activity, cardioprotective and antiobesity activity; the ferulic acid [phenolic acids] is a potent antioxidant and provides anticancer properties, and flavonoids have antioxidant activity). Though the genes responsible for pigment accumulation in cereals have been studied exclusively, only a few functional markers for genetic breeding have been reported. Therefore, more allele-specific markers should be explored so that the genetic engineering and metabolic strategies for improvising the pigmented cereals could be implemented further for ensuring food and nutritional security.

Numerous bioactive compounds have been detected in pigmented cereals and many others are still unidentified. The role of these secondary metabolites from the pigmented cereals concerning health benefits to humankind cannot be overemphasized. Many under-studied aspects of anthocyanins from pigmented cereals, such as genetics, chemistry, bioprocess technologies, and biological activities, should be explored further for the identification, isolation, and characterization of these compounds for their ultimate applications for the benefit of humankind. More research on pigmented cereal breeding is the need now to improvise the existing varieties on enhancing the product yield, quantity of secondary metabolites and their resistance to diseases and harsh climatic conditions. Several capacity-building training and sensitization programs should be encouraged to transfigure the negative public perception of the low yield of pigmented cereals compared to the conventional cereals for their adoption across the globe. Development and demonstration of optimized processing techniques of these pigmented cereals without losing their economic values has to be ensured for its earlier adoption for everyday use could contribute further to the development of new functional foods for improved nutrition and health benefits. Researchers should demonstrate the necessitate of pigmented cereals because of their higher nutrient content, antioxidant activities, and efficient utilization for the production of high value-added products. The promotion of pigmented cereals shall be facilitated at the grassroots level through nutritionists, food policymakers, and consumers.

Though there is an existing enormous genetic diversity trait of pigmented cereals for enriched flavonoids, the marketing niche for future functional foods should be developed in line with the growing public interest and consumer demand through the collaborative efforts of food scientists, dieticians and food industrialists. The role of functional bioactive components in pigmented cereals should be highlighted as attractive supplemental ingredients in retrofitting the food industry on conventional food products to create novel nutraceutical cereal-based foods. More metabolomic studies on fermented pigmented cereals should be carried out in the future to delineate the relationship between the bioactive components and their interaction with microorganisms. Additional validation research on gut-microbiomics should be promoted to ensure the health benefits of pigmented cereals. Further, the inclusion of emerging technologies such as extrusion-based three-dimensional technologies and micro/nano-encapsulation should be promoted to diversify pigmented cereals-based products without compromising their nutritional and/or sensory attributes. Promotion of wholegrain pigmented cereals consumption shall be leveraged to avert non-communicable diseases despite its poor shelf life and reduced palatability. Exploring the most efficient post-harvesting technologies for processing pigmented cereals without losing its bioactive components has to be encouraged further for the promotion of sustainable development goals of ensuring health security through the positioning of pigmented cereals as staple foods to meet the day-to-day human nutrition needs, which in return shall benefit both urban consumers' health concerns and address the malnutrition aspects of poorer sections of society. Therefore, pigmented cereals are important for developing new functional food products from the grain and obtaining natural colourants and antioxidants from the other parts of the plant.

Conflicts of Interest

The authors declare no conflict of interest to disclose.

Acknowledgements

The authors thanked the National Institute of Technology Rourkela for the necessary facilities.

References

1. R. Schoenlechner, S. Siebenhandl, and E. Berghofer, Pseudocereals, in *Gluten-free Cereal Products and Beverages*, ed. Arendt, E. K., and Bello, F. D., Academic Press, 2008, p. 149.
2. C. Wang, H. Lu, W. Gu, X. Zuo, J. Zhang, Y. Liu, Y. Bao and Y. Hu, *Holocene*, 2018, **28**(5), 738.
3. G. S. Khush, *Plant Mol. Biol.*, 1997, **35**(1), 25.
4. A. Menkir, W. Liu, W. S. White, B. Maziya-Dixon and T. Rocheford, *Food Chem.*, 2008, **109**, 521.
5. T. A. Brown, M. K. Jones, W. Powell and R. G. Allaby, *Trends Ecol. Evol.*, 2009, **24**(2), 103–109.
6. R. E. Evenson and D. Gollin, *Science*, 2003, **300**(5620), 758.
7. J. L. Weisdorf, *J. Econ. surv.*, 2005, **19**(4), 561.
8. R. Thomas, *Int. J. Hist. Archaeol.*, 2005, **9**(2), 71.
9. P. Barta, Feeding billions, a grain at a time, *Wall Street J.*, 2007, 28, A1.

10. P. D. Barsagade, P. Patil and M. J. Umekar, *World J. Pharm. Pharm. Sci.*, 2020, **9**(4), 683.
11. S. Hussain, T. Fujii, S. McGoey, M. Yamada, M. Ramzan and M. Akmal, *J. Anim. Plant Sci.*, 2014, **24**(5), 1504.
12. FAO, FAO Cereal Supply and Demand Brief | World Food Situation | Food and Agriculture Organization of the United Nations. 2022, retrieved from https://www.fao.org/worldfoodsituation/csdb/en/.
13. H. D. Jones, Introduction: Biotechnology of Major Cereals, *Biotechnology of Major Cereals*, 2016.
14. *Cereal Biotechnology*, ed. Morris, P. C., and Bryce, J. H., CRC Press, 2000.
15. S. O. Serna Saldivar *Encyclopedia of Food Sciences and Nutrition*, 2003, p. 1027.
16. G. Zamaratskaia, K. Gerhardt and K. Wendin, *Trends Food Sci. Technol.*, 2021, **107**, 114.
17. S. Muthayya, J. D. Sugimoto, S. Montgomery and G. F. Maberly, *Ann. N. Y. Acad. Sci.*, 2014, **1324**(1), 7.
18. B. T. Oleson. World wheat production, utilization and trade, in *Wheat*, Springer, Boston, MA, 1994, p. 1.
19. B. Ayalew and I. Sekar, *J. Dev. Agric. Econ.*, 2016, **8**(9), 193.
20. M. N. Nagendra Prasad, K. R. Sanjay, M. Shravya Khatokar, M. N. Vismaya and S. Nanjunda Swamy, *J. Nutr. Food Sci.*, 2011, **1**(3), 1.
21. T. S. Rathna Priya, A. R. L. Eliazer Nelson, K. Ravichandran and U. Antony, *J. Ethn. Foods*, 2019, **6**(1), 1–11.
22. F. M. Bhat and C. S. Riar, *Med. Aromat. Plants*, 2015, **4**, 198.
23. G. Pereira-Caro, S. Watanabe, A. Crozier, T. Fujimura, T. Yokota and H. Ashihara, Phytochemical profile of a Japanese black–purple rice, *Food Chem.*, 2013, **141**(3), 2821.
24. P. Beyer, Golden Rice and 'Golden' crops for human nutrition, *New Biotechnol.*, 2010, **27**(5), 478.
25. D. Fujita, A. Yoshimura and H. Yasui, *Theor. Appl. Genet.*, 2006, **113**(4), 567.
26. P. Kumar, R. K. Yadava, B. Gollen, S. Kumar, R. K. Verma and S. Yadav, *Life Sci. Med. Res.*, 2011, **22**(1), 1.
27. W. Li, F. Shan, S. Sun, H. Corke and T. Beta, *J. Agric. Food Chem.*, 2005, **53**(22), 8533.
28. E. S. M. Abdel-Aal and P. Hucl, *J. Agric. Food Chem.*, 2003, **51**(8), 2174.
29. F. S. Hosseinian, W. Li and T. Beta, *Food Chem.*, 2008, **109**(4), 916.
30. R. A. Graybosch, C. J. Peterson, P. S. Baenziger and D. R. Shelton, *J. Cereal Sci.*, 1995, **22**(1), 45.
31. R. Gupta, M. Meghwal and P. K. Prabhakar, *Trends Food Sci. Technol.*, 2021, **110**, 240.
32. G. J. J. Sanchez, M. M. Goodman and C. W. Stuber, *Econ. Bot.*, 2000, **54**(1), 43.
33. G. E. Orhun, *Int. J. Food Sci. Nutr. Eng.*, 2013, **3**(2), 13.
34. R. A. Malvar, P. Revilla, J. Moreno-González, A. Butrón, J. Sotelo and A. Ordás, *Crop Sci.*, 2008, **48**(4), 1373.

35. C. Capurso, *Nutrients*, 2021, **13**(8), 2540.
36. G. Fatma, M. J. Siddiqui, P. Wani and S. Nikhat, *World J. Pharm. Res.*, 2019, **8**(11), 396.
37. Y. Zeng, X. Pu, J. Yang, J. Du, X. Yang, X. Li, L. Li, Y. Zhou and T. Yang, *Oxid. Med. Cell. Longevity*, 2018, 3232080.
38. E. S. M. Abdel-Aal, T. M. Choo, S. Dhillon and I. Rabalski, *Cereal Chem.*, 2012, **89**(4), 198.
39. Y. Zhang, Y. Lin, L. Huang, M. Tekliye, H. A. Rasheed and M. Dong, *LWT–Food Sci. Technol.*, 2020, **125**, 109181.
40. S. Birhanu, *Int. J. Food Eng. Technol.*, 2021, **5**(1), 16.
41. M. J. McGinnis, and J. E. Painter, *Nutrition Today*, 2020, vol. 55(1), p. 38.
42. L. Dykes, L. M. Seitz, W. L. Rooney and L. W. Rooney, *Food Chem.*, 2009, **116**(1), 313.
43. S. Malik, *Int. J. Adv. Res. Innovative Ideas Educ.*, 2015, **1**(3), 414.
44. V. Tripathi, A. S. Mohd and T. Ashraf, *Pharmacogn. Rev.*, 2018, **12**(23), 66.
45. H. Adlercreutz, *Lancet Oncol.*, 2002, **3**(6), 364.
46. W. Dedio, R. D. Hill and L. E. Evans, *Can. J. Plant Sci.*, 1972, **52**(6), 981.
47. E. Rodríguez, J. L. Arqués, R. Rodríguez, M. Nuñez, M. Medina, T. L. Talanico, I. A. Casas, T. C. Chung, W. J. Dobrogosz, L. Axelsson, S. E. Lindgren, L. Kerkeni, P. Ruano, L. L. Thin, S. Picco, L. Villegas, F. Tonelli, M. Blackbird and M. Masuelli, *Intech*, 1989, **32**, 137.
48. S. Ghosh, T. Sarkar, A. Das and R. Chakraborty, *LWT–Food Sci. Technol.*, 2022, **153**, 112527.
49. P. D. Matthews, and E. T. Wurtzel *Food Colorants: Chemical and Functional Properties*, 2007, p. 34810.
50. K. Okada, T. Saito, T. Nakagawa, M. Kawamukai and Y. Kamiya, *Plant Physiol.*, 2000, **122**(4), 1045.
51. D. Trono, *Plants*, 2019, **8**(12), 551.
52. B. Enaru, G. Dreţcanu, T. D. Pop, A. Stǎnilǎ and Z. Diaconeasa, *Antioxidants*, 2021, **10**(12), 1967.
53. A. Francavilla and I. J. Joye, *Nutrients*, 2020, **12**(10), 2922.
54. M. Certik and S. Shimizu, *J. Biosci. Bioeng.*, 1999, **87**(1), 1.
55. T. Klempova, E. Basil, A. Kubatova and M. Certik, *Biotechnol. J.*, 2013, **8**(7), 794.
56. M. Čertík, Z. Adamechová and L. Guothová, *J. Biotechnol.*, 2013, **168**(2), 130.
57. D. Boros, *Biuletyn Instytutu Hodowli i Aklimatyzacji Roślin*, 2015, **277**, 7.
58. M. Nogala-Kałucka, A. Kawka, K. Dwiecki and A. Siger, *Acta Sci. Pol., Technol. Aliment.*, 2020, **19**(4), 405.
59. J. M. PihlavaSelected Bioactive Compounds in Cereals and Cereal Products–Their Role and Analysis by Chromatographic Techniques, *PhD Thesis*, University of Turku, Finland, 2017.
60. L. Paznocht, Z. Kotíková, M. Šulc, J. Lachman, M. Orsák, M. Eliášová and P. Martinek, *Food Chem.*, 2018, **240**, 670.
61. D. B. Ficco, V. De Simone, S. A. Colecchia, I. Pecorella, C. Platani, F. Nigro, F. Finocchiaro, R. Papa and P. De Vita, *J. Agric. Food Chem.*, 2014, **62**(34), 8686.

62. T. Oikawa, H. Maeda, T. Oguchi, T. Yamaguchi, N. Tanabe, K. Ebana, M. Yano, T. Ebitani and T. Izawa, *Plant Cell*, 2015, **27**(9), 2401.
63. B. O. Juliano, *Rice Chemistry and Quality*, Philippine Rice Research Institute, Manila, Philippines 2003, p. 25.
64. U. K. Kushwaha *Black Rice: Research, History and Development*. Springer, 2016.
65. V. C. Ito and L. G. Lacerda, *Food Chem.*, 2019, **301**, 125304.
66. R. Sompong, S. Siebenhandl-Ehn, G. Linsberger-Martin and E. Berghofer, *Food Chem.*, 2011, **124**(1), 132.
67. J. L. Mau, C. C. Lee, Y. P. Chen and S. D. Lin, *LWT–Food Sci. Technol.*, 2017, **75**, 434.
68. D. K. Verma and P. P. Srivastav, *Trends Food Sci. Technol.*, 2020, **97**, 355.
69. J. Li, S. S. Lim, J. Y. Lee, J. K. Kim, S. W. Kang, J. L. Kim and Y. H. Kang, *J. Nutr. Biochem.*, 2012, **23**(4), 320.
70. F. Colombo, C. Di Lorenzo, K. Petroni, M. Silano, R. Pilu, E. Falletta, S. Biella and P. Restani, *Foods*, 2021, **10**(8), 1770.
71. G. M. Borrelli, A. M. De Leonardis, C. Platani and A. Troccoli, *J. Cereal Sci.*, 2008, **48**(2), 494.
72. D. B. Ficco, A. M. Mastrangelo, D. Trono, G. M. Borrelli, P. De Vita, C. Fares, R. Beleggia, C. Platani and R. Papa, *Crop Pasture Sci.*, 2014, **65**(1), 1.
73. D. C. Knievel, E. S. Abdel-Aal, I. Rabalski, T. Nakamura and P. Hucl, *J. Cereal Sci.*, 2009, **50**(1), 113.
74. P. A. Parizad, J. Capraro, A. Scarafoni, F. Bonomi, M. Blandino, M. Marengo, D. Giordano, A. Carpen and S. Iametti, *Plant Foods Hum. Nutr.*, 2019, **74**(1), 128.
75. R. E. Asenstorfer, Y. Wang and D. J. Mares, *J. Cereal Sci.*, 2006, **43**(1), 108.
76. D. A. Urias-Lugo, J. B. Heredia, M. D. Muy-Rangel, J. B. Valdez-Torres, S. O. Serna-Saldívar and J. A. Gutierrez-Uribe, *Plant Foods Hum. Nutr.*, 2015, **70**(2), 193.
77. J. Li, S. S. Lim, J. Y. Lee, J. K. Kim, S. W. Kang, J. L. Kim and Y. H. Kang, *J. Nutr. Biochem.*, 2012, **23**(4), 320.
78. G. G. Bellido and T. Beta, *J. Agric. Food Chem.*, 2009, **57**(3), 1022.
79. E. S. M. Abdel-Aal, J. C. Young and I. Rabalski, *J. Agric. Food Chem.*, 2006, **54**(13), 4696.
80. M. Garg, M. Chawla, V. Chunduri, R. Kumar, S. Sharma, N. K. Sharma, N. Kaur, A. Kumar, J. K. Mundey, M. K. Saini and S. P. Singh, *J. Cereal Sci.*, 2016, **71**, 138.
81. X. Wei, and X. Huang, Origin, taxonomy, and phylogenetics of rice, in *Rice*, AACC International Press, 2019, vol. 1, p. 1.
82. T. L. Jeng, C. C. Lai, P. T. Ho, Y. J. Shih and J. M. Sung, *J. Cereal Sci.*, 2012, **56**(2), 425.
83. K. Murase, K. L. Morrison, P. Y. Tam, R. L. Stafford, F. Jurnak and G. A. Weiss, *Chem. Biol.*, 2003, **10**(2), 161.
84. P. C. Lee, B. N. Mijts and C. Schmidt-Dannert, *Appl. Microbiol. Biotechnol.*, 2004, **65**(5), 538.

85. J. Shipp and E. S. M. Abdel-Aal, *Open Food Sci. J.*, 2010, **4**(1), 7.
86. E. G. N. Mbanjo, T. Kretzschmar, H. Jones, N. Ereful, C. Blanchard, L. A. Boyd and N. Sreenivasulu, *Front. Genet.*, 2020, **11**, 229.
87. A. X. Cheng, X. J. Han, Y. F. Wu, and H. X. Lou, *Int. J. Mol. Sci.*, 2014, **15**(1), 1080.
88. A. Ghasemzadeh, M. T. Karbalaii, H. Z. Jaafar and A. Rahmat, *Chem. Cent. J.*, 2018, **12**(1), 1.
89. M. Rodríguez-Concepción, I. Ahumada, E. Diez-Juez, S. Sauret-Güeto, L. M. Lois, F. Gallego, L. Carretero-Paulet and A. Boronat, *Plant J.*, 2001, **27**(3), 213.
90. G. Farré, G. Sanahuja, S. Naqvi, C. Bai, T. Capell, C. Zhu and P. Christou, *Plant Sci.*, 2010, **179**, 28.
91. S. Zhai, X. Xia and Z. He, *Front. Plant Sci.*, 2016, **7**, 1197.
92. J. Gustafsson, C. Cederberg, U. Sonesson, and A. Emanuelsson, *The Methodology of the FAO study: Global Food Losses and Food Waste-Extent, Causes and Prevention"-FAO*, 2011.
93. M. Verni, C. G. Rizzello and R. Coda, *Front. Nutr.*, 2019, **6**, 42.

CHAPTER 11

Anthocyanins in Colored Cereals: Measurement and Composition

JASMEET KOUR(0000-0001-9182-7596)*[a], MALIHA AFREEN[b], LISA F. M. LEE NEN THAT(0000-0003-0808-4547)[c], İLKNUR UÇAK[d], NAVEEN ANAND[e] AND JESSICA PANDOHEE (0000-0003-4343-7192)[f]

[a]Padma Shri Padma Sachdev, Govt PG College for Women Gandhi Nagar, Jammu-180003, Jammu & Kashmir, India; [b]Department of Animal Production and Technologies, Nigde Omer Halisdemir University, Faculty of Agricultural Sciences and Technologies, 51240, Nigde, Turkey; [c]School of Science, RMIT University, PO Box 71, Bundoora, Victoria 3083, Australia; [d]Department of Animal Production and Technologies, Nigde Omer Halisdemir University, Faculty of Agricultural Sciences and Technologies, 51240, Nigde, Turkey; [e]Dean Academic affairs, Cluster University of Jammu, India; [f]Centre for Crop and Disease Management, School of Molecular and Life Sciences, Curtin University, Bentley, WA 6102, Australia
*E-mail: eru.jassi@gmail.com

11.1 Introduction

Anthocyanins are small biochemical metabolites belonging to a class of compounds called flavonoids. They are water soluble pigments and are responsible for the red, blue, purple and black colors in flowers, fruits, vegetables and several grains.[1] Foods that are rich in anthocyanins include berries

Food Chemistry, Function and Analysis No. 38
Pigmented Cereals and Millets: Bioactive Profile and Food Applications
Edited by Sneh Punia Bangar, Sajid Maqsood and Anil Kumar Siroha

Published by the Royal Society of Chemistry, www.rsc.org

and selected varieties of cereals. To date, over 500 anthocyanins have been reported.[2,3] These are mainly 3-glycoside anthocyanins. Plants produce four anthocyanins through a defence mechanism triggered by abiotic or biotic stress *via* the phenylpropanoid pathway.[5] In plants under environmental stress or pathogen attack, anthocyanins exhibit their antioxidative mechanism by entrapping reactive oxygen species.

Anthocyanins are believed to impart several health benefits.[6] Being vacuolar pigments pertaining to the class of flavonoids, anthocyanins derived from fruits and vegetables strongly correlate with various health benefits.[7] Considering the diversity of the anthocyanin group, little is known on the quantity needed to achieve beneficial health outcomes. Standard cereals have been consumed since the beginning of civilization for their high calorie content. They have been cultivated to a considerable extent, despite being less resistant to environmental and biotic challenges still producing a large amount of food for a rapidly growing population. Amid this, colored cereals are foods that will change the way we eat and look at cereal grains. While standard cereals are adequate in providing the daily energy requirement, they often lack phytochemicals and other bioactive compounds, such as anthocyanins. The consumption of colored cereals creates the opportunity to achieve the desired health benefits from anthocyanins as well as delivering the calories needed for daily energy requirements. Chemical profiling of colored cereals has been applied to characterize anthocyanin content apart from identification and quantification of anthocyanins. Various nutritional interventions investigating the effect of high anthocyanin-containing foods on disease have reported numerous valuable findings.

Anthocyanins are a group of phenolic compounds that are water soluble and give color to cereals, vegetables and fruits. Owing to consumers' unending demand for natural products, synthetic dyes are being replaced by natural colors imparted by these anthocyanins to a considerable extent in the food marketplace. According to a report in 2011, worldwide sales of natural dyes reached 600 million dollars annually, and for the first time their commercial worth increased more than that of artificial dyes.[8] The search for natural colorants is a challenge in the food industry. The high content of anthocyanin in colored cereals offers an appealing solution. These natural colorants are abundant in anthocyanins, meaning that the extract of purple corn is extensively used in coloring beverages and desserts such as mazamorra morada and chicha.[9] Blue, purple and black grains are produced for creating special foods and for ornamentation, owing to their colorful look.

Flavonoids are a group of anthocyanins which are water soluble, with a reddish to purple color being the primary colorants in these grains.[10–12] Anthocyanins have been known as health-promoting elements due to their antioxidant, anticancer, anti-inflammatory, and hypoglycemic properties.[13–18] Anthocyanin-containing cereals can be used as functional foods or as beneficial food colorants. Anthocyanins originate in nature in the form of methoxylated heterosides and polyhydroxylated, which are derived from the flavylium ion or 2-phenylbenzopyrilium.

The occurrence of many sugar molecules and the presence of hydroxyl groups on the sugar rings, make these compounds more feasible to dissolve in ethanol and water.[19] This chapter will shed light on various techniques that are used to measure and elucidate the composition analysis of anthocyanins in various colored cereals, apart from describing their benefits for human health.

11.2 Measurement

11.2.1 Extraction

Extraction is the removal of active components such as alkaloids, glycosides, phenolics, terpenoids and flavonoids of plant material including leaf, root, stem and fruit using selective techniques. For this purpose, both fresh and dried samples can be used. In most cases, the dried sample is preferred, as it is easier to handle than the fresh sample. The plant material under study is well ground using a mortar and pestle, electrical blender or ball mill, to reduce the particle size. It is then subjected to extraction to get the active components. Besides genetic factors, it is the physiology and environment in which cereals are grown, including the extraction and quantification methods employed, that deeply impact on anthocyanin composition.[20–22] In a few significant findings, formic acid solution (2%) has proven to be better solvent for the extraction of condensed and acylated anthocyanin pigments from maize.[21] Besides conventionally employed solvent extraction techniques, advanced methods such as ultrasound and microwave-assisted extractions (MAE) can also be exploited to extract these pigments.[23,24]

Suriano *et al.*[25] conducted a pivotal work on determining total phenolic content, flavonoids, anthocyanins, lipophilic phenolic antioxidants and carotenoids in red, yellow, blue and purple colored maize. This study characterized the anthocyanin profile as well as comparing the bioactive phytochemical substances along with the emergence of maize genotypes with new pigments. Cyanidin, petunidin, pelargonidin and peonidin-3-glucoside and their derivatives were the major anthocyanins detected, out of which the highest concentration was reported in purple kernel genotype. The extractability of anthocyanins from sorghum is a reflection of its composition, type of solvent used and the temperature used in extraction.[20]

Beta *et al.*[26] extracted phenolic compounds from purple wheat by using the methodologies that are used conventionally in the extraction of phenolic compounds from whole grains and are specifically applicable to colored wheat. This study concluded with the significant finding that aqueous alcohol (80% [v/v]), aqueous ethanol (80% [v/v]) and aqueous methanol can be used for the extraction of free soluble phenolic acids. Phenolic acids are polar (due to the carboxylic acid functional group) and hydroxyl substituents on the benzene ring, therefore it is expected that they will require polar organic solvents in order to dissolve and be extracted from the whole grain matrix.

Extraction of fat-free bran from wheat (*Triticum aestivum*) with a dark blue color was done by using ethanol followed by characterization by high-performance liquid chromatography (HPLC) and electrospray ionization mass spectrometry (ESI-MS). The separation of anthocyanins was done by phenolic acids, using solid phase extraction. The antioxidant activity of the fractions was evaluated by performing DPPH radical, ABTS radical scavenging tests and oxygen radical absorption assay. The most predominant anthocyanin reported was cyanidin-3-glucoside, apart from cyanidin-3-galactoside, pelargornidin-3-glucoside, and peonidin-3-glucoside. Some major simple phenolic acids were also identified with low concentration of *p*-hydroxybenzoic acid, caffeic acid, syringic acid and *p*-coumaric acid. The free radical scavenging capacity of dark blue wheat was attributed more to the anthocyanin content than the extractable phenolic acids.[27]

Apart from this, the extraction ability of anthocyanins from the brans of purple and black rice increased by up to ~50% by using an ultrasound-assisted process.[24] Accelerated solvent extraction has been reported to induce negligible structural changes in anthocyanins.[28] Varga *et al.*[29] evaluated the phenolic profile and antioxidant capacity of 20 oat genotypes with different colors of hull. Soluble phenolics were extracted from groats and hulls by using the methodology of Verardo *et al.*[30] with some modifications. For the extraction, sonication was done using methanol/water (3 ml) (80/20, v/v) as a solvent for 10 min at room temperature, followed by centrifugation at 8092 *g* for 5 min. The combined supernatants were evaporated by centrifugal evaporation. The hull fractions obtained exhibited an increased antioxidant activity (AOA) in comparison to groats.[30]

Suriano *et al.*[31] investigated free, soluble conjugated, and bound phenolics along with proanthocyanidins and carotenoids, in 20 genotypes of colored barley. The extraction of the total polyphenolic content was done by using the methodology of Hosseinian *et al.*[32] with slight modifications. The wholemeal samples (0.5 g) were subjected to extraction with methanol (8 ml) acidified with 1 N HCl (85 : 15; v/v) in centrifugation tubes followed by sonication for 18 min at room temperature in an ultrasonic bath. The maximum proportion of the extracts obtained were composed of insoluble bound phenolic acids with ferulic acid as the main conjugated phenolic acid. The free phenolic acids were composed of salicylic and gallic acids.

Li *et al.*[7] compared the distribution of anthocyanins in coproducts prepared from purple and blue corn from three conventional corn fractionation techniques. The extract, rich in total monomeric anthocyanin, obtained from purple corn was 4933.1 ± 43.4 mg of cyanidin-3-glucoside equivalent per kg dry corn. This content was found to be higher than in the blue corn, which had the highest total anthocyanin content in small grits. The anthocyanin content was the highest in steeping water for coproducts prepared from purple corn.

Abdel-Aal *et al.*[33] conducted significant work on extracting anthocyanins from anthocyanin-rich powder by processing bran fraction using ethanol-based extraction and column purification. Anthocyanins obtained were characterized with a triple quadrupole mass spectrometry, which reported a rich anthocyanin profile comprising five prominent pigments: cyanidin-3-glucoside, cyanidin-3-(6-malonyl glucoside), cyanidin-3-rutinoside, peonidin-3-glucoside and peonidin-3-(6-malonylglucoside). This constituted a significant proportion of total anthocyanin content. Extraction was carried out with acidified ethanol and the extracts were concentrated by using a rotary evaporator at 40 °C.

Shao *et al.*[34] extracted free phenolics from rice flour using the method of Qiu *et al.*[35] with few modifications using 80% methanol (20 mL) in duplicates. Defatting of the concentrated free fractions was done with hexane (10 mL). Extraction was carried out three times using ethyl acetate (70 mL). This significant work unfolded the prevalence of phenolic compounds and anthocyanin pigments in the endosperm, embryonic layer and bran portion of white, red and black colored rice. Bran reportedly exhibited the highest total phenolic content (TPC) (7.35 mg GAE per g), comprising the maximum proportion of phenolics in white, red and black rice.

Paucar-Menacho *et al.*[36] optimized time of germination and temperature to enhance the content of bioactives, as well as the antioxidant activity of purple corn (*Zea mays* L.), which was selected owing to its high anthocyanin content and rich phenolic profile. Proper cleaning of the seeds was done, followed by storage in dark conditions away from heat and light in polyethylene containers at 4–8 °C until they germinated. The samples were analysed by high performance liquid chromatography equipped with a reverse phase C18 column (5 mm, 150 mm, 4.6 mm id) at 35 °C, followed at 520 nm, according to Alcalde-Eon *et al.*[37] Different concentrations of cyanidin 3-*O*-glucoside, peonidin 3-*O*-glucoside and pelargonidin 3-*O*-glucoside were injected for the quantification of anthocyanins. Detection of twenty-one anthocyanin pigments was reported in raw and germinated purple corn. Prominent anthocyanin pigments were cyanidin, pelargonidin derivatives, as shown in their UV-vis spectra and mass spectral data. Some of the pigments included cyanidin 3,5-diglucose, cyanidin 3-glucoside, pelargonidin 3-glucoside, peonidin 3-glucoside and delphinidin 3-glucoside.

Defatting of the milled fractions, debranned fractions and cooked pasta followed by freeze drying was done by treatment with petroleum ether, which led to a high rate of extraction of anthocyanins. Extraction was done by employing the solvent extraction technique for a time period of 16 h at room temperature with continuous stirring. Solvent composition was ethanol (65 mL), water (35 mL) and HCl (0.1 mL). Centrifugation of the mixture was done at 10 000 *g* for a period of 10 min. The supernatant was obtained, while the residue was subjected to extraction twice. The three supernatants obtained in the three extraction steps were combined. The measurement of the total monomeric anthocyanins was achieved by using the pH differential method.[38]

Kurambhatti *et al.*[39] conducted a significant study on the determination of anthocyanin concentration from purple corn by using the method outlined by Hu *et al.*[27] and Wang *et al.*[40] with slight modifications. The sample was extracted with acidified ethanol (mixture of 95% v/v ethanol and 1 N HCl in 85 : 15 v/v ratio) with vortexing at room temperature for 1 h. Filtration of the mixture was done under vacuum by using Whatmann No. 4 filter paper followed by analysis for anthocyanin concentration. Solid residue was extracted using acidified ethanol for determining remaining anthocyanin concentration. Solids recovered from the second extraction were mixed with acidified ethanol to extract anthocyanins. The anthocyanins extracted were measured using the pH differential method.[41] The extracts obtained were subjected to dilution using buffer solution 0.25 M KCl and 0.4 M sodium acetate buffer and their absorbances were measured at 520 and 700 nm using UV/Vis microplate spectrophotometer. This study also shed light on the water-based extraction of anthocyanins from the pericarp of purple corn by treating the pericarp with deionized (DI) water at 32 °C, followed by centrifugation of the mixture at 150 rpm. Extraction was carried out at 1, 3, 6, 9, 12 and 24 h to evaluate the effect of residence time on extraction. To investigate the residence times, high sample loading experiments were carried out using a water-based extract rich in anthocyanins. The effect of pH on anthocyanin extraction was also evaluated by performing extraction at pH 1, 4 and 7 (adjusted with 1 M HCl) at 1, 3, 6, 9, 12 and 24 h under similar conditions.

Water-based anthocyanin recovery using ground purple corn (4 g) was mixed with 7 mL DI water (~33% w/w solid loading) at 32 °C with continuous mixing at 150 rpm. Anthocyanin extraction was performed for 1, 3, 6 and 12 h to determine optimum extraction time; anthocyanin measurement was performed similarly to previous treatments.

Teixeira *et al.*[42] conducted pressurized liquid extraction for obtaining the anthocyanin-rich extract using a self-assembled instrumented unit. In this procedure, 5 g of broken black beans (*Phaseolus vulgaris* L.) and 36 g of glass spheres were placed inside the extraction column, forming a fixed bed. Flow rate of the solvent was 100 bar for 26 min. Optimization of the processing conditions was done following a Box-Behnken design. A set of 15 experiments was conducted. The maximum yields of anthocyanins from ethanol:citric acid extracts of broken black bean hulls were obtained under the optimized conditions. Apart from this, the recovery rate of anthocyanin and other antioxidant compounds was on a higher level in comparison to maceration and ultrasound-assisted extraction. The predominant anthocyanin determined in anthocyanin-rich extracts was delphinidin-3-glucoside, which has exhibited various health-promoting activities, primarily antiradical and antidiabetic ones.

Hong *et al.*[43] extracted anthocyanins from three rows of kernels from each cob of purple pericarp sweetcorn. Extraction of the anthocyanins was done using 80 v/v % aqueous methanol acidified with 0.1 M hydrochloric acid as an extraction solvent. The sample was taken in a Falcon tube, with three technical replicates. Stock solution of delphinidin-3-glucoside (120 ml) was used as an internal standard. Sonication of the mixture was carried out at 4 °C for 10 min followed by vigorous shaking at 250 rpm min^{-1} for 10 min at 4 °C. Centrifugation was carried out at 4000 rpm for 10 min at 4 °C. Mobile phase A consisted of acetonitrile, deionized water and formic acid in the ratio of 92 : 7 : 1 (v/v/v), while formic acid (1%) in acetonitrile consisted of mobile phase B. The combined supernatants were filtered through a hydrophilic polytetrafluoroethylene syringe filter into a UPLC vial for further analysis.

Herrman *et al.*[44] determined the relative stability and extractability of sorghum 3-deoxyanthocyanins (3-DXA) *vs.* anthocyanins from maize and cowpea by using a MAE technique. The 3-DXA remained structurally stable to

MAE conditions up to 1200 W/100 °C/30 min. Powdered samples were given microwave energy at 300, 600 or 1200 W in a microwave-accelerated reaction system for 1–30 min. Solvent to sample ratio was kept at 10 : 1. Extraction was carried at 100 °C. The mixture was centrifuged, followed by the collection of supernatant and storage at −20 °C until further analysis. To determine the effect of microwave energy on the structure of 3-DXA, apigeninidin and luteolinidin standards (1.75 uM), along with the luteolinidin analog, cyanidin, were similarly subjected to microwave energy (1200 W) for up to 30 min at a maximum temperature of 100 °C. Compared with conventional extraction, microwave conditions resulted in significantly improved extraction efficiency (yield and speed) of 3-DXA from sorghum. In general, pigment yield increased with increasing extraction time at all MAE power levels used. For maize, MAE did not increase pigment yield compared with conventional extraction under any MAE condition used. In contrast to maize, for cowpea, relatively mild MAE conditions (300 W/5–10 min, 600 W/5 min, and 1200 W/1 min) significantly increased pigment yield compared with conventional extraction. In contrast to the sorghum 3-DXA profile, the anthocyanin profile of maize and cowpea was dramatically affected by MAE compared with conventional extraction. For example, in the maize pigment extract, an unusual presence of cyanidin aglycone was observed in the 1 min/1200 W MAE extract which was absent in the conventional extract. The poor stability of anthocyanin glycosides to MAE was prominently visible in the cowpea extract, where all the major anthocyanins in the conventional extract were glycosides. MAE appears to be uniquely suitable for significantly improving the extraction efficiency and yield of sorghum 3-DXA pigments. This is a new and highly relevant finding because attempts to improve extraction efficiency of anthocyanins (3-DXA analogs in other plants) with MAE have not been successful, largely due to their poor structural stability under MAE. The poor stability of anthocyanins is confirmed in our side-by-side comparative study. MAE effectively improved extractability. This particular method for extraction of anthocyanins and related phenolic compounds has experienced an upsurge in recent years. It is not very efficient in the extractability of anthocyanins from plant tissue due to the structural instability of anthocyanins to microwave energy. This study investigated the effect of microwave energy on structural stability and extractability of 3-DXA from sorghum grain in relation to anthocyanin analogs from monocot and dicots.

11.2.2 Separation of Anthocyanins

Anthocyanin concentration in cereals can vary, even in similar species. Therefore, to quantify the exact amount of anthocyanin in cereal varieties, many extraction methods can be used. The most popular is total anthocyanin content, in which pigments are extracted with precise acidic alcoholic solution and then quantity can be measured by spectrophotometer.[45] The pH differential technique was also mostly used to calculate anthocyanin in dark-colored maize.[46–48] Additionally, high-performance liquid chromatography and ultra-performance liquid chromatography methods have been used for quantification and composition of anthocyanin compounds.[49–51]

11.2.2.1 Total Anthocyanin Method

This method was first introduced in 1968,[52] and is still used today, for quick determination of anthocyanin, specifically in calculating anthocyanin quantity in cereals like in many types of maize cultivar.[53–56] Concisely, anthocyanin compounds could be extracted from powder of purple corn by using 85 : 15 (v/v) solution of 95% ethanol and 1.5 N HCl. Individual powders are soaked in the extraction solvent at 4 °C overnight. Then the solution is filtered to get extract in an identified volume and quantity of anthocyanin from extract is measured by using Shimadzu UV-2450, UV-visible spectrophotometer at a wavelength of 535 nm. Readings should be noted three times. This method is popular due to its simplicity. Anthocyanins can be calculated by the following formula:

$$\text{Total anthocyanins}\left(\text{mg per } 100\,\text{g}\right) = \frac{100 \times A \times \text{DF} \times V}{98.2 \times x}$$

where, 100/98.2 = A constant that takes the extinction coefficient and unit conversions into consideration
A = Absorbance of sample at 535 nm
DF = Dilution factor (for example, if 0.5 ml sample was diluted to 3 ml, DF = 3/0.5 = 6)
V = The known volume anthocyanin extract was made up to after extraction (ml)
x = The weight of purple corn powder for extraction (g).

11.2.2.2 High-performance Liquid Chromatography

High-performance liquid chromatography is a useful method for quantification of anthocyanin, as it gives the exact quantity of each separated compound. Although HPLC is a sophisticated technique, it is time-consuming compared with a simple spectrophotometric technique.[57] Preparation of the sample is done prior to the quantification of anthocyanin through HPLC. To quantify anthocyanins in purple corn, 2 ml of the crude extract of each purple corn powder was passed through C18 Sep-Pak cartridge to eliminate acids, sugars and extra non-phenolic compounds. When different quantities of cyanidin-3-glucoside chloride solution from purple corn were injected into HPLC, different quantities of colorants give different peaks and a standard curve is obtained. Then, anthocyanins are quantified in every sample by comparing peak regions of purple corncob with standard curve. Anthocyanin chromatogram of purple corn was obtained at a wavelength of 520 nm with mass spectrometer.

11.2.2.3 Ultra-performance Liquid Chromatography

This is another advanced method through which anthocyanin extraction can be done by using Waters Acquity Ultra Performance™ liquid chromatography system, comprising a quaternary pump system. Mobile phase consisted of aqueous formic acid (eluent A) and HPLC grade acetonitrile (eluent B). The chromatogram obtained revealed the composition of anthocyanins and the pattern of the peaks in advanced colored wheat lines. The compounds separated were pelargonidin 3-glucoside, cyanidin 3-glucoside, cyanidin 3-rutinoside, peonidin 3-galactoside and cyanidin chloride.[58]

11.2.3 Characterization of Anthocyanins

The amount of anthocyanin compounds varies considerably in red, purple, pink, blue, white and black grains. It was observed that red, brown and black sorghum contained significantly different amounts of anthocyanins, whereas red, brown and black sorghum have the highest anthocyanin concentrations.[59] Different types of black rice also have different concentrations of anthocyanins.[60] It was identified that black rice has the highest amount of anthocyanins among all other colored grains, as it is approximately thirty-five times more than red rice, while very few anthocyanins were observed in wild rice.

A broad range of anthocyanin content was observed in colored corn grains with pink, red, blue and purple colors with minimum value of 51 $\mu g\ g^{-1}$ and a maximum of 1277 $\mu g\ g^{-1}$. Higher anthocyanin content was found in purple corn compared with sweet scarlet red corn and shaman blue corn. Anthocyanin concentrations in blue wheat species was found to vary from 35–507 $\mu g\ g^{-1}$.[61]

11.3 Composition

11.3.1 Rice

The natural color of common rice is brown, owing to its bran, however, after it is polished the consumed grain is white. Reported colors of rice grains include red, black and purple.[62] Pramai and Jiamyangyuen[63] found that Thai black rice contained an anthocyanin content ranging between 21.5 and 441 mg per 100 g. Higher anthocyanin levels in black rice was also reported, compared with red rice. Additionally, the comparison of 25 purple rice cultivars identified cyanidin-3-glucoside, peonidin-3-glucoside, cyanidin-3-galactoside, cyanidin-3-rutinoside, cyanidin and peonidin as principal anthocyanins.[64] Similarly, LC-MS characterization of purple rice has reported that the main anthocyanins include cyanidin-3-glucoside, peonidin-3-glucoside and cyanidin chloride, with their range falling between 42 and 271 mg per 100 g.[65] Compared with other cereals, colored rice contains a lower proportion of anthocyanins.[53]

11.3.2 Wheat

Wheat exists in various colors such as white, yellow, red and purple. White and red wheat varieties do not contain anthocyanins. Interestingly, these are most commonly consumed varieties. In a comparative study, Liu *et al.*[66]

demonstrated that purple, red and yellow wheat had antioxidant properties, with purple wheat being the most bioactive, whereas white wheat had no antioxidant properties. Characterization of these wheat varieties proved that purple wheat has the most antioxidant properties due to these anthocyanins, while the other colored wheat did not contain anthocyanins except for phenolic acids, flavones and flavonols, which were also found in purple wheat. The main anthocyanin in purple wheat was cyanidin-3-glucoside (4 mg g^{-1}). Other anthocyanins identified in purple wheat include peonidin-3-glucoside (2 mg g^{-1}), cyanidin-malonyl-glucose (1 mg g^{-1}) and cyanidin-succinyl-glucose (1 mg g^{-1}), along with peonidin-malonyl-glucose, peonidin-succinyl-glucose and peonidin-malonyl-succinyl-glucose.[67] The total anthocyanin content of purple wheat varies from 10 to 305 mg kg^{-1}.[58,68] Anthocyanins have also been detected in black wheat varieties.[68] However, as black wheat is rare and the amount of anthocyanins was relatively low, anthocyanins in black wheat are under-studied. Garg *et al.*[69] identified cyanidin, delphinidin, pelargonidin, petunidin, malvidin and peonidin when studying black wheat's composition.

11.3.3 Barley

The color of barley varies from yellow to amber orange. Barley exists in various colors such as white, purple, blue and black.[70] Colored barley with antioxidant properties include black hulled, purple hulled, black unhulled, blue unhulled and purple unhulled barley.[71] Among these, purple and blue barley contain a higher level of anthocyanidins and proanthocyanidins than black barley. Interestingly, and similarly to wheat varieties, cyanidin 3-glucoside was the main anthocyanin in purple barley, whereas delphinidin 3-glucoside was the principal anthocyanin in black and blue barley.[72] Lee *et al.*[73] further confirmed that blue and purple barley consisted of anthocyanins with peonidin derivatives at a level of 77 mg kg^{-1} of grain. In general, it was found that the anthocyanin level in unhulled barley is higher than in hulled grains.[71]

11.3.4 Maize

Anthocyanin composition in maize has been well studied, identified and quantified.[33] A study by E. M. Abdel-Aal, Young and Rabalski[53] showed that blue, pink, purple and red maize contained 18 to 27 anthocyanin compounds. Red maize had 27 anthocyanin compounds, which was the highest number of anthocyanins present in pigmented maize, followed by purple maize. However, the highest quantity of anthocyanin was in purple maize (965 μg g^{-1}).[53,74] Other studies reported that blue and red maize had high anthocyanin content.[75] Anthocyanin content was too low or could not be detected in white, yellow and orange maize.[55,56] The anthocyanins detected in maize were cyanidin, delphinidin, peonidin, pelargonidin and malvidin.[22,76,77]

These anthocyanin compounds were mostly in acylated form as they had malonyl, succinyl derivatives or isomers.[53] The most abundant anthocyanin in pigmented maize was cyanidin-3-glucoside.[53,75] In purple maize, 6 anthocyanins were detected and cyanidin 3-*O*-β-D-glucoside was the main anthocyanin.[49] The 5 other anthocyanins included: pelargonidin 3-*O*-β-D-glucoside, peonidin 3-*O*-β-D-glucoside, cyanidin 3-*O*-β-D-(6-malonyl-glucoside), pelargonidin 3-*O*-β-D-(6-malonyl-glucoside) and peonidin 3-*O*-β-D-(6-malonyl-glucoside).[22,49,74,78] Derivatives of cyanidin represented 70% of the seed.[49] Nine anthocyanins were detected in blue maize and seven were malonyl or succinyl derivatives of cyanidin.[75] There were 2 sets of isomers of cyanidin disuccinyl-glucoside and cyanidin malonyl-glucoside (Scheme 11.1).

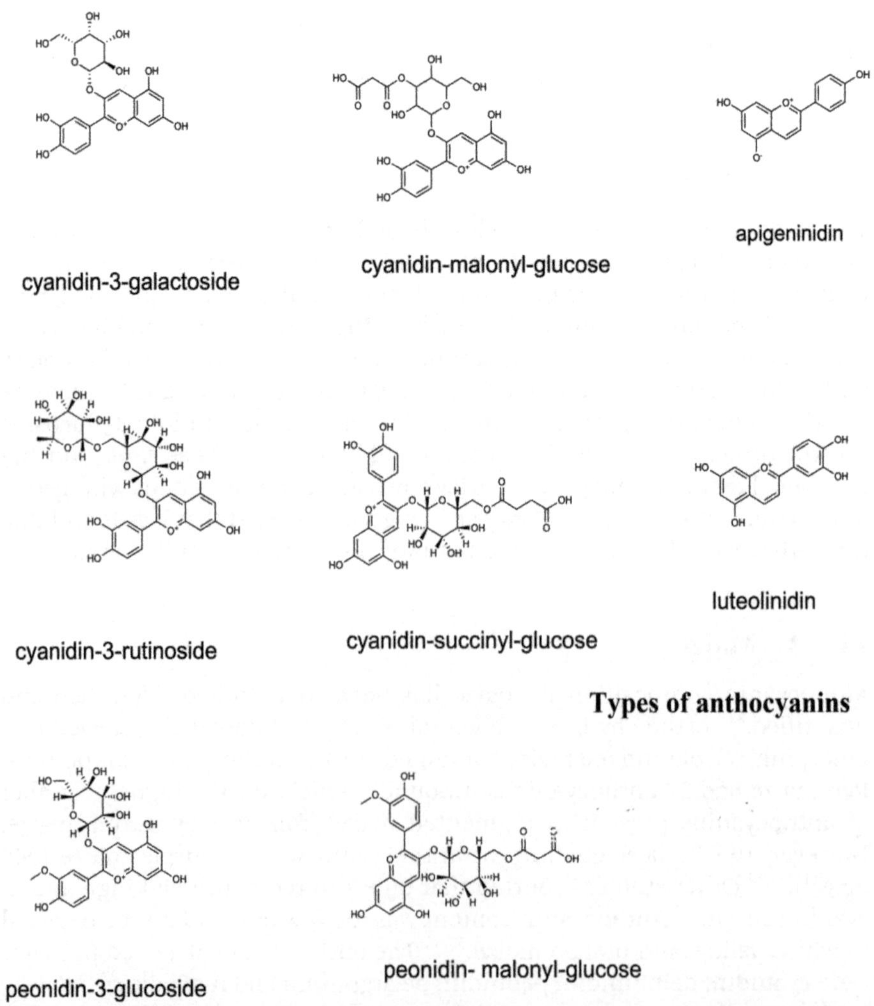

Scheme 11.1

11.3.5 Sorghum

The anthocyanin content of sorghum is unique, as it is not only rich in anthocyanins but also contains a group of deoxyanthocyanins.[59] These deoxyanthocyanins are lacking a hydroxyl group at the C3 position. While apigeninidin and luteolinidin are the main deoxyanthocyanins found in sorghum, their glucosides and methyl derivatives have also been detected in the cereal grain.[79,80,83] Other anthocyanins that have been identified in sorghum varieties include cyanidin, kuromanin, pelargonidin and fisetinidin.[80,81] Ashley *et al.*[82] reported that apigeninidin was not detected in sumac sorghum. Numerous studies have emphasized the various potential health benefits imparted by sorghum consumption, which are attributed to the bioactive compounds in it.

Several studies have reported that black sorghum (mean = 10.1 mg g^{-1}) had higher anthocyanin content than brown, red, sumac and Pakistani sorghum.[59,81,83] Indeed, anthocyanin content in black sorghum was 4.0 to 9.8 mg g^{-1} and both apigeninidin and luteolinidin represented approximately 50% of anthocyanins in black sorghum.[59] The anthocyanin content of pigmented (yellow, white and red) and non-pigmented (white, brown) sorghum grains were analysed by measuring the level of luteolinidin and apigeninidin.[84] The content ranged from 0.05 ± 0.03 to 0.22 ± 0.10 mg luteolinidin equivalent per g. The sorghum grains tested had a low quantity of anthocyanin monomers. In *Sorghum bicolor* leaves, apigeninidin represented 83.5% of total phenolic content (TPC) (29.87 ± 9.85 mg g^{-1} of dried leaf sheath) while TPC for luteolinidin accounted for 1.0% (0.34 ± 0.21).[85]

11.3.6 Millet

There have been very limited reports on the anthocyanins present in millets. No anthocyanins were present in two millet varieties from Pakistan.[81] However, Pradeep and Sreerama[86] analysed the total anthocyanin content of non-pigmented whole foxtail millet and little millet, and found they had a value of 43.53 and 67.52 mg of cyanidin-3-*O*-glucoside equivalents per g, respectively. Other studies looked at the anthocyanin content of ethyl acetate extract of several white, brown and dark brown varieties of finger millets.[87] The findings reveal that anthocyanin content was higher for brown grains than light-colored grains. Also, apigeninidin-type and luteolinidin-type compounds may be present in finger millets.

11.3.7 Rye

Few studies have analysed rye for its anthocyanin content. There were no anthocyanins detected in yellow rye.[88] Nevertheless, delphinidin-3-rutinoside was present in the aleurone layer of blue seeds. Cyanidin-3-rutinoside and cyanidin-3 glucoside were detected in the red coleoptile and first leaves of the coleoptile respectively. Dedio, Hill and Evans[89] reported that anthocyanins were identified in the coleoptile and pericarp of purple-seeded rye. The anthocyanin cyanidin-3-rutinoside was detected in the coleoptile. In the pericarp, cyanidin 3-glucoside, peonidin 3-glucoside and their corresponding acylated forms were also detected. Delphinidin 3-rutinoside was also present in the aleurone layer. A study looking at the anthocyanin content in a batch of pigmented seeds revealed that green rye seeds contained mainly delphinidin rutinoside or cyanidin rutinoside.[90] As for violet rye seeds, 6 varieties were tested and cyanidin or peonidin was the main aglycone. The highest content was peonidin rutinoside (115.37 mg kg^{-1} of dry matter) followed by cyanidin rutinoside (101.36 mg kg^{-1} of dry matter) in the violet grains. The anthocyanidin content consisted of cyanidin (dimalonyl) hexoside, cyanidin (malonyl) hexoside, cyanidin (sinapoyl) hexoside, cyanidin (succinyl) hexoside, cyanidin 3-*O*-glucoside, cyanidin rutinoside, pelargonidin rutinoside, peonidin (dimalonyl) hexoside, peonidin (malonyl) hexoside, peonidin 3-*O*-glucoside, peonidin rutinoside, malvidin 3-*O*-galactoside. Yellow and brown seeds did not contain any anthocyanins.

In wholegrain rye flour, one significant study revealed that delphinidin-3-rutinoside was the most abundant (2.9 ± 0.1 mg per kg dw) followed by delphinidin-3-glucoside, cyanidin-3-rutinoside, and petunidin-3-rutinoside.[91]

11.4 Benefits of Anthocyanins from Various Colored Cereals

There has been a wide array of beneficial health aspects (Table 11.1) that have been correlated with consuming the pigments derived from colored fruit and vegetables.[103] These health claims have led to a spike in the current exploitation of colored cereals as functional foods for human health promotion and other related applications.[28] The intake of anthocyanins on a daily basis in the human diet predicted on the consumption of various fruits and vegetables in the United States is estimated as 12.5 mg day^{-1}.[57] Anthocyanins have been proven to have multiple health benefits, including cholesterol reduction, and antioxidant, anti-inflammatory and anticarcinogenic potential.[104] These phenolic compounds act as potent antioxidants, and free radical scavengers, as well as metal chelators, diminishing peroxidation of fats.[105,106] The most intensively studied aspect of these phytochemicals is their antioxidative potential, which can be found in various forms in manifold food products.

These compounds have high antioxidative value, exhibiting predominant roles such as free radical scavenging, reducing agents, production of pro-oxidant metals (Adom and Liu, 2002).[107] Nevertheless, there is abundant documentation related to wheat, which has proven huge antioxidant activity derived from the phenolic acid-rich fractions.[107] Less information, however, is still available on the antioxidant potential of pigments of wheat. Besides anthocyanins, purple corn are also a potential store house of non-anthocyanin phenolic compounds exhibiting potential health benefits.[108]

A significant study has reported that a wheat variety with pigments exhibited high DPPH radical scavenging capacity in comparison with wheat devoid of pigments.[109] Among all colored cereals, it is black rice that has been significantly used *in vivo* and *in vitro* models owing to the various functional properties of the extracts. The exceptionally high antioxidant activity of black rice has proven its worth in imparting immense protection to the endothelial cells[110] as well as diminishing rates of cardiovascular diseases[111] and acting as potent anticancerous agents.[112]

Interest in colored cereals has upsurged in the last decade. Petroni *et al.*[113] highlighted the health-promoting potential of maize-derived anthocyanins as shown in the literature up to 2013. These coloring pigments of phytochemical origin have led to the prevention of various degenerative diseases, such as cancers. The yellow pigmentation of the endosperm of durum wheat is attributable to these carotenoids, whereas the red, purple or blue grain colors from anthocyanins are located in the aleurone or pericarp.[114] Purple wheat utilized on a commercial basis was evaluated as a source of functional ingredients primarily based on anthocyanin composition and its antioxidant properties.[33] Black rice is abundant in these anthocyanins and other pigmented cereal grains.[115,116] Bellido and Beta[116] identified prominent anthocyanin pigments such as delphinidin 3-glucoside, delphinidin 3-rutinoside, cyanidin 3-glucoside, petunidin 3-glucoside and cyanidin chloride in barley, whereas delphinidin 3-glucoside and cyanidin 3-glucoside were reported in yellow and purple barley, respectively. In comparison with other cereals, consumption of sorghum might have the potential in reducing the risk of the onset of various types of cancer in humans, owing to the high concentration of phytochemicals in the sorghum.[59]

Anthocyanins are predominantly absorbed in the small intestine, where they undergo hydrolysis. These anthocyanins have been quite successful in attracting the attention of researchers and consumers, owing to their pharmacology and health aspects. The anthocyanin content of pigmented and colored grains such as wheat (purple/blue), corn (purple/blue) and rice (red/black) is comparable to the anthocyanin content of fruits and vegetables, thereby making them a promising delivery agent for a plethora of functional ingredients.[53] These are the pigmented flavonols derived naturally, and pertain to a flavonoid group's subclass. However, there is a dearth of literature on the types and content of anthocyanins in cereals, owing to the belief that cereals are not a significant source of these coloring pigments.[117]

Interestingly, carotenoids, which are fat-soluble terpenoids, also provide a natural source of varied colors such as yellow, orange and red, with huge antioxidant potential.[118] These pigments have manifold applications in industry, mainly in the form of food colorants, due to their ability to produce bright colors, and more advanced benefits in the form of their health-promoting properties of these compounds, which have already been used in human diets for centuries.[119]

As per significant research, it was reported that sorghum fractions exhibited a higher anthocyanin activity *in vitro* in comparison with other cereals and fruits, and the health benefits imparted by it can be considered equivalent to those of fruits.[117] These compounds blessed with functional properties are biologically active substances of dietary origin. The consistent consumption of these phytochemicals leads to diminishing

Table 11.1 Anthocyanin profile of colored wheats and their health benefits.

Anthocyanin (characteristic color)	Health benefits	Derivatives	Chemical formula	Wheat type	Amount (mg kg^{-1})	Ref.
Cyanidin (orange-red)	Offers protection against heart and liver diseases, improves metabolic syndrome; improves spatial memory; suppression of ischemia/reperfusion liver oxidation damage; protective effect on DNA cleavage; inhibition of invasion and mortality of tumor cells; maintains meta-bolites in plasma; increase glucose uptake; improves metabolic profile and mitochondrial energy metabolism; acts as a barrier against inflammation induced permeabilization	Cyanidin-3-arabinoside	$C_{20}H_{19}ClO_{10}$	Purple	23.2–25.1	32
		Cyanidin-3-galactoside	$C_{21}H_{21}O_{11}^{+}$	Purple	0.98–72.0	12
		Cyanidin-3-glucoside	$C_{21}H_{20}O_{11}$	Blue	1.17–28.14	139 and 56
		Cyanidin di-glucoside	$C_{27}H_{31}O_{16}^{+}$	Purple	0.63–116.2	33
		Cyanidin-3-rutinoside	$C_{27}H_{30}O_{15}$	Purple	1.7	102
				Blue	7.12–16.8	
Delphinidin (blue-red)	Anti-inflammatory, bone protective, and neuroprotective effects; inhibits platelet activation and thrombosis; barrier against inflammation-induced permeabilization; inhibitory effect on human glyoxalase	Delphinidin-3-rutinoside	$C_{27}H_{30}O_{16}$	Blue	31.77–49.6	33
		Delphinidin-3-arabinoside	$C_{20}H_{19}ClO_{11}$	Purple	15.1–16.7	32
		Delphinidin-3-galactoside	$C_{21}H_{21}ClO_{12}$	Purple	38.3–36.4	102 and 53
		Delphinidin-3-glucoside	$C_{21}H_{20}O_{12}$	Blue	0.52–56.5	
Malvidin (blue-red)	Antioxidant, anticancer, anti-inflammatory activity in endothelial cells; alleviates postprandial hyperglycemia; cytotoxic to human leukemia cells	Malvidin-3-glucoside	$C_{23}H_{24}O_{12}$	Blue	10.26–13.8	102
		Malvidin-3-rutinoside	$C_{29}H_{34}O_{16}$	Red	0.06 0.43	

Pelargonidin (orange-red)	Have anti-inflammatory effects; alleviates progression of postprandial hyperglycemia; prevents neurodegenerative disorders; effective in treatment of cancer and vascular diseases	Pelargonidin-3-arabinoside Pelargonidin-3-galactoside	$C_{20}H_{19}O_9$+ $C_{21}H_{21}ClO_{10}$	Purple Purple	9.3–11.3 22.1–26.1	126 and 32
Peonidin (orange-red)	Prevention of vascular and cardiac diseases; inhibits growth of tumor cells; inhibitory and apoptotic effects on human breast cancer cells	Peonidin-3-arabinoside Peonidin-3-glucoside	$C_{21}H_{21}ClO_{10}$ $C_{22}H_{22}O_{11}$	Blue Purple	0.95–3.49 2.7–17.5	102 and 141
Petunidin (blue-red)	Stimulates osteoblastogenesis concomitant; reduces bone loss; anti-proliferative and anti-invasive properties; protects against liver cancer	Petunidin-3-rutinoside Petunidin 3-galactoside	$C_{28}H_{33}O_{16}$+ $C_{22}H_{23}O_{12}$+	Blue Purple	4.5 38.3–40.4	53 and 32

the risks of type II diabetes,[120] cancer,[121] cardiovascular disease[122] and ocular damage.[123] The extracts of purple corn and red rice rich in anthocyanins have proven to be a great functional ingredient and colorant agent derived from nature.[124] Nevertheless, although there is a variation in the composition of anthocyanins, the biological activities of anthocyanins derived from fruit and vegetables is equivalent to anthocyanins derived from cereals.[28]

The rich nutritional profile of these coloring pigments has attracted the attention of researchers across the globe. The major health benefits of anthocyanins cover their antioxidative, anticancerous, anti-ageing, and gut improving properties.[125] Wide variety of colored cereals that act as potential carriers for the developing of functional foods have been intensively studied for unleashing enormous health benefits. The claimed health promoting aspects of these pigments include anti-oxidative, anticancerous, glycemic, regulation of body weight, protection of retina, hepatoprotection, and anti-ageing.[28] Besides these, anthocyanins and carotenoids exhibit potent antioxidative properties in the human body.[126] The food extracts that are adequately rich in these coloring pigments are a rich source of health-promoting ingredients, which can pave the way for the development and production of functional foods, especially dietary food products.[119]

Anthocyanins have been credited with health-promoting potential in combating oxidative stress,[127–129] cancer[130–133] and diabetes.[134] Carotenoids serve as precursors in the production of vitamin A, leading to efficient normal function of the retina of the eye.[135] Of the various colored cereals, it is purple corn that is most rich in phenolics, primarily anthocyanins, which have been reported in animal studies to have cardioprotective potential,[136] to aid in weight management[134] and prevent diabetic nephropathy,[137] and to inhibit development of cancer of the mammary and prostate glands in animal studies.[99] Anthocyanins not only find use as nutrients, but also as components of various herbal medicines used traditionally to eradicate diseases such as hypertension, pyrexia, liver disorders, diarrhea, urinary problems and the common cold.[138]

11.5 Concluding Remarks and Future Perspectives

Various colored cereals have captured the attention of food growers in the last decade. Durum wheat acquires its color due to carotenoids, which impart a yellow hue, due to yellow pigmentation in the endosperm. On the other hand, grains' red, purple or blue colors originate from anthocyanins located in the aleurone or pericarp. In view of the high antioxidant and free radical scavenging potential against singlet oxygen, these coloring pigments can exert ample beneficial effects by regulating key enzymes, modulating nuclear receptors and cell pathways involved in proliferation, and averting DNA damage. Foods loaded with anthocyanins are not only food products, but can also be considered therapeutic food ingredients.

Over the years, anthocyanins have attracted the attention of researchers and consumers because of their pharmacological and health potential.

Anthocyanins possess antidiabetic, anticancer, anti-inflammatory, antimicrobial, and anti-obesity effects, as well as prevention of cardiovascular diseases.[2] Therefore, anthocyanins extracted from edible plants are potential pharmaceutical ingredients. Although their abundant health-promoting potential is backed by scientific evidence, further research is still needed to authenticate their pharmacokinetics in humans, along with the dose and the molecular mechanisms underlying their biological activities. Plant-based anthocyanins have been reported to have widespread medicinal value. Many anthocyanin extraction methods, such as conventional acidified water (CAW), ultrasound, microwave pre-treatment, supercritical fluid extraction and pulsed electric field (PEF) have been used. Among them, ultrasound and MAE are two putative methods for extracting anthocyanins from vegetables. They have significant advantages; they are cheap, easily manipulated, suitable for laboratory, domestic and large-scale industrial applications, less time-consuming, matrix independent, free sample particle size, less solvent used and long-term preservation. Supercritical fluid extraction and pressurized liquid extraction have proven effective in extracting these pigments from colored cereals. Thus far, the implementation of novel extraction technologies, using different mechanisms such as ultrasound, microwave energy, supercritical fluids and accelerated solvent extraction has been promoted;[3] the main objective of these methods is to reduce extraction time and energy consumption, which is reflected in the lowering of the final cost.

References

1. R. E. Wrolstad, R. W. Durst and J. Lee, *Trends Food Sci. Technol.*, 2005, **16**(9), 423–428.
2. M. T. Escribano-Bailón, C. Santos-Buelga and J. C. Rivas-Gonzalo, *J. Chromatogr. A*, 2004, **1054**(1–2), 129–141.
3. I. Ignat, I. Volf and V. I. Popa, *Food Chem.*, 2011, **126**(4), 1821–1835.
4. N. C. Veitch and R. J. Grayer, *Nat. Prod. Rep.*, 2008, **25**(3), 555–611.
5. T. Vogt, *Mol. Plant*, 2010, **3**(1), 2–20.
6. E. Pojer, F. Mattivi, D. Johnson and C. S. Stockley, *Compr. Rev. Food Sci. Food Saf.*, 2013, **12**(5), 483–508.
7. Q. Li, P. Somavat, V. Singh, L. Chatham and E. G. de Mejia, *Food Chem.*, 2017, **231**, 332–339.
8. Mintel and Leatherhead Food Research, *Food Colours: Market, Technical and Regulatory Insights*, Leatherhead Food Research, Leatherhead, 2013.
9. FAO, *Traditional High Andean Cuisine*, 2013, http://www.fao.org/docrep/018/i1466e/i1466e.pdf, accessed 26 May 2014.
10. Y. Salinas Moreno, G. S. Sánchez, D. R. Hernández and N. R. Lobato, *J. Chromatogr. Sci.*, 2005, **43**(9), 483–487.
11. G. Mazza and L. Gao, Blue and purple grains, in *Specialty Grains for Food and Feed*, ed. E.-S. M. Abdel-Aal, and P. J. Peter, American Association of Cereal Chemists, St. Paul, MN, 2005, pp. 45–67.
12. E. S. Abdel-Aal and P. Hucl, *J. Agric. Food Chem.*, 2003, **51**(8), 2174–2180.

13. S. H. Nam, S. P. Choi, M. Y. Kang, H. J. Koh, N. Kozukue and M. Friedman, *Food Chem.*, 2006, **94**(4), 613–620.
14. M. Philpott, K. S. Gould, C. Lim and L. R. Ferguson, *J. Agric. Food Chem.*, 2006, **54**, 1710–1715.
15. C. Zhao, M. M. Giusti, M. Malik, M. P. Moyer and B. A. Magnuson, *J. Agric. Food Chem.*, 2004, **52**, 6122–6128.
16. J. W. Hyun and H. S. Chung, *J. Agric. Food Chem.*, 2004, **52**, 2213–2217.
17. T. Tsuda, F. Horio and T. Osawa, *J. Nutr. Sci. Vitaminol.*, 2002, **48**, 305–310.
18. T. Tsuda, F. Horio, K. Uchida, H. Aoki and T. Osawa, *J. Nutr.*, 2003, **133**, 2125–2130.
19. F. Delgado-Vargas, A. R. Jiménez and O. Paredes-López, *Crit. Rev. Food Sci. Nutr.*, 2000, **40**(3), 173–289.
20. A. P. P. Kayodé, C. A. Bara, G. Dalodé-Vieira, A. R. Linnemann and M. J. R. Nout, *LWT–Food Sci. Technol.*, 2012, **46**, 49–55.
21. M. Paulsmeyer, L. Chatham, T. Becker, M. West, L. West and J. Juvik, *J. Agric. Food Chem.*, 2017, **65**, 4341–4350.
22. F. Lao, G. T. Sigurdson and M. M. Giusti, *Compr. Rev. Food Sci. Food Saf.*, 2017, **16**, 234–246.
23. E. S. M. Abdel-Aal, H. Akhtar, I. Rabalski and M. Bryan, *J. Food Sci.*, 2014, **79**, C138–C146.
24. A. B. Das, V. V. Goud and C. Das, *Ind. Crops Prod.*, 2017, **95**, 332–341.
25. S. Suriano, C. Balconi, P. Valoti and R. Redaelli, *LWT–Food Sci. Technol.*, 2021, **144**, 111257.
26. T. Beta, Y. Qiu, Q. Liu, A. Borgen and F. B. Apea-Bah, Purple Wheat (Triticum sp.) Seeds: Phenolic Composition and Antioxidant Properties, in *Nuts and Seeds in Health and Disease Prevention*, ed. V. R. Preedy, R. R. Watson and V. B. Patel, Academic Press, 2nd edn, 2020, pp. 103–125.
27. C. Hu, Y. Z. Cai, W. Li, H. Corke and D. D. Kitts, *Food Chem.*, 2007, **104**(3), 955–961.
28. F. Zhu, *Food Res. Int.*, 2018, **109**, 232–249.
29. M. Varga, R. Jojart, P. Fonad, R. Mihaly and A. Palagyi, *Food Chem.*, 2018, **268**, 153–161.
30. V. Verardo, C. Serea, R. Segal and M. F. Caboni, *J. Cereal Sci.*, 2011, **54**, 211–217.
31. S. Suriano, A. Iannucci, P. Codianni, C. Fares, M. Russo, N. Pecchioni, U. Marciello and M. Savino, *Food Res. Int.*, 2018, **113**, 221–233.
32. F. Hosseinian, W. Li and T. Beta, *Food Chem.*, 2008, **109**, 916–924.
33. E. S. M. Abdel-Aal, P. Hucl and I. Rabalski, *Food Chem.*, 2018, **254**, 13–19.
34. Y. Shao, F. Xu, X. Sun, J. Bao and T. Beta, *J. Cereal Sci.*, 2014, **59**(2), 211–218.
35. Y. Qiu, Q. Liu and T. Beta, *Food Chem.*, 2010, **121**(1), 140–147.
36. L. M. Paucar-Menacho, C. Martinez-Villaluenga, M. Dueñas, J. Frias and E. Peñas, *LWT–Food Sci. Technol.*, 2017, **76**, 236–244.
37. C. Alcalde-Eon, I. García-Estévez, R. Ferreras-Charro, J. C. Rivas-Gonzalo, R. Ferrer-Gallego and M. T. Escribano-Bailón, *J. Food Compos. Anal.*, 2014, **34**(1), 99–113.

38. M. Zanoletti, P. A. Parizad, V. Lavelli, C. Cecchini, P. Menesatti, A. Marti and M. A. Pagani, *LWT–Food Sci. Technol.*, 2017, **75**, 663–669.
39. C. Kurambhatti, D. Kumar, K. D. Rausch, M. E. Tumbleson and V. Singh, *Ind. Crops Prod.*, 2020, **158**, 112976.
40. Z. Wang, H. Huang, E. Gonzalez de Mejia, Q. Li and V. Singh, *Cereal Chem.*, 2016, **93**, 344–351.
41. J. Lee, R. W. Durst and R. E. Wrolstad, *J. AOAC Int.*, 2005, **88**, 1269–1278.
42. R. F. Teixeira, L. Benvenutti, V. M. Burin, T. M. Gomes, S. R. Ferreira and A. A. Zielinski, *Innovative Food Sci. Emerging Technol.*, 2021, **67**, 102587.
43. H. T. Hong, M. E. Netzel and T. J. O'Hare, *Data Brief*, 2020, **30**, 105495.
44. D. A. Herrman, J. F. Brantsen, S. Ravisankar, K. M. Lee and J. M. Awika, *Food Chem.*, 2020, **333**, 127494.
45. M. Leja, A. Mareczek and J. Ben, *Food Chem.*, 2003, **80**, 303–307.
46. P. Jing and M. M. Giusti, *J. Food Sci.*, 2007, **72**(7), C363–C368.
47. X. Zhao, M. Corrales, C. Zhang, X. Hu, Y. Ma and B. Tauscher, *J. Agric. Food Chem.*, 2008, **56**, 10761–10766.
48. F. Ramos-Escudero, A. M. Muñoz, C. Alvarado-Ortíz, Á. Alvarado and J. A. Yáñez, *J. Med. Food*, 2012, **15**, 206–215.
49. H. Aoki, N. Kuze and Y. Kato, *Foods Food Ingredients J. Jpn.*, 2002, **199**, 41–45.
50. P. Jing, V. Noriega, S. J. Schwartz and M. M. Giusti, *J. Agric. Food Chem.*, 2007, **55**, 8625–8629.
51. E. C. Montilla, S. Hillebrand, A. Antezana and P. Winterhalter, *J. Agric. Food Chem.*, 2011, **59**, 7068–7074.
52. T. Fuleki and F. J. Francis, *J. Food Sci.*, 1968, **33**, 72–77.
53. E.-S. M. Abdel-Aal, J. C. Young and I. Rabalski, *J. Agric. Food Chem.*, 2006, **54**, 4696–4704.
54. C. Y. Li, H. W. Kim, S. R. Won, K. J. Min, J. Y. Park, M. S. Ahn and H. I. Rhee, *J. Agric. Food Chem.*, 2008, **56**, 11413–11416.
55. L. X. Lopez-Martinez, R. M. Oliart-Ros, G. Valerio-Alfaro, C. H. Lee, K. L. Parkin and H. S. Garcia, *LWT–Food Sci. Technol.*, 2009, **42**(6), 1187–1192.
56. S. Žilić, A. Serpen, G. Akıllıoğlu, V. Gökmen and J. Vančetović, *J. Agric. Food Chem.*, 2012, **60**, 1224–1231.
57. X. Wu, G. R. Beecher, J. M. Holden, D. B. Haytowitz, S. E. Gebhardt and R. L. Prior, *J. Agric. Food Chem.*, 2006, **54**, 4069–4075.
58. S. Sharma, V. Chunduri, A. Kumar, R. Kumar, P. Khare, K. Kondepudi, M. Bishnoi and M. Garg, *PLoS One*, 2018, **13**(4), e0194367.
59. J. M. Awika, L. W. Rooney and R. D. Waniska, *Food Chem.*, 2004, **90**, 293–301.
60. S. N. Ryu, S. Z. Park and C.-T. Ho, *J. Food Drug Anal.*, 1998, **6**, 729–736.
61. E.-S. M. Abdel-Aal and P. Hucl, *Cereal Chem.*, 1999, **76**, 350–354.
62. B. K. Veni, *J. Pharmacogn. Phytochem.*, 2019, **SP2**, 303.
63. P. Pramai and S. Jiamyangyuen, *Songklanakarin J. Sci. Technol.*, 2016, **38**(5), 463–472.
64. W. Chen, D. Müller, E. Richling and M. Wink, *J. Agric. Food Chem.*, 2016, **61**, 3047–3053.

65. S. Yamuangmorn, B. Dell and C. Prom-U-Thai, *Rice Sci.*, 2018, **25**(5), 270–278.
66. Q. Liu, Y. Qiu and T. Beta, *J. Agric. Food Chem.*, 2010, **58**(16), 9235–9241.
67. E.-S. M. Abdel-Aal, T. M. Choo, S. Dhillon and S. Rabalski, *Cereal Chem.*, 2012, **89**, 198–204.
68. P. Saini, N. Kumar, S. Kumar, P. W. Mwaurah, A. Panghal, A. K. Attkan, V. K. Singh, M. K. Garg and V. Singh, *Crit. Rev. Food Sci. Nutr.*, 2021, **61**(19), 3197–3210.
69. M. Garg, M. Chawla, V. Chunduri, R. Kumar, S. Sharma, N. K. Sharma, N. Kaur, A. Kumar, J. K. Mundey and M. K. Saini, *et al.*, *J. Cereal Sci.*, 2016, **71**, 138–144.
70. E. K. Arendt and E. Zannini, *Cereal Grains for the Food and Beverage Industries*, Elsevier, Woodhead Publishing, 1st edn, 2013.
71. M. J. Kim, J. N. Hyun, J. A. Kim, J. C. Park, M. Y. Kim, J. G. Kim, S. J. Lee, S. C. Chun and I. M. Chung, *J. Agric. Food Chem.*, 2007, **55**(12), 4802–4809.
72. I. Diczházi and L. Kursinszki, *Cereal Chem.*, 2014, **91**(2), 195–200.
73. C. Lee, D. Han, B. Kim, N. Baek and B. K. Baik, *Int. J. Food Sci. Technol.*, 2013, **48**(5), 984–991.
74. F. Colombo, C. Di Lorenzo, K. Petroni, M. Silano, R. Pilu, E. Falletta, S. Biella and P. Restani, *Foods*, 2021, **10**, 1770.
75. D. A. Urias-Lugo, J. B. Heredia, M. D. Muy-Rangel, J. B. Valdez-Torres, S. O. Serna-Saldívar and J. A. Gutiérrez-Uribe, *Plant Foods Hum. Nutr.*, 2015, **70**, 193–199.
76. L. A. Bello-Pérez, G. A. Camelo-Mendez, E. Agama-Acevedo and R. G. Utrilla-Coello, *Agrocencia*, 2016, **50**(8), 1041–1063.
77. P. A. Parizad, J. Capraro, A. Scarafoni, F. Bonomi, M. Blandino, M. Marengo, D. Giordano, A. Carpen and S. Iametti, *Plant Foods Hum. Nutr.*, 2019, **74**, 128–134.
78. F. Fernandez-Aulis, L. Hernandez-Vazquez, G. Aguilar-Osorio, D. Arrieta-Baez and A. Navarro-Ocana, *J. Food Sci.*, 2019, **84**(5), 954–962.
79. W. K. Nip and E. E. Burns, *Cereal Chem.*, 1969, **46**, 490–494.
80. J. M. Awika and L. W. Rooney, *Phytochemistry*, 2004, **65**, 1199–1221.
81. F. Ahmad, I. Pasha, M. Saeed and M. Asgher, *Int. J. Food Prop.*, 2018, **21**(1), 1586–1597.
82. D. Ashley, D. Marasini, C. Brownmiller, J. A. Lee, F. Carbonero and S. Lee, *Nutrients*, 2019, **11**, 217.
83. L. Yang, K. F. Allred, B. Geera, C. D. Allred and J. M. Awika, *Nutr. Cancer*, 2012, **64**(3), 419–427.
84. S. C. Choi, J. M. Kim, Y. G. Lee and C. Kim, *Cereal Res. Commun.*, 2019, **47**(2), 228–238.
85. S. B. L. Makanjuola, A. O. Ogundaini, L. C. Ajonuma and A. Dosunmu, *Int. J. Rheum. Dis.*, 2018, **21**, 1487–1495.
86. P. M. Pradeep and Y. N. Sreerama, *J. Funct. Foods*, 2017, **35**, 682–693.
87. M. Siwela, J. R. N. Taylor, W. A. J. de Milliano and K. G. Duodu, *Food Chem.*, 2010, **121**, 443–449.

88. W. Dedio, P. J. Kaltsikes and E. N. Larter, *Phytochemistry*, 1969, **8**, 2351–2352.
89. W. Dedio, R. D. Hill and L. E. Evans, *Can. J. Plant Sci.*, 1972, **52**, 981–983.
90. P. A. Zykin, E. A. Andreeva, A. N. Lykholay, N. V. Tsvetkova and A. V. Voylokov, *Molecules*, 2018, **23**, 948.
91. J. Pihlava, J. Hellström, T. Kurtelius and P. Mattila, *J. Cereal Sci.*, 2018, **79**, 183–192.
92. Sugiyama Chemical Institute, Anthocyanin food coloring agent from purple corn, *Jpn. Patent*, 77130824, 1977.
93. B. A. Cevallos-Casals and L. Cisneros-Zevallos, *J. Agric. Food Chem.*, 2003, **51**, 3313–3319.
94. N. Ahmadiani, R. J. Robbins, T. M. Collins and M. M. Giusti, *J. Agric. Food Chem.*, 2014, **62**(30), 7524–7531.
95. S. Y. Wang and H. S. Lin, *J. Agric. Food Chem.*, 2000, **48**(2), 140–146.
96. S. E. Kulling and H. M. Rawel, *Planta Med.*, 2008, **74**(13), 1625–1634.
97. K. Fukamachi, T. Imada, Y. Ohshima, J. Xu and H. Tsuda, *Cancer Sci.*, 2008, **99**, 1841–1846.
98. P. Jing, J. A. Bomser, S. J. Schwartz, J. He, B. A. Magnuson and M. M. Giusti, *J. Agric. Food Chem.* 2008, **56**, 9391–9398.
99. N. Long, S. Suzuki, S. Sato, A. Naiki-Ito, K. Sakatani, T. Shirai and S. Takahashi, *Cancer Sci.*, 2013, **104**(3), 298–303.
100. M. L. Finkel, S. Sanchez, T. Mak, J. Granstein and A. Lefkowitz, *J. Evidence-Based Complementary Altern. Med.*, 2013, **18**, 237–242.
101. A. C. Zeven, *Euphytica*, 1991, **56**, 243–258.
102. D. B. M. Ficco, A. M. Mastrangelo, D. Trono, G. M. Borrelli, P. De Vita, C. Fares, R. Beleggia, C. Platani and R. Papa, *Crop Pasture Sci.*, 2013, **65**(1), 1–15.
103. T. Tsuda, *Mol. Nutr. Food Res.*, 2012, **56**, 159–170.
104. D. F. Birt, S. Hendrich and W. Wang, *Pharmacol. Ther.*, 2001, **90**(2–3), 157–177.
105. N. C. Cook and S. Samman, *J. Nutr. Biochem.*, 1996, **7**, 66–76.
106. M. Fattore, D. Montesano, E. Pagano, R. Teta, F. Borrelli, A. Mangoni, S. Seccia and S. Albrizio, *J. Food Compos. Anal.*, 2016, **53**, 61–68.
107. K. K. Adom and R. H. Liu, *J. Agric. Food Chem.*, 2002, **50**(21), 6182–6187.
108. R. Pedreschi and L. Cisneros-Zevallos, *Food Chem.*, 2007, **100**(3), 956–963.
109. W. Li, F. Shan, S. C. Sun, H. Corke and T. Beta, *J. Agric. Food Chem.*, 2005, **53**, 8533–8536.
110. M. W. Zhang, R. F. Zhang, B. J. Guo, J. W. Chi, Z. C. Wei and Z. H. Xu, *Acta Nutr. Sin.*, 2006, **28**, 216–220.
111. J. Tanaka, S. Nakamura, K. Tsuruma, M. Shimazawa, H. Shimoda and H. Hara, *Phytother. Res.*, 2011, **26**(2), 214–222.
112. V. Leardkamolkarn, W. Thongthep, P. Suttiarporn, R. Kongkachuichai, S. Wongpornchai and A. Wanavijitr, *Food Chem.*, 2011, **125**, 978–985.
113. K. Petroni, R. Pilu and C. Tonelli, *Planta*, 2014, **240**, 901–911.
114. S. Kruger and G. E. Morlock, *J. Chromatogr. A*, 2018, **1538**, 75–85.

115. S. Jang and Z. Xu, *J. Agric. Food Chem.*, 2009, **57**, 858–862.
116. G. G. Bellido and T. Beta, *J. Agric. Food Chem.*, 2009, **57**, 1022–1028.
117. H. Davis, X. Su, Y. Shen, J. Xu, D. Wang, J. S. Smith, F. Aramouni and W. Wang, Phenotypic diversity of colored phytochemicals in sorghum accessions with various pericarp pigments, in *Polyphenols Plants*, ed. R. R. Watson, Academic Press, 2019, pp. 123–131.
118. L. Paznocht, Z. Kotíkova, M. Sulc, J. Lachman, M. Orsak, M. Eliášová and P. Martinek, *Food Chem.*, 2018, **240**, 670–678.
119. A. Calderaro, D. Barreca, E. Bellocco, A. Smeriglio, D. Trombetta and G. Laganà, Colored phytonutrients: role and applications in the functional foods of anthocyanins, in *Phytonutrients in Food*, Woodhead Publishing, 2020, pp. 177–195.
120. P. L. Lutsey, D. R. Jacobs Jr, S. Kori, E. Mayer-Davis, S. Shea, L. M. Steffen, M. Szklo and R. Tracy, *Br. J. Nutr.*, 2007, **98**(2), 397–405.
121. I. C. W. Arts and P. C. H. Hollman, *Am. J. Clin. Nutr.*, 2005, **81**, 317s–325s.
122. P. B. Mellen, T. F. Walsh and D. M. Herrington, *Nutr., Metab. Cardiovasc. Dis.*, 2008, **18**(4), 283–290.
123. E. J. Johnson, *Nutr. Clin. Care*, 2002, **5**, 56–65.
124. S. Siebenhandl, H. Grausgruber, N. Pellegrini, D. Del Rio, V. Fogliano, R. Pernice and E. Berghofer, *J. Agric. Food Chem.*, 2007, **55**(21), 8541–8547.
125. J. M. Kong, L. S. Chia, N. K. Goh, T. F. Chia and R. Brouillard, *Phytochemistry*, 2003, **64**, 923–933.
126. J. Lachman, P. Martinek, Z. Kotíková, M. Orsák and M. Šulc, *J. Cereal Sci.*, 2017, **74**, 145–154.
127. B. J. Glove and C. Martin, *Curr. Biol.*, 2012, **22**, R147–R150.
128. M. P. Kähkönen and M. Heinonen, *J. Agric. Food Chem.*, 2003, **51**, 628–633.
129. G. Mazza, C. D. Kay, T. Cottrell and B. J. Holub, *J. Agric. Food Chem.*, 2002, **50**, 7731–7737.
130. N. R. Song, H. Yang, J. Park and J. Y. Kwon, *et al.*, *Food Chem.*, 2012, **133**, 658–664.
131. K. D. Vu, H. Carlettini, J. Bouvet and J. Côté, *et al.*, *Food Chem.*, 2012, **132**, 959–967.
132. W. Liu, J. Xu, S. Wu and Y. Liu, *et al.*, *PLoS One*, 2013, **8**, e81586.
133. S. Lim, J. Xu, J. Kim and T.-Y. Chen, *et al.*, *Mol. Nutr. Food Res.*, 2013, **57**, 1908–1917.
134. T. Tasuda, H. Fumihiko, K. Uchida, H. Aoki and T. Osawa, *J. Nutr.*, 2003, **133**, 2125–2130.
135. Z. Kotíková, M. Šulc, J. Lachman, V. Pivec, M. Orsák and K. Hamouz, *Food Chem.*, 2016, **197**, 992–1001.
136. M. C. Toufektsian, M. De Lorgeril, N. Nagy, P. Salen, M. B. Donati and L. Giordano, *et al.*, *J. Nutr.*, 2008, **138**(4), 747–752.
137. M. K. Kang, S. S. Lim, J. Y. Lee, K. M. Yeo and Y. H. Kang, *PLoS One*, 2013, **8**(11), e79823.
138. I. Konczak and W. Zhang, *J. Biomed. Biotechnol.*, 2004, **5**, 239–240.
139. T. H. Gamel, A. J. Wright, M. Pickard and E. S. M. Abdel-Aal, *Cereal Chem.*, 2020, **97**(1), 34–38.

CHAPTER 12

Pigmented Cereals as a Source of Carotenoids

D. TRONO*[a]

[a]Research Centre for Cereal and Industrial Crops, Council for Agricultural Research and Economics, Foggia, Italy
*E-mail: daniela.trono@crea.gov.it

12.1 Introduction

Cereals play an important role in agriculture worldwide. According to the World Food and Agriculture Organization's statistics, global cereal production in 2021/2022 has been estimated at 2815 million tonnes.[1] Globally, rice, maize, and wheat are the three most important cereal crops, which together comprise at least 75% of grain production.[2] However, the utilization pattern of these cereal grains differs between developed and developing countries. In the former, more than 70% of total cereal production is fed to animals, whereas in the latter, almost all cereal production is used for human consumption.[3] Cereals are considered a rich source of carbohydrate, protein, and dietary fiber, with rice, maize, and wheat providing more than 90% of total calories from cereals.[4] In addition to high amounts of carbohydrates and proteins, cereal grains also contain many health-promoting components such as vitamins, minerals, and phytochemicals.[5] 'Phytochemical' is a broad term that refers to chemically diverse compounds that occur naturally in plants and that exert protective or disease-preventing effects in humans.[6] Some of these compounds are pigments that impart colour to those plant tissues in which they are accumulated in large amounts.

Food Chemistry, Function and Analysis No. 38
Pigmented Cereals and Millets: Bioactive Profile and Food Applications
Edited by Sneh Punia Bangar, Sajid Maqsood and Anil Kumar Siroha

Published by the Royal Society of Chemistry, www.rsc.org

In cereals, coloured phytochemicals can accumulate in different tissues of the kernel. This gives the unique characteristic colouration of pigmented cereals, which vary from purple and blue to yellow and orange, and is mainly due to two classes of pigments: anthocyanins and carotenoids.[7,8] The blue colour of grains is due to the accumulation of anthocyanins in the aleurone layer, whereas the purple colouration comes from anthocyanins located in the pericarp layer.[9] Carotenoids are responsible for the characteristic yellow to orange colours of pigmented cereals. They mainly accumulate in the starchy endosperm, although significant levels have been also detected in the germ and aleurone layer.[10] Thanks to the accumulation of carotenoid pigments in the endosperm, a significant amount of this class of phytochemicals can be found in refined flour, which is utilized in the formulation of numerous products. In contrast, the accumulation of anthocyanins in the outer layers of grains limits their presence to end-products made with wholemeal flour.

The bright yellow colour given by carotenoid accumulation is important for the aesthetic quality of some cereal-based end-products. For instance, the high carotenoid content in durum wheat, and therefore in semolina, results in pasta products with a more attractive yellow colour for consumers.[11] Also, yellow maize is preferred to white maize as livestock feed, as it confers a yellowish colour to poultry meat and egg yolk.[12] In addition to their role as an important aesthetic parameter, carotenoids also have important nutritional and health roles. Like many other phytochemical compounds (*e.g.* flavonoids, anthocyanins, tocols, and tocopherols) all the carotenoids show antioxidant activity in reducing the oxidative stress generated by reactive oxygen species (ROS).[13] As ROS and oxidative damage to biomolecules are involved in the onset and progression of chronic degenerative diseases,[14] carotenoids can be considered vital prophylactic strategic molecules. Moreover, specific carotenoids have the peculiar capacity to act as precursors of vitamin A, which provides protection from ocular diseases.[15]

In the light of the important health benefits of carotenoids and the high cereal *per capita* consumption in the world, the improvement of carotenoid content in pigmented cereal grains and the shifting of consumption from white to yellow, orange, and dark-orange grains would be helpful to reduce the risk of chronic diseases in industrialized countries and mitigate vitamin A deficiency in rural areas of developing countries.

12.2 Carotenoids: Chemical Structures and Biological Properties

Carotenoids are among the most abundant families of pigments found in nature. They are widely distributed in plants, fungi, algae, and some bacteria that naturally exhibit yellow, orange, and red colours.[16] Nearly all animals, including humans, do not have the endogenous enzymes required for the *de*

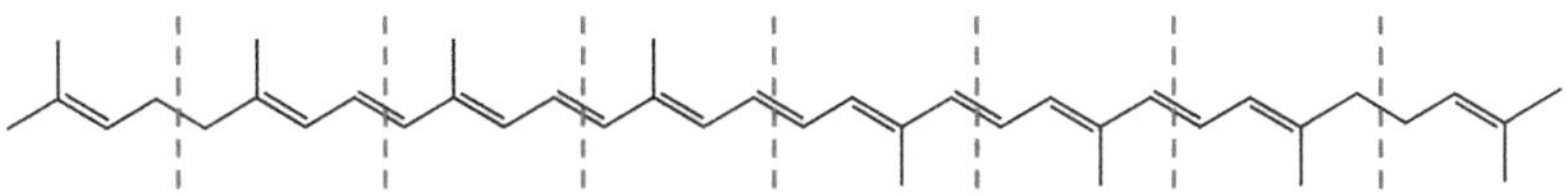

Figure 12.1 Basic chemical structure of carotenoids. The carotenoid general structure consists of eight C5 isoprene units (separated by red dashed lines) joined to form a C40 polyene chain.

novo synthesis of carotenoid pigments and, therefore, they rely on diet for their supply of carotenoids.[17] More than 700 carotenoids have been identified in nature, and they are characterized by a basic structure consisting of eight C5 isoprenoid units that result in a C40 backbone (Figure 12.1). The isoprenoid units are joined so that the order of the units is reversed at the center of the molecule.

The typical features of this group of compounds, therefore, include a long chain of conjugated double bonds that give carotenoids their ability to absorb visible light between 400 and 500 nm, which is responsible for their characteristic colouration in the yellow to red range.[18] The main group of carotenoids can be divided into two subgroups: carotenes, which are formed exclusively by carbon and hydrogen atoms, and xanthophylls, which also contain oxygen in their epoxy, carbonyl, hydroxyl, methoxy, or carboxylic acid functional group.[18] Carotenoids can also be further classified into provitamin A and non-provitamin A carotenoids. Provitamin A carotenoids can be cleaved by the β-carotene-15,15′-dioxygenase (BCO) to form retinal, which is reversibly reduced to retinol (vitamin A) or irreversibly oxidized to retinoic acid[19] (Figure 12.2). BCO is considered the key enzyme in the retinoid metabolism; it is present not only in the intestinal mucosa but also in the retinal pigment epithelium, where it may provide a local supplement of retinoids to photoreceptors.[20]

The ability of carotenoids to act as precursors of vitamin A is linked in part to the number of β-rings present in the molecular structure. Due to the presence of two β-rings in its molecule (Figure 12.3), β-carotene has the highest provitamin A activity, since it can give rise to two retinal molecules.[21] All the other provitamin A carotenoids, including α-carotene and β-cryptoxanthin, have one β-ring, thus they form only one retinal molecule. Carotenoids without β-rings, such as lycopene, have no provitamin A activity (Figure 12.3).

The interest in carotenoids found in plants is not only due to their provitamin A activity but also to their antioxidant activity. Carotenoids can scavenge two of the most dangerous ROS, the singlet oxygen and the peroxyl radicals; in addition, they are effective deactivators of electronically excited molecules that are involved in the generation of ROS and singlet oxygen.[13] The antioxidant activity of carotenoids is closely related to the number of conjugate double bonds in their structure. The longer the polyene chain that may stabilize the radical, the higher the antioxidant ability. There are three mechanisms of interaction between carotenoids and radical species; namely, electron

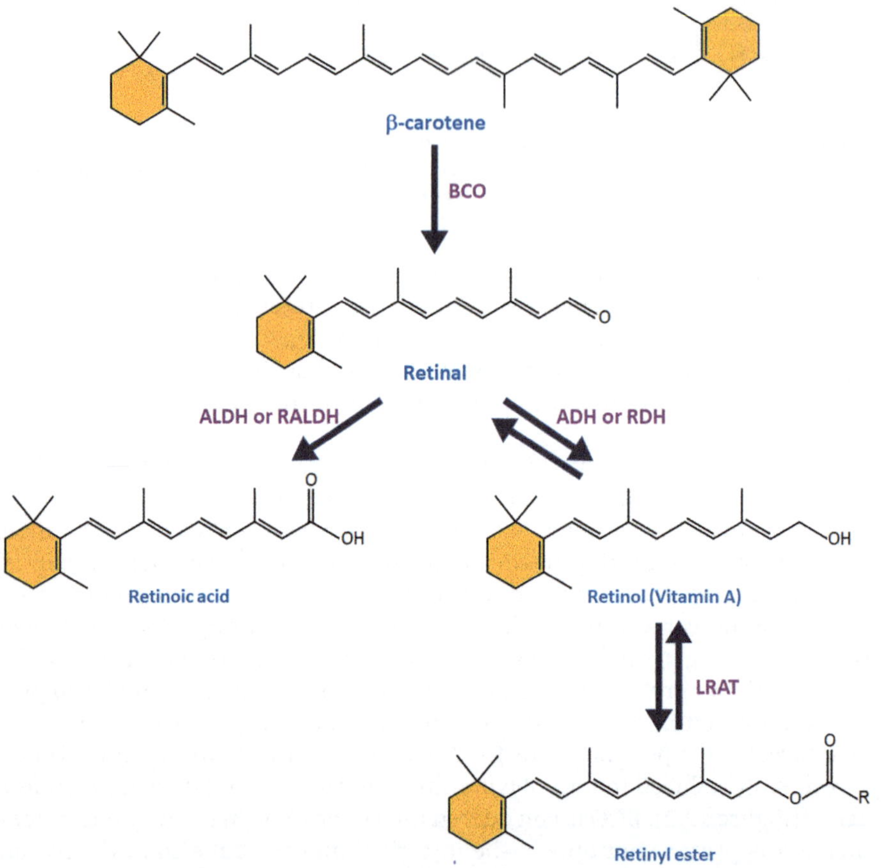

Figure 12.2 Main steps in retinoid formation from β-carotene in animals. β-Carotene-15,15′-dioxygenase (BCO) catalyzes the cleavage of β-carotene that yields two molecules of retinal. Subsequently, retinal can be irreversibly oxidized into retinoic acid by members of the aldehyde dehydrogenase family (ALDH or RALDH), or reversibly reduced into retinol by different forms of alcohol dehydrogenase (ADH) and retinol dehydrogenases (RDH). Retinol can be further esterified into retinyl esters by the enzyme lecithin:retinol acyltransferase (LRAT).

transfer, which results in the formation of the carotenoid (CAR) radical cation (eqn (1)) or CAR radical anion (eqn (2)), radical adduct formation (eqn (3)), and hydrogen atom transfer, which leads to the formation of the neutral CAR radical (eqn (4)):[22]

$$R^{\bullet} + CAR \rightarrow R^{-} + CAR^{+\bullet} \tag{1}$$

$$R^{\bullet} + CAR \rightarrow R^{+} + CAR^{-\bullet} \tag{2}$$

$$R^{\bullet} + CAR \rightarrow R\text{–}CAR^{\bullet} \quad (3)$$

$$R^{\bullet} + CAR \rightarrow RH + CAR^{\bullet} \quad (4)$$

The interruption of the conjugated double bond system either by cleavage or by addition to one of the double bonds – which occurs due to the carotenoid exposition to radical or oxidizing species – leads to the colour loss of the carotenoid molecule, a phenomenon known as carotenoid 'bleaching'.[22]

Evidence exists about a cooperative interaction between carotenoids and other antioxidant molecules. *In vitro* studies have demonstrated that β-carotene regenerates α-tocopherol from the α-tocopheroxyl radical, being the resulting carotenoid radical subsequently repaired by vitamin C.[23] Also, compared with single antioxidants, β-carotene, vitamin E, and vitamin C were found to act synergistically in the scavenging of reactive nitrogen species,[24] and a combination of both β-carotene and α-tocopherol resulted in an inhibition of lipid peroxidation significantly greater than the sum of the inhibition of the two individual antioxidants.[25]

12.3 Metabolic Pathways of Carotenoids

The colour intensity in yellow, orange, and dark orange cereal grains depends on the amount of carotenoid pigments accumulated, which, in turn, is the net result of biosynthesis, degradation, and stable accumulation.[26] Therefore, identifying the rate-limiting steps that regulate these metabolic pathways and the characterization of genes/alleles associated with higher carotenoid content, may be helpful for more efficient selection of new, highly pigmented genotypes.

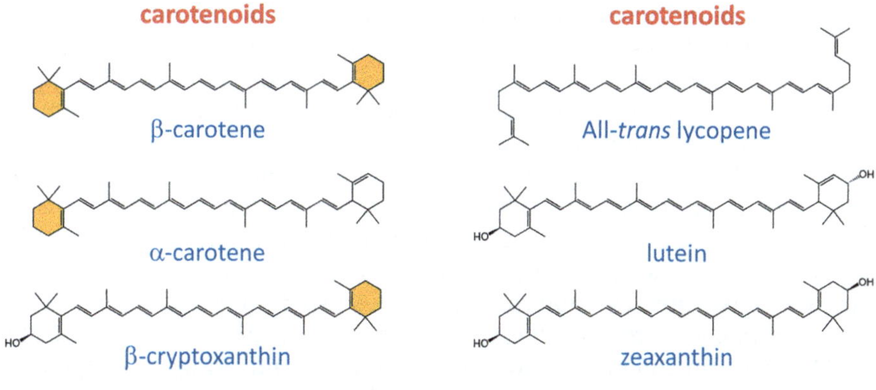

Figure 12.3 Chemical structures of some provitamin A and non-provitamin A carotenoids. The β-ring in provitamin A carotenoids is highlighted in orange.

12.3.1 Biosynthesis

The biosynthetic pathway of carotenoids has been thoroughly investigated in maize, primarily because of the availability of the *viviparous* mutants, which include mutants associated with reduced or suppressed carotenoid accumulation in the endosperm.[27] The biosynthesis of carotenoids starts in the plastid membranes with the isopentenyl pyrophosphate (IPP) synthesized through the methylerythritol 4-phosphate (MEP) pathway (Figure 12.4). Evidence has been reported that the reaction catalyzed by the enzyme 1-deoxy-D-xylulose 5-phosphate synthase (DXS), which supplies isoprenoid precursors from the MEP pathway, is a key bottleneck that may limit the flux throughout the entire pathway in cereal grains.[28,29] The IPP is isomerized to dimethylallyl pyrophosphate (DMAPP) by the enzyme isopentenyl pyrophosphate isomerase (IPPI) and the subsequent addition of three IPP molecules to the DMAPP leads to the formation of the geranylgeranyl pyrophosphate (GGPP) through the action of the geranylgeranyl pyrophosphate synthase (GGPPS) (Figure 12.4).

The first step in the carotenoid pathway is catalyzed by the phytoene synthase (PSY, CrtB in bacteria) and consists of the condensation of two GGPP molecules to form a 15-*cis* isomer of phytoene (15-*cis*-phytoene) (Figure 12.4). This reaction represents the first rate-limiting step in the biosynthetic pathway of carotenoids in the endosperm of cereal grains.[26] Evidence has been reported that the transcript abundance of the *PSY1* gene correlated with carotenoid content in maize endosperm,[30] and that allelic differences in the *PSY1* gene were associated with differences in grain yellow pigment content in wheat.[31]

The second step includes four desaturation and isomerization reactions, which lead to the conversion of the colourless 15-*cis*-phytoene to the red-coloured all-*trans*-lycopene (Figure 12.4). The desaturation reactions are catalyzed by the enzymes phytoene desaturase (PDS) and ζ-carotene desaturase (ZDS), whereas the carotenoid isomerase (CrtISO) is responsible for the isomerization of the *cis* configuration of lycopene into the *trans* configuration (Figure 12.4). A single enzyme (CrtI) is responsible for both the desaturation and isomerization reactions in bacteria. The activity of CrtISO in maize endosperm may control the rate of carotenoid accumulation in grains by affecting the availability of all-*trans*-lycopene.[32]

The third step is the cyclization of lycopene, which represents the first branching point in carotenoid biosynthesis. In this step, lycopene β-cyclase (LCYB) and lycopene ε-cyclase (LCYE) (CrtY in bacteria) compete for the substrate lycopene to form either β-carotene or α-carotene. In one branch LCYB catalyzes a two-step reaction that introduces two β-rings at each end of the lycopene molecule to form β-carotene, whereas in the other branch LCYB and LCYE act in conjunction to introduce both β- and ε-rings to lycopene to form α-carotene (Figure 12.4). The cyclization of lycopene has a major role in the determination of the β,β/β,ε branching ratio. In maize, LCYE played a key role in controlling this ratio, since the levels of predominant provitamin A compounds β-carotene and β-cryptoxanthin were highly associated with variation at the LCYE locus.[33]

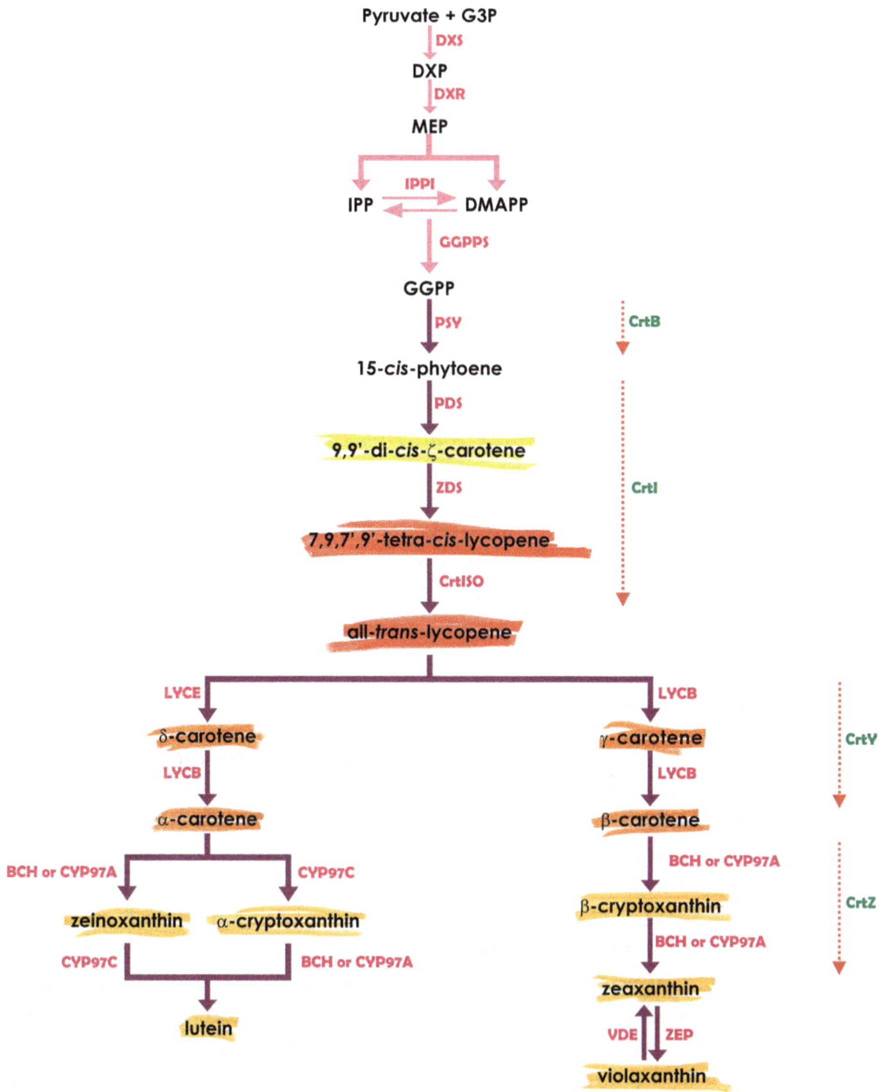

Figure 12.4 Carotenoid biosynthesis pathway in higher plants. G3P, glyceraldehyde 3-phosphate; DXS, 1-deoxy-D-xylulose 5-phosphate synthase; DXP, 1-deoxy-D-xylulose 5-phosphate; DXR, 1-deoxy-D-xylulose 5-phosphate reductoisomerase; MEP, methylerythritol 4-phosphate; IPP, isopentenyl pyrophosphate; IPPI, isopentenyl pyrophosphate isomerase; DMAPP, dimethylallyl pyrophosphate; GGPPS, geranylgeranyl pyrophosphate synthase; GGPP, geranylgeranyl pyrophosphate; PSY, phytoene synthase; PDS, phytoene desaturase; ZDS, ζ-carotene desaturase; CrtISO carotenoid isomerase; LYCB, lycopene β-cyclase; LYCE, lycopene ε-cyclase; BCH, non-heme di-iron carotene hydroxylases (also abbreviated to CHYB, HYD or CrtRB); CYP97A and CYP97C, heme-containing cytochrome P450 carotene β- and ε-ring hydroxylases; VDE, violaxanthin de-epoxidase; ZEP, zeaxhantin epoxidase; CrtB, bacterial phytoene synthase; CrtI, bacterial phytoene desaturase/isomerase; CrtY, bacterial lycopene β-cyclase; CrtZ, bacterial β-carotene hydroxylase.

The fourth step is the hydroxylation of α-carotene and β-carotene to xanthophylls. Hydroxylation of cyclic carotenes is catalyzed by two types of enzymes: the non-heme di-iron carotene hydroxylases (BCH, also abbreviated to CHYB, HYD or CrtRB, CrtZ in bacteria), which catalyze the hydroxylation of β-rings, and the cytochrome P450 carotene hydroxylases (CYP97 type), which catalyze the hydroxylation of both β- and ε-rings. The hydroxylation of the two β-rings of β-carotene leads to the formation of zeaxanthin, whereas the hydroxylation of the β-ring and the ε-ring of α-carotene leads to the formation of lutein (Figure 12.4). The BCH (CHYB, HYD or CrtRB) type of hydroxylases are mainly responsible for the conversion of β-carotene to zeaxanthin in cereals.[34] Consistently, in maize endosperm the *HYD3* alleles associated with reduced transcript expression correlated with higher β-carotene concentration.[35] Also, the suppression of the *HYD1* gene in durum wheat induced an increase of more than 70% of β-carotene compared to the wild-type.[36]

While lutein represents the end-product of the β,ε branch, zeaxanthin can be converted to violaxanthin by the enzyme zeaxanthin epoxidase (ZEP) (Figure 12.4). Violaxanthin is the precursor of the hormone abscisic acid (ABA) that in seeds plays a key role in dormancy, embryo maturation, and germination.[37]

12.3.2 Degradation

Carotenoid degradation also affects the amount of carotenoids accumulated in cereal grains. Carotenoids can undergo both specific cleavage and non-specific degradation. The enzymes responsible for the specific cleavage of carotenoids include the carotenoid cleavage dioxygenases (CCDs) that cleave the 9,10 double bonds of their carotenoid substrates, and the 9-*cis*-epoxycarotenoid dioxygenases (NCEDs) that cleave 11,12 double bonds.[38] The specific cleavage of carotenoids gives rise to various apocarotenoids that play important roles in many physiological processes of plants. In particular, the NCEDs are involved in the biosynthesis of the ABA hormone, whereas the carotenoid cleavage by CCDs produces apocarotenoids that are important components of plant aroma and colour and play a role in attracting pollinators.[38] In the last decade, evidence has accumulated on the role of CCDs in the production of strigolactones, a class of plant hormones that stimulate branching and colonization of roots by symbiotic arbuscular mycorrhizal fungi.[39] Evidence exists that CCDs can negatively regulate β-carotene accumulation in cereal grains. In wheat, a correlation analysis between the expression of *CCD* genes and the carotenoid content revealed that *CCD1* genes had a predominant role in the degradation of β-carotene in the endosperm.[40]

The non-specific mechanisms of carotenoid degradation include enzymatic and non-enzymatic oxidation. Enzymatic oxidation is triggered by lipoxygenases (LOXs). These are non-heme iron-containing dioxygenases that catalyze the hydroperoxidation of polyunsaturated fatty acids containing 1,4-*cis*,*cis* pentadiene structures to give the corresponding conjugated diene. The peroxy radicals produced in the intermediate steps of this reaction cleave carotenoids by the random attack of the polyene chromophore and lead to the consequent 'bleaching' of the carotenoid pigments.[41] LOX activity is mainly responsible for the oxidative degradation of carotenoids that occurs during processing of durum wheat semolina, leading to discolouration of its end-products.[42–44]

The non-enzymatic carotenoid degradation is strictly linked to the antioxidant activity of this class of compounds. The attack of the carotenoid molecule by ROS may lead to the generation of epoxy- and peroxyl-derivatives, which decompose into apocarotenoids. Since all the double bonds present in the carotenoid backbone can be attacked by radicals, the number of apocarotenoid species formed non-enzymatically far exceeds the number of those produced enzymatically.[45]

12.3.3 Accumulation

Except for proplastids, all plastids in fruits, flower, tubers, and seeds can accumulate carotenoids. These include chromoplasts, amyloplasts, leucoplasts, etioplasts, and elaioplasts, and their capacity to sequester and stably store carotenoids contributes to the final level of carotenoid accumulation in plant tissues.[46] Compared with the other plastids, chromoplasts accumulate high levels of carotenoids.[47] These organelles have developed a unique mechanism for the sequestration of massive amounts of carotenoids, which relies on the hydrophobic nature of carotenoids that allows these compounds to associate with lipid–protein complexes.[48] These carotenoid-lipoprotein structures probably serve as a metabolic sink to sequester carotenoids, thus preventing their accumulation in the chromoplast membranes, the site of carotenoid biosynthesis. The existence of such a metabolic sink enhances biosynthesis and determines a high storage capacity of carotenoids.[46]

Amyloplasts are specialized plastids for the storage of starch granules in the endosperm of starchy seeds like cereal grains. Unlike chromoplasts, amyloplasts accumulate low to medium levels of carotenoids.[32] The reduced ability of amyloplasts to accumulate carotenoids, compared with chromoplasts is linked to the lack of appropriate lipoprotein sequestering substructures in the amyloplasts, which results in the accumulation of carotenoids in the envelope membranes of these organelles.[49] This reduces the rate of carotenoid biosynthesis and limits the capacity of the amyloplasts to accumulate carotenoids.[50]

12.4 Carotenoids and Human Health

ROS are continuously generated as a by-product of normal aerobic metabolism. However, when the amount of ROS increases, it may cause damage to cell structures, in particular membrane lipids, proteins, and DNA.[51] Oxidative stress has been recognized as a major factor contributing to the onset and progression of several degenerative disorders, representing a growing cause of disability in developed countries. Carotenoids, with their ability to scavenge singlet oxygen and peroxy radicals and interact synergistically with other antioxidants, may lower the risk for several degenerative disorders. Indeed, data from various epidemiological studies and clinical trials have strongly supported the observation that adequate intake of carotenoid-rich foods may significantly reduce the risk of several chronic disorders, including age-related macular degeneration, cardiovascular diseases, neurodegenerative diseases, and certain types of cancer.[17,52] Moreover, dietary intake of provitamin A carotenoids has been shown to play a critical role in maintaining healthy vision, immune response, growth and differentiation, fertility, and fetal development.[53] Vitamin A deficiency remains a major public health problem in developing countries. As the populations of these areas strongly depend on cereal grains for their nutritional needs, the introduction of cereals with high levels of provitamin A carotenoids in their diet can provide a sustainable solution to eliminating vitamin A deficiency in these areas of the world.

12.4.1 Cardiovascular Diseases

Over the past decades, several epidemiological studies have supported the role of carotenoids in the prevention of cardiovascular diseases (CVD), such as atherosclerosis, hypertension, micro-angiopathy, coronary artery disease, myocardial infarction, and stroke.[54] Oxidatively modified low-density lipoproteins (LDL) are responsible for the onset and progression of CVD, and β-carotene and lycopene that are transported in LDL have a central role in preventing LDL oxidation. Consistently, carotenoid intake was found to be inversely associated with the concentration of plasma oxidized LDL.[55,56] In addition to their antioxidant properties, carotenoids can prevent cardiovascular diseases by reducing other CVD risk factors, such as hyperlipidemia, C-reactive protein, and total homocysteine concentrations. Studies indicated that higher serum carotenoid concentrations were inversely associated with some of these biomarkers. In a large population (more than 14 000 individuals), low serum levels of α-carotene, β-carotene, β-cryptoxanthin, lycopene, and lutein + zeaxanthin were found to be positively correlated with elevated C-reactive protein levels.[57] Also, a significant inverse correlation was observed between LDL-cholesterol levels and dietary intake of β-carotene and lutein + zeaxanthin, as well as between total homocysteine levels and dietary intake of β-carotene, lycopene, and total carotenoids.[58] A positive correlation was also found between dietary intake of lutein + zeaxanthin and HDL-cholesterol levels.[58]

12.4.2 Cancer

Evidence exists that oxidative stress is involved in the onset and progression of cancer, by increasing DNA mutations and inducing genomic instability if the oxidative DNA damage is not repaired prior to DNA replication.[59]

With their antioxidant activity, carotenoids have been believed to decrease the risk of certain types of cancer, such prostate, lung, and colon cancer. Among carotenoids, lycopene seems to have the most significant effects in reducing the risk of prostate cancer. A case-control study demonstrated the existence of an inverse, dose–response association between dietary lycopene and prostate cancer risk, whereas no association was found with other dietary carotenoids, such as α-carotene, β-carotene, β-cryptoxanthin, lutein, and zeaxanthin.[60] Also, a high-lycopene dietary intervention significantly reduced prostate oxidative DNA damage in patients diagnosed with prostate cancer.[61] An inverse association was observed between dietary lutein and colon cancer, whereas the association with other carotenoids was unremarkable.[62] High serum levels of zeaxanthin were associated with decreased rates of polyps and colorectal cancer development in males, whereas zeaxanthin, α-carotene, and β-carotene showed significant inverse association with colorectal cancer development in females.[63] Consistently, the same authors found an inverse association between the consumption of green–yellow vegetables and fruits and the development of polyps in males and colorectal cancer in females.[63] Several studies have examined associations between carotenoid intake and lung cancer, but the results are controversial. High dietary intakes of carotenoid-rich fruits and vegetables, as well as β-carotene, α-carotene, β-cryptoxanthin, and lycopene, were found to be associated with a decreased risk of lung cancer.[64,65] Conversely, chemoprevention trials showed that β-carotene alone or in combination with vitamin A or vitamin E increased lung cancer incidence and mortality in heavy smokers and asbestos workers.[66] This is probably linked to the instability of the β-carotene molecule in the lung exposed to cigarette smoke. Oxidized β-carotene metabolites may act as pro-oxidants, thus enhancing lung carcinogenesis.[67]

12.4.3 Neurodegenerative Disease

Alzheimer's disease is the principal cause of dementia throughout the world. Evidence has been reported that serum levels of lutein, zeaxanthin, β-cryptoxanthin, α-carotene, β-carotene and lycopene were significantly lower in patients affected by Alzheimer's disease than cognitively intact controls.[68] Consistently, high serum levels of lycopene and lutein + zeaxanthin were associated with a reduced risk of Alzheimer's disease mortality in adult patients.[69] Contrary to what was observed for Alzheimer's disease, data obtained on a possible association between increased intake of carotenoids and the reduced risk of Parkinson's disease are controversial. A meta-analysis carried out on data reported in epidemiological studies assessing the association between vitamin A and/or carotenoids and this pathology indicated only a weak and not significant inverse association between both α- and β-carotene and the risk of Parkinson's disease.[70] A meta-analysis was also carried out on nine studies on no demented subjects, and the results obtained revealed a significant effect of carotenoid intervention on cognitive outcomes.[71] Very recently, Tanprasertsuk and coworkers[72] constructed *a posteriori* the nutritional pattern of fat-soluble nutrients measured in brain tissues acquired post-mortem from 47 centenarians. Their results revealed that the brain nutrient pattern explained mainly by carotenoid concentrations correlated with cognitive function among subjects with no dementia.[72]

12.4.4 Visual Disorders

Vitamin A and its derivatives play a crucial role in vision and eye health. Visual pigments are composed of a transmembrane protein (opsin) covalently linked to the chromophore 11-*cis* retinal.[73] The eye's supply of vitamin A depends on dietary intake of preformed vitamin A from animal food, or in the form of provitamin A carotenoids from fruit and vegetables.[74] Individuals suffering hypovitaminosis A can develop night blindness, xerophthalmia, Bitot's spots, corneal ulcerations, and lesions.[75] Consistently, several cross-sectional studies carried out on young children in developing countries showed an association between low consumption of carotene-rich vegetables and fruits, and increased risk of xerophthalmia.[76] In the same manner, intervention studies in disadvantaged areas suggested that the vitamin A status of the populations was improved by promoting the cultivation and/or consumption of carotenoid-rich foods.[76]

A role in maintaining a healthy vision is also played by non-provitamin A carotenoids. Lutein and zeaxanthin are the major constituents of macular pigments[77] and numerous epidemiological studies have reported a relationship between dietary intake of these xanthophylls and the reduced risk of light-induced retinal damage and age-related macular degeneration,[78] the leading cause of blindness among the elderly in the Western world.[79] These carotenoid pigments exert their protective role on the macula through a simple light-filter effect and their antioxidant activity. They absorb light between 400 and 500 nm with a peak around 460 nm. These wavelengths are known to induce light-mediated damage to photoreceptors and the pigmentary epithelium of the retina. Both lutein and zeaxanthin may act as broadband filters, thus reducing the intensity of blue light and protecting the underlying retinal structures such as the photoreceptors.[80] The antioxidant properties of lutein and zeaxanthin may also protect the macula from oxidative stress.[81] The retina is susceptible to oxidative stress because of its high demand for oxygen, its aerobic metabolism and its high level of polyunsaturated fatty acids.[82] The various ROS produced under the aerobic metabolism may cause the peroxidation of the fatty acids in the membrane of retina photoreceptors. Due to their antioxidant properties, lutein and zeaxanthin can reduce changes in membrane permeability through scavenging ROS and free radicals.

12.5 Carotenoids in Pigmented Cereals

Compared with most fruits and vegetables, cereal grains have a lower carotenoid content.[83] However, cereals such as maize, wheat, and rice are staple foods in the diet of many populations, especially in underdeveloped and developing countries, and pigmented grains may represent an important source of carotenoids. On the other side, in developed countries, the demand for so-called 'functional' foods is increasing, driven by evidence of possible health benefits associated with the regular intake of food products rich in phytochemicals. In this regard, food products derived from pigmented cereals are of particular interest, and breeding of these crops for high carotenoid levels in grains could have a huge impact on human health.

Different strategies have been used to improve carotenoid levels and composition in maize, wheat, and rice. The germplasm of maize and durum wheat shows high genetic variability for the yellow to orange pigmentation of grains. Therefore, traditional breeding has been successfully applied to these crops to improve their carotenoid levels. Conversely, the low genetic variability of carotenoid content in bread wheat, as well as the absence of carotenoid accumulation in rice, warrants the adoption of genetic engineering strategies to manipulate carotenoid accumulation in these crops and obtain yellow to orange grains. Carotenoids in cereals are almost exclusively represented by xanthophylls, mainly lutein and zeaxanthin, whereas provitamin A carotenoids are present at low levels. Moreover, the accumulation of provitamin A carotenoids has very little or no naturally occurring variation in cereal grains. This prompted scientists to identify favourable alleles responsible for redirecting carotenoid pathway toward enhanced provitamin A carotenoid accumulation (mainly β-carotene), and breeders to use these alleles to select new genotypes with enhanced levels of provitamin A carotenoids through marker-assisted breeding and metabolic engineering approaches. In this regard, one of the most successful breeding interventions is the one conducted by HarvestPlus, a global interdisciplinary partnership programme that aims to develop and promote biofortified crops for a better nutrition, which has set a target of 15 μg of β-carotene in cereal grains to help in alleviating vitamin A deficiency in humans.[84] With this aim in mind, HarvestPlus researchers have developed orange maize varieties containing significant amounts of provitamin A carotenoids.[85] These varieties are currently cultivated in Zambia and their consumption has been demonstrated to improve total body vitamin A stores,[86] and significantly improve visual function in vitamin A-deficient children.[87]

Carotenoids are also present, although at a lower level, in anthocyanin-rich grains with a purple/blue pigmentation. It is generally agreed that the synergistic action of phytochemicals has numerous health benefits due to their antioxidant potential.[88] In light of this, the consumption of these pigmented cereals has great potential in promoting human health and in developing nutraceuticals and functional foods.

12.5.1 Maize

Maize is one of the most widely produced and consumed cereal crops worldwide. It represents the preferred staple food for 900 million poor consumers and is the most important food crop in sub-Saharan Africa.[89] Maize possesses the greatest natural variation in the colour of grains. It is possible to find a wide range of grain pigmentations, such as red, purple, blue, brown, or black, but the yellow to orange pigmentation determined by the carotenoid accumulation in the endosperm is the most common basic colour of maize grains. Among cereals, maize has the highest carotenoid content in grains,[32] as well as the highest variability in grain carotenoid content and carotenoid profile.[33,90] Indeed, the analysis of genotypes from different areas worldwide revealed that carotenoid content varied greatly, reaching values even higher than 60 μg g^{-1} [91–101] (Table 12.1). The yellow to orange colour of grains was determined almost exclusively by the accumulation of xanthophylls, which were mainly represented by lutein and zeaxanthin, each commonly accounting for 30–50% of total carotenoids, whereas the provitamin A carotenoids, α-carotene, β-carotene and β-cryptoxanthin, altogether constituted only 10–20% of total carotenoids (Table 12.1). In particular, β-carotene, which presents the highest provitamin A activity, reached a maximum of 6 μg, which is far below the daily requirement of 15 μg g^{-1} prescribed by HarvestPlus for humans.[84]

The high variability in carotenoid content has been extensively used by maize breeders to develop new pigmented genotypes with higher carotenoid levels and, particularly, with high β-carotene content. In this context, studies have been carried out to assess the possible association between the intensity of colour pigmentation of maize grains and the carotenoid content and profile. Several studies revealed that grain colour, when visually scored, was associated with the total carotenoid content.[33,96,102] Sivaranjani and coworkers[96] analyzed a panel of 111 Indian and exotic inbred lines and found a total carotenoid content among the deep orange inbred lines of 31.1 μg g^{-1}, whereas the same was found to be 28.03 μg g^{-1}, 23.97 μg g^{-1}, and 16.13 μg g^{-1} among the orange, yellow, and pale-yellow lines, respectively (Table 12.1). These findings indicate that the visual score of grain pigmentation should respond favourably to phenotypic selection and could represent a simple, low-cost and effective screening in breeding programs aimed at improving the total carotenoids in the maize grains. Results on a possible correlation between grain colour and provitamin A carotenoids are contrasting. In Brazilian maize genotypes, a significant correlation was found between the colour intensity of grains and the content of α-carotene and β-carotene.[98] Genotypes with light yellow grains presented α-carotene + β-carotene levels ranging from 2.79 μg g^{-1} to 4.50 μg g^{-1}, whereas higher levels, ranging from 4.62 μg g^{-1} to 8.63 μg g^{-1}, were detected in dark orange grains[98] (Table 12.1). Conversely, Harjes and coworkers[33] found a low correlation between visual selection of grain pigmentation and the levels of β-carotene and β-cryptoxanthin. Based on these findings, the authors suggested combining phenotypic selection with marker-assisted selection (MAS) to increase the proportion of provitamin A carotenoids in maize grains.

In this regard, favourable *LCYE* alleles with reduced functionality have been identified that may divert more lycopene to the β,β branch, thus improving the production of β-carotene.[33] However, most of the β-carotene

Table 12.1 Carotenoid content (μg g^{-1}) and composition in pigmented maize grains.

Genotype	Pigmentation	Total carotenoids	Lutein	Zeaxanthin	β-Cryptoxanthin	β-Carotene	α-Carotene	Ref.
Inbred lines from USA	Yellow-orange	26.32–38.23	8.64–13.67	13.10–20.75	2.17–3.51	1.22–2.42	—	91
and derived hybrids		17.9–51.4	4.7–17.5	8.9–30.7	1.1–4.9	0.5–3.4		
Genotypes from CYMMIT	Yellow-dark orange	—	1.33–32.31	0.38–34.88	0.00–6.13	0.00–5.81	0.00–2.31	92
Inbred lines from Africa	Yellow	—	0.1–19.0	0.3–32.4	0.3–5.5	0.3–4.7	0.0.1.9	93
Inbred lines from China	Yellow	0.096–22.495	0.040–17.504	0.022–6.728	0.012–3.66 6	0.016–1.726	0.004–0.852	94
Inbred lines from India	Yellow	5.58–63.90	—	—	—	0.12–4.74	—	95
Inbred lines from India	Deep orange	31.10	—	—	—	—	—	96
	Orange	28.03						
	Yellow	23.97						
	Pale yellow	16.13						
Inbred lines from Italy	Yellow	20.89–37.06	7.14–11.73	10.50–19.00	1.51–5.81	0.38–1.83	n.d.–0.28	97
and derived hybrids		24.74–33.81	6.96–13.93	12.47–17.07	3.12–6.64	0.61–1.60	n.d.–0.32	
Genotypes from Brazil	Light yellow	15.30–31.66	9.07–21.07[a]		3.45–6.10	2.79–4.50[b]		98
	Dark orange	30.05–44.45	20.57–27.95[a]		4.86–9.80	4.62–8.63[b]		
Inbred lines from USA	Light yellow-dark orange	9.55–62.96	1.23–23.93	1.44–32.40	0.13–5.17	0.31–3.27	0.45–2.65	99
Inbred lines from India	Yellow-orange	—	1.35–15.75	2.65–22.76	0.30–4.37	0.27–2.77	—	100
Hybrids from Africa	Orange	—	1.27–7.59	4.89–26.71	0.96–7.01	1.11–6.08	0.09–1.18	101
MAS-derived inbred lines	Orange	—	—	—	>15[c]		—	85
and derived hybrids					7–11[c]			
MAS-derived inbred lines	Yellow-orange	—	—	—	—	8.6–17.5	—	104
and derived hybrids						10.5–21.7		
MAS-derived inbred lines	Yellow-orange	—	—	—	7.38–13.59[c]		—	105
and derived hybrids					9.25–12.88[c]			
MAS-derived inbred lines	Yellow-orange	—	—	—	—	7.06–9.23	—	106
and derived hybrids						8.24–9.05		

(*continued*)

Table 12.1 (*continued*)

Genotype	Pigmentation	Total carotenoids	Lutein	Zeaxanthin	β-Cryptoxanthin	β-Carotene	α-Carotene	Ref.
Transgenic lines (*CrtB* + *CrtI*)	Yellow-orange	—	2.93–6.68	3.41–6.21	0.26–0.74	1.54–8.64	0.28–4.65	107
Transgenic lines (*ZmPSY1*)	Yellow	58.21	14.95	18.25	2.49	7.10	0.93	108
Transgenic lines (*ZmPSY1* + *CrtI*)	Orange-red	156.14	9.76	25.36	5.97	57.35	6.10	108 and 109
Genotypes from USA/Mexico border region	Red	3.21	—	—	—	—	—	111
	Blue	3.29						
	Pink	3.50						
	Brown	4.17						
	Purple	4.45						
	Yellow	8.36						
	Orange	8.48						
Genotypes from USA	Blue	—	0.05	0.14	0.03	0.23	—	112
	Red		1.22	1.12	0.13	0.20		
	Yellow		4.06	3.53	0.19	0.34		
Genotypes from China	Black	8.55	3.65	1.42	n.d.	—	—	113
	Yellow	20.16	9.65	4.14	0.40			
	Golden yellow	22.55	10.44	4.22	0.38			
Genotypes from Serbia	Dark blue	—	5.34	—	—	0.60	—	114
	Red-yellow		9.59			1.15		
	Red		13.89			1.68		
	Yellow		5.91			0.70		
	Orange		11.14			2.42		
Genotypes from Brazil	Purple	6.74–11.72	0.85–1.86	4.82–10.70	0.02–0.10	n.d.–0.06[b]		115
	Yellow-orange	1.49–10.86	0.48–3.69	0.07–7.05	n.d.–0.06	0.04–0.07[b]		

[a]These values corresponded to the sum of lutein plus zeaxanthin.
[b]These values corresponded to the sum of α-carotene plus β-carotene.
[c]These values corresponded to the provitamin A concentration, calculated as the sum of β-carotene plus half β-cryptoxanthin.
n.d. not detectable.

is hydroxylated to produce β-cryptoxanthin, with provitamin A activity 50% lower than that of β-carotene, and zeaxanthin, with no provitamin A activity. Therefore, blocking the hydroxylation reaction can potentially increase β-carotene levels in maize endosperm. Natural allelic variation in the *CrtRB1* gene responsible for β-carotene hydroxylation has been reported[35] and favourable alleles with reduced functionality have been identified that are able to induce from 2- to 10-fold increase in β-carotene concentration in maize endosperm.[103] Inbred lines and hybrids were developed using MAS targeting the favourable alleles of the *CrtRB1* and/or *LCYE* genes. HarvestPlus and its partners identified maize genotypes carrying the favourable *CrtRB1* allele in homozygous form with β-carotene concentrations up to 26 μg g^{-1} and total provitamin A as high as 30 μg g^{-1}.[85] These genotypes were used in MAS programs to develop orange inbred lines with more than 15 μg g^{-1} provitamin A carotenoids[85] that were in turn used to reconstitute orange maize hybrids with a provitamin A content ranging between 7 μg g^{-1} and 11 μg g^{-1} (Table 12.1), three of which were released in Zambia in 2015.[85] Yellow to orange inbred lines with significant enhancement of provitamin A carotenoids up to 17 μg g^{-1} were obtained through MAS approaches also by other research groups,[104–106] and values even higher (>20 μg g^{-1}) were reached in the derived hybrids[104] (Table 12.1).

Attempts have been made to improve the carotenoid content and profile also through transgenic approaches. Maize lines were metabolically engineered by overexpressing the bacterial *CrtB* and *CrtI* genes under the control of an endosperm-specific promoter.[107] The T2 seeds obtained were characterized by a yellow/orange pigmentation of grains and a β-carotene content that exceeded 8 μg g^{-1}[107] (Table 12.1). Carotenogenic genes, under the control of different endosperm-specific promoters, were differently combined by Zhu and coworkers[108] to transform a South African elite white maize variety. An extraordinary increase in the grain carotenoid content was obtained, and was associated with an intense pigmentation of grains. In particular, transgenic lines expressing the maize *PSY1* (*ZmPSY1*) gene were characterized by yellow-pigmented grains and a 53-fold increase in total carotenoids over the wild-type genotype (58.21 μg g^{-1} *vs.* 1.10 μg g^{-1})[108] (Table 12.1). A further increase in total carotenoids and a more intense pigmentation of grains was observed in lines transformed with the *ZmPSY1* gene and the bacterial *CrtI* gene, which showed a 142-fold increase in total carotenoid content compared to the wild-type genotype (156.14 μg g^{-1} *vs.* 1.10 μg g^{-1}) and orange to red pigmentation of grains[108] (Table 12.1). Interestingly, transformants carrying the *ZmPSY1* and the *CrtI* genes produced very high β-carotene levels (on average 57.35 μg g^{-1})[108] (Table 12.1). One of these lines became the first-generation of Carolight® variety, which is a biofortified line representing an excellent and sustainable way to alleviate vitamin A deficiency.[109] The pathway was extended even further to ketocarotenoids and this led to a 133-fold increase in total carotenoid content and to the accumulation of high amounts of ketocarotenoids, such as adonixanthin (22.36 μg g^{-1}) and astaxanthin (4.46 μg g^{-1}), which determined the red pigmentation of grains in these transformants.[108] These compounds are scarce in plants but are abundant in some microalgae, bacteria, and yeast, and are characterized by a strong antioxidant activity.[110]

Anthocyanin-rich coloured maize grains with a red, purple, or blue pigmentation also showed significant carotenoid levels. In most cases, these were lower than those observed in yellow and orange maize grains, although some exceptions have been reported. Ryu and coworkers[111] analyzed landraces from the USA/Mexico border region with diverse grain pigmentation. They found that the average carotenoid content in purple, blue, brown, red, and pink grains ranged from 3.21 μg g^{-1} to 4.45 μg g^{-1} and was significantly lower compared with that of yellow (8.36 μg g^{-1}) and orange (8.48 μg g^{-1}) grains[111] (Table 12.1). Similar findings were obtained by de la Parra and coworkers,[112] who found the highest amounts of xanthophylls and β-carotene in yellow grains, followed by red and blue grains (Table 12.1). Black grains also presented lower levels of total carotenoids, mainly lutein and zeaxanthin, compared with yellow and golden-yellow grains (8.55 *vs.* 20.16 and 22.55 μg g^{-1}, respectively)[113] (Table 12.1). Zilić and coworkers[114] found significantly lower levels of lutein in dark blue grains compared with yellow and orange grains (5.34 *vs.* 5.91 and 11.14 μg g^{-1}, respectively), whereas comparable or even higher levels were observed in red-yellow and red grains (9.59 and 13.89 μg g^{-1}, respectively) (Table 12.1). The authors also observed that dark blue and red-yellow grains presented β-carotene levels significantly higher compared with yellow grains (1.68 and 1.18 μg g^{-1}, respectively, *vs.* 0.70 μg g^{-1})[114] (Table 12.1). Two purple-pigmented Brazilian landraces were also identified by Kuhnen and coworkers,[115] which showed great potential as a source of carotenoids due to their high concentration of zeaxanthin, which was higher than that observed in yellow to orange grains (4.82–10.70 μg g^{-1} *vs.* 0.07–7.05 μg g^{-1}) (Table 12.1).

12.5.2 Wheat

Wheat is the third most important cereal crop worldwide after maize and rice, but it represents the most important cereal crop in the temperate regions, and is the stable food for 40% of the world's population.[116] Bread wheat accounts for 95% of the world's wheat production, whereas the other 5% is durum wheat.[116] Bread wheat flour is produced primarily for making bread, whereas durum wheat semolina is mainly used for making pasta, but is also an important ingredient for couscous and bulgur, and other typical local products. Both bread and durum wheat represent an important source of starch and are the leading source of vegetable protein in the human diet. Moreover, wheat grains represent a rich source of bioactive phytochemicals including carotenoids.

The yellow colour determined by the accumulation of carotenoid pigments in durum wheat grains is a desirable quality trait for semolina and pasta and, therefore, a high yellow pigment concentration is an important target in durum wheat breeding programs.[11] Several studies have been carried out to assess the carotenoid content and profile in different durum wheat genotypes[117–122] (Table 12.2). These studies described a high variability in the total carotenoids (from 0.5 μg g^{-1} to 11.8 μg g^{-1}) that are represented almost exclusively by lutein (60–80%) followed by zeaxanthin, whereas the provitamin A carotenoids, β-cryptoxanthin, β-carotene, and α-carotene are present only in traces[117–120] (Table 12.2). Carotenoid pigments are variably distributed in durum wheat grains. Lutein is accumulated almost exclusively in the endosperm, whereas zeaxanthin and β-carotene are concentrated near the outer layers of grains.[123,124] The interest of breeders in increasing the carotenoid content and, consequently, the yellowness of durum wheat grains, is evidenced by the fact that modern varieties present significantly higher total carotenoid content than old ones. Digesù and coworkers[117] analyzed a collection of 102 tetraploid accessions and found that durum wheat varieties released after 1991 had a significantly higher carotenoid content than those released before 1991 (1.857–4.416 μg g^{-1} *vs.* 1.178–3.532 μg g^{-1}) (Table 12.2). Over the past several years, durum wheat breeders have developed different techniques for a fast evaluation of the yellow pigmentation of semolina. Among these, the reflectance spectrophotometric measurement of semolina yellowness, known as Yellow Index (YI), was found to be the most reliable since it was found to be highly correlated with both lutein ($r = 0.95$) and β-carotene ($r = 0.91$).[125] Therefore, YI is currently used as a rapid and accurate method for the screening of yellow-pigmented grains in large-scale durum wheat breeding programs.

In the last decade, evidence has been accumulated that grain yellowness in durum wheat is genetically controlled. Allelic variation at the *Psy1-A1* and *Psy1-B1* loci was found to explain most of the variability in the yellow pigmentation of durum wheat grains[126,127] and molecular markers suitable for the selection of yellow-grained durum wheat varieties were developed. The *Psy-A1SSRe* allele associated with high yellow pigment content was introgressed into two Indian durum wheat varieties with low yellow pigment using marker-assisted backcrossing (MAB).[127] The MAB-derived durum wheat lines carrying the *Psy-A1SSRe* allele showed an 89% to 98% increase in yellow pigment content over recurrent parents (5.00–7.70 μg g^{-1} *vs.* 3.26–3.57 μg g^{-1}), which indicated the robustness of this marker[127] (Table 12.2). However, high carotenoid content in the endosperm does not guarantee an equally high content in pasta products because oxidative degradation of carotenoids by LOX enzyme can occur during pasta processing, leading to carotenoid loss and discolouration of the end-product. The LOX-encoding gene family *Lpx-B1* is mainly responsible for the LOX activity in durum wheat endosperm and evidence has been reported that durum wheat genotypes that carry the deleted *Lpx-B1.1c* allele are characterized by very low LOX activity in the endosperm.[44] Therefore, this allele has been included in the list of markers used in the MASWheat project financed by the US Department of Agriculture[128] to select new durum wheat lines characterized by low LOX activity in semolina and, consequently, by high yellow pigment content in pasta products.

Table 12.2 Carotenoid content (μg g^{-1}) and composition in pigmented wheat grains.

Genotype	Pigmentation	Total carotenoids	Lutein	Zeaxanthin	β-Cryptoxanthin	β-Carotene	α-Carotene	Ref.
Durum wheat varieties from Italy	Yellow							117
Released before 1970		1.178–3.345	0.721–2.395	0.109–0.184	0.006–0.011	0.007–0.020	0.006–0.018	
Released between 1971 and 1990		1.352–3.532	0.843–2.405	0.088–0.178	0.002–0.014	0.007–0.021	0.003–0.018	
Released between 1991 and 2008		1.857–4.416	1.332–3.078	0.805–0.176	0.006–0.013	0.008–0.033	0.007–0.029	
Durum wheat varieties from Italy	Yellow	1.95–6.33	1.64–5.50	0.14–0.39	—	0.06–0.44	—	118
Durum wheat varieties from North America and Europe	Yellow	0.5–8.8	0.3–8.0	0.1–2.0	—	—	—	119
Durum wheat varieties and breeding lines from different countries	Yellow	—	1.99–7.56	0.27–0.39	—	0.03–0.13	0.00–0.17	120
Durum wheat varieties from Italy	Yellow	4.766–11.838	—	—	—	—	—	121
Durum wheat landraces from Italy	Yellow	1.96–4.79	—	—	—	—	—	122
MAB-derived durum wheat lines from India	Yellow	5.00–7.46						127
Bread wheat varieties from Europe	White	0.43–1.74	—	—	—	—	—	129
Bread wheat variety 'Chinese Spring'	White	0.603	0.391	0.213	—	—	—	130
Bread wheat varieties from China	White	0.349–1.159	0.183–0.939	0.049–0.120	—	0.000–0.369	—	131
Transgenic bread wheat lines (*ZmPSY1* + *CrtI*)	Yellow	2.31–4.96	—	—	—	—	—	133
Transgenic bread wheat lines (*CrtB* + *CrtI*)	Yellow-dark red	0.64–4.76	0.20–0.36	0.12–0.53	0.02–0.29	0.12–3.21	0.02–0.34	134
Transgenic bread wheat lines (*CrtB* + *HYD silencing*)	Yellow-red	3.26–9.32	0.34–2.15	0.39–0.59	0.13–0.16	1.52–5.05	0.29–0.36	135

Eikorn wheat accessions from Mediterranean countries	Yellow	5.72–13.64	3.92–12.64	—	—	n.d.–0.65	n.d.–1.74	136
Eikorn wheat variety 'Monlis'	Yellow	8.09	6.75	0.29	0.10	0.96[a]		137
Eikorn wheat accessions from Mediterranean countries	Yellow	4.73–8.22	3.26–6.38	0.52–1.02	0.17–0.42	0.35–0.72[a]		138
Wild barley	Yellow	6.14	5.38	0.74	—	0.02	—	139
Primary tritordeum accessions	Yellow	2.6–8.8	—	—	—	—	—	143
Secondary tritordeum accessions	Yellow	3.7–9.4						
Wheat genotypes from Czech Republic and Austria	Purple	—	1.67	0.34	—	—	—	144
	Blue		1.24	0.29				
	Yellow		3.62	0.34				
Wheat genotypes from Czech Republic	Purple	3.60	2.91	0.59	—	0.06	n.d.	145
	Blue	2.41	1.76	0.53		0.03	n.d.	
	Yellow	7.04	5.16	1.17		0.22	0.05	
Wheat genotypes from Czech Republic	Purple	1.15–2.72	—	—	—	—	—	146
	Blue	1.16–2.07						
	Yellow	3.54–7.10						

[a]These values corresponded to the sum of α-carotene plus β-carotene.
n.d. not detectable.

Unlike durum wheat, white grain genotypes are usually selected for bread wheat because consumers prefer white bread made from bread wheat. Therefore, the levels of carotenoid pigments accumulated in bread wheat endosperm are very low (ranging between 0.3 and 1.7 μg g^{-1})[129–131] (Table 12.2). However, the increasing weight of nutritional aspects in the definition of food quality has increased the importance of improving individual carotenoids, particularly provitamin A carotenoids, not only in durum wheat but also in bread wheat. Therefore, transgenic approaches have been exploited to increase the carotenoid content and confer the typical yellow pigmentation to bread wheat grains. He and coworkers[132] referred to these pigmented lines as 'golden wheat', in analogy with the more famous 'golden rice' and 'golden potato' lines also obtained with transgenic approaches and characterized by yellow to orange pigmentation of endosperm and tubers, respectively. Transgenic bread wheat lines were obtained by overexpressing the *ZmPSY1* gene, under the control of an endosperm-specific promoter, and the bacterial *CrtI* gene, under the control of a constitutive promoter.[133] An increase in the total carotenoid content was observed in the transgenic lines up to 10.8-fold as compared with the wild-type genotype (4.96 μg g^{-1} *vs.* 0.46 μg g^{-1}) (Table 12.2). Consistently, the transformed grains showed a yellow colour of the endosperm, which indicated carotenoid accumulation, whereas the endosperm of the wild-type grains was bright white.[133] Wang and coworkers[134] substituted the *ZmPSY1* gene with the bacterial *CrtB* and obtained transgenic bread wheat lines with yellow-dark red grains and a maximum increase in carotenoid levels of 8-fold compared with the wild-type genotype (4.76 μg g^{-1} *vs.* 0.59 μg g^{-1}) (Table 12.2). Interestingly, the most abundant carotenoid in these transgenic lines was β-carotene, which accounted for up to 67% of total carotenoids (3.21 μg g^{-1})[134] (Table 12.2). A similar result was obtained by Zeng and coworkers,[135] who simultaneously overexpressed the bacterial *CrtB* gene and silenced the *HYD* gene in bread wheat endosperm. This approach resulted in a significant increase in the total carotenoid content compared to the wild-type genotype (up to 9.32 μg g^{-1} *vs.* 1.22 μg g^{-1}), with β-carotene accounting for up to 54% of total carotenoids (5.05 μg g^{-1}) (see Table 12.2). Consistently, grains of these transgenic lines presented a yellow to red pigmentation confirmed by measurements of the YI, which showed higher values than the wild-type genotype and was significantly correlated with lutein and β-carotene content.[135]

Among triticae, high carotenoid levels can be found in some ancient and wild species, which may represent an interesting gene pool to improve yellow pigment content in wheat. In einkorn, a total carotenoid content ranging from 4.73 μg g^{-1} to 13.64 μg g^{-1} was detected[136–138] (Table 12.2). As in durum wheat, lutein was the most common carotenoid (on average 90%), but several einkorn accession had up to 25–33% of α-carotene + β-carotene, which, instead, were present at very low or even undetectable levels in durum and bread wheat.[136] High carotenoid levels were also found in the wild barley. Mellado-Ortega and Hornero-Mendez[139] reported an average total carotenoid content of 6.14 μg g^{-1} in the small grains of the wild barley, with lutein as the major component (5.38 μg g^{-1}) and β-carotene only in trace amounts (0.02 μg g^{-1}) (Table 12.2). Up to 55% of lutein in wild barley grains was found to be esterified with palmitic and linoleic acids. This finding is of particular interest, since the esterification of xanthophylls with fatty acid extends the stability and preserves the antioxidant activity of these pigments.[140] This wild species was crossed with durum wheat to generate the hexaploid tritordeum,[141] a species that could be used as a vector to transfer genes for increasing carotenoid content in wheat.[142] Primary tritordeum accessions, directly obtained from the chromosome doubling of the hybrid obtained from the cross, exhibited a total carotenoid content ranging between 2.6 μg g^{-1} and 8.8 μg g^{-1}[143] (Table 12.2). Higher values were detected in secondary tritordeum accessions obtained from the tritordeum breeding programs, which showed a total carotenoid content ranging between 3.7 and 9.4 μg g^{-1}[143] (Table 12.2).

Carotenoids are also present in purple and blue wheat grains, but their accumulation in these pigmented grains was found to be lower compared with yellow grains, with purple grains generally exhibiting higher levels compared with blue ones (1.15–3.60 μg g^{-1} *vs.* 1.16–2.41 μg g^{-1})[144–146] (Table 12.2). Lutein was the main carotenoid, followed by zeaxanthin, whereas α-carotene and β-carotene were present at very low or even indetectable levels[145] (Table 12.2). Fractionation of the milled samples revealed that carotenoids are distributed differently in purple and blue wheat grains. Only 42% of lutein + zeaxanthin was recovered in the flour fraction of purple wheat, whereas the flour fraction of blue wheat contained more than double the amount in the bran, thus indicating a higher concentration of carotenoid pigments in the endosperm.[147]

12.5.3 Rice

Rice represents an important food crop worldwide. More than 90% of the world's rice is cultivated in Asia, where it represents a staple food that accounts for 35–75% of the calories consumed by the population.[148] Determinations of the carotenoid pigments in raw rice grains highlighted much lower levels than in common non-rice cereals. The major carotenoids are β-carotene and lutein (roughly 100 ng g^{-1}), while zeaxanthin levels are lower (roughly 30 ng g^{-1}).[149] These pigments are accumulated almost exclusively in the bran and in the outer endosperm, whereas they are absent in the core endosperm.[149]

The absence of carotenoids in the endosperm of rice grains prompted scientists to adopt genetic engineering strategies to manipulate carotenoid accumulation in this tissue and develop new genotypes characterized by a yellow to orange pigmentation of grains. In particular, the efforts have been directed towards generating new rice lines characterized by a high β-carotene content in the endosperm. Ye and coworkers[150] developed the first genetically modified rice with high carotenoid content. The authors introduced simultaneously, in the background of the *Japonica* variety Taipei309, the *Narcissus pseudonarcissus PSY* (*NpPSY*) gene, under the control of an endosperm-specific promoter, and the bacterial *CrtI* gene, under a constitutive promoter. The genetically modified rice was named 'Golden Rice' due to the bright yellow pigmentation of its grains. The best line presented 1.6 μg g^{-1} total carotenoids and 0.7 μg g^{-1} β-carotene (Table 12.3). Successively, new transgenic lines were developed by Syngenta using as background the *Javanica* variety Cocodrie and a modified vector in which the bacterial *CrtI* gene was under an endosperm-specific promoter control.[151] These new yellow lines were named 'Golden Rice 1' and showed significantly higher carotenoid contents, up to 6.0 μg g^{-1}, than the 'Golden Rice'[151] (Table 12.3). But the highest increase in carotenoids was obtained by Paine and coworkers.[152] The authors hypothesized that the NpPSY was the limiting step in the carotenoid accumulation in 'Golden Rice 1'. Therefore, they tested PSY from different sources and identified the ZmPSY as the one able to increase the carotenoid accumulation in the transgenic lines. The introduction of the *ZmPSY* gene together with the bacterial *CrtI* gene generated the 'Golden Rice 2'. The transgenic lines presented a marked yellow to orange pigmentation of grains and a carotenoid content that ranged from 8.8 μg g^{-1} to 36.7 μg g^{-1} (Table 12.3). Interestingly, the highest levels of carotenoid accumulation were found to be associated with a very high proportion (80–90%) of β-carotene, between 7.1 μg g^{-1} and 30.9 μg g^{-1}[152] (Table 12.3). Since the overexpression of biosynthetic enzymes may lead to the depletion of biosynthetic precursors, limiting the accumulation of the desired product,[153] the *ZmPSY* and the bacterial *CrtI* genes were co-expressed with the *Arabidopsis thaliana DXS* (*AtDXS*) gene, which supplies metabolic precursors from the MEP pathway (genotype D), or the *Arabidopsis thaliana ORANGE* (*AtOR*) gene, which promotes the formation of a metabolic sink by inducing chromoplast differentiation (genotype O).[29] Transgenic rice seeds were obtained characterized by an orange endosperm and a carotenoid content that ranged from 14.94 μg g^{-1} to 31.78 μg g^{-1} in genotypes D, and from 11.53 μg g^{-1} to 25.83 μg g^{-1} in genotypes O[29] (Table 12.3). Both genotypes were characterized by high levels of the provitamin A carotenoids β-carotene and α-carotene, with values of 7.50–16.61 μg g^{-1} and 3.60–9.70 μg g^{-1}, respectively, in genotype D, and 5.87–10.52 μg g^{-1} and 3.20–9.70 μg g^{-1}, respectively, in genotypes O[29] (Table 12.3). Very recently, an optimized CRISPR-Cas9-based method was used to achieve the targeted insertion of the carotenoid biosynthesis cassette previously used to obtain the 'Golden Rice 2'. Rice plants were obtained that were characterized by yellow pigmented grains and a carotenoid content that ranged between 7.74 μg g^{-1} and 8.12 μg g^{-1}[154] (Table 12.3).

Transgenic rice lines producing diverse carotenoids beyond β-carotene have also been generated. Four genes were synthesized with codon optimization for rice based on the protein sequences of ZmPSY1, bacterial CrtI, and algal HYD and β-carotene ketolase (BKT), and used to promote the accumulation in rice endosperm of the ketocarotenoid astaxanthin. The transgenic lines obtained (called aSTARice) presented grains with an orange-red pigmentation and astaxanthin levels that ranged from 9.76 μg g^{-1} to 16.23 μg g^{-1}, accounting for nearly 73% of the total carotenoid content (13.50–21.94 μg g^{-1})[155] (Table 12.3).

Table 12.3 Carotenoid content (μg g^{-1}) and composition in pigmented rice grains.[a]

Genotype	Pigmentation	Total carotenoids	β-Carotene	α-Carotene	Lutein	Zeaxanthin	β-Cryptoxanthin	Astaxanthin	Ref.
Transgenic 'Golden Rice' (*NpPSY* + CrtI)	Yellow	1.6	0.7	—	—	—	—	—	150
Transgenic 'Golden Rice 1' (*NpPSY* + *CrtI*)	Yellow-orange	6.0	—	—	—	—	—	—	151
Transgenic 'Golden Rice 2' (*ZmPSY* + *CrtI*)	Yellow-orange	8.8–36.7	7.1–30.9	1.1–5.1	—	—	0.1–0.4	—	152
Transgenic lines (*ZmPSY* + *CrtI* + *AtDXS*)	Orange	14.94–31.78	7.50–16.61	3.60–9.70	0.26–0.69	0.13–0.33	0.39–0.41	—	29
Transgenic lines (*ZmPSY* + *CrtI* + *AtOR*)	Orange	11.53–25.83	5.87–10.52	3.20–9.70	0.55–1.73	0.23–0.57	0.19–1.13	—	29
CRISPR-Cas9 edited lines (*ZmPSY* + *CrtI*)	Yellow	—	7.74–8.12	—	—	—	—	—	154
Transgenic lines 'aSTARice' (*ZmPSY1*+*CrtI* + *HYD* + *BKT*)	Orange-red	13.50–21.94	n.d.	n.d.	0.73–3.52	n.d.	n.d.	9.76–16.23	155
Korean varieties	Black	4.82	0.21–0.48	—	4.41–6.43	0.04–0.16	—	—	157
	Red	0.52	0.06–0.07		0.22–0.23	0.04–0.05			
Indian variety 'Kavuni'	Brownish-black	—	0.462	—	2.216	—	—	—	158
Japan variety 'Asamurasaki'	Black-purple	3.9	0.14	—	3.05	0.65	—	—	159
Camargue's varieties	Black	6.6	0.2	—	4.3	1.9	—	—	159
	Red	0.5	n.d.		0.4	0.1			
	Brown	0.08	n.d.		0.07	0.01			

[a]n.d. not detectable.

Although rice grains with yellow or orange pigmentation do not exist in nature, evidence has been reported that rice grains with other pigmentations, *e.g.* purple, black and red, possess relatively high levels of different carotenoids. Similarly to white rice varieties, these pigmented varieties accumulated most of the carotenoid pigments in the outer layers of grains, as demonstrated by Nakornriab and coworkers.[156] Marked differences in the carotenoid content were observed that depended upon grain colour. Kim and coworkers[157] assessed the carotenoid content in grains of black and red Korean varieties and reported the highest carotenoid content in black varieties with an average value 4.82 μg g^{-1}, whereas red varieties showed an average value of 0.52 μg g^{-1} (Table 12.3). Lutein was the most abundant carotenoid, with values that ranged between 4.41 μg g^{-1} and 6.43 μg g^{-1} in black rice, whereas much lower values ranging between 0.22 μg g^{-1} and 0.23 μg g^{-1} were detected in red varieties (Table 12.3). Provitamin A β-carotene levels were also found to be significantly higher in black rice than in red rice varieties (0.21–0.48 μg g^{-1} *vs.* 0.06–0.07 μg g^{-1}) (Table 12.3). Similarly, the Indian brownish-black rice variety 'Kavuni' was reported to possess detectable levels of β-carotene (0.462 μg g^{-1}) and significantly higher levels of lutein (2.216 μg g^{-1})[158] (Table 12.3). The Japanese black-purple rice variety 'Asamurasaki' contained 3.9 μg g^{-1} of total carotenoids, mainly consisting of lutein (3.05 mg g^{-1}), with small amounts of zeaxanthin and β-carotene that occurred as minor components (0.65 μg g^{-1} and 0.14 μg g^{-1}, respectively)[159] (Table 12.3). In line with these findings, the same authors analyzed different pigmented rice grown in the Camargue region of France and reported a total carotenoid content of 6.6 μg g^{-1}, 0.5 μg g^{-1}, and 0.08 μg g^{-1} in black, red, and brown rice grains, respectively[160] (Table 12.3).

12.6 Conclusions

The high daily consumption of cereal-based foods around the world, and the beneficial effect of carotenoids in the prevention and management of many diseases, will ensure interesting future prospects for pigmented cereals in the field of human nutrition and nutraceuticals. Foods based on pigmented grains will have an excellent chance of meeting the favour of consumers since their introduction in the diet will not involve a drastic change in eating habits, but only a change in food composition. In the light of this, it is feasible that the development and delivery of pigmented cereals could offer immense health benefits to human beings by alleviating micronutrient deficiencies in the world's poorest people and reducing the risk of chronic diseases linked to the lifestyle typical of the industrialized countries. Cereal breeders are continuously engaged in selecting new genotypes of pigmented cereals with improved properties, including high yield, good technological quality and higher phytochemical content. In most cases, traditional breeding approaches have been successfully used to enhance the carotenoid levels in cereal grains, sometimes drawing on wild relatives of common cereals, which represent an excellent source of variability for carotenoid content. Moreover, in the last decade, significant advances have been made in understanding carotenoid metabolism and regulation, and favourable alleles for higher carotenoid content and better carotenoid composition have been identified. Consequently, MAS and genetic engineering have emerged as important breeding strategies for selecting new pigmented genotypes with higher levels of nutritionally relevant carotenoids. Recently, new technologies have quickly evolved that will aid breeders in achieve this goal in the future. The genome editing tools, such as transcription activator-like effector nucleases (TALENs) and clustered regularly interspaced palindromic repeat-associated protein 9 (CRISPR/Cas9), provide the means to modify target genes rapidly and precisely, thus making possible to introduce desirable traits directly into elite varieties and avoid the time-consuming crossing cycles typical of traditional breeding approaches. It is feasible that in the future the genome-editing technology, in combination with other breeding techniques, will be successfully applied to select more colourful and nutritious cereal grains that will make our lives healthier.

References

1. FAO Cereal Supply and Demand Brief, available at https://www.fao.org/worldfoodsituation/csdb/en/, accessed on 5 February 2022.
2. L. Cordain, *World Rev. Nutr. Diet.*, 1999, **84**, 19.
3. O. O. Olugbire, S. Olorunfemi and D. O. Oke, *J. Trop. Agric. Food Environ. Ext.*, 2021, **20**(1), 9.
4. S. O. S. Saldivar, *Encyclopedia of Food Sciences and Nutrition*, ed. B. Caballero, L. Trugo and P. Finglas, Academic Press, London, 2nd edn, 2003, p. 1027.
5. B. McKevith, *Nutr. Bull.*, 2004, **29**, 111.
6. C. Leitzmann, *Forsch. Komplementmed.*, 2016, **23**(2), 69.
7. D. B. M. Ficco, A. M. Mastrangelo, D. Trono, G. M. Borrelli, P. De Vita, C. Fares, R. Beleggia, C. Platani and R. Papa, *Crop Pasture Sci.*, 2014, **65**, 1.
8. S. H. Arunima, G. S. Kumar, R. Krishnan and T. Mohammed, *Pharma Innovation*, 2021, **10**(5), 317.
9. E. S. M. Abdel-Aal, J. C. Young and I. Rabalski, *J. Agric. Food Chem.*, 2006, **54**(13), 4696.
10. V. U. Ndolo and T. Beta, *Food Chem.*, 2013, **139**(1–4), 663.
11. D. Trono, *Plants*, 2019, **8**(12), 551.
12. J. C. McCann, *Maize and Grace: Africa's Encounter with a New World Crop 1500-2000,* Harvard University Press, Cambridge, 1st edn, 2005.
13. A. J. Young and G. L. Lowe, *Antioxidants*, 2018, **7**(2), 28.
14. I. Liguori, G. Russo, F. Curcio, G. Bulli, L. Aran, D. Della-Morte, G. Gargiulo, G. Testa, F. Cacciatore, D. Bonaduce and P. Abete, *Clin. Interventions Aging*, 2018, **13**, 757.
15. J. D. Ribaya-Mercado and J. B. Blumberg, *J. Am. Coll. Nutr.*, 2004, **23**(6 Suppl), 567S.

16. H. E. Khoo, K. N. Prasad, K. W. Kong, Y. Jiang and A. Ismail, *Molecules*, 2011, **16**, 1710.
17. A. V. Rao and L. G. Rao, *Pharmacol. Res.*, 2007, **55**, 207.
18. G. Britton, S. Liaaen-Jensen and H. Pfander, *Carotenoids HandBook*, Birkhäuser, Basel, 2004.
19. H. K. Biesalski, G. R. Chichili, J. Frank, J. von Lintig and D. Nohr, *Vitam. Horm.*, 2007, **75**, 117.
20. W. Yan, G. F. Jang, F. Haeseleer, N. Esumi, J. Chang, M. Kerrigan, M. Campochiaro, P. Campochiaro, K. Palczewski and D. J. Zack, *Genomics*, 2001, **72**(2), 193.
21. T. Grune, G. Lietz, A. Palou, A. C. Ross, W. Stahl, G. Tang, D. Thurnham, S. A. Yin and H. K. Biesalski, *J. Nutr.*, 2010, **140**(12), 2268S.
22. N. I. Krinsky and K. J. Yeum, *Biochem. Biophys. Res. Commun.*, 2003, **305**(3), 754.
23. F. Böhm, R. Edge, E. J. Land, D. J. McGarvey and T. G. Truscott, *J. Am. Chem. Soc.*, 1997, **119**, 621.
24. F. Böhm, R. Edge, D. J. McGarvey and T. G. Truscott, *FEBS Lett.*, 1998, **436**, 387.
25. P. Palozza and N. I. Krinsky, *Arch. Biochem. Biophys.*, 1992, **297**, 184.
26. S. Zhai, X. Xia and Z. He, *Front. Plant Sci.*, 2016, **7**, 1197.
27. D. Robertson, *Genetics*, 1955, **40**, 745.
28. W. B. Suwarno, K. V. Pixley, N. Palacios-Rojas, S. M. Kaeppler and R. Babu, *Theor. Appl. Genet.*, 2015, **128**, 851.
29. C. Bai, T. Capell, J. Berman, V. Medina, G. Sandmann, P. Christou and C. Zhu, *Plant Biotechnol. J.*, 2016, **14**(1), 195.
30. B. Buckner, P. S. Miguel, D. Janick-Buckner and J. L. Bennetzen, *Genetics*, 1996, **143**, 479.
31. W. Zhang and J. Dubcovsky, *Theor. Appl. Genet.*, 2008, **116**(5), 635.
32. C. A. Howitt and B. J. Pogson, *Plant, Cell Environ.*, 2006, **29**(3), 435.
33. C. E. Harjes, T. R. Rocheford, L. Bai, T. P. Brutnell, C. B. Kandianis, S. G. Sowinski, A. E. Stapleton, R. Vallabhaneni, M. Williams, E. T. Wurtzel, J. Yan and E. S. Buckler, *Science*, 2008, **319**(5861), 330.
34. Q. Li, G. Farre, S. Naqvi, J. Breitenbach, G. Sanahuja, C. Bai, G. Sandmann, T. Capell, P. Christou and C. Zhu, *Transgenic Res.*, 2010, **19**(6), 1053.
35. J. Yan, C. B. Kandianis, C. E. Harjes, L. Bai, E. H. Kim, X. Yang, D. J. Skinner, Z. Fu, S. Mitchell, Q. Li, M. G. Salas Fernandez, M. Zaharieva, R. Babu, Y. Fu, N. Palacios, J. Li, D. Dellapenna, T. Brutnell, E. S. Buckler, M. L. Warburton and T. Rocheford, *Nat. Genet.*, 2010, **42**, 322.
36. M. D. Garcia Molina, E. Botticella, R. Beleggia, S. Palombieri, P. De Vita, S. Masci, D. Lafiandra and F. Sestili, *Theor. Appl. Genet.*, 2021, **134**(12), 4013.
37. M. P. Maluf, I. N. Saab, E. T. Wurtzel and M. M. Sachs, *J. Exp. Bot.*, 1997, **48**(311), 1259.
38. M. E. Auldridge, D. R. McCarty and H. J. Klee, *Curr. Opin. Plant Biol.*, 2006, **9**, 315.

39. K. P. Jia, L. Baz and S. Al-Babili, *J. Exp. Bot.*, 2018, **69**(9), 2189.
40. N. Thakur, Flowerika, N. Thakur, S. Khan, A. K. Pandey and S. Tiwari, *3 Biotech*, 2021, **11**(5), 221.
41. G. M. Borrelli and D. Trono, *Int. J. Mol. Sci.*, 2016, **17**(7), 1177.
42. D. Trono, D. Pastore and N. Di Fonzo, *J. Cereal Sci.*, 1999, **29**, 99.
43. D. Pastore, D. Trono, L. Padalino, S. Simone, D. Valenti, N. Di Fonzo and S. Passerella, *J. Cereal Sci.*, 2000, **31**, 41.
44. A. Verlotta, V. De Simone, A. M. Mastrangelo, L. Cattivelli, R. Papa and D. Trono, *BMC Plant Biol.*, 2010, **10**, 263.
45. P. Schaub, F. Wüst, J. Koschmieder, Q. Yu, P. Virk, J. Tohme and P. Beyer, *J. Agric. Food Chem.*, 2017, **65**, 6588.
46. C. I. Cazzonelli and B. J. Pogson, *Trends Plant Sci.*, 2010, **15**(5), 266.
47. J. T. Kirk and R. A. Tiliney-Bassett, *The Plastids: Their Chemistry, Structure, Growth and Inheritance*, ed. S. T. Kirck and R. A. Tiliney-Bassett, Elsevier/North Holland Biomedical Press, Amsterdam, 1978, p. 217.
48. L. Li and H. Yuan, *Arch. Biochem. Biophys.*, 2013, **539**(2), 102.
49. A. B. Lopez, J. Van Eck, B. J. Conlin, D. J. Paolillo, J. O'Neill and L. Li, *J. Exp. Bot.*, 2008, **59**, 213.
50. L. Li, Y. Yang, Q. Xu, K. Owsiany, R. Welsch, C. Chitchumroonchokchai, S. Lu, J. Van Eck, X. X. Deng, M. Failla and T. W. Thannhauser, *Mol. Plant.*, 2012, **5**(2), 339.
51. J. Fiedor and K. Burda, *Nutrients*, 2014, **6**(2), 466.
52. A. F. G. Cicero and A. Colletti, *Curr. Pharm. Des.*, 2017, **23**(17), 2422.
53. F. T. Johra, A. K. Bepari, A. T. Bristy and H. M. Reza, *Antioxidants*, 2020, **9**(11), 1046.
54. T. Gori and T. M. Nzel, *Ann. Med.*, 2011, **43**, 259.
55. S. Kiokias and M. Gordon, *Eur. J. Clin. Nutr.*, 2003, **57**, 1135.
56. P. G. Cocate, A. J. Natali, R. G. Alfenas and A. de Oliveira, *Br. J. Nutr.*, 2015, **114**, 257.
57. E. S. Ford, S. Liu, D. M. Mannino, W. H. Giles and S. J. Smith, *Eur. J. Clin. Nutr.*, 2003, **57**, 1157.
58. Y. Wang, S. J. Chung, M. L. McCullough, W. O. Song, M. L. Fernandez, S. I. Koo and O. K. Chun, *J. Nutr.*, 2014, **144**(7), 1067.
59. J. E. Klaunig, L. M. Kamendulis and B. A. Hoceva, *Toxicol. Pathol.*, 2010, **38**(1), 96.
60. D. Van Hoang, N. M. Pham, A. H. Lee, D. N. Tran and C. W. Binns, *Nutrients*, 2018, **10**, 70.
61. L. Chen, M. Stacewicz-Sapuntzakis, C. Duncan, R. Sharifi, L. Ghosh, R. van Breemen, D. Ashton and P. E. Bowen, *J. Natl. Cancer Inst.*, 2001, **93**, 1872.
62. M. L. Slattery, J. Benson, K. Curtin, K. N. Ma, D. Schaeffer and J. D. Potter, *Am. J. Clin. Nutr.*, 2000, **71**(2), 575.
63. Y. Okuyama, K. Ozasa, K. Oki, H. Nishino, S. Fujimoto and Y. Watanabe, *Int. J. Clin. Oncol.*, 2014, **19**(1), 87.
64. Y. Takata, Y. B. Xiang, G. Yang, H. Li, J. Gao, H. Cai, Y. T. Gao, W. Zheng and X. O. Shu, *Nutr. Cancer*, 2013, **65**(1), 51.

65. M. Shareck, M. C. Rousseau, A. Koushik, J. Siemiatycki and M.-E. Parent, *Front. Oncol.*, 2017, **7**, 23.
66. G. S. Omenn, G. E. Goodman, M. D. Thornquist, J. Balmes, M. R. Cullen, A. Glass, J. P. Keogh, F. L. Meyskens, B. Valanis, J. H. Williams, S. Barnhart and S. Hammar, *N. Engl. J. Med.*, 1996, **334**, 1150.
67. R. M. Russell, *Pure Appl. Chem.*, 2002, **74**(8), 1461.
68. K. Mullan, M. A. Williams, C. R. Cardwell, B. McGuinness, P. Passmore, G. Silvestri, J. V. Woodside and G. J. McKay, *Alzheimer's Dementia*, 2017, **3**(3), 432.
69. J. Y. Min and K. B. Min, *Dementia Geriatr. Cognit. Disord.*, 2014, **37**(3–4), 246.
70. A. Takeda, O. P. Nyssen, A. Syed, E. Jansen, B. Bueno-de-Mesquita and V. Gallo, *Neuroepidemiology*, 2014, **42**(1), 25.
71. S. Davinelli, S. Ali, V. Solfrizzi, G. Scapagnini and G. Corbi, *Antioxidants*, 2021, **10**(2), 223.
72. J. Tanprasertsuk, T. M. Scott, A. K. Barbey, K. Barger, X. D. Wang, M. A. Johnson, L. W. Poon, R. Vishwanathan, N. R. Matthan, A. H. Lichtenstein, G. Ferland and E. J. Johnson, *Front. Nutr.*, 2021, **8**, 704691.
73. R. K. Crouch, G. J. Chader, B. Wiggert and D. R. Pepperberg, *Photochem. Photobiol.*, 1996, **64**, 613.
74. S. L. Booth, T. Johns and H. V. Kuhnlein, *Food Nutr. Bull.*, 1992, **14**(1), 1.
75. J. F. Faustino, A. Ribeiro-Silva, R. F. Dalto, M. M. Souza, J. M. Furtado, G. de Melo Rocha, M. Alves and E. M. Rocha, *Arq. Bras. Oftalmol.*, 2016, **79**, 56.
76. S. Y. Hess, D. I. Thurnham and R. F. Hurrell, Influence of Provitamin A Carotenoids on Iron, Zinc, and Vitamin A Status, *HarvestPlus Technical Monograph*, 2005, vol. 6, p. 1.
77. J. T. Landrum and R. A. Bone, *Arch. Biochem. Biophys.*, 2001, **385**, 28.
78. J. Wu, E. Cho, W. C. Willett, S. M. Sastry and D. A. Schaumberg, *JAMA Ophthalmol.*, 2015, **133**(12), 1415.
79. N. M. Bressler, *J. Am. Med. Assoc.*, 2004, **291**, 1900.
80. A. Kijlstra, Y. Tian, E. R. Kelly and T. T. Berendschot, *Prog. Retinal Eye Res.*, 2012, **31**(4), 303.
81. F. M. Barker 2nd, D. M. Snodderly, E. J. Johnson, W. Schalch, W. Koepcke, J. Gerss and M. Neuringer, *Invest. Ophthalmol. Visual Sci.*, 2011, **52**(7), 3934.
82. S. Carpentier, M. Knaus and M. Y. Suh, *Crit. Rev. Food Sci. Nutr.*, 2009, **49**, 313.
83. R. D. Graham and J. M. Rosser, *Food Nutr. Bull.*, 2000, **21**(4), 404.
84. H. E. Bouis, C. Hotz, B. McClafferty, J. V. Meenakshi and W. H. Pfeiffer, *Food Nutr. Bull.*, 2011, **32**(1 Suppl), S31.
85. E. Simpungwe, T. Dhliwayo, M. Palenberg, V. Taleon, E. Birol, A. Oparinde, A. Saltzman and M. T. Diressie, *Afr. J. Food, Agric., Nutr. Dev.*, 2017, **17**, 11973.
86. B. Gannon, C. Kaliwile, S. A. Arscott, S. Schmaelzle, J. Chileshe, N. Kalungwana, M. Mosonda, K. Pixley, C. Masi and S. A. Tanumihardjo, *Am. J. Clin. Nutr.*, 2014, **100**(6), 1541.

87. A. C. Palmer, K. Healy, M. A. Barffour, W. Siamusantu, J. Chileshe, K. J. Schulze, K. P. West Jr. and A. B. Labrique, *J. Nutr.*, 2016, **146**(12), 2551.
88. L. F. Călinoiu and D. C. Vodnar, *Nutrients*, 2018, **10**(11), 1615.
89. O. Ekpa, N. Palacios-Rojas, G. Kruseman, V. Fogliano and A. R. Linnemann, *Global Food Secur.*, 2018, **17**, 48.
90. K. Pixley, N. Palacious, R. Babu and A. Menkir, Maize HarvestPlus: Bio-fortying Maize with Provitamin A carotenoids, in *Addressing Climate Change Effects and Meeting Maize Demand for Asia. Book of Extended Summaries of the 11th Asian Maize Conference*, ed. P. H. Zaidi, R. Babu, J. Cairns, D. Jeffers, L. Q. Kha, G. K. Krishna, V. Krishna, A. McDonald, G. Ortiz-Ferrara, N. Palacios, K. Pixley, B. M. Prasanna, Z. Rashid, T. Tefera, T. P. Tiwari, M. T. Vinayan, V. Vengadessan, F. Xingming, Y. Xu, C. Wei-dong, S. Zhang, and B. S. Vivek, Nanning, China, 2011, pp. 250–252.
91. C. O. Egesel, J. C. Wong, R. J. Lambert and T. R. Rocheford, *Crop Sci.*, 2003, **43**, 818.
92. J. I. Ortiz-Monasterio, N. Palacios-Rojas, E. Meng, K. Pixley, R. Trethowan and R. J. Penã, *J. Cereal Sci.*, 2007, **46**, 293.
93. A. Menkir, W. Liu, W. S. White, B. Maziya-Dixon and T. Rocheford, *Food Chem.*, 2008, **109**, 521.
94. S. Chander, Y. Meng, Y. Zhang, J. Yan and J. Li, *J. Agric. Food Chem.*, 2008, **56**(15), 6506.
95. T. Safawo, N. Senthil, M. Raveendran, S. Vellaikumar, K. N. Ganesan, G. Nallathambi, S. Saranya, V. G. Shobhana, B. Abirami and E. Vijaya Gowri, *Electron. J. Plant Breed.*, 2010, **1**(4), 548.
96. R. Sivaranjani, B. M. Prasanna, F. Hossain and I. M. Santha, *Indian J. Agric. Sci.*, 2013, **83**(4), 431.
97. M. Alfieri, A. Hidalgo, N. Berardo and R. Redaelli, *J. Cereal Sci.*, 2014, **59**, 181.
98. S. de Almeida Rios, M. C. D. Paes, W. S. Cardoso, A. Borém and F. F. Teixeira, *Am. J. Plant Sci.*, 2014, **5**, 857.
99. B. F. Owens, A. E. Lipka, M. Magallanes-Lundback, T. Tiede, C. H. Diepenbrock, C. B. Kandianis, E. Kim, J. Cepela, M. Mateos-Hernandez, C. R. Buell, E. S. Buckler, D. DellaPenna, M. A. Gore and T. Rocheford, *Genetics*, 2014, **198**(4), 1699.
100. V. Muthusamy, F. Hossain, N. Thirunavukkarasu, N. Pandey, A. K. Vishwakarma, S. Saha and H. S. Gupta, *Food Biotechnol.*, 2015, **29**, 276.
101. E. O. Alamu, B. Maziya-Dixon, A. Menkir, E. A. Irondi and O. Olaofe, *Front. Nutr.*, 2021, **8**, 640563.
102. K. Chandler, A. E. Lipka, B. F. Owens, H. Li, E. S. Buckler, T. Rocheford and M. A. Gore, *Crop Sci.*, 2013, **53**, 189.
103. R. Babu, N. P. Rojas, S. Gao, J. Yan and K. Pixley, *Theor. Appl. Genet.*, 2013, **126**(2), 389.
104. V. Muthusamy, F. Hossain, N. Thirunavukkarasu, M. Choudhary, S. Saha, J. S. Bhat, B. M. Prasanna and H. S. Gupta, *PLoS One*, 2014, **9**(12), e113583.

105. R. U. Zunjare, F. Hossain, V. Muthusamy, A. Baveja, H. S. Chauhan, J. S. Bhat, N. Thirunavukkarasu, S. Saha and H. S. Gupta, *Front. Plant Sci.*, 2018, **9**, 178.
106. S. Natesan, T. Duraisamy, B. Pukalenthy, S. Chandran, J. Nallathambi, K. Adhimoolam, D. Manickam, V. Sampathrajan, S. J. Muniyandi, L. J. Meitei, N. Thirunavukkarasu, G. Kalipatty Nalliappan and R. Rajasekaran, *Front. Nutr.*, 2020, **7**, 134.
107. M. Aluru, Y. Xu, R. Guo, Z. Wang, S. Li, W. White, K. Wang and S. Rodermel, *J. Exp. Bot.*, 2008, **59**(13), 3551.
108. C. Zhu, S. Naqvi, J. Breitenbach, G. Sandmann, P. Christou and T. Capell, *Proc. Natl. Acad. Sci. U. S. A.*, 2008, **105**(47), 18232.
109. C. Zhu, G. Farré, D. Zanga, J. Lloveras, A. Michelena, J. P. Ferrio, J. Voltas, G. Slafer, R. Savin, R. Albajes, M. Eizaguirre, C. Lopez, C. Cantero-MartÍnez, J. DÍaz-Gómez, C. Nogareda, J. A. Moreno, E. Angulo, J. Estany, R. N. Pena, M. Tor, M. Portero-Otin, N. Eritja, G. Arjó, J. C. E. Serrano, X. Matias-Guiu, R. M. Twyman, G. Sandmann, T. Capell and P. Christou, *Phytochem. Rev.*, 2018, **17**, 195.
110. Y. M. Naguib, *J. Agric. Food Chem.*, 2000, **48**(4), 1150.
111. S. H. Ryu, L. Werth, S. Nelson, J. C. Scheerens and R. C. Pratt, *Econ. Bot.*, 2013, **67**(2), 98.
112. C. de la Parra, S. O. Saldivar and R. H. Liu, *J. Agric. Food Chem.*, 2007, **55**(10), 4177.
113. Q. P. Hu and J. G. Xu, *J. Agric. Food Chem.*, 2011, **59**(5), 2026.
114. S. Zilić, A. Serpen, G. Akıllıoğlu, V. Gökmen and J. Vančetović, *J. Agric. Food Chem.*, 2012, **60**(5), 1224.
115. S. Kuhnen, P. M. Lemos, L. H. Campestrini, J. B. Ogliari, P. F. Dias and M. Maraschin, *J. Sci. Food Agric.*, 2011, **91**(9), 1548.
116. J. Peng, Z. Liu, X. Liu, J. Yan, D. Sun, and E. Nevo, Evolutionary agriculture domestication of wild emmer wheat, in *New Horizons in Evolution*, ed. S. P. Wasser and M. Frenkel-Morgenstern, Academic Press, 2021, ch. 8, pp. 193–255.
117. A. M. Digesù, C. Platani, L. Cattivelli, G. Mangini and A. Blanco, *J. Cereal Sci.*, 2009, **50**, 210.
118. A. Fratianni, M. Irano, G. Panfili and R. Acquistucci, *J. Agric. Food Chem.*, 2005, **53**(7), 2373.
119. S. Growth, R. Wittmann, C. Friedrich, H. Longin and V. Böhm, *Eur. Food Res. Technol.*, 2020, **246**, 2079.
120. A. Ramachandran, C. J. Pozniak, J. M. Clarke and A. K. Singh, *J. Cereal Sci.*, 2010, **52**, 30.
121. R. Beleggia, C. Platani, F. Nigro and L. Cattivelli, *J. Cereal Sci.*, 2010, **52**, 106.
122. A. Brandolini, A. Hidalgo and S. Moscaritolo, *J. Cereal Sci.*, 2008, **48**, 599.
123. V. Hentschel, K. Kranl, J. Hollmann, M. G. Lindhauer, V. Böhm and R. Bitsch, *J. Agric. Food Chem.*, 2002, **50**(23), 6663.
124. G. M. Borrelli, A. M. De Leonardis, C. Platani and A. Troccoli, *J. Cereal Sci.*, 2008, **48**(2), 494.

125. R. Acquistucci and L. A. Pasqui, *Nahrung*, 1991, **35**, 345.
126. C. J. Pozniak, R. E. Knox, F. R. Clarke and J. M. Clarke, *Theor. Appl. Genet.*, 2007, **114**(3), 525.
127. R. Patil, M. Oak, A. Deshpande and S. Tamhankar, *Mol. Breed.*, 2018, **38**, 136.
128. MASWheat-Marker Assisted Selection in Wheathttps://maswheat.ucdavis.edu/protocols, available at , accessed on 2 February 2022.
129. F. Leenhardt, B. Lyan, E. Rock, A. Boussard, J. Potus, E. Chanliaud and C. Remesy, *Eur. J. Agron.*, 2006, **25**, 170.
130. M. D. Rey, M. C. Calderón, M. J. Rodrigo, L. Zacarías, E. Alós and P. Prieto, *PLoS One*, 2015, **10**(8), e0134598.
131. W. Li, S. Zhai, H. Jin, W. Wen, J. Liu, X. Xia and Z. He, *Front. Agric. Sci. Eng.*, 2016, **3**(2), 124.
132. Y. He, Q. Wang, J. Zeng, T. Sun, G.-X. Yang and G.-Y. He, *J. Integr. Agric.*, 2015, **14**(3), 438.
133. L. Cong, C. Wang, L. Chen, H. J. Liu, G. X. Yang and G. Y. He, *J. Agric. Food Chem.*, 2009, **57**, 8652.
134. C. Wang, J. Zeng, Y. Li, W. Hu, L. Chen, Y. Miao, P. Deng, C. Yuan, C. Ma, X. Chen, M. Zang, Q. Wang, K. Li, J. Chang, Y. Wang, G. Yang and G. He, *J. Exp. Bot.*, 2014, **65**, 2545.
135. J. Zeng, X. Wang, Y. Miao, C. Wang, M. Zang, X. Chen, M. Li, X. Li, Q. Wang, K. Li, J. Chang, Y. Wang, G. Yang and G. He, *J. Agric. Food Chem.*, 2015, **63**, 9083.
136. A. Hidalgo, A. Brandolini, C. Pompei and R. Piscozzi, *J. Cereal Sci.*, 2006, **44**, 182.
137. A. Hidalgo and A. Brandolini, *J. Sci. Food Agric.*, 2014, **94**(4), 601.
138. E. Giambanelli, F. Ferioli, B. Koçaoglu, M. Jorjadze, I. Alexieva, N. Darbinyan and L. F. D'Antuono, *J. Sci. Food Agric.*, 2013, **93**(14), 3490.
139. E. Mellado-Ortega and D. Hornero-Mendez, *J. Cereal Sci.*, 2015, **62**, 15.
140. H. Fu, B. Xie, G. Fan, S. Ma, X. Zhu and S. Pan, *Food Chem.*, 2010, **122**, 602.
141. A. Martín and E. Sanchez-Monge Laguna, *Euphytica*, 1982, **31**, 261.
142. A. Martín, J. A. Alvarez, L. M. Martín, F. Barro and J. Ballesteros, *J. Cereal Sci.*, 1999, **30**, 85.
143. S. G. Atienza, J. Ballesteros, A. Martín and D. Hornero-Méndez, *J. Agric. Food Chem.*, 2007, **55**(10), 4244.
144. D. Giordano, M. Locatelli, F. Travaglia, M. Bordiga, A. Reyneri, J. D. Coïsson and M. Blandino, *Food Chem.*, 2017, **233**, 483.
145. L. Paznocht, Z. Kotíková, M. Šulc, J. Lachman, M. Orsák, M. Eliášová and P. Martinek, *Food Chem.*, 2018, **240**, 670.
146. M. Zrckova, I. Capouchova, M. Eliášová, L. Paznocht, K. Pazderů, P. Dvořák, P. Konvalina, M. Orsák and Z. Štěrba, *Plant, Soil Environ.*, 2018, **64**, 530.
147. S. Siebenhandl, H. Grausgruber, N. Pellegrini, D. Del Rio, V. Fogliano, R. Pernice and E. Berghofer, *J. Agric. Food Chem.*, 2007, **55**(21), 8541.
148. G. S. Khush, *Plant Mol. Biol.*, 2005, **59**(1), 1.
149. L. Lamberts and J. A. Delcour, *J. Agric. Food Chem.*, 2008, **56**(24), 11914.

150. X. Ye, S. Al-Babili, A. Klöti, J. Zhang, P. Lucca, P. Beyer and I. Potrykus, *Science*, 2000, **287**, 303.
151. S. Al-Babili and P. Beyer, *Trends Plant Sci.*, 2005, **10**, 565.
152. J. A. Paine, C. A. Shipton, S. Chagger, R. M. Howles, M. J. Kennedy, G. Vernon, S. Y. Wright, E. Hinclie, J. L. Adams, A. L. Silverstone and R. Drake, *Nat. Biotechnol.*, 2005, **23**, 482.
153. G. Giuliano, *Curr. Opin. Biotechnol.*, 2017, **44**, 169.
154. O. X. Dong, S. Yu, R. Jain, N. Zhang, P. Q. Duong, C. Butler, Y. Li, A. Lipzen, J. A. Martin, K. W. Barry, J. Schmutz, L. Tian and P. C. Ronald, *Nat. Commun.*, 2020, **11**(1), 1178.
155. Q. Zhu, D. Zeng, S. Yu, C. Cui, J. Li, H. Li, J. Chen, R. Zhang, X. Zhao, L. Chen and Y.-G. Liu, *Mol. Plant*, 2018, **11**, 1440.
156. M. Nakornriab, T. Sriseadka and S. Wongpornchai, *J. Food Lipids*, 2008, **15**, 488.
157. J. K. Kim, S. Y. Lee, S. M. Chu, S. H. Lim, S. C. Suh, Y. T. Lee, H. S. Cho and S. H. Ha, *J. Agric. Food Chem.*, 2010, **58**(24), 12804.
158. R. Valarmathi, M. Raveendran, S. Robin and N. Senthil, *J. Plant Biochem. Biotechnol.*, 2015, **24**(3), 305.
159. G. Pereira-Caro, S. Watanabe, A. Crozier, T. Fujimura, T. Yokota and H. Ashihara, *Food Chem.*, 2013, **141**(3), 2821.
160. G. Pereira-Caro, G. Cros, T. Yokota and A. Crozier, *J. Agric. Food Chem.*, 2013, **61**(33), 7976.

CHAPTER 13

Pigmented Cereal: Dietary and Health Importance

OSENI KADIRI*[a], BEATRICE M. FASOGBON[b], DESMOND O. ONOBUN[a], MOMOH EMOSHIOKE[a], RUTH N. UKEJEH[c] AND PRATIK NAYI[d]

[a]Department of Biochemistry, Faculty of Basic Medical Sciences, Edo State University Uzairue, Nigeria; [b]Department of Biotechnology & Food Technology, Faculty of Science, University of Johannesburg, South Africa; [c]Department of Food Sciences and Technology, Faculty of Technology, Obafemi Awolowo University, Ile-Ife, Nigeria; [d]Department of Tropical Agriculture and International Cooperation, National Pingtung University of Science and Technology, 1 Shuefu Road, Neipu, Pingtung, 91201, Taiwan
*E-mail: kadirioseni@yahoo.com

13.1 Introduction

Pigmented grains are an attractive cereal group because, in addition to fibre, they may include a high concentration of some potential bioactive components. Pigments from plant sources are secondary metabolites that are nutritionally, therapeutically edible and also possess anti-cancer, anti-obesity, and anti-hypercholesterolemia activity.[1] There has been renewed public attention to, and acceptance of, these food sources, as a result of these positive health benefits. Some food products that are formulated from pigmented cereals include breakfast cereals, baked foods, snack foods, non-alcoholic beverages, and extruded snacks.[2] The advantages of using pigmented cereal grains or their bran in developing and producing novel functional and nutraceutical

Food Chemistry, Function and Analysis No. 38
Pigmented Cereals and Millets: Bioactive Profile and Food Applications
Edited by Sneh Punia Bangar, Sajid Maqsood and Anil Kumar Siroha

Published by the Royal Society of Chemistry, www.rsc.org

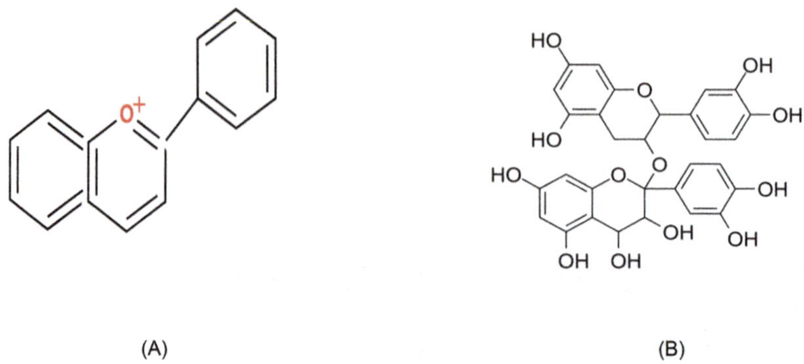

Figure 13.1 Structure of (A) anthocyanin and (B) proanthocyanidin.

cereal-based products of improved nutritional value and health benefits have been on the increase in recent years.

Rice, sorghum, wheat, maize, and rice have varieties of pigmented grains that are rich in polyphenols and bioactive compounds. Bioactive compounds such as flavonoids and polyphenols are some of the reasons for the significant antioxidant activities reported in these grains and their roles in combating the onset of degenerative diseases such as cancer.[2]

Anthocyanins are a group of water-soluble phenolic pigments belonging to the flavonoid class, which are responsible for the purple, blue, red, or black color of pigmented cereals (Figure 13.1).[1] Antioxidant activity increases significantly as pigment concentration increases.[3] A higher concentration of anthocyanins and other nutrients in pigmented wheat, barley, corn, sorghum, and rice is also responsible for an increase in the nutritional content and prolonged shelf life of food products formulated from them.[2] Anthocyanins present in pigmented cereal such as purple rice have been demonstrated to have beneficial health effects,[4] such as improving gut microflora,[5] antioxidant, anti-inflammatory, anti-obesity, anti-atherosclerosis, anti-adipogenic, anti-angiogenesis, anti-metastatic, and pro-apoptotic potential in *in vitro* and *in vivo* experiments.[6–8] The consumption of pigmented corn has been confirmed as safe and suitable for celiac patients, because they are rich sources of soluble phenolic compounds (anthocyanins with known antioxidant capacity).[9] Pigmented sorghum has the highest polyphenolic content and antioxidant activity, which is beneficial in preventing and treating neurodegenerative diseases.[10] Ferulic acid in pigmented sorghum makes the grain a suitable ingredient for functional antioxidant and anti-aging foods.[11]

13.1.1 Dietary Pigmented Cereal Species

Cereals, edible seeds, and grains belonging to the grass family Gramineae are commonly grown because of their higher nutritional characteristics.[12–14] The Gramineae family comprises several crops, including rice, flaxseed, wheat, psyllium, barley, millets *etc.* Several types of cereal offer enormous nutritional value for both humans and livestock and have been used for centuries. Among the various sources of food, cereals have been considered the most important as cereal-based foods as important sources of protein, energy, vitamin B, and minerals.[12] Generally, these cereals are relatively inexpensive to produce. They are also easily stored and transported, as long they are kept dry in order to prevent them from deteriorating. Rice and wheat jointly account for about 50% of global cereal production. Sap-soluble flavonoids among the >5000 flavonoids found in nature have been related to the observed colors of several plants.[14] A specific important flavonoid called anthocyanin has been reported as the major constituent impacting colors such as red, purple, black, blue, and even a mixture of different colors on cereals such as maize, barley, wheat, and rice.[15–18]

*13.1.1.1 Rice (*Oryza sativa*)*

Rice is a major diet of the diets of about half of the world's population. While white rice is widely consumed worldwide, several Asian nations also consume colored cereals such as red rice, black rice, reddish-brown rice, purple-black rice, and others.[19] Rice is a key cereal crop that provides sustenance for over half of the world's population, earning it the nickname 'the grain of life'. It is the second most widely consumed cereal in the world, after wheat.[20] In the many colors of different kernels, rice shows its great diversity just like maize. Anthocyanin compounds mainly present inside the bran layer of dehusked rice kernels. Anthocyanins content varies with the different parts of the rice kernel. More than 2 billion people depend on rice for 80% of their energy in Asia alone. Until recent times, rice was considered the sole source of carbohydrates due to its starchy nature, although with a minute amount of proteins. Rice has a great nutritional value despite the low quantity of protein.[21] Rice is found in white, brown, and colored forms. These colored forms of rice have been associated with several health benefits that have been proven. The presence of polyphenols, which are known to have various health-promoting qualities such as anti-cancer, antioxidant, and hypoglycemic capabilities, has been linked to the health advantages of these colored rice variants.[14] Colored rice is becoming increasingly popular worldwide, with black rice being combined with non-colored rice before cooking, to improve its taste.[22] As reported in this study, there are many food items like bread, ice cream, and liquors in which red rice can also be added in such products.[23] In black rice, the anthocyanin contents of bran, embryo, and endosperm were reported as 6.3, 0.34, and 0.0 mg CGE per g, respectively.[24] With the help of a high-efficiency transgene stacking technique, a novel and highly unique rice genotype with purple endosperm rich in anthocyanins content was recently developed. This developed rice kernel contains up to 1 mg g^{-1} (db) of polished grain, making it the biggest source of anthocyanins.[25]

13.1.1.2 *Wheat (*Triticum aestivum*)*

Wheat is essential in the human diet. It is a species that belongs to the Gramineae (wild grass) family and is widely present in regions of western Asia. Wheat is not only consumed for its nutritional benefits, but also for its nutraceutical and functional attributes. It is a major constituent of the diet of many people worldwide and, as such it has a significant impact on human nutrition and health.[13] Wheat is found in different pigmented colors such as black, blue, purple, red, and yellow. Wheat's black, blue, and purple colors are caused by anthocyanins (Figure 13.1), while the red and yellow colors are caused by phlobaphene and carotenoids.[14] The colored varieties are of higher nutritional quality and offer more health benefits when compared with non-pigmented wheat varieties, and this has led to their utilization in the production of items such as biscuits, bread, and noodles.[26] The health benefits of colored wheat types include cholesterol absorption inhibition, high antioxidant activity, anti-aging, and neuroprotection activities.[14] Black wheat was developed by crossing blue and purple wheat. The black color of the wheat grain is caused by anthocyanin, which is found in the outer layer. In addition to the essential nutrients, the principal useful substances found in wheat include phenolic acids (PAs), flavonoids, anthocyanins, and carotenoids (carbohydrates, protein, and fat). Phenolic acids (PAs) bioavailability is mainly influenced by the dietary matrix and their chemical interactions with carbs, proteins, and lipids. These chemicals are mostly contained in the outer layer of wheat seed, therefore, removing the bran and germ decreases antioxidant activity. Furthermore, antioxidant activity is affected by the genotype and the growth state of the grain. Because of some internal changes like enzymatic activity and external factors mainly UV light, high temperature, oxidation, and extreme pH range.[27]

Black wheat is rich in total phenolic contents (TPC), antioxidant activity, and total flavonoid content (TFC) compared with purple wheat.[28] Black wheat has the highest antioxidant activity, total flavonoid content (TFC), and total phenolic compounds (TPC), whereas light purple wheat has the lowest. According to the study, TPC and TFC concentrations in black wheat were found to be 659.8 μg gallic acid equivalents and 319.3 μg rutin equivalents per g^{-1}, respectively.[29]

13.1.1.3 *Maize (*Zea mays*)*

Maize is generally known as corn, and is widely produced around the world. It is closely related to rice, wheat, and oats.[13] It is a major part of the diets of individuals across different socioeconomic classes. Although pigmented maize is not produced in large quantities in the world, they are often used for ornamental purposes and for making special foods. Several compounds, such as phenolics and carotenoids have been reported for colored maize.[14] Flavonoids and phenolic acid are the most commonly found phenolics in maize and they exist either in insoluble bound or soluble free forms.

Maize possesses diverse pigmented colors such as orange, crimson, scarlet, mauve, blue, violet, yellow, cream, and ivory, for which flavonoids are responsible.[17] The amount of anthocyanins in colored maize samples have shown a greater genetic diversity. Purple maize kernel has maximum anthocyanin [891-3312 CGE (cyanidin 3-glucoside equivalent) μg g^{-1}, db (dry basis)] than blue (up to 540 CGE μg g^{-1}, db), red/blue (up to 368 CGE μg g^{-1}, db), and red (up to 127 CGE μg g^{-1}, db) samples.[30] The anthocyanin levels of 398 colored maize genotypes ranged from 23 to 252 CGE mg kg^{-1}.[31] In one study, 14 maize cobs were studied, and it was reported that anthocyanin content ranged from 4.3 to 117 C3G mg g^{-1}.[32]

13.1.1.4 *Barley (*Hordeum vulgare*)*

Barley is a cereal that is cultivated and consumed for its nutritional and health benefits.[13] This kind of crop can be grown in different climatic conditions around the world. Although barley contains protein, crude fibre, fat, phosphorus, and calcium, it is a major carbohydrate source.[33] Pigmented barleys come in various colors, including black, red, blue, and purple, and include substances such as anthocyanin, melanin, phenolic acids, and carotenoids.[14] The anthocyanin content and genetic variability in barley kernel color – mainly purple, black, and blue – has been reported. The different barley kernels were studied in seven groups of barley samples, and it was found that variation in color values ranged from 60 to 350 μg g^{-1}. Generally, anthocyanin content is available in the outer layer of barley kernels, similar to other cereals. In one genotype of barley, up to 1.66 g kg^{-1} anthocyanins content was found in the bran layer. From several research studies, the different ranges of anthocyanins compounds have been measured in barley kernels. It was mainly peonidin 3-(6″-succinyl)glucoside, delphinidin 3-glucoside, cyanidin 3-(6″-succinyl)glucoside, cyanidin 3-glucoside, and delphinidin 3-malonylglucoside.[34] The content of these anthocyanins varies greatly amongst barley genotypes.[25]

13.1.2 Relationship of Pigmented Cereal Composition to Diet and Health

13.1.2.1 *Protein*

Protein content is unevenly distributed in colored cereal grains. Protein content in wheat cultivars has been found to range between 10 and 15% of dry weight when grown under field conditions. The nutritional quality of protein found in cereals is defined by the amount of essential amino acids present, which must be supplied to human diets. When one essential amino acid is deficient, the other essential amino acids are broken down and excreted, resulting in decreased human growth and nitrogen loss from the diet.[35] Adults require a lower amount of essential amino acids as they need them solely for maintenance purposes compared with children who require more, as they need them not only for maintenance but also for growth. Protein is the second most important component in cereals, after starch, and it impacts cereal's eating and nutritional quality. Pigmented wheat has been shown to have a high glutamine and proline content but a low lysine content. Pigmented rice is very rich in lysine, with a content of 3.5–4.0% making it the cereal with the highest amount of protein.[36,37] It was reported that rice has highly digestible protein (93%), good biological value (74%), and a good protein efficiency ratio (2.02–2.04%) because of its lysine content.[38,39]

13.1.2.2 *Carbohydrates*

Carbohydrates are the most abundant components in cereals and they are mostly present in the form of starch (amylose and amylopectin).[39,40] It has been reported that wheat contains about 85% carbohydrate, of which 80% appears in the form of starch once the wheat is mature. For humans, cereal starch is the most important source of dietary carbohydrates.[35] The carbohydrate content of cereals includes monosaccharides such as D-glucose, D-fructose, as well as small quantities of their phosphorylated forms, oligosaccharides such as fructooligosaccharides, and starch. Fructans derived from cereals are useful in mitigating the onset of diseases such as ulcerative colitis and Crohn's disease.[41,42] It has been reported that fructans can help with calcium absorption.[43,44]

13.1.2.3 *Phytochemicals*

Pigmented cereals and their derived food items provide a very good source of phytochemicals. These phytochemicals, which are non-nutritional components of foods have been proven both clinically and epidemiologically to benefit human and animal health.[45] The phytochemicals present in cereals include, but are not restricted to, vitamin E (tocotrienols and tocopherols), γ-oryzanols, and several phenolic compounds as well as carotenoids, and these compounds have all been reported to possess several health benefits.[46] Phenolic compounds are the most common phytochemicals found in pigmented cereals, with anthocyanins and proanthocyanidins being the most common.[47–50] Cereals contain phenolic compounds such as benzoic and cinnamic acids, flavonoids, flavonols, flavones, lignans, avenanthramides, and alkylresorcinols. Other phytochemicals in pigmented cereals include different carotenoids such as α- and β-carotene, β-cryptoxanthin, lutein, and zeaxanthin, phytate, numerous terpenes, and terpenoids.[51] According to Siebenhandl *et al.*,[52] ferulic acid, vanillic acid, and *p*-coumaric acid were found in purple wheat, blue wheat, and black barley. These phytochemicals are capable of combating a variety of degenerated ailments such as diabetes, obesity, cardiovascular disease, *etc.*[51,53,54]

13.1.2.4 Dietary Fibre

Dietary fibre (DF) is an important nutritive component of cereals and other foods. Dietary fibres have been reported to increase the bulk of faeces, and increase food viscosity, reducing the access of starch granules to digestive enzymes.[55] They also reduce the rate of carbohydrate assimilation and decrease postprandial glycemic and insulinemic responses.[56] The soluble dietary fibre content of rice was reported to contain arabinoxylans, β-D-glucan, xylose, rhamnose, galactose, mannose, and glucose, while the dietary fibre reported to be insoluble contained cellulose, insoluble β-glucan, hemicellulose, and arabinoxylans.[37,57] In maize, the dietary fibres were primarily composed of arabinoxylans, heteroxylans, and cellulose. Dietary fibres have been studied and reported by many researchers, and it was reported that health benefits such as controlling diabetes and cardiovascular diseases and these activities have been linked with the ability of these fibres to increase mineral absorption, improve the function of the colon, and decrease colorectal cancer.[58] Arabinoxylan can transport phenolic compounds such as ferulic and coumaric acid, as well as oligomers, to the colon, where they exhibit antioxidant activity and promote health.[59] Polysaccharides such as hemicellulose, pectin, and cellulose are dietary fibre components that are not digested by gastrointestinal enzymes. These fibres assist in raising the viscosity of the food items, which slows the accessibility of starch granules to digesting enzymes found in the human digestive organs. Black wheat (BW) grain has around 1.15% total dietary fibres (TDF) by total weight. The insoluble type of fibre dominates this TDF, providing volume and body to the faeces, stimulating intestinal transit, and preventing constipation.[27]

13.1.2.5 Vitamins

Several vitamins have been reported for different pigmented cereals. Cereals are considered good sources of vitamin A, and B vitamins such as thiamine (B1), riboflavin (B2), niacin (B3), pantothenic acid (B5), pyridoxine (B6), and folate (B9); and vitamin E and vitamin K, but they are generally deficient in cobalamin (B12), vitamin C, and vitamin D.[12,60,61] Similarly, vitamin B5 can be beneficial in the metabolism of carbs, fat, protein, and hemoglobin synthesis. Vitamin E (tocopherol) can prevent hemolysis of red blood corpuscle and is also responsible for women's infertility. The formation of blood cells is directly associated with vitamin B5 (pantothenic acid) and its helpful in hormone release, conversion of food into energy, and gland functioning. A considerable amount of vitamins B3, B5, and vitamin E is available in black wheat, as reported by Tian *et al.*[62] The presence of 19 vitamins in black wheat was also reported by Wang *et al.*[63] In a black grained wheat (BGW 76), concentration of vitamin K (which is responsible for bone metabolism and blood clotting) was found to be 11.47 mg kg^{-1}, which is 1.6-fold higher than in conventional wheat (Jinchun 9).[27]

Tocols, often known as vitamin E, is a term that is used for a set of four tocopherols (α, β, γ, and δ) and four tocotrienols (α, β, γ, and δ). α-Tocopherol has the highest biological activity, as reported by several researchers. Sookwong *et al.*[64] reported that total vitamin E in pigmented glutinous rice variety (Leum Pua) was 32.61 ± 1.6 (μg g^{-1}) and similarly for the pigmented non-glutinous rice varieties such as HomPayom, rice berry, and Hua Bon was found to be 3.25 ± 0.1, 28.77 ± 0.8, and 1.95 ± 0.2 μg g^{-1}, respectively.[19]

13.1.2.6 Bio-functional Properties

Pigmented cereals are known to have significant levels of phenolic and flavonoid compounds in various parts of the kernel, and these compounds are typically responsible for the colors of these cereals.[23,65] Several studies have linked flavonoid anthocyanin, as well as certain phenolic chemicals, to antioxidant and anti-inflammatory properties.[65–67] Through modification of NF-κB driven pathways, bioactive substances can interfere with endogenous cell signaling pathways, resulting in decreased inflammatory responses.[68,69] In studies, anthocyanins and polyphenols have been demonstrated to reduce the activity of enzymes, including pancreatic α-amylase and brush-border α-glucosidase. Reduced rate of digestion and carbohydrate absorption is reported as a result of intestinal α-glucosidase inhibition, which results in lower postprandial glycemia and is regarded as a good strategy for controlling type 2 diabetes.[70]

Anthocyanins, a subtype of the much broader flavonoid superfamily, are the major bioactive components in pigmented grains.[71] Anthocyanins are also beneficial in treating certain disorders, such as cardiovascular disease and hypertension owing to their antioxidant and anti-inflammatory qualities. Several studies have found that anthocyanins as a whole may decrease the oxidation of low-density lipoproteins. In addition to this 'direct' process, pro-anthocyanidins from grape seeds have been shown in cell cultures to influence the activity and gene expression of antioxidant enzymes, such as glutathione *S*-transferase, glutathione reductase, and glutathione peroxidase.

Anthocyanin consumption also offered health advantages in metabolic disorders such as dyslipidemia and type 2 diabetes.[72] Microbes are countered *via* cells by a damaging membrane, intercellular matrix, and cell wall with the help of anthocyanins. Vitamin E is known as tocopherol or tocotrienol and vitamins B1, B2, B3, B6, and B9, namely thiamine, riboflavin, niacin, pyridoxine, and folate, respectively, are rich in pigmented wheat. Besides these vitamins, it also contains a minimum amount of provitamin A, D, and K, namely β-carotene, calciferol, and phylloquinone, respectively. The amount of Zn, Fe, and magnesium ranged from 108.54–142.68%, 8.57–42.86%, and 5.31–40.63%, respectively in pigmented wheat, which is higher than in normal wheat.[27]

13.1.3 Nutrient Contribution of Pigmented Cereals to Diet and Health

Although the body's own antioxidant defense mechanism is effective, it can get overwhelmed at times, necessitating the use of exogenous antioxidants. Antioxidant molecules such as flavonols and isoflavones can help restore oxidative equilibrium.[73] In one study, people who included anthocyanin-rich fruits, vegetables, and whole grains in their daily diet had a lower risk of chronic diseases such as cancer, diabetes, and cardiovascular disease (CVD).[74] Exogenous antioxidant substances consumed in various amounts from plant sources are responsible for most of these positive health effects.[75] Anthocyanins are a type of bioactive molecule that is known for their health benefits. Anthocyanins are water-soluble pigments that give cereal grains their blue, purple, and red hues. Anthocyanins are one of the most studied flavonoids in cereal grains because of their antioxidant effects. As reported in previous studies, they have anti-cancer, anti-aging, retinal protective, antioxidant, and gut improving benefits for human health.[76,77] Anthocyanins are 2-phenyl benzo pyrylium salts (flavylium) polyhydroxy derivatives or glycosides of poly-methoxy (Figure 13.1). Some researchers reported that certain chronic disorders can be prevented by consuming food products rich in anthocyanin.[78] Anthocyanins can be found in abundance in pigmented cereal grains.[25] They can be transformed into functional components that can be used to create unique food products with health benefits.[79] Food additives and natural anthocyanins with high colorant stability can be isolated from pigmented cereal grains (*e.g.* anthocyanins from sorghum).[80] Understanding the composition of anthocyanins in various pigmented cereal grains is important for human nutrition and food processing.[30]

The key health benefits and bioactive components are shown in Figure 13.2. Pigmented cereals can be helpful in many ways, mainly through

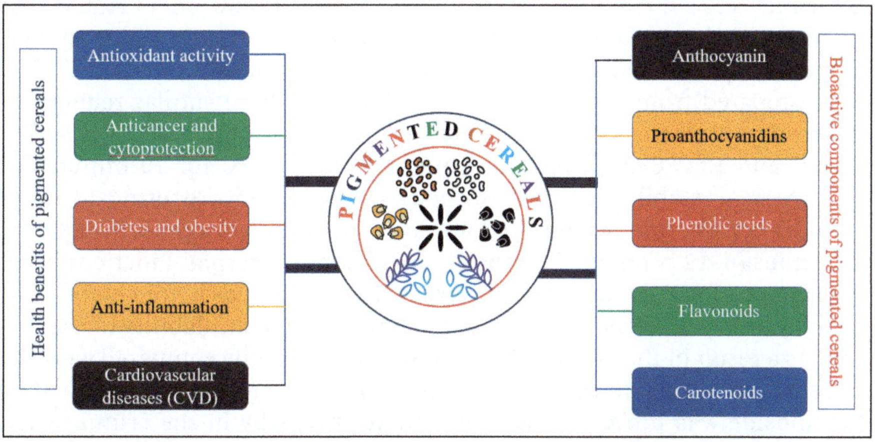

Figure 13.2 Major health benefits and bioactive components of pigmented cereals.

their antioxidant, anti-cancer, cytoprotection, anti-diabetes, anti-obesity, anti-inflammation, and cardiovascular disease preventive (CVD) activities, due to bioactive compounds like anthocyanin, proanthocyanidins, phenolic acids, flavonoids, and carotenoids. Extracts of anthocyanin-rich color cereals have been linked to a variety of health benefits. *In vivo* and *in vitro* experiments carried out by various researchers have confirmed its anti-cancer, antioxidant, glycaemic, retinal protection, and body weight control among many benefits. These claims are consistent with anthocyanins from fruits and vegetables, as well as pure anthocyanin molecules. The *in vitro* findings are mostly consistent with those of the *in vivo* investigations, particularly regarding retinal protection and antioxidant activity. The information provided by pigmented cereals could, for instance, assist in a tailored diet with focused nutritional applications. These anthocyanin-rich pigmented grains could be utilized to partially replace artificial colorants in a variety of food applications. Pigmented cereals could be employed as functional food ingredients to aid in treating a variety of chronic conditions.

There has been much research interest in the attractive nutritional values of pigmented cereal grains in recent years. Pigmented cereals have also been used in functional food production due to the beneficial effects of their ingredients on human health. Several phytochemicals are found in cereal grains, some of which substantially impact grain color. Accumulation of anthocyanins results in a combination of hues, such as purple or blue coloration. Flavonoids are found in the grain's outer layer. Typical flavonoids found in these layers include the yellow C-glycosides of flavonols, flavones, proanthocyanidin, flavones, and the reddish color of phlobaphenes, whereas carotenoids are found in the endosperm and are responsible for their yellow color. As a result, the accumulation of these pigments in grain may be a key target in breeding efforts aimed at enhancing bioactive component concentrations in grains and their associated products. After experimenting on two varieties of wheat, Bartl and coworkers stated that purple wheat has greater acetylated anthocyanins, which are prone to thermodynamic deterioration when compared with those of blue wheat.[81] Biscuits prepared from blue maize were observed to have similar reductions in anthocyanin concentration, where the losses are roughly 11% (baked at 179 °C and 182 °C for 4 minutes) and 25% (at 204 °C for 10 minutes).[82] The thermal stability of two anthocyanins (cyanidin-3-rutinoside and cyanidin-3-glucoside) were tested at increasing temperatures (200–240 °C) and times (0–12 minutes).[83] In both cases, non-isothermal kinetic models were created for the crumb and the crust. The crumb has higher stability of anthocyanin, which could be due to the lower oxygen availability in the system. As a result of the crumb's lower temperature profile compared with the crust, anthocyanins in the crumb were less thermally damaged. Cyanidin-3-rutinoside was proven to be more stable thermally in the crumb when compared with cyanide-3-glucoside.

13.1.4 Genetic Basis of Nutritional Quality Improvement of Pigmented Cereals

Variation in bioactive components of pigmented grains exists due to differences in the geographical locations, cultivar, associated climate conditions (levels of salinity, water, and temperature), extraction, and storage techniques. Genetic variability of anthocyanins in pigmented cereals has been shown across species and genotypes.[84] Anthocyanins are accumulated in a genotype-specific manner in pigmented grains,[85] particularly in the grains' outer layer, making it possible for cereals to show improved resistance to economically relevant plant diseases and other contaminations such as mycotoxins.[86] The large genetic component associated with the accumulation of anthocyanins in the grain could contribute to developing breeding programs to increase phenolic contents in non-pigmented grains. These could lead to increasing micronutrient concentrations in the grain and the nutritional value of the final products.[85] The changes from non-pigmented to pigmented grains occur through the spontaneous and chemically induced mutation, controlled by Ra, Rc, and Rd genes.[87] These genes are responsible for producing the grains' purple, brown, and red pericarp.

13.1.5 Pigmented Cereals as Functional Foods and Nutraceuticals

Pigmented cereals all have functional bioactive components, which is a reason for their significantly higher antioxidant activity. Antioxidants have been reported to prevent the onset of degenerative diseases like atherosclerosis, type 2 diabetes, and cancer.[88] Pigmented cereal grains are used to produce baked products and confectionaries. They are also used in the enhancement of traditional food products, as well as retrofitting. According to recent research studies, fermented pigmented cereals can improve their nutritional properties by reducing their anti-nutrient compounds, and boosting their bioactive compounds. Higher levels of antioxidants, anthocyanins, and phenolics with lower digestible starch have been reported in a traditional fermented Peruvian beverage known as Chicha.[89]

In pigmented barley, many researchers has reported a nearly 10-fold increase in gallic acid, catechin, vanillin, and resorcinol.[90] Dela Rosa and Medina[91] reported the application of fermented black rice in snacks and beverages formulation. Because of the richness of phytochemicals in pigmented cereals, they have been used in sourdough bread and have been shown to have anti-inflammatory properties.[92] Alteration of fermentation conditions improves polyphenol and antioxidants composition in previous studies. For instance, metabolomics investigations of fermented black rice revealed a decrease in phenolic compounds after 60 hours, which could be due to molecular breakdown and bacterial consumption.[93] Metabolomics analysis of fermented pigmented grains could shed fresh light on phenolic variety and its interactions with microbes. Natural pigments found in color grains, such as anthocyanins and proanthocyanidins (Figure 13.1), have been employed to boost the nutritional value of foods while also extending their shelf life. Grains may be converted into food additives due to the low cost of broken-color cereal.[94]

Due to their nutritional benefits and simplicity of eating, extruded color cereals have received a lot of attention. The total phenolic content was reduced during processing, but the bioavailability of these compounds was raised in a study by Hole *et al.*[95] They found that extruded puffed rice had more extractable phenolic acids, such as gallic and protocatechuic acids, but there was a substantial decrease in cyanidin-3-glucoside. The level of chemical composition such as phenolic, anthocyanin, antioxidant activity, and vitamin E was significantly higher in black rice-based products than those of white rice.[96]

13.2 Conclusion

As discussed in this chapter, pigmented cereal grains have a wide range of physiologically active chemicals that can be recovered and employed in several nutraceutical applications. This aligns with a growing understanding of the main beneficial components found in pigmented cereals, namely anthocyanin and proanthocyanidin. Pigmented cereal applications in dietary therapy, functional foods, cosmetics use, and other high-value uses have received a lot of attention. There is a need to identify and verify bioactive compounds of new varieties of pigmented cereal. This presents a novel research interest suitable for future studies. However, substantial research on nutritional values, functional properties, product making characteristics, and processing features of diverse cereal grains variations is required. The main bioactive components, namely anthocyanins, flavonoids, and carotenoids, should be further studied and more research should be carried out.

References

1. R. K. Upadhyay, *Int. J. Green Pharm.*, 2018, **12**, 93.
2. D. Kasote, R. N. Tiozon Jr, K. J. Sartagoda, H. Itagi, P. Roy, A. Kohli, A. Regina and N. Sreenivasulu, *Front. Plant Sci.*, 2021, **12**, 771276.
3. K. Hosoda, M. Miyaji, H. Matsuyama, S. Haga, H. Ishizaki and K. Nonaka, *Livest. Sci.*, 2012, **145**, 266.
4. P. Kaur, S. S. Purewal, K. S. Sandhu and M. Kaur, *Bioresour. Bioprocess.*, 2019, **6**, 1.
5. M. V. Selma, J. C. Espin and F. A. Tomas-Barberan, *J. Agric. Food Chem.*, 2009, **57**, 6485.
6. M. Carcea, *Agriculture*, 2021, **11**, 720.
7. S. Rao, L. J. Schwarz, A. B. Santhakumar, K. A. Chinkwo and C. L. Blanchard, *Cereal Chem.*, 2018a, **95**, 589.

8. D. K. Verma and P. P. Srivastav, *Trends Food Sci. Technol.*, 2020, **97**, 355.
9. F. Colombo, C. Di Lorenzo, K. Petroni, M. Silano, R. Pilu, E. Falletta, S. Biella and P. Restani, *Foods*, 2021, **10**, 1770.
10. M. C. B. Santos, N. Barouh, E. Durand, B. Baréa, M. Robert, V. Micard, V. Lullien-Pellerin, P. Villeneuve, L. Claudio Cameron, E. P. Ryan, M. Simões Larraz Ferreira and C. Bourlieu-Lacanal, *Metabolites*, 2021, **11**, 110.
11. V. U. Ndolo, T. Beta and R. G. Fulcher, *Food Res. Int.*, 2013, **52**, 109.
12. B. McKevith, *Nutr. Bull.*, 2004, **29**, 111.
13. N. Chaturvedi, P. Sharma, K. Shukla, R. Singh and S. Yadav, *J. Appl. Pharm. Sci.*, 2011, **1**, 6.
14. S. H. Arunima, G. S. Kumar, R. Krishnan and T. Mohammed, *Pharma Innovation J.*, 2021, **10**, 317.
15. M. J. Kim, J. N. Hyun, J. A. Kim, J. C. Park, M. Y. Kim, J. G. Kim, ... and I. M. Chung, *J. Agric. Food Chem.*, 2007, **55**, 4802.
16. M. K. Kim, H. A. Kim, K. Koh, H. S. Kim, Y. S. Lee and Y. H. Kim, *Nutr. Res. Pract.*, 2008, **2**, 46.
17. S. Žilić, D. Dodig, J. Vančetović, N. Grčić, V. Perić, P. Titan and V. Maksimović, *Pol. J. Food Nutr. Sci.*, 2019, **69**, 137.
18. M. Garg, M. Chawla, V. Chunduri, R. Kumar, S. Sharma, N. K. Sharma, ... and S. P. Singh, *J. Cereal Sci.*, 2016, **71**, 138.
19. D. Samyor, A. B. Das and S. C. Deka, *Int. J. Food Sci. Technol.*, 2017, **52**, 1073.
20. N. L. Kent, *Kent's Technology of Cereals: An Introduction for Students of Food Science and Agriculture*, Elsevier, 1994.
21. R. C. Chaudhary and D. V. Tran, Speciality rices of the world – a prologue, in *Specialty Rices of the World: Breeding, Production, and Marketing*, ed. Chaudhary R. C. and Tran D. V., FAO, Rome, Italy, and Oxford & IBH Publishing Co. Pvt. Ltd, New Delhi, India, 2001, pp. 3–14.
22. H. Yoshida, Y. Tomiyama and Y. Mizushina, *Food Chem.*, 2010, **123**, 210.
23. E. S. Abdel-Aal, J. C. Young and I. Rabalski, *J. Agric. Food Chem.*, 2006, **54**, 4696.
24. Y. Shao, F. Xu, X. Sun, J. Bao and T. Beta, *J. Cereal Sci.*, 2014, **59**, 211.
25. F. Zhu, *Food Res. Int.*, 2018, **109**, 232.
26. P. Saini, N. Kumar, S. Kumar, P. W. Mwaurah, A. Panghal, A. K. Attkan, ... and V. Singh, *Crit. Rev. Food Sci. Nutr.*, 2021, **61**, 3197.
27. S. Dhua, K. Kumar, Y. Kumar, L. Singh and V. S. Sharanagat, *Trends Food Sci. Technol.*, 2021, **112**, 780.
28. R. Gupta, M. Meghwal and P. K. Prabhakar, *Trends Food Sci. Technol.*, 2021, **110**, 240.
29. Y. Li, D. Ma, D. Sun, C. Wang, J. Zhang, Y. Xie and T. Guo, *Crop J.*, 2015, **3**, 328.
30. A. Collison, L. Yang, L. Dykes, S. Murray and J. M. Awika, *J. Agric. Food Chem.*, 2015, **63**, 5528.
31. M. Paulsmeyer, L. Chatham, T. Becker, M. West, L. West and J. Juvik, *J. Agric. Food Chem.*, 2017, **65**, 4341.
32. F. Lao and M. M. Giusti, *Food Anal. Methods*, 2016, **9**, 1367.

33. B. K. Baik and S. E. Ullrich, *J. Cereal Sci.*, 2008, **48**, 233.
34. I. Diczházi and L. Kursinszki, Anthocyanin content and composition in winter blue barley cultivars and lines, *Cereal Chem.*, 2014, **91**, 195.
35. P. R. Shewry and S. J. Hey, *Food Energy Secur.*, 2015, **4**, 178.
36. B. O. Eggum, in *Proceedings of Workshop on Chemical Aspects of Rice Grain Quality*, 1979, pp. 91–111.
37. T. S. Rathna Priya, A. R. Eliazer Nelson, K. Ravichandran and U. Antony, *J. Ethn. Foods*, 2019, **6**, 1.
38. B. O. Juliano, *Rice, in Human Nutrition*, International Rice Research Institute, 1993.
39. D. K. Verma and P. P. Srivastav, *Rice Sci.*, 2017, **24**, 21.
40. L. A. Bello-Pérez, G. A. Camelo-Mendez, E. Agama-Acevedo and R. G. Utrilla-Coello, *Agrociencia*, 2016, **50**, 1041.
41. P. R. Gibson and S. J. Shepherd, *J. Gastroenterol. Hepatol.*, 2010, **25**, 252.
42. J. G. Muir and P. R. Gibson, *Gastroenterol. Hepatol.*, 2013, **9**, 450.
43. S. A. Abrams, I. J. Griffin and K. M. Hawthorne, *J. Nutr.*, 2007, **137**, 2524S.
44. S. A. Abrams, K. Hawthorne, O. Aliu, P. Hicks, Z. Chen and I. Griffin, *FASEB J.*, 2007, A175.
45. R. H. Liu, *J. Nutr.*, 2004, **134**, 3479S.
46. K. K. Adom and R. H. Liu, *J. Agric. Food Chem.*, 2002, **50**, 6182.
47. K. K. Adom, M. E. Sorrells and R. H. Liu, *J. Agric. Food Chem.*, 2005, **53**, 2297.
48. N. Okarter and R. H. Liu, *Crit. Rev. Food Sci. Nutr.*, 2010, **50**, 193.
49. M. N. Irakli, V. F. Samanidou, D. N. Katsantonis, C. G. Biliaderis and I. N. Papadoyannis, *Cereal Res. Commun.*, 2016, **44**, 98.
50. E. Idehen, Y. Tang and S. Sang, *J. Food Drug Anal.*, 2017, **25**, 148.
51. D. P. Belobrajdic and A. R. Bird, *Nutr. J.*, 2013, **12**, 1.
52. S. Siebenhandl, H. Grausgruber, N. Pellegrini, D. Del Rio, V. Fogliano, R. Pernice and E. Berghofer, *J. Agric. Food Chem.*, 2007, **55**, 8541.
53. J. C. Stoclet, T. Chataigneau, M. Ndiaye, M. H. Oak, J. El Bedoui, M. Chataigneau and V. B. Schini-Kerth, *Eur. J. Pharmacol.*, 2004, **500**, 299.
54. Y. Curin and R. Andriantsitohaina, *Pharmacol. Rep.*, 2005, **1**(57), 97.
55. A. Angioloni and C. Collar, *LWT–Food Sci. Technol.*, 2011, **44**, 747.
56. K. A. Harris and P. M. Kris-Etherton, *Curr. Atheroscler. Rep.*, 2010, **12**, 368.
57. A. O. Oko and S. C. Onyekwere, *Int. J. Biotechnol. Biochem.*, 2010, **6**, 949.
58. F. Saeed, I. Pasha, F. M. Anjum and M. T. Sultan, *Crit. Rev. Food Sci. Nutr.*, 2011, **51**, 467.
59. P. Vitaglione, A. Napolitano and V. Fogliano, *Trends Food Sci. Technol.*, 2008, **19**, 451.
60. N. Shaheen, L. Bari, and M. A. Mannan, *Food Composition Table for Bangladesh*, 2013.
61. B. A. Acosta-Estrada, J. A. Gutiérrez-Uribe, and S. O. Serna-Saldivar, in *Corn*, AACC International Press, 1 January 2019, pp. 369–403.
62. S. Q. Tian, Z. C. Chen and Y. C. Wei, *J. Cereal Sci.*, 2018, **83**, 63.

63. X. Wang, X. Zhang, H. Hou, X. Ma, S. Sun, H. Wang and L. Kong, *Food Res. Int.*, 2020, **138**, 109711.
64. P. Sookwong, P. Suttiarporn, P. Boontakham, P. Seekhow, S. Wangtueai and S. Mahatheeranont, *Food Chem.*, 2016, **211**, 140.
65. Q. Liu, Y. Qiu and T. Beta, *J. Agric. Food Chem.*, 2010, **58**, 9235.
66. E. S. M. Abdel-Aal, A. A. Abou-Arab, T. H. Gamel, P. Hucl, J. C. Young and I. Rabalski, *J. Agric. Food Chem.*, 2008, **56**, 11171.
67. K. Masisi, T. Beta and M. H. Moghadasian, *Food Chem.*, 2016, **196**, 90.
68. B. Romier, Y. J. Schneider, Y. Larondelle and A. During, *Nutr. Rev.*, 2009, **67**, 363.
69. S. Vendrame and D. Klimis-Zacas, *Nutr. Rev.*, 2015, **73**, 348.
70. V. Gowd, Z. Jia and W. Chen, *Trends Food Sci. Technol.*, 2017, **68**, 1.
71. J. Lachman, P. Martinek, Z. Kotíková, M. Orsák and M. Šulc, *J. Cereal Sci.*, 2017, **74**, 145.
72. C. Bladé, G. Aragonès, A. Arola-Arnal, B. Muguerza, F. I. Bravo, M. J. Salvadó, L. Arola and M. Suárez, Proanthocyanidins in Health and Disease, *Biofactors*, 2016, **42**, 5.
73. D. M. Kasote, S. S. Katyare, M. V. Hegde and H. Bae, *Int. J. Biol. Sci.*, 2015, **11**, 982.
74. K. B. Pandey and S. I. Rizvi, *Oxid. Med. Cell. Longevity*, 2009, **2**, 270.
75. J. He and M. M. Giusti, *Annu. Rev. Food Sci. Technol.*, 2010, **1**, 163.
76. S. Kamiloglu, E. Capanoglu, C. Grootaert and J. Van Camp, *Int. J. Mol. Sci.*, 2015, **16**, 21555.
77. F. J. Olivas-Aguirre, J. Rodrigo-García, N. D. R. Martínez-Ruiz, A. I. Cárdenas-Robles, S. O. Mendoza-Díaz, E. Álvarez-Parrilla, G. A. González-Aguilar, L. A. De la Rosa, A. Ramos-Jiménez and A. Wall-Medrano, *Molecules*, 2016, **21**, 1264.
78. M. T. Escribano-Bailón, C. Santos-Buelga and J. C. Rivas-Gonzalo, Anthocyanins in cereals, *J. Chromatogr. A*, 2004, **1054**, 129.
79. D. B. M. Ficco, V. De Simone, A. M. De Leonardis, V. Giovanniello, M. A. Del Nobile, L. Padalino, L. Lecce, G. M. Borelli and P. De Vita, *Food Chem.*, 2016, **205**, 187.
80. B. Geera, L. O. Ojwang and J. M. Awika, *J. Food Sci.*, 2012, **77**, C566.
81. P. Bartl, A. Albreht, M. Skrt, B. Tremlová, M. Ošťádalová, K. Šmejkal, I. Vovk and N. P. Ulrih, *Int. J. Food Sci. Nutr.*, 2015, **66**, 514.
82. J. Li, C. E. Walker and J. M. Faubion, *J. Sci. Food Agric.*, 2011, **91**, 38.
83. X. Sui, P. Y. Yap and W. Zhou, *Food Bioprocess Technol.*, 2015, **8**, 983.
84. S. Rao, A. B. Santhakumar, K. A. Chinkwo, T. Vanniasinkam, J. Luo and C. L. Blanchard, *Nutr. Cancer*, 2018, **70**, 913.
85. R. Beleggia, D. Ficco, F. M. Nigro, V. Giovanniello, S. A. Colecchia, I. Pecorella and P. De Vita, *Agronomy*, 2021, **11**, 591.
86. S. Iametti, P. Abbasi Parizad, F. Bonomi, and M. Marengo, Pigmented Grains as a Source of Bioactives, in *Reference Module in Food Science*, 2019, p. 261.
87. V. C. Ito and L. G. Lacerda, *Food Chem.*, 2019, **301**, 125304.

88. O. Kadiri, *Int. J. Food Prop.*, 2017, **20**, S798.
89. D. Vargas-Yana, B. Aguilar-Morón, N. Pezo-Torres, K. Shetty and L. G. Ranilla, *J. Ethn. Foods*, 2020, **7**, 1.
90. S. P. Bangar, K. S. Sandhu, S. S. Purewal, M. Kaur, P. Kaur, A. K. Siroha, K. Kumari, M. Singh and M. Kumar, *J. Food Process. Preserv.*, 2021, e15543.
91. J. G. L. Dela Rosa and P. M. B. Medina, *Food Sci. Technol.*, 2022, **v42**, e45120.
92. S. Luti, V. Galli, M. Venturi, L. Granchi, P. Paoli and L. Pazzagli, *Appl. Sci.*, 2021, **11**, 3291.
93. J. L. Mau, C. C. Lee, Y. P. Chen and S. D. Lin, *LWT–Food Sci. Technol.*, 2017, **75**, 434.
94. I. Sapna and A. Jayadeep, *J. Food Sci. Technol.*, 2021, **58**, 2089.
95. A. S. Hole, N. P. Kjos, S. Grimmer, A. Kohler, P. Lea, B. Rasmussen, ... and S. Sahlstrøm, *J. Agric. Food Chem.*, 2013, **61**, 2739.
96. T. B. Bagchi, K. Chattopadhyay, M. Sivashankari, S. Roy, A. Kumar, T. Biswas and S. Pal, *J. Cereal Sci.*, 2021, **100**, 103263.

CHAPTER 14

The Role of Biotechnology in the Production of Pigmented Cereals

ALI RAZA ISHAQ*[a], RABIA BATOOL[b], SHOUWEN CHEN[a], DONGBO CAI[a] AND ASFANDYAR[a]

[a]State Key Laboratory of Biocatalysis and Enzyme Engineering, Environmental Microbial Technology Center of Hubei Province, College of Life Science, Hubei University, 430062 Wuhan, China; [b]Department of Zoology, Government College University Faisalabad, Punjab, Pakistan
*E-mail: alirazaattari476@gmail.com

14.1 Introduction

Nature provides several beneficial compounds for the betterment of human health. Carotenoid pigments are also essential natural antioxidant compounds owing to a broad spectrum of biological activities, including anticancer, antitumor, anti-inflammatory, reactive oxygen species (ROS) scavengers *etc.*[1] Lycopene is an acyclic carotene acting as a metabolite regulator *via* modulating carotenoid intermediates (β-carotene, torulene). Carotenoids are used as food colorants and nutritional supplements in the food industry. They have multiple preventive actions to cure various diseases, such as Alzheimer's disease, oxidative stress-related disorders, atherosclerosis, and diabetes. Carotenoids also manage cell death, mutations, and protect from heart disease and cancer.[2]

Food Chemistry, Function and Analysis No. 38
Pigmented Cereals and Millets: Bioactive Profile and Food Applications
Edited by Sneh Punia Bangar, Sajid Maqsood and Anil Kumar Siroha

Published by the Royal Society of Chemistry, www.rsc.org

By 2050, the global population will increase by approximately 2.3 billion, bringing the total to ~9.8 billion. Along with human population growth, environmental changes have arisen from the last four decades that pose a substantial human-made threat to farming and food safety.[3]

The call for foodstuffs is therefore responsive to advanced developments such as vegetal oils, livestock, and dairy foods. By 2050, the demand for nutrition will be increased by about 70% due to human population growth. Emerging states would need to double their agronomic production. According to recent research, harvesting of important crops such as maize, soybean, wheat, and rice has increased by roughly 2.4% due to overpopulation as well as global village demand.[4,5] Based on a meta-analysis of ~1700 model simulations, it was found that worldwide mean crop production is predicted to reduce by about 3–10% per degree of warming.[4]

This rise in production was accompanied by a healthier capacity of crops to manage with abiotic and biotic pressures.[76] These advances have been knowing genetic variations, molecular and cellular pathways, and exploring the crop production. The characteristics that can markedly improve crop plant yield comprise the features of resistance peptides, such as identifying the signs from invasive pathogens and resisting pathogens, the existence of vegetation under submerged conditions, improved root structure to grow in restricted water environments, augmented metal stress as well as salt stress tolerance, and improved photosynthetic efficacy with better light-harvesting and CO_2 fixing potential.[6] Flora with such characteristics are being developed through advances in genetic technologies, for example sequencing, genotyping and genomic assisted breeding,[7] and synchronized approaches that unite conventional breeding, genomics, and bio-technological methods.[8]

Carotenoids have various favourable physiological actions in mammalian cells due to their various properties, such as antioxidant, anti-mutagenic, and photoprotective potential. Due to restrictions, synthetic organic carotenoids cannot be used in food, pharmaceutical, or medicinal applications. Thus, fermentation and genetic engineering are the best methods to produce pigmented cereals that are widely used in the food, medical and pharmaceutical industries, as shown in Figure 14.1.

Mycotoxins are herbal, secondary metabolites produced with the aid of numerous fungal species.[9] These mycotoxins are extremely harmful and can enter into the human body *via* food to cause metabolic dysfunction. Thus, the risks to human and animal health posed by these herbal materials may be greater than those produced through people because of pesticide use in plant safety, veterinary drugs used in the protection of animal health, preservatives, or food additives.[8] The simplest way to maintain the nutritional profile of food is to apply correct agricultural techniques, create a specific version of crops, utilize plant varieties resistant to fungal infections, and use fungicides. Regulations on synthetic fungicides have brought about the expanded use of natural fungicides, including metabolites from flowers, together with phenolic compounds and essential oils.[10,11]

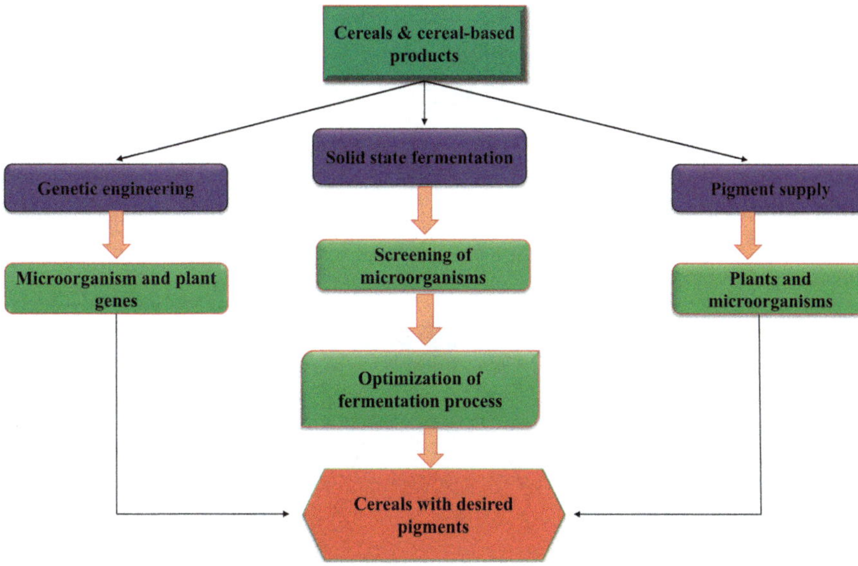

Figure 14.1 Diagrammatic representation of methods employed to boost the productivity of pigmented cereals.

Genetic varieties are being discovered at nucleotide-scale accuracy through genome-wide association revisions and additional gene-mapping approaches, corresponding with innovative phenotyping systems. Recent studies have revealed that loci directly engaged in trait expression, molecular-marker-assisted breeding, and the emergence of novel variations in the next generation. Complementary to breeding methods, developments in the spatial and sequential regulation of engineered genetic factors and paths are progressively enhanced by the battered editing of genomics using the CRISPR–Cas method.[6]

This chapter endorses for the current advances in various biotechnological strategies to be combined with traditional plant breeding techniques to produce crops with an improved nutritional profile and benefits for human health.

14.2 Nutritional Significance of Pigmented Cereals

Pigmented cereal grains have a greater ratio of flavonoids and phenolic compounds, including anthocyanins, which are found in different plant tissue types of the kernel and are involved in the pigmentation of these grains. The latest investigations have explained that phenolic, flavonoid and anthocyanin compounds have potential biological activities including cardioprotective, antidiabetic, anti-inflammatory, anti-allergic, anticancer, and antioxidant, as shown in Figure 14.2. These chemicals and their derivatives can act on intrinsic cellular signaling pathways and attenuate inflammatory responses *via*

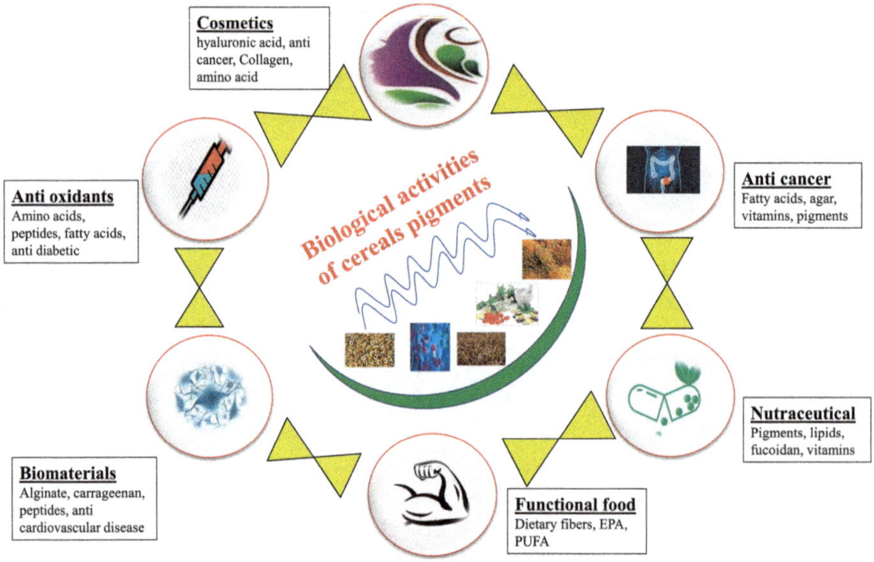

Figure 14.2 Outline of the possible biological activities of cereal pigments.

influencing NF-κB-mediated pathways.[1] Polyphenols and anthocyanins have also been significantly lowered pancreatic α-amylase and brush-border glucosidase activities. They inhibit intestinal α-glucosidases, limiting the digestion and absorption of postprandial glycemia, which is an efficacious method of controlling type 2 diabetes.[12] Likewise, rice brans have a diverse nutrition profile, including proteins, lipids and phytochemicals that actively induce intestinal microbiota, inflammatory response, production of secondary metabolites and intestinal immunity against bacterial infections. Pigmented rice brans have caffeic, isoferulic, syringic, vanillic, sinapic and pro-anthocyanidins in their nutritional chemistry. There is a further need to improve the nutritional chemistry of rice brans and production at an industrial scale.[13]

Vegetables have been used as a primary resource in the human diet since ancient times. The cultivation of cereals and other vegetables improves and diversifies the diet by being an important source of nutrients, vitamins, secondary metabolites, and other compounds that contribute to health and nutrition.[14] Cereals, such as wheat, maize, or rice are consumed all over the world and represent an essential source of carbohydrates and nutrients for the diet.[15] Cereals are also a fundamental part of the livestock diet, including maize and other coarse grains, such as sorghum, barley, or oats, and, together with those used in human nutrition, represent a powerful economic engine for many countries and the livelihood of their families. Other vegetable crops, such as lettuce, zucchini, tomato, eggplant, cucumber, onion, peas, or potatoes are widely consumed as an important source of starch and cellulose, containing a wide variety of vitamins and minerals.[16]

The physiological connection between lactic acid bacteria (LAB) yeasts is responsible for producing "sourdough bread" that contains following ingredients: cereal flour, water, and complex microbial biological system. The nutritional chemistry of sourdough bread is dependent on metabolic pathways of species and strains. Like, some strains can produce essential amino acids or vitamins such as vitamin E and folates, bioactive compounds, amino acid derivatives, enzymes, and anti-nutritional factors.[17]

Beetroot is a dispensable reservoir of components with various human health benefits, such as flavonoids, carotenoids, nitrate, and iron. Beetroot has diverse biological activities due to its antioxidant, anti-diabetic, anti-cancer, cardiovascular disease, hypertensive, hepatoprotective and wound healing benefits.[18] Thus, it is a big challenge for researchers to develop a functional transgenic food along with beetroot gene implantation and find a suitable expression system. Corn and rice are gluten-free products with a high number of anthocyanins to scavenge free radicals as well as behaving like an antiradical. Astaxanthin is a natural pigment known as a carotenoid that provides salmon its distinct orange red colour.

Violacein is an herbal compound formed by condensation of tryptophan molecules, which turned into a potential novel pharmaceutical reagent due to its large biological properties including antitumoral.[19] antibacterial,[20] antiviral,[21] and antioxidant activities.[22] Violacein biosynthesis starts with L-tryptophan and is catalyzed with the aid of enzymes VioA, B, E, D, and C successively, which were encoded through the vioABCDE operon.[23] Several Gram-negative micro-organisms were said to provide violacein as their secondary metabolites together with *Chromobacterium violaceum*, *Alteromonas luteoviolacea*, *Pseudoalteromonas luteoviolacea*, B2, and so forth. Some of the natural manufacturers have been studied to enhance their violacein manufacturing, such as *C. violaceum* and *Duganella* sp. B2.[24,25] However, fermentation by natural producers have been constrained by low productiveness, bacterial pathogenicity, and problems in huge scale culture.[26]

Carotenoids are a large array of isoprenoid-derived compounds that might be synthesized using flowers, micro-organism, fungi, and animals. In flora, carotenoids are crucial additives of the photosynthetic apparatus and are responsible for the orange, and yellow colour of many vegetation and fruit. In addition to their flora roles, they essentially make contributions to human fitness.[27]

Viola Galli *et al.*[28] examined the antioxidant activity of different conventional cereals compared with sourdoughs made with pigmented wheat and barley. Despite being inoculated with the identical LAB strains, the antioxidant activity of sourdoughs prepared using pigmented wheat and barley was found to be more than that of those manufactured with ordinary wheat flour in this investigation.

14.3 Fermentation a Strategy for Pigmented Cereal Production

Fermentation is defined as the incomplete oxidation of macromolecules in the presence of microbial culture (fungi and bacteria) to produce desired products. Fermentation technology has the potential to design novel biotechnology and modify the processes used in existing ones by keeping in mind numerous specialized, financial, and ecological factors. Given the financial importance and continuous challenges in the biotech industry, multidisciplinary design developments are preparing to become important tools with the development of current innovation and advancement.[29] It is the most effective method to modify nutrient contents (such as the introduction of PUFAs and pigments) in cereal-based food items for the better health of human beings. Solid state fermentation (SSF) is frequently used to grow microbes on wet solid media to achieve the desired result. The reasons to use SSF are that the polymeric nature of wet solid media is unable to dissolve in water and it acts as the source of various valuable nutrients like hydrogen, nitrogen, carbon, water, electrolytes, and other metabolites for microbial cultures.[30,31] The most superficial benefit of SSF is to trigger the process of fermentation, meaning degradation of agro-industrial wastes like wheat, hull, bran, and other cereals for additional bioproducts. SSF is going to open new doors at a commercial level in different industries such as food, animal feed, pharmaceutical, diary, environmental industries.

14.3.1 Features of Ideal Microbial Strain–Substrate Selection

Previous investigations revealed that filamentous fungi are widely employ on substrates to increase the yield of pigmented cereals. *Monascus purpureus, M. pilosus, M. anka, M. ruber* and some *Mucor* strains are the best producers of pigmented cereals. Fermented food sources and refreshments result from the enzymatically changed food parts achieved by various microorganisms. Fermented food sources have increased in prevalence of late because of their supposed medical advantages. Biogenic amines, bioactive peptides, antinutrient decrease, and polyphenol change to physiologically dynamic synthetics are medical advantages of fermented items. Ethiopian-fermented food varietiesare for the most part handled utilizing unconstrained fermentation process. Injera is one of the fermented food items consumed in all areas of the country and sourdough fermentation could be accomplished utilizing different LAB and yeast strains. Additionally, the type of the substrate and the type of microbial vegetation, as well as temperature, air supply, and pH, all impact the aging system of injera.[32] Figure 14.3 represent the mechanism of microbial strains digesting the substrate for value pigments. There are some unique features of microbes for the effective production of pigmented cereals; (1) large surface area of substrate for optimal growth of microbes (2) have an adaptive structure to frequently utilize the substrate like fungal hyphae or digestive enzymes (3) have a mechanism to inhibit the production of anti-nutrient products (4) the microbial strain must grow in dehydrated conditions to stunt bacterial contamination. Algae, yeast, and fungi are the primary producers of β-carotene or lycopene at a commercial level. In the kingdom of fungi, the members of *Zygomycetes* have the ability to produce both PUFAs and carotenes along with the bulk of bioactive compounds. Milan Certík *et al.*[33] employed SSF on cereals to increase the pigment yield along with

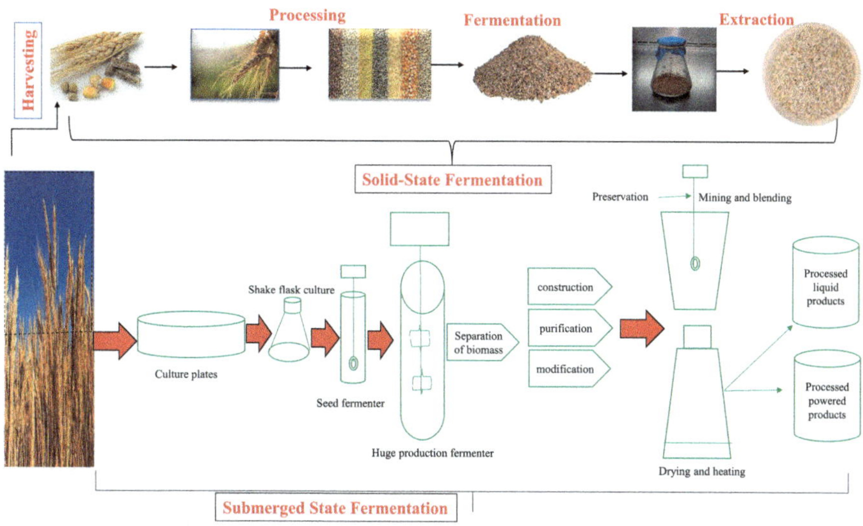

Figure 14.3 The detailed mechanism involved in the enhancement of pigmented cereals *via* microbial action.

PUFAs *via* fungal growth. *M. circinelloides* CCF-2617 cultures on a 3:1 ratio of mixtures of rye bran–spent malt grain, respectively, were used to boost the pigment yield at stationary phase. Wheat bran, rye bran oat flakes and barley groats can be digested by various *Mucor* strains including *M. petrinsularis, M. dimorphosporus, M. circinelloides and M. hiemalis* to give off an augmented number of bioactive compounds.

14.3.2 Pigment Isolation Methods

Pigment isolation is an important step in fermentation due to their reversible nature, **converting back** into intermediate products. Various types of acid are used to extract pigments. Paulsmeyer *et al.*[34] utilized the acidic medium to isolate the anthocyanin from maize kernels *via* 2% formic acid solution. The main drawback of acidic extraction is that hydrogen ions can alter the chemical composition of the isolated pigments. In a strong acidic medium, anthocyanin converts into anthocyanidin and consequently decomposes into by-products. So, we can conclude that the extraction of pigments depends on the following: type of pigment, temperature, source of pigment. Temperature and solvent readily affect the ability to extract anthocyanins from sorghum.[35] Solvent systems dissolve the other solutes as well as metabolites. Methanol, ethanol andacetone with the addition of different acids (0.01 *vs.* 1% HCl, TFA, formic acid) are the best solvent systems for extraction of pigments.

Despite the solvent applications in pigment extraction, there is another beneficial method in extracting pigments known as ultrasound- and microwave-assisted extraction. Das *et al.*, 2017 proved the 50% reliability of this method by extracting anthocyanin from rice bran *via* an ultrasound-assisted process.[36] Abdel-Aal *et al.*[37] noticed that any extraction method can change the chemistry of the pigment. They found that the microwave-based method does not affect the composition of anthocyanins of wheat, rice, and maize samples. Lao & Giusti[38] proved that the yield of purple maize anthocyanins depends on physiochemical factors including pH, and HPLC-UVV with HCl-hydrolysis. The results show that the total anthocyanins has the highest value compared with the other three methods.[38] In conclusion, selecting a method for isolating pigments is also critical for fermenting the cereal products.

14.3.3 Fermentation and Pigment Production

Nature provides us with approaches to treat agro-industrial wastes for the production of beneficial products. Fermentation is one of best methods to produce hormones, antiviral proteins, enzymes, bioactive compounds, growth hormones and medicinal drugs. Researchers provide optimal conditions like temperature, pH, nutrient supplements, microbial cultures *etc.*, to the fermenter for escalating the production at an industrial level. The duration of SSF also influences the yield of pigments such as co-production of GLA and β-carotene is enhanced by 96% and 88%, respectively, during 140 hours of SSF. Microbial culture efficiently digests the specific amount of supplemental substrate to increase the production of pigments. Table 14.1 represent the culture conditions for microbes to employ biomass for escalating the production of pigmented cereals. The substrate amount, kind, and duration of inoculation greatly affect the yield of bio-products. For example, fungal culture consumes only 5% olive oil to augment the production of enriched fermented cereal with β-carotene about 1–2 fold. Not all oils as co-substrates with specified concentrations can increase the production of pigmented cereals. The previous investigation proved that sunflower and linseed oil completely stunt the growth of fungi on the substrate, indicating the negative feedback mechanism on β-carotene production. Recent investigations show the significant increase in carotenoids production because of inserting supplemented oils in various microbial cultures such as *Blakes leatrispora* and *Phaffia rhodozyma*.[39] Tatiana Klempová *et al.*[40] provide an alternative narrative that the inoculation of oil to the substrate has no prominent effect on the co-production of β-carotene under solid state fermentation.

The species of zygomycetes are the factories for the production of β-carotene and other carotenoids. β-Carotene in zygomycetes varies from strain-to-strain fermenter conditions as well. Tatiana Klempová *et al.*[40] investigated the chemical composition of crude pigment extract that possess the 75–90% β-carotene content along with other accessory pigments like gamma-carotene, lycopene, and *cis*-β-carotene. *Umbelopsis isabellina CCF 2412* has carotene productivity ranging between 3.2 and 40.7 mg L^{-1} next to 3.2 mg g^{-1} cells in culture. Klempova *et al.*, 2013, showed experimental data that

Table 14.1 Solid state fermentation conditions for pigmented cereals production.

Microbes	Culture conditions	Methods of pigment analysis	Substrate	Reference
Monascus purpureus	Potato dextrose agar slant + SSF	TLC and (LC. MS/MS)	Rice, DSG, WSG, Corn, SB *Monascus purpureus*	4
M. circinelloides	Agar slants Yeast extract (2.5 g L^{-1})	HPLC	Rye bran/spent malt grains	41
Mucor wosnessenskii	Potato Dextrose Agar slant	HPLC	Cereals including (corn, wheat, rye, rice)	40
	SMF		Legumes	
Umbelopsis isabelline	Sabourad agar slants	HPLC	Biosynthesis from soil fungi	42
Mucor azygosporus MTCC 414	YpSS (Yeast phosphate soluble starch) agar slants at 25 °C	HPLC, UV spectrophotometric	Cheese whey	43
Lactobacillus sp.	LAB-strain/flour sourdough fermentation	RP-HPLC/DAD	Wheat, Barley	28
Aspergillus oryzae NK	*Aspergillus oryzae* NK (*f*SSE)	HPLC	Sorghum	44

indicates that the yield of β-carotene was 0.2–9.4 mg L^{-1} in *M. azygosporus*.[40,42] Comparative analysis of these results indicate that the production of β-carotene from *Umbelopsis isabelline* is 10-times more than *M. azygosporus* under minimum cell cultures.

M. azygosporus cultivates on whey containing 3.5% lactose to produce a maximum yield of β-carotene (318 μg g^{-1} dcw). So, this investigation indicates that fungal species are also lactose tolerant. For instance, further increasing the content of lactose to 4.0% (w/v) significantly decreased the production of β-carotene (224 μg g^{-1} dcw). Here, we noticed a relationship between biomass and carotene production which means the exponential growth of fungi hyphae resulted in the escalation of β-carotene biosynthesis.[43] Previous research has not found any links between abiomass aggregation, carotenoid production and lactose content *via* yeast co-culture with lactic acid.[45]

There is no significant positive effect of nitrogen sources in terms of β-carotene yield in *M. azygosporus MTCC 414*. Further, organic nitrogen inhibits the secretion of β-carotene in contrast to nitrogen free cultures. Moreover, *M. azygosporus* show maximum cell division under the nitrogen-containing culture medium.[43] β-Carotene shows a slope-like growth graph which means that the highest yield (374 μg g^{-1} dcw) noticed after 5 d of incubation is consequently gradually downregulated.[43] *Blakeslea trispora* was cultivated on cheese whey and an optimum output of β-carotene was noticed after 8 days of incubation in submerged fermentation.[46]

14.3.4 Factors Affecting Fermented Food

Most fermented substances contain no less than 106 microbial cells in every gram, with the value fluctuating given various factors like the item's topography, age, and season of investigation/utilization.[32] The encompassing food network seems to assume a basic part in probiotic strain endurance by buffering and preparing for gastrointestinal circumstances (*e.g.*, low pH and bile acids). Various investigations indicate that microorganisms from fermented food sources can infiltrate the gastrointestinal tract. Be that as it may, this fluctuates by item, and their presence in the stomach seems, by all accounts, to be temporary.[47] These microorganisms might in any case have the option to give a physiological benefit in the stomach by contending with hurtful microscopic organisms and creating insusceptible administrative and neurogenic maturation side-effects.[48]

14.3.4.1 Aeration

Aeration is a significant factor that influences large-scale fermentation systems. The impacts could be the Custer impact, which restrains development in anaerobic circumstances, or the Pasteur impact, which represses maturation in oxygen-consuming circumstances. In oxygen-consuming circumstances, *D. bruxellensis* is seriously smothered by *S. cerevisiae* in cluster maturations.[32,49] The presence of high air circulation causes a diminishing in the last ethanol yield. Meanwhile, yeast has the ability to produce acidic corrosive ethanol in the presence of air. Aeration prompts the formation of corrosive acid and influences the total ethanol yield, so restricted air is the superior fermentation condition to get a successful result.[50] The oxygen movement coefficient, oxygen movement rate, and broken down oxygen pressure should all be regulated for ethanol production. KLa values in the scope of 2.3 to 5.9 h1 were viewed as having the ideal balance between efficiency and yield. Regarding DOT, *S. stipitis* delivered more ethanol when the immersion level was under 1%.[32]

14.3.4.2 pH

pH is a factor that could influence cell advancement as well as ethanol yield.[51] The impact on yeast development is in all probability because of changes in toxicity after hydrolysis instead of the direct impact of pH since it isn't exposed to extreme circumstances.[52] Variations in the pH of the ethanol production system can influence the basic fermentation process. It is fundamental to keep the pH between 4.0 and 5.0, however, in the event that it exceeds this range, the resulting products, for example, acidic and butyric corrosive substances can interact with the substrates, reducing the efficiency of ethanol fermentation.[53]

Physical factors like pH, temperature, and moisture content are also major contributors to increasing the yield of bioproducts. pH is the main factor in terms of changing the concentration of hydrogen ions in the fermenter, resulting in the reduction of bioproducts into oligomers. The optimum yield β-carotene (365 μg g^{-1} dcw) was seen when the medium had a pH of 5.5.[43] Previous investigations Martelli *et al.* and Frengova *et al.* indicate that the ideal pH of the medium must be 5.05.5 for the maximum production of β-carotene specifically with lactose as a co-substrate.[45,54]

An increase in pH ultimately inactivates the lactase enzyme to decrease the utilization of the co-substrate as a nutrient source, resulting in a lower yield of bio-products. Despite this, Seon-won *et al.*[55] proved that *B. trispora* has the ability to secrete β-carotene under alkaline pH. The experimental result prove that β-carotene was produced maximally (48.0 mg I-carotene/l) at pH 10. Sheetal Choudhari *et al.*[46] indicated that the production of β-carotene (99 ± 1 mg L^{-1}) was achieved at pH 6–7 as compared to pH 4–6 or pH 8–10. Almost all fungi have a similar incubation period, which ranges between 25 and 35 °C for the optimal production of pigmented cereals. Table 14.2 shows the biosynthetic pathway and production of pigmented cereals.

14.3.4.3 Temperature

Temperature is a significant factor for microbial development; all microorganisms have a specific ideal range where they can develop.[56] It is even possible that a sequential temperature has a considerable effect with some of the other variables. As the temperature climbs, the maturation time diminishes, however, a significantly higher temperature restrains cell development and therefore fundamentally decreases aging because the higher temperature changes the vehicle action or level of dissolvable mixtures and solvents in the cells, possibly expanding the collection of poisons like ethanol inside cells. On the other handt the cells, show decreased explicit development rates at lower temperatures, which could be because of their low resilience to ethanol at lower temperatures.[57]

The fermentation temperature and the baking temperature of the fermented teff batter may likewise significantly affect the nature of injera. Grain crushes with a pH of 5–6.2, which are abundant in fermentable starches, will be specially fermented by LAB to a pH under 4, and below this point, acid tolerant yeasts overwhelm the fermented the pH of injera to be 3.4.[58] The ideal temperature range for yeasts is 20–30 °C. Most lactic acid microscopic organisms flourish at temperatures somewhere in the range of 18 and 22 °C, while *Lactobacillus* species flourish at temperatures over 22 °C.[59]

14.3.4.4 Substrate Concentration and Type

Different fermentation designs happened to rely upon the substrates. As indicated by Assefa,[32] the super fermentable sugars in grain wheat and wheat-red sorghum-injera were maltose and glucose, with a transient aggregation of maltose. Substrate groupings of 20 to 300 kg m^{-3} impact ethanol age. When

Table 14.2 Pigments and their yield, biosynthesis pathway, biological activity.

Microbes	Pigments	Pigment yield	Biosynthesis pathway	Biological activity of cereals pigment	Reference
Monascus purpureus	Rubropunctamine Monascorurin Rubropunctatin Monascin Ankaflavin Xanthomonascin A,B Monascopyridine A,B Yellow II Monapilol B		Fatty acid metabolism Upregulation of acetyl-CoA pathway	Food colorants Food supplement Functional food Anti-inflammatory Anticancer Antimicrobial	4
M. circinelloides	β-carotene	8.5 mg	Stationary fungal growth	Anti-*Helicobacter pylori* (*H. pylori*), anti-human immunodeficiency virus (HIV), cytotoxic, multidrug resistance (MDR)	41
Mucor wosnessenskii	β-carotene	261.5 mg	—		40
Umbelopsis isabelline	β-carotene	40.7 mg L^{-1}	—		42
Mucor azygosporus	β-carotene	385 μg g^{-1} dcw	Proteolysis of wheat		43
Lactobacillus sp.	Bioactive compounds in grains flour like anthocyanin, carotenioids	—	By promoting the protein and phytic acid degradation	Antioxidant and anti-inflammatory	28
Aspergillus oryzae NK	*p*-coumaric acid	3.82 *f* SSE	—	Anti-tyrosinase activity and suppression of MMP-1, -2	44

the pH was not controlled, greater initial glucose fixations over 80 kg m^{-3} at 30 °C required a longer incubation period, however when the pH was regulated, a longer incubation time period was expected for greater initial glucose concentrations than 80 kg m^{-3} at 30 °C.[60]

Besides, when the pH value was not changed, larger starting glucose concentrations, like 300 kg m^{-3}, may have diminished ethanol transformation proficiency since the higher substrate and product concentrations might have obstructed the ethanol fermentation interaction.[32]

14.3.4.5 Chemical Attributes

As well as creating acidity, microbiological digestion affects the composition of substrate of the aging teff batter. The angular shape of starch granules is harmed during fermentation because of enzymatic breakdown, as shown in light and electron microscopy investigation of matured teff.[61] Fermentation has an effect on grain and immature organism pieces, cell dividers, and protein bodies. During fermentation, microorganisms formed fibrillar strings that connected flour particles together.[62] Exopolysaccharide, which is created by lactic acid microorganisms, is the most probable source of this fibrillar material.

Parapouli *et al.*[62] detailed that *Candida guilliermondii* was isolated from fermented teff and they showed that it was responsible for starch hydrolysis and an expansion in the centralization of reducing sugars in the early stage.

14.4 Commercialization in Fermentation of Pigmented Cereals

Interest in nutraceuticals and functional foods continues to increase because of growing public interest. As a result, food scientists, dietitians, and food industrialists are participating in creating new functional food materials.[63] Identifying important post-harvest processing strategies to retain extra polyphenolic compounds through conventional cooking and baking processes must be recommended to maintain the human health benefits. More than one hundred fifty million lots of wheat bran and 76 million tons of rice bran nutritional material are produced as a by-product of milling and generally discarded. It can be channeled into technological advancements in micronization, microfluidization, ultrasound processing, nixtamalization, pulsed electric fields, and micro/nano-encapsulation to extract bioactive phenolic compounds and produce diverse useful by-products with widespread human health benefits.[64]

Furthermore, bioprocessing techniques can be used to produce unique functional food products from germination sprouts, fermentation and enzymatic treatments of bran and pigmented grains to enhance the bioaccessibility of phenolic compounds. Implementing holistic strategies to discover donor traces with enriched bioactives such as the usage of untargeted metabolite profiling to discover novel phenolic compounds with higher antioxidant ability and deploying contemporary processing packages to provide final food products can be beneficial in growing novel opportunities to place cereals as staple meals to fulfill human vitamin needs. Long-time period scientific studies for the whole variety of useful food objects produced from cereals are required to decide lengthy-time period advantages.[64,65]

14.5 Genetic Engineering: a Pioneering Method for Transgenic Cereal Production

Genetic engineering is the manipulation of genes with genes of interest to inject a cell with a desirable trait. Due to the demand for pigments in pharmaceutical industry, researchers are looking to the practical application of CRISPR-Cas9 technology for the formulation of pigment-rich seeds.[66] Over the last twenty year, a significant development has been manipulate genes from different sources and insert them into cereal-based products. Transgenic plants are so-called green factories due to their high nutrition profile as well as sustainable and cheap availability in the market (Figure 14.4).

Cen Li *et al.*[67] developed a visualized multigene editing system *via* an unusual tRNA-guide RNA array comprising the gRNAs of a pigment gene *albA* and target genes. The scientific community succeeded in manipulating metabolic pathways in terms of inserting genes into a system and silencing other negative genes.

The anticipated population increase by 2050 will be coupled with increased food demand. To attain maintainable food supplies to feed the world's

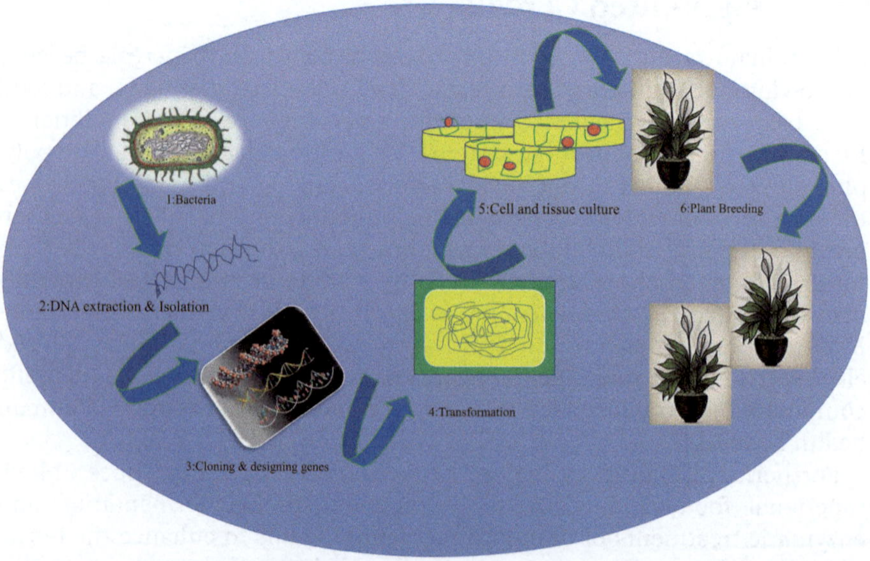

Figure 14.4 Representation of genetic manipulation in cereals *via* introducing selected genes.

inhabitants by 2050, approximately a 2.4% increase in the main crop yield is mandatory. It is noteworthy that yield-improved crops are a positive consequence of genomic variation and knowledge of molecular markers, numerical trait loci, genetic factors, and pathways related to advanced yields and improved tolerance to biotic and abiotic pressures. Developments in genetic strategies are helpful to plant breeders and geneticists to boost yield plants with better-quality agronomic methods. This distinct problem is an effort to report genetic advances by adapting genetic methods and genetic assortment.[4]

Speed breeding permits researchers to explore gene bank accessions and distorted collections for an un-paralleled swift gene detection and gene placement. A combination of speed breeding and leading-edge shrub breeding strategies with planned global corporations can probably attain the mandatory genetic gain goals to bring our future harvests.[7]

14.6 Prospects

14.6.1 Solid State Fermentation

Solid-state fermentation is a cost-effective bioproduct synthesis process that takes place on a solid space in the near absence of free water.[68] This system has been efficiently used to increase the protein content, while also dropping undesirable elements such as tannins, phytic acid and fiber. The frequently used fermentation microorganisms such as *Aspergillus niger, B. subtilis*, and *C. utilis* were approved and are usually regarded as safe and legal animal feed in China.[69]

14.6.1.1 Advantages

Secondary metabolites of microbial organisms are the by-products or the products of metabolism that are obtained *via* a fermentation process. These are produced through the stationary phase and show a key role in competition, growth and self-defense mechanisms.[70] Fungal fermentation had similar effects on the contents of several hydrolytic amino acids and improved the production of bitter and umami components.[71]

14.6.1.2 Disadvantages

The solid state fermentation process with *Aspergillus niger* improved the nutritive value and flavour of tea dregs but further animal and toxicity trials are mandatory to regulate the safety of fermented tea as animal feed.[71]

14.6.1.3 Demand

There is demand for *S. cerevisiae* in the fermentation industry due to the rapid production of CO_2 and more consumption of sugar as an energy source to give off bread. It is essential to develop a microbial strain having both features at the same time with genetic manipulation.

14.6.2 Genetic Engineering

Genetic engineering is a molecular biology technique used to change the DNA sequence in the genome using different methods such as the homologous recombination technique used to mark specific sequences in the embryonic stem cell genome or other cultured cells.[72]

14.6.2.1 *Advantages*

Genomic selection is a breeding strategy that has been successfully applied on crops to report the limitations of phenotypic assortment, such as decreasing biased marker outcome evaluation and allowing scientists to stop additional differences due to the minor effect of quantitative trait *loci*.[4,73] That can be attained by incorporating genome-wide indicator data in a breeding rate prediction model. Genomic selection has a vital role in complementing population genetics and computational genetics, which had extensive benefits in the development of sequencing and genomics. Therefore, the application of genomic selection in cereals and beans could be valuable[74] to forecast the antioxidant production of new and un-phenotypical sorghum genotypes.[74]

The developments in molecular markers and genomic sequencing assisted the delimitation of the genetic basis of agronomical characteristics. Mainly, the advances in genome sequencing have pushed forward the cause of genome-wide association studies that helped in the finding of significant genes linked to the aptitude of a crop plant to endure biotic and abiotic pressures. The Green Revolution aided a steep rise in the yields of key staple grain crops to meet the caloric needs of a growing global population. This was empowered through leading diversity breeding, hybrid yield growth, fertilizer tender, and organizational development through substantial public assets. The subsequent rise in food safety advanced many areas of the ecosphere and upgraded agricultural advances (mainly in India and south-east Asia), reducing poverty and starvation.[6,75]

14.6.2.2 *Disadvantages*

Corynebacterium glutamic is an old food-grade engineering micro-organism, in which well-organized endotoxin recombinant polypeptide protein expression is emerging in recent years.[1] Vertical drainage is used for monitoring the expanding groundwater stages in some chunks of Australia and helps to decrease the extent of water logging in India.[2] Nevertheless, the maintenance and operative costs are higher than parallel drainage systems as it includes high energy to work a system of tube wells. Gene silencing by RNA interference is a significant tool in studying protein sequence role in polyploid cereals from which the groups of natural or induced mutants are inadequate.[3]

14.6.2.3 Demand

Therefore, the standardization of extraction methods for anthocyanins in cereals remains to be developed so that the results from different studies may be compared directly. Global initiatives on method comparison and the involvements of international agencies such as AOAC are needed for such purposes.

14.7 Conclusion

Cereals and its by-products are major natural food resources in the human diet generated by various processing chains. While cereal processing industry by-products can potentially give important mixtures, their utilization in the diet is restricted by their difficult industrial resources. Hence, the improvement of creative biotechnology is fundamental for redesigning these results, possibly prompting the plan for novel food sources. Fermentation has been demonstrated as a truly achievable choice to improve the innovative, tactile, and particularly healthy and useful elements of cereal industry by-products. The key aspect of an effective use of cereal products in food tenders is using proper bioprocessing skill, including a fermentation process with selected starters. The manipulation of cereal-based products for the extraction of their functional components, whether for food, make up, or pharmacological manufacturing, suggests an auspicious alternative to manufacturing the mixtures and it is a growing trend.

References

1. A. R. Ishaq, T. Younis, A. Noor, F. Jabeen and C. Shouwen, Natural Polyphenols a New Paradigm in Treatment of Various Diseases, *Polyphenols-based Nanotherapeutics for Cancer Management*, Springer, 2021, p. 17.
2. G. A. Engwa Free radicals and the role of plant phytochemicals as antioxidants against oxidative stress-related diseases. *Phytochemicals: Source of Antioxidants and Role in Disease Prevention BoD–Books on Demand*, 2018, vol. 7, p. 49.
3. D. K. Ray, N. D. Mueller, P. C. West and J. A. Foley, *PLoS One*, 2013, **8**, e66428.
4. M. A. Nawaz and G. Chung, *Genes*, 2020, **11**(11), 1255.
5. H. C. J. Godfray, J. R. Beddington, I. R. Crute, L. Haddad, D. Lawrence, J. F. Muir, J. N. Pretty, S. Robinson and C. Toulmin, *Science*, 2010, **327**(5967), 812.
6. J. Bailey-Serres, J. E. Parker, E. A. Ainsworth, G. E. Oldroyd and J. I. Schroeder, *Nature*, 2019, **575**(7781), 109.
7. H. Li, A. Rasheed, L. T. Hickey and Z. He, *Trends Plant Sci.*, 2018, **23**(3), 184.

8. R. K. Varshney, M. Thudi, M. K. Pandey, F. Tardieu, C. Ojiewo, V. Vadez, A. M. Whitbread, K. H. M. Siddique, H. T. Nguyen, P. S. Carberry and D. Bergvinson, *J. Exp. Bot.*, 2018, **69**(13), 3293.
9. M. Bryła, A. Waśkiewicz, E. Ksieniewicz-Woźniak, K. Szymczyk and R. Jędrzejczak, *Molecules*, 2018, **23**(4), 963.
10. R. H. Esper, E. Gonçalez, M. O. Marques, R. C. Felicio and J. D. Felicio, *Front. Microbiol.*, 2014, **5**, 269.
11. Y. Tian, Y. Tan, N. Liu, Y. Liao, C. Sun, S. Wang and A. Wu, *Front. Microbiol.*, 2016, **7**, 395.
12. P. A. Parizad, J. Capraro, A. Scarafoni, F. Bonomi, M. Blandino, M. Marengo, D. Giordano, A. Carpen and S. Iametti, *Plant Foods Hum. Nutr.*, 2019, **74**(1), 128.
13. M. C. B. Santos, N. Barouh, E. Durand, B. Baréa, M. Robert, V. Micard, V. Lullien-Pellerin, P. Villeneuve, L. C. Cameron, E. P. Ryan, M. S. L. Ferreira and C. Bourlieu-Lacanal, *Metabolites*, 2021, **11**(2), 110.
14. R. Agregán, N. Echegaray, M. López-Pedrouso, R. M. Aadil, C. Hano, D. Franco and J. M. Lorenzo, *Molecules*, 2021, **26**(16), 4924.
15. P. R. Shewry and S. J. Hey, *Food Energy Secur.*, 2015, **4**(3), 178.
16. K. D. Kaur, A. Jha, L. Sabikhi and A. K. Singh, *J. Food Sci. Technol.*, 2014, **51**(8), 1429.
17. M. Palla, M. Blandino, A. Grassi, D. Giordano, C. Sgherri, M. F. Quartacci, A. Reyneri, M. Agnolucci and M. Giovannetti, *Sci. Rep.*, 2020, **10**(1), 1.
18. A. Panghal, K. Virkar, V. Kumar, S. B. Dhull, Y. Gat and N. Chhikara, *Curr. Res. Nutr. Food Sci. J.*, 2017, **5**(3), 257–262.
19. C. J. Balibar and C. T. Walsh, *Biochemistry*, 2006, **45**(51), 15444.
20. S. Y. Choi, S. Kim, S. Lyuck, S. B. Kim and R. J. Mitchell, *Sci. Rep.*, 2015, **5**(1), 1.
21. C. R. Andrighetti-Fröhner, R. V. Antonio, T. B. Creczynski-Pasa, C. R. M. Barardi and C. M. O. Simões, *Mem. Inst. Oswaldo Cruz*, 2003, **98**(6), 843.
22. M. Konzen, D. De Marco, C. A. Cordova, T. O. Vieira, R. V. Antonio and T. B. Creczynski-Pasa, *Bioorg. Med. Chem.*, 2006, **14**(24), 8307.
23. H. Sun, D. Zhao, B. Xiong, C. Zhang and C. Bi, *Microb. Cell Factories*, 2016, **15**(1), 1.
24. T. Hoshino, *Appl. Microbiol. Biotechnol.*, 2011, **91**(6), 1463.
25. A. L. Rodrigues, N. Trachtmann, J. Becker, A. F. Lohanatha, J. Blotenberg, C. J. Bolten, C. Kornelia, A. O. de Souza Limac, L. M. Portod, G. A. Sprenger and C. Wittmann, *Metab. Eng.*, 2013, **20**, 29.
26. A. Fabiyi, P. L. Calcagno, J. L. Sever, T. Antonovych and F. Wolman, *Nature*, 1967, **215**(5096), 88.
27. H. Park, S. S. Kreunen, A. J. Cuttriss, D. DellaPenna and B. J. Pogson, *Plant Cell*, 2002, **14**(2), 321.
28. V. Galli, M. Venturi, S. Guerrini, M. Blandino, S. Luti, L. Pazzagli and L. Granchi, *Foods*, 2020, **9**(5), 640.
29. S. Zhuang, N. Renault and I. Archer, *J. Biotechnol.*, 2021, **339**, 32.

30. G. Suarato, M. Contardi, G. Perotto, H. G. Jose'A, F. Fiorentini, L. Ceseracciu, C. Pignatelli, D. Debellis, R. Bertorelli and A. Athanassiou, *Mater. Sci. Eng. C*, 2020, **116**, 111151.
31. M. Čertík, T. Klempová, L. Guothová, D. Mihálik and J. Kraic, *Eur. J. Lipid Sci. Technol.*, 2013, **115**(11), 1247.
32. Y. Mengesha, A. Tebeje and B. Tilahun, *Int. J. Food Sci.*, 2022, 4419955.
33. T. Klempová, O. Slaný, M. Šišmiš, S. Marcinčák and M. Čertík, *J. Biotechnol.*, 2020, **311**, 1.
34. M. Paulsmeyer, L. Chatham, T. Becker, M. West, L. West and J. Juvik, *J. Agric. Food Chem.*, 2017, **65**(21), 4341.
35. A. P. P. Kayodé, C. A. Bara, G. Dalodé-Vieira, A. R. Linnemann and M. J. R. Nout, *LWT--Food Sci. Technol.*, 2012, **46**(1), 49.
36. A. B. Das, V. V. Goud and C. Das, *Ind. Crops Prod.*, 2017, **95**, 332.
37. E. S. M. Abdel-Aal, P. Hucl and I. Rabalski, *Food Chem.*, 2018, **254**, 13.
38. F. Lao and M. M. Giusti, *J. Cereal Sci.*, 2018, **80**, 87.
39. E. H. Papaioannou and M. Liakopoulou-Kyriakides, *Food Bioprod. Process.*, 2010, **88**(2–3), 305.
40. T. Klempová, O. Slaný, M. Šišmiš, S. Marcinčák and M. Čertík, *J. Biotechnol.*, 2020, **311**, 1.
41. M. Čertík, Z. Adamechová and L. Guothová, *J. Biotechnol.*, 2013, **168**(2), 130.
42. T. Klempova, E. Basil, A. Kubatova and M. Certik, *Biotechnol. J.*, 2013, **8**(7), 794.
43. W. Azmi, M. Thakur and A. Kumar, *Acta Microbiol. Immunol. Hung.*, 2011, **58**(3), 189.
44. L. E. E. Da Eun, J. E. Kwon and E. Choung, *J. Cosmet. Sci.*, 2017, **68**, 271.
45. G. I. Frengova, E. D. Simova, D. M. Beshkova and Z. I. Simov, *Z. Naturforsch. C*, 2002, **57**(9–10), 805.
46. S. Choudhari and R. Singhal, *Bioresour. Technol.*, 2008, **99**(4), 722.
47. P. Tsafrakidou, A. M. Michaelidou and C. G Biliaderis, *Foods*, 2020, **9**(6), 734.
48. E. Dimidi, S. R. Cox, M. Rossi and K. Whelan, *Nutrients*, 2019, **11**(8), 1806.
49. L. De Vuyst, A. Comasio and S. V. Kerrebroeck, *Crit. Rev. Food Sci. Nutr.*, 2021, 1.
50. A. I. Kaneda, *Indoor Air Monitoring of Ethanol and Benzene in a Pilot Winery Using Active Sampling*, 2019.
51. N. Hossain, J. H. Zaini and T. M. I. Mahlia, *Int. J. Technol.*, 2017, **8**, 5.
52. B. Singh, P. Kumar, A. Yadav and S. Datta, *Chem. Eng. J.*, 2019, **361**, 1152.
53. P. T. Sekoai, A. Ghimire, O. T. Ezeokoli, S. Rao, W. Y. Ngan, O. Habimana, Y. Yao, P. Yang, A. H. Y. Fung, K. O. Yoro, M. O. Daramola and C. H. Hung, *Renewable Sustainable Energy Rev.*, 2021, **143**, 110971.
54. H. L. Martelli, I. M. Da Silva, N. O. Souza and D. Pomeroy, *World J. Microbiol. Biotechnol.*, 1992, **8**(6), 635.
55. S. W. Kim, W. T. Seo and Y. H. Park, *Biotechnol. Lett.*, 1996, **18**(11), 1287.
56. M. I. Khan, J. H. Shin and J. D. Kim, *Microb. Cell Factories*, 2018, **17**(1), 1.

57. C. Auesukaree, *J. Biosci. Bioeng.*, 2017, **124**(2), 133.
58. W. D. Attuquayefio, *Influence of Processing Parameters on Eye Size and Elasticity of Tef-based Injera*. 2014.
59. H. B. Moonga, S. E. Schoustra, A. R. Linnemann, J. van den Heuvel, J. Shindano and E. J. Smid, *LWT--Food Sci. Technol.*, 2021, **136**, 110350.
60. J. O. Westman, R. Wang, V. Novy and C. J. Franzén, *Biotechnol. Biofuels*, 2017, **10**(1), 1.
61. A. Mezgebe, *Sorghum Waxy and High Protein Digestibility Traits and Their Relationship with Malting and Dough-based Product Making Quality*, University of Pretoria, 2018.
62. M. Parapouli, A. Vasileiadis, A. S. Afendra and E. Hatziloukas, *AIMS Microbiol.*, 2020, **6**(1), 1.
63. M. AlAli, M. Alqubaisy, M. N. Aljaafari, A. O. AlAli, L. Baqais, A. Molouki, A. Abushelaibi, K.-S. Lai and S. H. E. Lim, *Molecules*, 2021, **26**(9), 2540.
64. D. Kasote, R. N. Tiozon Jr, K. J. D. Sartagoda, H. Itagi, P. Roy, A. Kohli, A. Regina and N. Sreenivasulu, *Front. Plant Sci.*, 2021, **12**, 771276.
65. Z. Zeng, C. Liu, S. Luo, J. Chen and E. Gong, *PLoS One*, 2016, **11**(8), e0161086.
66. W. A. Ansari, S. U. Chandanshive, V. Bhatt, A. B. Nadaf, S. Vats, J. L. Katara, H. Sonah and R. Deshmukh, *Int. J. Mol. Sci.*, 2020, **21**(11), 4040.
67. C. Li, J. Zhou, S. Rao, G. Du and S. Liu, *ACS Synth. Biol.*, 2021, **10**(10), 2607.
68. H. Shi, B. Su, X. Chen and R. Pian, *PeerJ.*, 2020, **8**, e10358.
69. B. Sun, Y. Zhang, M. Ding, Q. Xi, G. Liu, Y. Li, D. Liu and X. Chen, *J. Anim. Physiol. Anim. Nutr.*, 2018, **102**(1), 194.
70. V. Kumar, V. Ahluwalia, S. Saran, J. Kumar, A. K. Patel and R. R. Singhania, *Bioresour. Technol.*, 2021, **323**, 124566.
71. Y. Cui, J. Li, D. Deng, H. Lu, Z. Tian, Z. Liu and X. Ma, *PLoS One*, 2021, **16**(11), e0260045.
72. T. M. Lanigan, H. C. Kopera and T. L. Saunders, *Genes*, 2020, **11**(3), 291.
73. L. Chen, Q. Wu, T. He, J. Lan, L. Ding, T. Liu, Q. Wu, Y. Pan and T. Chen, *Genes*, 2020, **11**(2), 177.
74. E. Habyarimana and M. Lopez-Cruz, *Genes*, 2019, **10**(11), 841.
75. J. Ramalingam, S. Palanisamy, G. Alagarasan, V. G. Renganathan, A. Ramanathan and R. Saraswathi, *Genes*, 2020, **11**(11), 1266.
76. S. M. Manik, G. Pengilley, G. Dean, B. Field, S. Shabala and M. Zhou, *Front. Plant Sci.*, 2019, **10**, 140.

CHAPTER 15

Potential Use of Pigmented Cereals and Millets as a New Ingredient for Developing New Bakery Products

YASH D. JAGDALE[a], GULZAR AHMAD NAYIK*[b] AND MOHAMMAD JAVED ANSARI[c]

[a]MIT School of Food Technology, MIT ADT University, Pune, Maharashtra-412201, India; [b]Department of Food Science & Technology, Govt. Degree College, Shopian-192303, J&K, India; [c]Department of Botany, Hindu College Moradabad (Mahatma Jyotiba Phule Rohilkhand University Bareilly), India-244001
*E-mail: gulzarnaik@gmail.com

15.1 Introduction

Pigmented cereal grains and millet contain a high concentration of flavonoids and phenolics (mainly anthocyanins),[1–3] which are found in diverse anatomical locations of the kernel and contribute to the grain's color. Blue, purple, or a combination of colors are produced when anthocyanins (AC) are concentrated in the pericarp or aleurone layer. Flavonoids, such as flavones, reddish-colored phlobaphenes, yellow *C*-glycosides of proanthocyanidins, and flavanonols are concentrated in the grain's outer layer, but carotenoids, which contribute to the grain's yellow color, are concentrated in the endosperm.[4] Numerous publications have linked AC and different phenolic

Food Chemistry, Function and Analysis No. 38
Pigmented Cereals and Millets: Bioactive Profile and Food Applications
Edited by Sneh Punia Bangar, Sajid Maqsood and Anil Kumar Siroha

Published by the Royal Society of Chemistry, www.rsc.org

compounds to antioxidant activities.[2,3,5] Although the AC level of some types of colored cereals may approach that of some berries and grapes,[6] research on grains is restricted.[5,7,8] In the case of cereals, effective debranning techniques can yield anthocyanin-rich parts prior to milling.[9] AC exhibit genetic diversity between species and genotypes. In the case of pigmented wheat, the highest AC content is seen in blue-aleurone soft wheat, followed by red- and purple-pericarp durum wheat.[1,10] The standard roller method of milling removes most of their composition, because these chemicals are present in the grains' outer layer.[1] Wheat's primary end products are pasta and bread. These are consumed daily, and one approach to enhance their health benefits may be by including wholemeal in their manufacture,[11] owing to its high content of bioactive components. On the other hand, wholemeal goods contain bran, which may have a negative effect on their technological and sensory aspects, restricting their consumption. Indeed, numerous studies have established that bran has a typically negative impact on the quality of pasta and bread.[12–14] A solution to this problem could be found by utilizing technological procedures that enable the recovery of proper milling fractions, or contemplating debranning to preserve external bioactive chemicals as a pre-milling step. Additionally, such processes must consider consumer safety, which may be jeopardized by the presence of pollutants.[15,16]

A cortical fraction produced from colored grain debranning has been used to enrich pasta. Different cereal-based foods manufactured similarly prompted concerns regarding the susceptibility of grain bioactives to standard processing procedures. As an alternative to synthetic colorants, AC pigmented grains can be utilized to manufacture foods either whole or after the natural colorants are removed from the anthocyanin-rich grain fractions. Since AC is concentrated in the kernel's outer layers, their extraction may be facilitated.[4] Recently, there has been a substantial surge in interest in the functional characteristics of foods, with antioxidants capable of scavenging free radicals playing a crucial role. In the nineteenth century, Ethiopians recognized wheat with AC coloring. Numerous red, purple, black, and blue-grained wheat cultivars have been developed in recent years.[17] Due to the beneficial effects of pigmented cereal and millets, bakery products enhanced with phenolic compounds have been extensively explored.[18,19] However, adding phenolic components to bakery products can negatively impact various technological qualities, including color, texture, and specific volume.[20] Phenolic molecules are reactive and thermally sensitive and may react with other ingredients and chemicals produced during baking.[20] Although the stability of phenolic components due to heat treatment is unclear, Abdel-Aal and Rabalski[21] observed that baking improves the number of free phenolics in bread items while decreasing the number of bound phenolics. Slavin *et al.*[22] also demonstrated that baking black soybean crackers at optimum temperatures for a brief period preserves the AC and other phenolic components.

Healthy bread is defined as baked goods that improve humans' health and well-being, and sourdough fermentation significantly improves the product's shelf life, nutritional value, and sensory attributes, while also naturally inducing the creation of bioactive peptides with anti-inflammatory and antioxidant properties.[23] Certain lactic acid bacteria (LAB) strains have been demonstrated to create doughs with better nutritional and functional characteristics than others.[24–26] Most peptides formed during fermentation are sourced from the flour's primary proteins.[25] Therefore, it is critical to carefully select flours, LAB, and yeast strains to generate bread rich in beneficial components, such as bioactive peptides, polyphenols, and free amino acids and derivatives.[26–29] Additionally, flours from colored cereal grain varieties may be new potential ingredients that might be utilized in gluten-free food production systems because of their higher dietary phenolic content, which may have an impact on starch digestibility through interaction with starch components and inhibiting digestive enzymes, thereby inhibiting amylolytic attack.[30] When cooked pigmented rice variants are compared with their unpigmented counterparts, higher resistant starch concentrations (on average +30%) have been recorded.[31] Similarly, Hernández-Uribe *et al.*[32] observed that tortillas made entirely of blue corn (BC) had a lower *in vitro* GI than their white corn counterparts, regardless of storage period.

Nowadays, consumers' increasing health consciousness has led to increased demand for healthier snack foods. As a result, the food industry has concentrated its efforts on developing enhanced calorie-reduced and functional snack formulations with palatable flavors. According to a recent survey, 39% of customers choose healthier snack options.[33] In a healthy diet, the RDA consumption of dietary fiber should be between 8 and 20 g per 1000 kcal nutrient density, and foods labeled 'source of fiber' or 'high fiber' should contain at least 3% or 6% fiber, respectively. Furthermore, the consumption of a bioactive-rich diet prevents most chronic diseases.[34] Consumers are aware of the favorable effects of dietary fiber (DF) and whole-grain products on their health.[35] The bran fraction is high in fibers (*e.g.*, lignin, cellulose, and hemicellulose), proteins, phenols, carotenoids, and AC.[36–38] The phytochemical concentration is more remarkable in colored wheat varieties than in standard wheat.[39] According to the food industry trend and consumer expectations of incorporating non-wheat cereals into bakery products to obtain products with numerous functional benefits, the use of emmer, einkorn, barley, and pigmented wheat cultivars and spelt is increasing worldwide.[16,40,41] This chapter focuses on applying different cereals and millets such as wheat, corn, rice, and sorghum to develop different bakery products that are nutritionally enhanced and superior in their overall characteristics compared with their traditional counterparts.

15.2 Wheat

Wheat is the world's most cultivated cereal crop on the planet, and wheat bread is essential for human nourishment. Whole-grain wheat products have been linked to a range of health benefits, which may be due to the presence of different phytochemical components. These exhibit considerable variation amongst wheat varieties and lines, as evidenced by investigations of HEALTH

GRAIN bread wheat lines.[42–44] The pigments in grain and grain products vary according to genetic variables, growth conditions, and technological procedures. Colored wheat types may contain AC and carotenoids employed in manufacturing various baking items, most of which are made from whole grain. Certain colored whole grains, in particular, offer novel and essential raw ingredients for the manufacture of functional meals.[4] Along with phenolic acid composition, these pigmented wheat grains contain considerable amounts of carotenoids and AC, beneficial chemicals that contribute to their distinctive yellow-orange and blue-purple colors, and increased nutritional composition.[1,45–47] Different researchers have used them to enhance the nutritional characteristics of baked goods such as bread, pasta, and biscuits. Thus, pigmented cereals and millets are an excellent source of bioactive components, minerals, and DF; however, some bioactive components are lost when the seed's outer layers are separated from the endosperm.[29,48] The presence of natural AC content at higher concentrations in blue, purple, and black grained wheat distinguishes them from white and red grained kinds of wheat.[4]

One of the most popular pigmented wheat varieties is 'black wheat', which is generated from blue and purple wheat crossing. The AC found in the outer layer gives wheat grain its black color. It contains various nutritious and bioactive elements, including dietary fibers, AC, phenolic compounds, phytic acid, carotenoids, minerals, and vitamins. It is used to manufacture functional meals and associated natural colors.[1] Garg *et al.*[49] also noted the potential for improved functional and nutritional qualities in new products derived from unconventional pigmented wheat. However, few studies have been done to date on black wheat, with the majority focusing on its qualitative analysis and breeding, total phenolic content, protein characteristics, antioxidant activity, flavonoid content, and free radical scavenging activity. In addition, studies have focused on the development of black wheat flour, its ability to create dough and bread, and intake studies of its various by-products to improve glycemic management and reduce inflammatory activity in patients.[50] The outer pericarp layer (aleurone) contains high-value compounds (such as AC).[49] Other cereal flours (such as soy flour, green gram flour, and chickpea flour) and conventional wheat flour can be combined with the bran fractions to generate a more bioactive enriched flour that can be used to produce functional foods.

Because black wheat (BW) flour lacks optimal baking qualities, it can be combined with regular wheat flour to create good baked goods. Fortification of black wheat is likewise a critical area that requires immediate attention. Kumari *et al.*[51] investigated the influence of biofortification on chapatti quality. They reported that the black wheat-enriched chapatti had a soft texture and an excellent sensory profile. Additionally, it contains a more significant concentration of DF, protein, insoluble, and soluble phenolic compounds and AC, and a lower carbohydrate concentration. Chapattis made with black wheat flour had a low stickiness, a low sodium dodecyl sulphate sedimentation (SDSS) value, were easy to roll out, and had good color and flavor.

It received an overall score of 7.90, comparable to chapattis made from regular wheat flour. Additionally, no reduction in fat amount occurred due to the heat used to make chapatti.[51] The study found that colorful wheat varieties (mainly BW) have significant characteristics for product development and their utilization in industries, primarily determined by the product manufacturing method and AC extraction. In terms of additional products, as well as instant noodles, Li and Beta[52] reported that BW might be utilized to create vinegar and breakfast cereals. However, other colored wheat products such as bran muffins, soy sauce, fine dry noodles, and antho-beer have yet to be tested with black wheat.[53]

In an experiment, a barley flour (hull-less), which has more β-glucans,[54] and flour made up of blue grain (cv Skorpion), which contains high concentrations of AC,[9] were compared with a typical red wheat (cv Aubusson). The results indicated that the flour of Skorpion and barley flour had greater total phenolic acids, whereas barley flour possessed the maximum antioxidant activity. These findings corroborate earlier studies[55,56] and emphasize the health benefits of using barley high in bioactive components and wheat cultivars in bread.[29] The bread's nutritional quality is also determined by its β-content, which hull-less barley contains in abundance; as a result, it has been advocated for use in the production of functional foods.[57] As a result, determining the appropriate combination of functional-value flour with selected lactic acid bacteria (LAB) might facilitate the manufacturing of nutrient-rich baked products. In a recent study on the selection of LAB strains, several lactobacilli strain with the best peptidase and proteolytic activity, antioxidant activity, and optimal pro-technological characteristics were identified using sourdoughs made with cv Skorpion flour (blue grain wheat), cv Aubusson flour (red-grained wheat), and cv Rondo flour (hull-less barley).[29]

Pigmented wheat has been the subject of numerous studies. In purple-grained wheat, the AC fraction is dominated by peonidin 3-*O*-glucoside and cyanidin 3-*O*-glucoside, but in blue varieties of wheat, delphinidin 3-*O*-glucoside and delphinidin 3-Orutinoside predominate.[1,2,10,58] The antioxidant activity of purple cultivars was tested using 2,2-diphenyl-1-picrylhydrazyl (DPPH) radical scavenging activity and ranged between 5.87 and 8.57 lmol Trolox per g,[2,59] although dark blue wheat has shown greater values.[46] Additionally, the distribution of grain pigments was also studied, demonstrating that the pericarp contains a higher concentration of purple pigments, and the aleurone layer contains blue pigments.[17] Little effort has been made to examine the manufacturing performance of pigmented wheat, despite manufacturers' desire to obtain whole-grain food specialties from pigmented wheat with potentially beneficial properties. A single study (Li *et al.*[59]) evaluated the influence of processing on the antioxidant capabilities of muffins made from purple wheat bran. Purple- and red-pericarp durum wheat and blue-aleurone soft wheat have been recently re-evaluated for their AC content.[10] As a result, they can be used as functional foods or as critical functional components in wheat-based products.[41] Because the AC in blue and purple wheat is found in the grains' outer layers, so bran-enriched or wholemeal flour can produce colored food products.[60] The higher content of AC in purple and blue wheat varieties can enhance the AC concentration of bakery items.[61]

Pasqualone *et al.*[41] produced and analyzed biscuits made from purple wheat. For comparison, control biscuits were utilized with the same recipe but made from non-colored wheat. Total AC was greater in the purple wheat biscuits (13.86 mg kg^{-1} cyanidin-3-glucoside). Purple biscuits had a lower concentration of carboxylic acids (lipid-derived) and a higher concentration of aldehydes and alcohols than control biscuits, suggesting that the lipid component had been subjected to limited oxidative degradation. The (alcohols (lipid-derived) + aldehydes)/acids ratio was 5.9 in purple biscuits and 3.0 in control biscuits. These compounds are found in bran concentrates, which can be used as nutraceutical additives or food coloring agents.[4] AC-rich food extracts have been integrated and developed into nutritional supplements. For instance, fermented red rice is promoted as a cholesterol-lowering dietary supplement called Cholestin. Additionally, Shipp *et al.* reported that blue wheat bran could be used to create blue wheat powder (anthocyanin-rich) as a dietary supplement.[62] Pasqualone *et al.*[63] created AC-enriched biscuits with olive oil, durum wheat semolina, and AC derived from black and red berries, grapes, squeezed grape marc, and sour cherries. Thepla is an Indian flatbread (unleavened) cooked with vegetables, whole wheat flour, and spices. The effect of edible powdered fruit by-products obtained from waste of the fruit industry, such as papaya peels, apple pomaces, and watermelon rinds, on GI-lowering activity, has been investigated.[64] The GI of the control thepla was 68.1, according to the authors. However, when watermelon rinds (3%), papaya peels (9%), and apple pomaces (9%) were added, the GI decreased dramatically to 63.38, 55.24, and 55.29, respectively.[65]

In an experiment by Li *et al.*,[59] bread made from purple-colored wheat flour had a total anthocyanin content of 78 mg kg^{-1}. The bread made from purple and blue wheat baked at a temperature of 180 °C for 31 minutes and 240 °C for 21 minutes, respectively, had significant changes in the concentration of AC.[66] The bread made with blue wheat contained 8.33 mg kg^{-1} AC (a drop of 7.1%), while 5.2 mg per kg AC was observed in purple wheat bread (a decrease of 61%). At 180 °C, bread made with blue wheat contained an AC of about 5.31 mg kg^{-1} (40.8% drop), whereas bread made with purple-grained wheat contained 3.63 mg kg^{-1} AC (72.8% drop). The AC loss varied according to the baking time and was more significant after 31 minutes at 180 °C than after 21 minutes at 240 °C. As the AC content loss was greater in purple wheat bread, Bartl *et al.*[66] concluded that purple wheat's acetylated AC is more vulnerable to the thermal degradation process of baking than blue wheat. A Similar trend of AC reductions was observed in blue corn (BC) biscuits, with losses of approximately 11% (baked at a temperature of 179 and 182 °C for 4 minutes) and 25% (temperature of 204 °C for 10 minutes) (Li *et al.*[67]).

15.3 Corn/Maize

Maize is a commonly farmed cereal crop that is a staple diet for people of all socioeconomic classes worldwide. Corn is a gluten-free grain and, together with rice, the most widely farmed cereal globally.[68] Among cereals, the most abundant source of AC is colored maize. Due to their appearance, colored maize kernels are produced in limited quantities throughout the world for use in decoration and the production of specialty foods. Maize that has been colored contains numerous secondary metabolites such as carotenoids and phenolic components. Flavonoids and phenolic acids are the most abundant phenolic chemicals discovered in entire maize kernels, either free or bonded insolubly. Flavonoid chemicals are responsible for most mauve, scarlet, orange, crimson, blue, and violet colors in maize types and provide yellow, ivory, and cream colors. Whether acylated or simple, AC is primarily found in the pericarp or aleurone layer of the corn endosperm, where they have a substantial effect on the antioxidant capacity and color of the kernels,[7] which can be red, purple, blue, or black.[69] Corn is dubbed the 'queen of cereals' due to its genetic yield prospects, which are the greatest of all cereals. It is currently the world's third most widely cultivated food cereal, trailing only wheat and rice,[70] and ranges in numerous colors ranging from purple, yellow, white, and red.[71] Del Pozo-Insfran *et al.*[72] found that pigmented corn (containing high levels of carotenoids and AC) with a floury endosperm had higher nutritional value than most dent equivalents.

Carotenoids and AC have been discovered in native pigmented maize, contributing to their potential health benefits.[69,73] Carotenoids contribute to various red, orange, and yellow colors of various grains, vegetables, and fruits. There are over 600 carotenoids (naturally occurring), and up to 60 are found in foods; the most abundant are xanthophylls (zeaxanthin and lutein), as well as carotenes (β- and α-carotene) in specific genotypes of colored maize.[74,75] Carotenoids generated from maize kernels have been associated with several functional roles, including increased antimutagenic activity, lowering oxidative stress, cancer prevention, and decreased risk of age-related macular degeneration.[76,77] However, corn kernels must be processed before being consumed by humans, altering or degrading their phytochemicals. More research is needed on the effect of processing conditions on the carotenoid profiles of pigmented corn, especially in terms of the antioxidant chemicals' fate in nixtamalized products.[78]

Over the last few decades, scientific study has focused on colored corn cultivars, due to AC's well-known health benefits.[79] As a result, breeding procedures have been identified to enhance the AC content of uncolored cultivars in various ways, most notably those derived from those traditionally grown in Central and South America.[80] Purple corn (PC) is indigenous to South America's Andean region, where it has been used for millennia to prepare traditional foods and beverages. It was recently regarded as a good source of natural dyes for the textile industry[80] due to its high AC content; however, it also contains additional phenolic components with health-promoting properties. Purple variants grabbed researchers' attention due to their possible usage in the cosmetic and food supplement industries. AC is abundant in the aleurone part of red corn (RC) and black corn varieties, whereas it is present in the pericarp of blue corn (BC) varieties.[81] Numerous research has been conducted on the rheological behavior of dough and the variables impacting such doughs (*i.e.*, temperature, size of the particle, shear rate, and water absorption), all of which are critical for selecting the technological conditions, have recently increased corn flour's use in the manufacturing of gluten-free products. The dough's rheological properties and the chemical characteristics of flours were significantly affected by corn types and milling conditions; indeed, when using the same milling process, the particle size of PC and yellow corn (YC) flours was relatively smaller than that of white flour, and both corn starch reduction and an increase in water absorption.[79] Furthermore, elastic behavior took precedence over viscous behavior, resulting in lower viscoelastic moduli values. The results of the rheological analysis demonstrated a close link between dough attributes, processing, and raw maize feature.[82] Additionally, purple wheat genotypes had a greater gelatinization enthalpy than non-purple genotypes due to AC increasing the fragility of starch granules during cooking (these genotypes possess a lower amylose concentration than non-purple genotypes), as Sebastián Mansilla *et al.*[83] demonstrated for PC flour from Argentina.

According to prior studies,[69] RC and BC kernels contain a greater range of total AC; the most abundant is C3G. However, because maize-based products must be thermally treated before consumption, thermal loss of different phenolic components and AC is a significant issue when colored corn is used as a raw material in the confectionery and baking industries. Cookies could easily be supplemented with polyphenols as a bakery product for the prevention of diseases and to improve health. Refined wheat flour has been partially substituted by anthocyanin-rich fruits and tea extract.[84,85] Functional cookies were prepared using RC and BC flours in this work. This research aimed to examine the impact of baking parameters such as temperature and time on the concentration of color, antioxidant property, and phenolic compounds of cookies produced with AC-rich maize flour. The primary purpose was to determine the influence of citric acid on AC stability in maize cookies to determine the organic acid's suitability as a bakery food ingredient.[7]

PC or BC, which originated in Peru, has been produced throughout the New World for centuries by the Aztec, Maya, and Inca peoples. Numerous civilizations produce BC items. Through nixtamalization, tortilla chips, tortillas, and other items are still created from BC. Regrettably, the technique, also known as lime cooking, significantly increases the pH of the food, resulting in a loss of total AC.[67] Numerous elements such as temperature, pH, oxygen, light, enzymes, sugars, metallic ions, and their breakdown products all affect the color stability of AC pigments. They change color from red at low pH to bluish at middle pH to colorless at high pH.[67] In tiny villages throughout Mexico, pigmented corn tortillas are cooked and devoured because they have a relatively softer texture and a sweeter flavor than yellow or white maize tortillas.[86] Traditionally, colored corn tortillas were cooked from locally grown maize and were used to make a variety of traditional foods, including tlacoyos, quesadillas, and tacos. Due to the growing popularity of multi, colored tortillas, manufacturers of nixtamalized corn flours produced BC flour to create a colored nixtamalized corn with better nutraceutical properties from which new products with improved visual qualities can be manufactured.[32] The most important staple for the Mexican people is maize tortillas, made by the nixtamalization method. Tortillas are a vital source of protein, energy, fiber, and calcium for Mexicans and other Latin Americans.[87] Corn tortillas and derivatives such as taco shells, corn chips, and tortilla chips have expanded their commercial significance. Unfortunately, the nixtamalization process and other tortilla-making procedures results in the loss of antioxidants and phenolic compounds.[88] De la Parra *et al.*[89] have proven that in five varieties of corn (yellow, red, high carotenoid, white, and blue), lime-cooking dramatically lowered the antioxidant activity and phytochemical concentration of the final products. Thus, after nixtamalization, adding phenolic components could be a way to improve antioxidant and phytochemical activity in final products.

Small communities in which BC and RC are grown use them to make traditional foods such as tlacoyos, tortillas, atoles, gorditas, and tostadas. These products were in high demand because of their soft texture and delicious flavor. Furthermore, because of their calcium concentration, nixtamalized products can benefit human diet. Due to current legal and health tendencies toward eliminating synthetic colorants, natural pigmented grains are of tremendous interest to industrial manufacturers, traditional farmers, and consumers. Several authors have previously reported using pigmented corn in various foods.[86] A few researchers outlined some of the properties of BC extruded goods and indicated that expanded snacks might be created successfully from BC. Plascencia-González *et al.*[90] produced tortilla and corn chips with excellent sensory characteristics from lime-cooked BC (texture, color, and flavor). Using BC fortified with calcium hydroxide, Zazueta-Morales, Martinez-Bustos *et al.*[91,92] developed useful functional extruded products. The nixtamalization method produces tortillas, tortilla chips, baked tortilla chips, and other products from BC. The process, also known as lime cooking, begins with the corn being heated in a 1–2% lime (CaO) solution for 5 to 50 minutes, then steeping (14 hours to overnight), and finally, washing. The nixtamal is then stone-crushed into a maize dough (masa), combined, cut, baked, fried, or extruded into tortilla chips and tortillas. Nixtamalization enhances niacin bioavailability and calcium content, eliminates the majority of the pericarp, and decreases mycotoxins in maize kernels considerably.[89] Additionally, it is responsible for forming flavor and color compounds that give the final products their distinct organoleptic characteristics.[89] In supermarkets and health food stores, various breakfast meals, novel snacks, and other items, such as muffins, cornflakes, pancakes, and various extruded snacks made from BC flour, are now accessible. Most of them are manufactured using organic farm-grown components and sold as organic food at a premium price.[91]

A negative relationship was found between AC content and amylopectin retrogradation enthalpy, most likely due to associations created during starch gelatinization. Camelo-Méndez *et al.*[93] obtained intriguing results when they evaluated the incorporation of varying amounts of BC flour to composite pasta made with unripe plantain and chickpeas to develop a gluten-free product. Even after extrusion (20% loss) and cooking (30% loss), the finished product had a dark color and a high polyphenol content (especially pasta with 75% BC flour), as well as enhanced adhesiveness and decreased chewiness and hardness when compared to white corn-based pasta. Due to the starch disorganization during the extrusion process, blue flour pasta had a lower peak viscosity than white flour pasta; however, a second peak viscosity was observed upon cooling for BC pasta, indicating a reorganization of starch compounds that may contribute to polyphenol retention. Another application of BC in extruded snacks uses orange bagasse (the principal waste product of juice production) as a fiber source. Extrudates with many holes per area and tiny pores were likewise toughened by orange bagasse (as shown by a drop in expansion index).[79] It was also reported that PC flours might make gluten-free eggless muffins without gum or yellow and white accessions. Despite their variances in physicochemical, pasting, and antioxidant qualities, all examined species had a low breakdown viscosity, indicating that they were all thermally stable. There was a positive link between cohesiveness and chewiness, but this was not negatively correlated with paste viscosities. Muffins made with PC accessions were less palatable than those made with yellow corn. These preparations are promising, primarily because they are egg-free, and their use could assist in reducing the rise in egg allergy cases.[79,94] Li *et al.*[67] examined the total AC content retention in cookies made from BC flour following baking, as affected by the interactions of lactic and citric acids, as well as glucono-δ-lactone, using three different oven types with varying heat transfer coefficients (convection oven 182 °C/4 min, reel oven 204 °C/10 min, and impingement oven 179 °C/4 min). The authors demonstrated that cookies prepared with citric acid in a convection oven would preserve the highest total AC content (227 ± 3 mg kg^{-1}). Mora-Rochin *et al.*[88] showed a drop in total antioxidant capacity (TAC) during BC processing into tortillas, and Li *et al.*[59] noticed a decrease in TAC during the baking of muffins containing purple wheat bran. Blue wheat TAC has been proven to be more thermally stable at pH 1, but raising from 65 to 95 °C causes higher degradation at pH 5.[58] After HPLC analysis, a high concentration of C3G was discovered in wholemeal flour and biscuits. The AC profile of biscuits was equivalent to wholemeal flour, despite the decline in AC observed after processing.

Utrilla-Coello *et al.*[95] examined the *in vitro* digestibility of starch, acceptability, and proximate composition of three cereal bars containing BC, unripe banana flours (UBF), or white maize. Whole cereal grains are low in saturated fat, and high in slowly digested starch (SDS), DF, minerals, vitamins (especially B type vitamins), and phytochemicals linked to improved long-term health. Due to indigestible components like resistant starch, unripe bananas benefit human health. UBF has shown nutraceutical/nutritional potential as a source of antioxidants and indigestible carbs. SDS is also found in pigmented maize varieties, which constitute a one-of-a-kind supply. Hernández-Uribe *et al.*[32] demonstrated that tortillas prepared from nixtamalized BC have a much lower projected glycemic index (pGI) value than those made from ordinary white maize varieties. The bar containing UBF had the optimal soluble/insoluble indigestible fraction ratio, *i.e.*, closer to one. This may be physiologically significant, considering the favorable effect of indigestible, soluble components, particularly indigestible carbs. The three bars have a slow rate of digestion. The lowest pGI was found in the UBF bar, followed by the BC and white corn bars. The three bars could be considered low glycemic index foods.[95] Depending on the formulation, snack bars with a slower rate of starch digestion and a higher concentration of resistant starch can be prepared, which may benefit people with varying nutritional needs in their diet.

15.4 Rice

For a long period in Asia, rice cultivars with purple, red, and black bran layers have been cultivated.[96] This rice has a long and illustrious history, and one of its strains is known as 'Imperial rice', which was once designated solely for the consumption of Emperors. Another variant, dubbed 'forbidden rice', has been consumed in Asia for many years; the Chinese aristocracy only consumed it for decades. Royal families and kings ate these specialized rice varieties to maintain their health and emphasize their wealth.[96] Additionally, these rice types were believed to be effective medicines capable of 'curing' a variety of ailments, as well as sources of nutritious staple foods with a pleasant taste, smell, and fluffiness. When these cereals were more widely available, their beneficial properties were found, and black rice became nutritionally and functionally significant.[97] Black rice (BR) is a pigmented rice that originated in Asia and was frequently utilized in traditional medicine,[98] Due to its climate instability, photosensitivity, and incompatibility, it was only introduced into Italy in the last 25 years. Venere rice was the first BR variety cultivated in Italy; it was created by a hybridization process involving a BR variety from Southeast Asia and a native variation. Although colored rice production is still restricted in Italy, it has increased due to growing customer appreciation. Currently, the Italian Varietal Register has 23 colored varieties (black or red) (the crop year 2019/2020). Artemide rice, a natural hybrid between white Indica rice and the Venere rice variety, has high phenolic concentration, especially in AC antioxidant activity.[99]

Rice is a vital commodity for Asian countries economically; thus, improving rice production to boost health benefits is quite tempting. Purple rice is used in Thailand and China to make desserts and snacks such as coconut black rice pudding and consumed as cooked rice grains.[100] The consumption of colored rice as a staple meal is expanding rapidly due to its perceived health benefits and status as a functional food additive. Numerous colored rice variants have garnered increased interest due to their diverse biological activity. Ferulic acid, anthocyanidins, AC, polymeric proanthocyanidins, and diferulates have been identified as phenolic chemicals. AC is primarily found in the rice kernel bran layers, whereas phenolic acids are primarily found in the rice bran layers, where they exist in their bound, conjugated, and free forms.[101] Pigmented rice is an important food ingredient in some Asian countries due to its multiple beneficial health properties such as anti-cancer, antioxidant, and anti-inflammatory characteristics. The usage of bran in foods and its incorporation into functional foods is rising.[101] Pigmented rice contains antioxidants such as proanthocyanidins, flavones, and phenolic acids which add to its nutritional profile.[1,102] ‘Artemide’ is an Italian BR type that resulted from natural hybridization between Indica ecotype (narrow and long grain, white pericarp) and Venere rice (medium-sized grain with a black pericarp). Compared with other Italian colored varieties, it contains a more significant concentration of polyphenols, particularly AC (C3G as the primary ingredient), and significant antioxidant activity.[99] Colored rice grain is also available in purple, reddish, and blackish colors. The black and red paddy varieties have about 18% more crude fiber, 38% more protein, and are higher in vitamin B1, lysine, and other minerals than the traditional varieties. Colored rice varieties have valuable features due to the antioxidant compounds they contain, which can suppress the generation of reactive cell-damaging free radicals and high fiber, vitamin, and protein content.[101]

Due to the multiple health benefits of black rice, particularly its anti-cancer effects, it has recently acquired popularity among rice consumers. Cyanidin-3-Glucosides (C3G) is anthocyanin's principal bioactive component in black pericarp rice. Rice with black and red pericarp pigments contain a higher concentration of polyphenols.[103,104] Itani *et al.*[105] found a clear correlation between antioxidant activity and the overall phenolic content of the grain. As a result, it decreases oxidative stress, cancer, blood lipids, and linked disorders and thus protects against cardiovascular problems and diabetes complications.[106] Food intolerance has developed into a significant public health issue, necessitating the development of effective preventative techniques. Gluten is the cause of celiac disease, which is an allergic illness. This intolerance can occur at any age, from infancy to old age. Due to the lack of gluten protein in rice, rice-based bread products have several advantages over wheat-based products. BR is colored rice with a high market value due to its anthocyanin content. Nowadays, people are extremely health conscious and seek out high-quality food products typically enhanced with nutraceutical characteristics, such as colored rice.[106]

Compared with brown and white rice, BR is higher in protein and lower in calories and carbs. Because BR is typically consumed unpolished or in brown rice form, gelatinizing it through cooking may be difficult due to its high fiber content. As a result, brown rice should not be consumed alone, but rather in combination with polished red or white rice, which may be healthy for humans. Brown rice may be unappealing to senior citizens, so a polished and brown rice blend may be preferred. It will improve cooking conditions and provide more flavor than just brown-black rice. Given this, it was assessed that the loss of anthocyanins, particularly C3G, during cooking and proposed a strategy for maintaining a significant concentration of C3G by combining white or red pericarp rice with it. As a result, current research has focused on the physicochemical characterization of available BR cultivars and evaluating C3G as a potential antioxidant.[106] Sui *et al.*[107] baked bread using anthocyanin-rich BR powder and tested the stability of two distinct AC, C3G and cyanidin-3-rutinoside (C3R), throughout baking at various temperatures (200, 220, and 240 °C) and durations (0, 2, 4, 6, 8, 10, and 12 min). Non-isothermal kinetic models were developed for both the bread crust and the crumb. Crumb has higher stability of AC, which could be due to the system's lesser oxygen availability. Because AC is thermally unstable, those in the crumb were less thermally damaged than those in the crust due to the crumb's lower temperature profile. C3R was shown to be a more thermally stable AC disaccharide in the crumb than C3G.[4]

Purple rice (PR) (*Oryza sativa* L. var. Sanpatong) is a frequently cultivated crop in Northern Thailand. It has gained popularity due to its unique color, pleasant flavor, and lower price than standard wheat flour.[108] PR has also grown in popularity due to its greater concentration of beneficial compounds, including AC and phenolic acids. Furthermore, it contains a higher level of essential fiber than wheat flour.[109] These bioactive substances confer significant antioxidant activity on the grain. The natural antioxidant content of red, black, and purple rice is well established.[41,109] PR has a higher vitamin, mineral, and protein content than white rice, including zinc, iron, phosphorus, and manganese, among other nutrients. This rice's bran layer contains a significant concentration of antioxidant components, including flavonoids, oryzanols, phenolic acids, and anthocyanins. Purple rice is substantially higher in anthocyanins than pigmented barley, wheat, and corn.[110] Kim *et al.*[111] established that PR powders enhanced traditional Korean rice wine's (takju) anthocyanin and antioxidant properties. When consumed in moderation, these anthocyanins and phenolic acids are expected to lessen the incidence of colon cancer.[112] Additionally, pigments derived from PR bran can be utilized in place of synthetic colorants in beverages and foods.[113] These pigments have been utilized to enhance the nutritional value of various breakfast cereals, cakes, and snacks.[110]

Red rice (RR) was once thought to be a weed in traditional rice cultivation; however, as science progressed, it was discovered that RR contains functional and nutritional components such as phenolic acids, proanthocyanidins, γ-oryzanol, and gamma-aminobutyric acid (GABA) and is high in dietary fibers.[1,114] Rice contains insoluble (bonded) and soluble phenolic compounds; in black and red pericarp grains, the soluble form accounts for around 8% of the total polyphenol content.[115] GABA has grabbed the interest of researchers due to its multiple health benefits and is recognized as a promising biofunctional compound.[116] RR, commonly known as 'arroz da terra', is a naturally occurring variety of *Oryza sativa* with a unique texture, nutritional composition, flavor, and high zinc and iron content that has sparked consumer attention. The inclusion of proanthocyanin, which enhances digestibility and has antioxidant effects that may assist the cardiovascular system, gives it its red color.[117] Furthermore, RR is a significant diet component of many northeastern Brazilian states (Paraiba, Cerará, Pernambuco, Alagoas, and Bahia) due to its economic and social importance.

RR flour is considered a promising cereal ingredient for manufacturing gluten-free products in place of wheat flour due to its hypoallergenic, bland, and natural taste. Because rice has a low prolamin concentration, additives such as emulsifiers and gums must be used to increase the system's viscosity or consistency.[118] Gusmão *et al.*[119] developed probiotic bread (gluten-free) from flour made of RR flour. The addition of microbial transglutaminase (MTgase) and its elevation improved bread-specific volume, chewiness, and firmness. It was able to keep MTgase enzyme levels between 0.5 and 1.0% with a fermentation duration of 60–80 mins to cover the most optimized attributes. The softness attribute in gluten-free bread needs to be improved, as consumers prefer softer products, and the CP formulation (1.0% MTgase and 80 min TF) was the best defined by sensory characteristics (flavor, softness, and overall impression) in the multivariate analysis, a fact confirmed by the higher scores obtained in all of these attributes, indicating greater acceptance and a better product. These findings suggest that celiacs and other consumers seeking higher physical and sensory quality products can employ RR flour, MTgase enzyme, and prebiotics in gluten-free bread recipes. Cappa *et al.*[120] examined potato-based pasta's (gnocchi) nutritional, technical, and sensory aspects after adding RR or buckwheat flour. The researchers discovered that gnocchi made with buckwheat flour had better texturizing properties and might be considered a 'source of fiber'. On the other hand, Wholemeal buckwheat causes an adverse sensory reaction, reducing the acceptability of the enriched gnocchi.

15.5 Barley

Barley (*Hordeum vulgare* L.) is an ancient grain that has been grown in temperate climates for centuries and is primarily malted for beer brewing. Like ordinary bread wheat, barley has a low level of total carotenoids compared with other cereals. The most common variety of yellow barley grain has 2.25 g of total carotenoids per gram, whereas the purple form has more than double that amount (4.54 gram per gram).[121] Additionally, these authors stated that zeaxanthin is the primary carotenoid found in barley. The colored barley kernel is a source of anthocyanin, carotenoids, phenolic acids, and other phytochemicals produced during secondary metabolism. The blue, red, purple, and black varieties of barley are pigmented due to the presence of both anthocyanins and melanins. According to several studies, these compounds act as a powerful antioxidant, metal chelators, and free radical scavengers, contributing to the prevention of lipid peroxidation.[122]

Martínez-Subirà *et al.*[123] studied purple barley biscuits with increased bioactive compounds. The study evaluated the anthocyanins, arabinoxylans, β-glucans, and other phenolic components of biscuits made with various proportions of whole purple barley flour and pearling fractions and compared them with biscuits made entirely of refined and whole wheat flour. The results indicated that purple barley-based biscuits had a higher antioxidant capacity and a lower glycemic index GI than whole and refined wheat flour biscuits. The physical properties of the biscuits were comparable to those of refined wheat biscuits but slightly inferior to those of whole wheat biscuits. The baking process did not affect the amount of -glucans and arabinoxylans in the food matrix, favoring the release of phenolic components and this increased antioxidant activity. The anthocyanins in purple barley were thermally unstable and decreased in concentration following baking, but this was improved by adding tartaric acid. The biscuit made entirely of purple barley contains more than 0.75 g of β-glucans, and thus one serving of four biscuits meets the daily requirement of 3 g of β-glucans, which helps to lower blood cholesterol and the risk of heart-related diseases (Table 15.1).[123]

15.6 Conclusion

Pigmented cereals and millets represent a new class of grains, capable of bringing a revolution in the bakery industry by developing novel bakery products. The major pigmented cereals and millets utilized in the world are wheat, corn, rice, and barley. Pigmented grains are popularly known for their rich content of phytochemicals and anthocyanins and can exhibit effective antioxidant activity and many other beneficial health activities by preventing numerous diseases. Several bakery products such as bread, biscuits, cookies, *etc.*, can be produced through pigmented cereals and millets, which are superior in terms of nutritional, texture, and sensorial properties to the products prepared from non-pigmented grains. However, there are only limited studies on the utilization of these colored grains in developing different bakery products. More research studies usually focus on developing bakery products by including pigmented cereal flour with non-pigmented ones. Also, pigmented cereals are cultivated in only selective areas of the world and many people are unaware of the potential of pigmented grains in bakery products. Hence, more and more research should be carried out in including the different categories of pigmented millets and cereals and major emphasis should also be given to developing traditional bakery products through pigmented grains.

Table 15.1 Application of different millets/cereal in the development of nutritionally enhanced bakery products.

Sr. No.	Cereal or millet varieties	Origin	Type of bakery product	Research results	Reference
1.	A red wheat (cv Aubusson), hull-less spring barley (cv Rondo), and a blue wheat (cv Skorpion)	Red variety (Italy), blue variety (Czech Republic), and barley (Italy)	Bread	The mix of flours, LAB, and yeasts utilized in this study resulted in the manufacture of bakery items that benefited cell viability and the inflammatory and oxidative state.	124
2.	Yellow- and blue-grained wheat varieties (cv Bona Vita and cv Skorpion) and red wheat (cv Aubusson)	Italy and the Czech Republic	Bread	Compared to standard wheat bread, the brans (pre-fermented) incorporation resulted in greater digestibility of protein (up to 79%) and a significant decrease in the fortified bread's anticipated glycemic index (about 8%).	57
3.	Blue-aleurone soft wheat and purple-pericarp durum wheat.	Italy	Bread	Bread enriched with purple durum wheat (medium-fine fraction) had a more pleasant taste and aroma.	45
4.	Blue wheat (cv Skorpion) and red wheat (cv Aubusson).	Italy and the Czech Republic	Sourdough	Despite using the identical LAB strains, sourdoughs created with pigmented barley and wheat demonstrated more antioxidant properties than those made with regular wheat flour.	29
5.	Purple wheat (Triticum turgidum ssp.)	Ethiopia	Biscuit	Although the biscuit quality parameters were somewhat lower in purple wheat samples compared to control samples, this was offset by increased antioxidant activity, total phenolic content (TPC), and total anthocyanin content (TAC).	41
6.	Blue standard corn, dark-red and blue popping corn.	Belgrade, Serbia	Cookies	The cookies made with 0.5 g per 100 g citric acid and cooked at 150 °C for 12 minutes had 60% more total flavonoids and 70% more anthocyanins than the control cookies heated at 200 °C for 7 minutes.	7

7.	Purple rice (*Oryza sativa* L. var. Sanpatong)	New Zealand	Biscuit	The study discovered that the greater percentage of purple rice (PR) flour resulted in the more excellent protein digestibility of the final biscuit. The biscuits composed solely of 100% PR flour had the lowest anticipated glycemic index (pGI) of all.	125
8.	Burma black rice and Jyothi red rice.	Karnataka, India	Flour	Brown rice flours obtained from various pulverizers, and toasting unpolished rice broken from pigmented rice species are possible sources of nutraceuticals such as phenolic components and can be used to make convenience foods, according to the findings.	126
9.	Italian black rice (*Oryza sativa* L., var. Artemide)	Italy	Biscuit	Polyphenols (from Artemide black rice) spray-dried with gum arabic and maltodextrin (50 : 50, w/w) resulted in the most stable component for storage and partial protection against heat damage during baking in a model biscuit.	127
10.	Red rice paddy variety, namely Megh II	Meghalaya, India	Rice cake	According to rheological tests, rice cakes manufactured from rice flour derived after heat-moisture treatment (HMT) of paddy had a greater volume increase and acceptability, and their batters had a higher viscosity.	128
11.	Black, brown, and red rice grains	—	Cake	Colored rice varieties, notably red rice, are potential gluten-free cake alternatives with superior technological qualities and a high concentration of bioactive compounds.	129
12.	White, red, and black pericarp rice	Bangladesh	—	Incorporating C3G-enriched black rice (25%) with white or red pericarp rice (75%) may improve the nutritional quality of rice-based bakery products under both baking and cooking conditions.	106
13.	Black glutinous rice flour	Thailand	Cake	The black glutinous rice flour-substituted cakes had higher firmness, gumminess, and chewiness values than wheat cake.	130

(*continued*)

Table 15.1 (*continued*)

Sr. No.	Cereal or millet varieties	Origin	Type of bakery product	Research results	Reference
14.	Chinese black, black Japonica, Purple Sticky, Himalayan Red, Short Grain Brown, Long, Grain Brown, French Camargue, Wild and Wehani.	USA	Flatbreads	The acrylamide concentration of sixteen examined flatbreads ranged from 0 to 49.1 μg kg^{-1}, indicating that these flatbreads may serve as low-acrylamide functional foods.	131
15.	Black rice (*Oryza sativa* L.)	Taiwan	Chiffon cake	The antioxidant potential and bioactive components in the chiffon cake (CC) made with up to 60% black rice (BR) powder instead of wheat flour were higher than in the CC control cake.	132
16.	Purple rice	Thailand	Biscuit	The partial substitution of purple rice flour for wheat flour alters both the physical and chemical properties of the biscuits (*e.g.*, hardness).	110
17.	Black rice	China	Bread	Another technique to make bread with a slower digestion rate and increased health advantages is to fortify it with anthocyanins.	133
18.	Peeled red pericarp rice grain.	Brazil	Cupcake	Red rice flour (made with 40 hours of germination) could be utilized to manufacture cupcakes, as suggested by the acceptability index derived from untrained panelists.	116
19.	Black rice (*Oryza sativa* L. ssp. *Japonica*)	Italy	Muffins	The muffins manufactured with black rice (BR) flour exhibited a more outstanding AC content and antioxidant potential than the control sample made with wheat flour.	134
20.	Brown rice (K-332 and Khosar)	Kashmir, India	Cracker	Crackers prepared with 9% pomace have higher antioxidant levels, notably dietary fiber, and polyphenols.	135

21.	Purple rice	New Zealand	Biscuits	The inclusion of increasing purple rice flour in the mixture had a substantial influence on enhancing the *in vitro* digestibility of protein, which can decrease the starch digestion rate in human metabolism.	125
22.	Red and blue maize kernels.	Mexico	tortillas	The findings show indisputably that the proposed lime-cooking extrusion approach contributed significantly to the retention of phytochemicals and antioxidants in all tortillas, particularly ferulic acid.	88
23.	Red, yellow, black, and purple maize.	Mexico	Tortilla chips	Black and red maize tortilla chips contained less acrylamide than yellow, purple, or commercial maize tortilla chips.	136
24.	Eight contrasting pigmented (red and yellow) kernels	Mexico	Tortillas	When compared to raw kernels, extruded and traditional nixtamalization tortillas from yellow corn retained and contained a higher level of carotenoids and lipophilic antioxidant activity.	75
25.	Red, yellow, and blue phenotypes of maize	Turkey	Functional baked snack	According to the results, snacks produced from pigmented corn had higher total, bound, and free phenolic levels.	34
26.	Black rice	Galaţi	Biscuits	The black rice flour biscuits contained a high protein and fiber content, implying that their functionality is primarily derived from their high concentration of biologically active compounds.	137
27.	Yellow, white, and purple corn	New Delhi, India.	Muffin	Muffins made with the yellow maize had the highest phenolic compounds and antioxidant activity, making them the most acceptable.	138
28.	Yellow and red-pigmented corn varieties	Himachal Pradesh, India	Chapatti	The chapatti made from standard (yellow) corn received a greater acceptability score from consumers, whereas the chapatti made from waxy corn received a low score.	139

(*continued*)

Table 15.1 (*continued*)

Sr. No.	Cereal or millet varieties	Origin	Type of bakery product	Research results	Reference
29.	Blue corn	Peru	Cookies	Increased oven heat transfer efficiency through shorter baking time and incorporation of an acidulent blue corn cookie results in increased anthocyanin retention. Cookies baked in a convection oven with citric acid retained their maximum TAC.	67
30.	White and red sorghum	—	Cookies	Malted sorghum cookies demonstrated significantly higher chemical scores (98% for red sorghum flour cookies), protein digestibility, and bio-functional compounds (free phenolic acid and GABA) than cookies produced with native sorghum flours.	140
31.	White and red sorghum	USA	Sorghum crisps	While the chemical composition and thermal characteristics of red and white sorghum flours were similar, discrepancies in the flour's chemical composition and thermal properties were identified.	141
32.	White and red sorghum	Australia	Flatbread	Because flatbreads made from sorghum have lower RDS levels than refined wheat flour flatbreads, integrating refined wheat flour into whole-grain sorghum flour flatbreads may have the ability to reduce blood glucose levels (post-meal) in humans.	142

References

1. E.-S. M. Abdel-Aal, J. C. Young and I. Rabalski, *J. Agric. Food Chem.*, 2006, **54**, 4696.
2. Q. Liu, Y. Qiu and T. Beta, *J. Agric. Food Chem.*, 2010, **58**, 9235.
3. E.-S. M. Abdel-Aal, A. A. Abou-Arab, T. H. Gamel, P. Hucl, J. C. Young and I. Rabalski, *J. Agric. Food Chem.*, 2008, **56**, 11171.
4. J. Lachman, P. Martinek, Z. Kotíková, M. Orsák and M. Šulc, *J. Cereal Sci.*, 2017, **74**, 145.
5. K. Masisi, T. Beta and M. H. Moghadasian, *Food Chem.*, 2016, **196**, 90.
6. X. Wu, G. R. Beecher, J. M. Holden, D. B. Haytowitz, S. E. Gebhardt and R. L. Prior, *J. Agric. Food Chem.*, 2006, **54**, 4069.
7. S. Žilić, T. Kocadağlı, J. Vančetović and V. Gökmen, *LWT--Food Sci. Technol.*, 2016, **65**, 597.
8. D. Angelino, M. Cossu, A. Marti, M. Zanoletti, L. Chiavaroli, F. Brighenti, D. D. Rio and D. Martini, *Food Funct.*, 2017, **8**, 2368.
9. D. Giordano, M. Locatelli, F. Travaglia, M. Bordiga, A. Reyneri, J. D. Coïsson and M. Blandino, *Food Chem.*, 2017, **233**, 483.
10. D. B. M. Ficco, V. De Simone, S. A. Colecchia, I. Pecorella, C. Platani, F. Nigro, F. Finocchiaro, R. Papa and P. De Vita, *J. Agric. Food Chem.*, 2014, **62**, 8686.
11. E.-S. M. Abdel-Aal and I. Rabalski, *Open Agric. J.*, 2008, **2**, 7.
12. A. Pasqualone, B. Laddomada, I. Centomani, V. M. Paradiso, D. Minervini, F. Caponio and C. Summo, *LWT*, 2017, **78**, 151.
13. T. Steglich, D. Bernin, A. Moldin, D. Topgaard and M. Langton, *Cereal Chem.*, 2015, **92**, 617.
14. S. Hemdane, P. J. Jacobs, E. Dornez, J. Verspreet, J. A. Delcour and C. M. Courtin, *Compr. Rev. Food Sci. Food Saf.*, 2016, **15**, 28.
15. R. Ciccoritti, F. Taddei, I. Nicoletti, L. Gazza, D. Corradini, M. G. D'Egidio and D. Martini, *Food Chem.*, 2017, **225**, 77.
16. M. Zanoletti, P. Abbasi Parizad, V. Lavelli, C. Cecchini, P. Menesatti, A. Marti and M. A. Pagani, *LWT*, 2017, **75**, 663.
17. A. C. Zeven, *Euphytica*, 1991, **56**, 243.
18. M. Gómez and M. M. Martinez, *Crit. Rev. Food Sci. Nutr.*, 2018, **58**, 2119.
19. F. Zhu, R. Sakulnak and S. Wang, *Food Chem.*, 2016, **194**, 1217.
20. J. Ou, M. Wang, J. Zheng and S. Ou, *Food Chem.*, 2019, **284**, 99.
21. E.-S. M. Abdel-Aal and I. Rabalski, *J. Cereal Sci.*, 2013, **57**, 312.
22. M. Slavin, Y. Lu, N. Kaplan and L. (Lucy) Yu, *Food Chem.*, 2013, **141**, 1166.
23. M. Gobbetti, M. De Angelis, R. Di Cagno, M. Calasso, G. Archetti and C. G. Rizzello, *Int. J. Food Microbiol.*, 2019, **302**, 103.
24. V. Galli, L. Mazzoli, S. Luti, M. Venturi, S. Guerrini, P. Paoli, M. Vincenzini, L. Granchi and L. Pazzagli, *Int. J. Food Microbiol.*, 2018, **286**, 55.
25. S. Luti, L. Mazzoli, M. Ramazzotti, V. Galli, M. Venturi, G. Marino, M. Lehmann, S. Guerrini, L. Granchi, P. Paoli and L. Pazzagli, *Food Chem.*, 2020, **322**, 126710.

26. V. M. Koistinen, O. Mattila, K. Katina, K. Poutanen, A.-M. Aura and K. Hanhineva, *Sci. Rep.*, 2018, **8**, 5684.
27. B. Păcularu-Burada, L. A. Georgescu, M. A. Vasile, J. M. Rocha and G.-E. Bahrim, *Microorganisms*, 2020, **8**, 643.
28. P. Petrova and K. Petrov, *Nutrients*, 2020, **12**, 1118.
29. V. Galli, M. Venturi, S. Guerrini, M. Blandino, S. Luti, L. Pazzagli and L. Granchi, *Foods*, 2020, **9**, 640.
30. G. A. Camelo-Méndez, E. Agama-Acevedo, M. M. Sanchez-Rivera and L. A. Bello-Pérez, *Food Chem.*, 2016, **211**, 281.
31. G. Deepa, V. Singh and K. A. Naidu, *J. Food Sci. Technol.*, 2010, **47**, 644.
32. J. P. Hernández-Uribe, E. Agama-Acevedo, J. J. Islas-Hernández, J. Tovar and L. A. Bello-Pérez, *J. Sci. Food Agric.*, 2007, **87**, 2482.
33. S. Grasso, *Trends Food Sci. Technol.*, 2020, **99**, 284.
34. H. Çetin-Babaoğlu, N. Yalım, E. Kale and S. A. Tontul, *Food Biosci.*, 2021, **44**, 101349.
35. V. S. Mialon, M. R. Clark, P. I. Leppard and D. N. Cox, *Food Qual. Prefer.*, 2002, **13**, 1.
36. K. Katina, A. Laitila, R. Juvonen, K.-H. Liukkonen, S. Kariluoto, V. Piironen, R. Landberg, P. Åman and K. Poutanen, *Food Microbiol.*, 2007, **24**, 175.
37. M. C. Messia, A. Reale, L. Maiuro, T. Candigliota, E. Sorrentino and E. Marconi, *J. Cereal Sci.*, 2016, **69**, 138.
38. K. K. Adom and R. H. Liu, *J. Agric. Food Chem.*, 2002, **50**, 6182.
39. G. R. Carson and N. M. Edwards, *Wheat: Chemistry and Technology*, 2009, p. 97.
40. S. Bartłomiej, R.-K. Justyna and N. Ewa, *Food Sci. Technol. Int.*, 2012, **18**, 559.
41. A. Pasqualone, A. M. Bianco, V. M. Paradiso, C. Summo, G. Gambacorta, F. Caponio and A. Blanco, *Food Chem.*, 2015, **180**, 64.
42. P. R. Shewry and S. Hey, *J. Cereal Sci.*, 2015, **65**, 236.
43. P. R. Shewry and J. L. Ward, *Food Energy Secur.*, 2012, **1**, 47.
44. P. R. Shewry, G. Charmet, G. Branlard, D. Lafiandra, S. Gergely, A. Salgó, L. Saulnier, Z. Bedő, E. N. C. Mills and J. L. Ward, *Trends Food Sci. Technol.*, 2012, **25**, 70.
45. D. B. M. Ficco, G. M. Borrelli, V. Giovanniello, C. Platani and P. De Vita, *J. Cereal Sci.*, 2018, **79**, 118.
46. C. Hu, Y.-Z. Cai, W. Li, H. Corke and D. D. Kitts, *Food Chem.*, 2007, **104**, 955.
47. S. Siebenhandl, H. Grausgruber, N. Pellegrini, D. Del Rio, V. Fogliano, R. Pernice and E. Berghofer, *J. Agric. Food Chem.*, 2007, **55**, 8541.
48. V. Sovrani, M. Blandino, V. Scarpino, A. Reyneri, J. D. Coïsson, F. Travaglia, M. Locatelli, M. Bordiga, R. Montella and M. Arlorio, *Food Chem.*, 2012, **135**, 39.
49. M. Garg, M. Chawla, V. Chunduri, R. Kumar, S. Sharma, N. K. Sharma, N. Kaur, A. Kumar, J. K. Mundey, M. K. Saini and S. P. Singh, *J. Cereal Sci.*, 2016, **71**, 138.

50. S. Dhua, K. Kumar, Y. Kumar, L. Singh and V. S. Sharanagat, *Trends Food Sci. Technol.*, 2021, **112**, 780.
51. A. Kumari, S. Sharma, N. Sharma, V. Chunduri, P. Kapoor, S. Kaur, A. Goyal and M. Garg, *Molecules*, 2020, **25**, 5071.
52. W. Li and T. Beta, in *Flour and Breads and their Fortification in Health and Disease Prevention*, ed. V. R. Preedy, R. R. Watson and V. B. Patel, Academic Press, San Diego, 2011, p. 59.
53. J. Lachman, K. Hejtmánková and Z. Kotíková, *J. Cereal Sci.*, 2013, **57**, 207.
54. M. Blandino, M. Locatelli, A. Gazzola, J. D. Coïsson, S. Giacosa, F. Travaglia, M. Bordiga, A. Reyneri, L. Rolle and M. Arlorio, *J. Cereal Sci.*, 2015, **65**, 48.
55. A. K. Holtekjølen, C. Kinitz and S. H. Knutsen, *J. Agric. Food Chem.*, 2006, **54**, 2253.
56. A. K. Holtekjølen, A. B. Bævre, M. Rødbotten, H. Berg and S. H. Knutsen, *Food Chem.*, 2008, **110**, 414.
57. E. Pontonio, C. Dingeo, R. Di Cagno, M. Blandino, M. Gobbetti and C. G. Rizzello, *Int. J. Food Microbiol.*, 2020, **313**, 108384.
58. E.-S. M. Abdel-Aal and P. Hucl, *J. Agric. Food Chem.*, 2003, **51**, 2174.
59. W. Li, M. D. Pickard and T. Beta, *Food Chem.*, 2007, **104**, 1080.
60. S. N. Syed Jaafar, J. Baron, S. Siebenhandl-Ehn, T. Rosenau, S. Böhmdorfer and H. Grausgruber, *Plant Breed.*, 2013, **132**, 546.
61. Y. Li, D. Ma, D. Sun, C. Wang, J. Zhang, Y. Xie and T. Guo, *Crop J.*, 2015, **3**, 328.
62. J. Shipp and E.-S. M. Abdel-Aal, *Open Food Sci. J.*, 2010, **4**, 7.
63. A. Pasqualone, A. M. Bianco and V. M. Paradiso, *CyTA--J. Food*, 2013, **11**, 301.
64. A. G. Waghmare and S. S. Arya, *J. Food Process. Preserv.*, 2014, **38**, 1198.
65. N. Rawat and D. Indrani, *Food Rev. Int.*, 2015, **31**, 125.
66. P. Bartl, A. Albreht, M. Skrt, B. Tremlová, M. Ošťádalová, K. Šmejkal, I. Vovk and N. P. Ulrih, *Int. J. Food Sci. Nutr.*, 2015, **66**, 514.
67. J. Li, C. E. Walker and J. M. Faubion, *J. Sci. Food Agric.*, 2011, **91**, 38.
68. E. de la Hera, M. Talegón, P. Caballero and M. Gómez, *J. Sci. Food Agric.*, 2013, **93**, 924.
69. S. Žilić, A. Serpen, G. Akıllıoğlu, V. Gökmen and J. Vančetović, *J. Agric. Food Chem.*, 2012, **60**, 1224.
70. Y. Wang, J. M. King, Z. Xu, J. Losso and A. Prudente, *J. Agric. Food Chem.*, 2008, **56**, 7942.
71. N. Singh, S. Singh and K. Shevkani, in *Flour and Breads and their Fortification in Health and Disease Prevention*, ed. V. R. Preedy, R. R. Watson and V. B. Patel, Academic Press, San Diego, 2011, pp. 89–99, DOI: 10.1016/B978-0-12-380886-8.10009-1.
72. D. Del Pozo-Insfran, C. H. Brenes, S. O. Serna Saldivar and S. T. Talcott, *Food Res. Int.*, 2006, **39**, 696.
73. L. X. Lopez-Martinez, R. M. Oliart-Ros, G. Valerio-Alfaro, C.-H. Lee, K. L. Parkin and H. S. Garcia, *LWT--Food Sci. Technol.*, 2009, **42**, 1187.

74. E. G. Kean, B. R. Hamaker and M. G. Ferruzzi, *J. Agric. Food Chem.*, 2008, **56**, 9918.
75. A. B. Corrales-Bañuelos, E. O. Cuevas-Rodríguez, J. A. Gutiérrez-Uribe, E. M. Milán-Noris, C. Reyes-Moreno, J. Milán-Carrillo and S. Mora-Rochín, *J. Cereal Sci.*, 2016, **69**, 64.
76. S. Mendoza-Díaz, M. del C. Ortiz-Valerio, E. Castaño-Tostado, J. de D. Figueroa-Cárdenas, R. Reynoso-Camacho, M. Ramos-Gómez, R. Campos-Vega and G. Loarca-Piña, *Plant Foods Hum. Nutr.*, 2012, **67**, 442.
77. R. Reynoso-Camacho, G. Guerrero-Villanueva, J. de Dios Figueroa, M. A. Gallegos-Corona, S. Mendoza, G. Loarca-Piña and M. Ramos-Gomez, *Plant Foods Hum. Nutr.*, 2015, **70**, 146.
78. B. Nayak, R. H. Liu and J. Tang, *Crit. Rev. Food Sci. Nutr.*, 2015, **55**, 887.
79. R. Colombo, L. Ferron and A. Papetti, *Molecules*, 2021, **26**, 199.
80. K. Petroni, R. Pilu and C. Tonelli, *Planta*, 2014, **240**, 901.
81. L. A. Chatham, M. Paulsmeyer and J. A. Juvik, *Theor. Appl. Genet.*, 2019, **132**, 2927.
82. R. Moreira, F. Chenlo, S. Arufe and S. N. Rubinos, *J. Food Sci. Technol.*, 2015, **52**, 7954.
83. P. S. Mansilla, M. C. Nazar and G. T. Pérez, *Int. J. Biol. Macromol.*, 2020, **146**, 311.
84. S. Mildner-Szkudlarz, R. Zawirska-Wojtasiak, W. Obuchowski and M. Gośliński, *J. Food Sci.*, 2009, **74**, S362.
85. A. Pasqualone, A. M. Bianco, V. M. Paradiso, C. Summo, G. Gambacorta and F. Caponio, *Food Res. Int.*, 2014, **65**, 385.
86. A. Cortés-Gómez, E. S. Martín-Martínez, F. Martínez-Bustos and G. M. Vázquez-Carrillo, *J. Food Eng.*, 2005, **66**, 273.
87. A. J. Palacios-Fonseca, C. Vazquez-Ramos and M. E. Rodríguez-García, *J. Food Eng.*, 2009, **93**, 45.
88. S. Mora-Rochin, J. A. Gutiérrez-Uribe, S. O. Serna-Saldivar, P. Sánchez-Peña, C. Reyes-Moreno and J. Milán-Carrillo, *J. Cereal Sci.*, 2010, **52**, 502.
89. C. de la Parra, S. O. Serna Saldivar and R. H. Liu, *J. Agric. Food Chem.*, 2007, **55**, 4177.
90. G. Plascencia-González, F. Martínez-Bustos, Y. Salinas-Moreno, F. Sánchez-Sinencio and A. Hernández-Montes, in *Proceedings of the 10th World Congress of Food Science & Technology*. Sydney, Australia, 1999.
91. J. J. Zazueta-Morales, F. Martínez-Bustos, N. Jacobo-Valenzuela, C. Ordorica-Falomir and O. Paredes-López, *J. Sci. Food Agric.*, 2001, **81**, 1379.
92. J. Zazueta-Morales, F. Martinez-Bustos, N. Jacobo-Valenzuela, C. Ordorica-Falomir and O. Paredes-Lopez, *J. Food Sci.*, 2002, **67**, 3350.
93. G. A. Camelo-Méndez, J. Tovar and L. A. Bello-Pérez, *J. Food Sci. Technol.*, 2018, **55**, 2739.
94. S. Khampas, K. Lertrat, K. Lomthaisong, S. Simla and B. Suriharn, *Turkish J. Field Crop.*, 2015, **20**, 15.

95. R. G. Utrilla-Coello, E. Agama-Acevedo, P. Osorio-Díaz, J. Tovar and L. A. Bello-Pérez, *Starch-Stärke*, 2011, **63**, 416.
96. T. Oikawa, H. Maeda, T. Oguchi, T. Yamaguchi, N. Tanabe, K. Ebana, M. Yano, T. Ebitani and T. Izawa, *Plant Cell*, 2015, **27**, 2401.
97. V. C. Ito and L. G. Lacerda, *Food Chem.*, 2019, **301**, 125304.
98. G.-F. Deng, X.-R. Xu, Y. Zhang, D. Li, R.-Y. Gan and H.-B. Li, *Crit. Rev. Food Sci. Nutr.*, 2013, **53**, 296.
99. M. Bordiga, S. Gomez-Alonso, M. Locatelli, F. Travaglia, J. D. Coïsson, I. Hermosin-Gutierrez and M. Arlorio, *Food Res. Int.*, 2014, **65**, 282.
100. S. Kraithong, S. Lee and S. Rawdkuen, *J. Cereal Sci.*, 2018, **79**, 259.
101. F. M. Bhat, S. R. Sommano, C. S. Riar, P. Seesuriyachan, T. Chaiyaso and C. Prom-u-Thai, *Agronomy*, 2020, **10**, 1817.
102. S. S. Dipti, C. Bergman, S. D. Indrasari, T. Herath, R. Hall, H. Lee, F. Habibi, P. Z. Bassinello, E. Graterol, J. P. Ferraz and M. Fitzgerald, *Rice*, 2012, **5**, 16.
103. S. Tian, K. Nakamura and H. Kayahara, *J. Agric. Food Chem.*, 2004, **52**, 4808.
104. Z. Zhou, K. Robards, S. Helliwell and C. Blanchard, *Food Chem.*, 2004, **87**, 401.
105. T. Itani, H. Tatemoto, M. Okamoto, K. Fujii and N. Muto, *Nippon Shokuhin Kagaku Kogaku Kaishi*, 2002, **49**, 540.
106. H. B. Shozib, M. M. Islam, S. A. S. Mahmud, M. N. Bari, N. Akter, S. Jahan, S. Hosen, M. N. Hossain, A. H. M. N. Nabi, M. A. Siddiquee and Md. M. Haque, *Saudi J. Biol. Sci.*, 2021, **28**, 7472.
107. X. Sui, P. Y. Yap and W. Zhou, *Food Bioprocess Technol.*, 2015, **8**, 983.
108. S. Jang and Z. Xu, *J. Agric. Food Chem.*, 2009, **57**, 858.
109. R. Yawadio, S. Tanimori and N. Morita, *Food Chem.*, 2007, **101**, 1616.
110. W. Klunklin and G. Savage, *Adv. Food Sci. Eng.*, 2018, **2**, 81.
111. C. Kim, S. Kikuchi, Y. Kim, S. Park, U. Yoon, G. Lee, J. Choi, Y. Kim and S. Park, *Biochip J.*, 2010, **4**, 247.
112. D. Sumczynski, E. Kotásková, H. Družbíková and J. Mlček, *Food Chem.*, 2016, **211**, 339.
113. Y. Shao and J. Bao, *Food Chem.*, 2015, **180**, 86.
114. A. Gunaratne, K. Wu, D. Li, A. Bentota, H. Corke and Y.-Z. Cai, *Food Chem.*, 2013, **138**, 1153.
115. N. V. M. de Mira, I. L. Massaretto, C. de S. C. I. Pascual and U. M. Lanfer Marquez, *J. Food Compos. Anal.*, 2009, **22**, 405.
116. C. P. Müller, J. F. Hoffmann, C. D. Ferreira, G. W. Diehl, R. C. Rossi and V. Ziegler, *Int. J. Gastron. Food Sci.*, 2021, **25**, 100379.
117. H. Li, S. Prakash, T. M. Nicholson, M. A. Fitzgerald and R. G. Gilbert, *Food Chem.*, 2016, **196**, 702.
118. A. B. Das and S. Bhattacharya, *LWT*, 2019, **102**, 197.
119. T. A. S. Gusmão, R. P. de Gusmão, H. V. Moura, H. A. Silva, M. E. R. M. Cavalcanti-Mata and M. E. M. Duarte, *J. Food Sci. Technol.*, 2019, **56**, 2949.

120. C. Cappa, M. Laureati, M. C. Casiraghi, D. Erba, M. Vezzani, M. Lucisano and C. Alamprese, *Foods*, 2021, **10**, 91.
121. V. U. Ndolo and T. Beta, *Food Chem.*, 2013, **139**, 663.
122. L. Paznocht, Z. Kotíková, M. Šulc, J. Lachman, M. Orsák, M. Eliášová and P. Martinek, *Food Chem.*, 2018, **240**, 670.
123. M. Martínez-Subirà, M. P. Romero, E. Puig, A. Macià, I. Romagosa and M. Moralejo, *LWT*, 2020, **131**, 109582.
124. S. Luti, V. Galli, M. Venturi, L. Granchi, P. Paoli and L. Pazzagli, *Appl. Sci.*, 2021, **11**, 3291.
125. W. Klunklin and G. Savage, *J. Food Qual.*, 2018, **2018**, e8052847.
126. I. Sapna, M. Kamaljit, R. Priya and P. A. Jayadeep, *J. Food Sci. Technol.*, 2019, **56**, 273.
127. V. A. Papillo, M. Locatelli, F. Travaglia, M. Bordiga, C. Garino, M. Arlorio and J. D. Coïsson, *Food Chem.*, 2018, **269**, 603.
128. P. M. S. Pathaw, C. L. Mahanta and S. Bhattacharya, *J. Food Process. Preserv.*, 2021, **45**, e15206.
129. G. H. Lang, D. H. Kringel, T. dos S. Acunha, C. D. Ferreira, Á. R. G. Dias, E. da R. Zavareze and M. de Oliveira, *Food Chem.*, 2020, **318**, 126480.
130. P. Itthivadhanapong and A. Sangnark, *Int. Food Res. J.*, 2016, **23**, 1190.
131. X. Li, T. Kahlon, S. C. Wang and M. Friedman, *Foods*, 2021, **10**, 2909.
132. J.-L. Mau, C.-C. Lee, Y.-P. Chen and S.-D. Lin, *LWT*, 2017, **75**, 434.
133. X. Sui, Y. Zhang and W. Zhou, *Food Chem.*, 2016, **196**, 910.
134. C. Croitoru, C. Mureșan, M. Turturică, N. Stănciuc, D. G. Andronoiu, L. Dumitrașcu, V. Barbu, E. Enachi (Ioniță), G. Horincar (Parfene) and G. Râpeanu, *Molecules*, 2018, **23**, 3047.
135. S. A. Mir, S. J. D. Bosco, M. A. Shah, S. Santhalakshmy and M. M. Mir, *J. Saudi Soc. Agric. Sci.*, 2017, **16**, 25.
136. R. M. Delgado, G. Arámbula-Villa, G. Luna-Bárcenas, V. Flores-Casamayor, J. J. Veles-Medina, E. Azuara, R. Salazar, R. M. Delgado, G. Arámbula-Villa, G. Luna-Bárcenas, V. Flores-Casamayor, J. J. Veles-Medina, E. Azuara and R. Salazar, *Rev. Mex. Ing. Quim.*, 2016, **15**, 69.
137. C.-A. Bolea, M. Turturică, E. Enachi, C. Vizireanu and N. Stănciuc, *Ann. Univ. Dunarea Jos Galati, Fasc. VI*, 2021, **45**, 48.
138. S. Trehan, N. Singh and A. Kaur, *J. Food Sci. Technol.*, 2018, **55**, 2334.
139. S. Thakur, N. Singh and A. Kaur, *J. Food Sci. Technol.*, 2017, **54**, 3285.
140. A. G. Garzón, M. Erben, C. A. Osella and S. R. Drago, *J. Food Process. Preserv.*, 2020, **44**, e14571.
141. J. G. Pezzali, A. Suprabha-Raj, K. Siliveru and C. G. Aldrich, *PLoS One*, 2020, **15**, e0234940.
142. A. Yousif, D. Nhepera and S. Johnson, *Food Chem.*, 2012, **134**, 880.

Subject Index